Concepts and Strategies in Plant Sciences

This book series highlights the spectacular advances in the concepts, techniques and tools in various areas of plant science. Individual volumes may cover topics like genome editing, phenotyping, molecular pharming, bioremediation, miRNA, fast-track breeding, crop evolution, IPR and farmers' rights, to name just a few. The books will demonstrate how advanced strategies in plant science can be utilized to develop and improve agriculture, ecology and the environment. The series will be of interest to students, scientists and professionals working in the fields of plant genetics, genomics, breeding, biotechnology, and in the related disciplines of plant production, improvement and protection.

Interested in editing a volume? Please contact Prof. Chittaranjan Kole, Series Editor, at ckoleorg@gmail.com

Luis María Vaschetto
Editor

Plant Transcriptomics and Epitranscriptomics

Decoding the RNA Landscape

Springer

Editor
Luis María Vaschetto
National Coalition of Independent Scholars
South Bend, IN, USA

ISSN 2662-3188 ISSN 2662-3196 (electronic)
Concepts and Strategies in Plant Sciences
ISBN 978-981-95-5182-8 ISBN 978-981-95-5183-5 (eBook)
https://doi.org/10.1007/978-981-95-5183-5

This Springer imprint is published by the registered company Springer Nature Singapore Pte Ltd.
The registered company address is: 152 Beach Road, #21-01/04 Gateway East, Singapore 189721, Singapore

Preface

Transcriptomics can be defined as the study of all known types of RNA transcripts present in a cell, tissue, or organism under specific environmental conditions. As the direct product of gene expression, the transcriptome provides a dynamic and comprehensive view of which genes are active and their relative levels. High-throughput RNA-Sequencing (RNA-Seq) technologies are now revolutionizing this research area by enabling the quantitative analysis of gene expression on a genome-wide scale, allowing researchers to identify novel genes, study alternative splicing, and map entire gene regulatory networks. Transcriptomics research offers thus invaluable insights into the molecular mechanisms that govern plant growth, development, and adaptation.

Epitranscriptomics is a field that complements transcriptomics by investigating the chemical modifications on RNA molecules after transcription. Much like epigenetics, which studies heritable changes in gene expression not caused by DNA sequence changes, epitranscriptomics explores gene regulation through the "writing," "erasing," and "reading" of RNA modifications. This discipline is now revealing a vast and dynamic regulatory world, with over 100 different chemical modifications identified on various types of RNA. Modifications like N6-methyladenosine (m6A) and 5-methylcytosine (m5C) can influence RNA stability, splicing, localization, and translation without altering the underlying RNA sequence. Consequently, while transcriptomics is a well-established field that has advanced rapidly in recent years due to the emergence of high-throughput technologies, epitranscriptomics is expanding even faster, offering crucial new insights into how plants fine-tune their gene expression to respond to environmental cues.

Plant Transcriptomics and Epitranscriptomics: Decoding the RNA Landscape provides a comprehensive, updated, and integrated vision of plant gene regulation by holistically covering transcriptomic and epitranscriptomic profiles. The book is a vital resource that connects regulatory RNA omics layers to real-world outcomes, being this particularly valuable in crop improvement. I am deeply grateful to the researchers whose contributions have made this book a unique source of information for years to come.

South Bend, IN, USA Luis María Vaschetto

Contents

Editor and Contributors

About the Editor

Luis María Vaschetto obtained his Ph.D. in Biological Sciences from the Universidad Nacional de Córdoba (Argentina). He was a postdoctoral researcher at the Consejo Nacional de Investigaciones Científicas y Técnicas (CONICET) and a Research Associate at the South Dakota State University (SDSU). He has first-authored over ten academic articles in top-tier journals, and also published several book chapters. He has edited and authored ten (10) international books. His research mainly focuses on understanding the epigenetic mechanisms involved in plant and animal development, as well as their effects on human health and diseases.

Contributors

Maria Almira-Casellas Plant Physiology Laboratory, Bioscience Faculty, Universitat Autònoma de Barcelona, Barcelona, Spain

Silvia Busoms Plant Physiology Laboratory, Bioscience Faculty, Universitat Autònoma de Barcelona, Barcelona, Spain
Plant Physiology Laboratory, Department of Environmental Sciences, Universitat de Girona, Girona, Spain

María Jesús Cañal Department of Organisms and Systems Biology, University of Oviedo, Oviedo, Spain
Biotechnology Institute of Asturias (IUBA), University of Oviedo, Oviedo, Spain

Régis L. Corrêa Institute of Integrative Systems Biology (I2SysBio, CSIC-UV), Paterna, Spain
Department of Genetics, Federal University of Rio de Janeiro (UFRJ), Rio de Janeiro, Brazil

Andrés J. Cortés Facultad de Ciencias Agrarias – Departamento de Ciencias Forestales, Universidad Nacional de Colombia – Sede Medellín, Medellin, Colombia

Pelayo G. de Lena Department of Organisms and Systems Biology, University of Oviedo, Oviedo, Spain
Health Research Institute of Asturias (ISPA), Oviedo, Spain
Foundation for Biomedical Research and Innovation in Asturias (FINBA), Oviedo, Spain

Mónica Escandón Department of Organisms and Systems Biology, University of Oviedo, Oviedo, Spain
Biotechnology Institute of Asturias (IUBA), University of Oviedo, Oviedo, Spain

Glòria Escolà Plant Physiology Laboratory, Bioscience Faculty, Universitat Autònoma de Barcelona, Barcelona, Spain

Leonardo Furci Plant Epigenetics Unit, OIST, Okinawa, Japan

Antoni Garcia-Molina Centre for Research in Agricultural Genomics (CRAG), CSIC-IRTA-UAB-UB, Edifici CRAG, Barcelona, Spain

Ruolin Gong State Key Laboratory of Crop Stress Resistance and High-Efficiency Production, College of Agronomy, Northwest A&F University, Yangling, Shaanxi, China

Jihong Hu State Key Laboratory of Crop Stress Resistance and High-Efficiency Production, College of Agronomy, Northwest A&F University, Yangling, Shaanxi, China

Sarath Chandra Janga Department of Biomedical Engineering and Informatics, Luddy School of Informatics, Computing, and Engineering, Indiana University Indianapolis, Indianapolis, IN, USA
Center for Computational Biology and Bioinformatics, Indiana University School of Medicine, Health Information and Translational Science (HITS), Indianapolis, IN, USA
Department of Medical and Molecular Genetics, Indiana University School of Medicine, Medical Research and Library Building, Indianapolis, IN, USA

Alexander Krohannon Department of Biomedical Engineering and Informatics, Luddy School of Informatics, Computing, and Engineering, Indiana University Indianapolis, Indianapolis, IN, USA

Vijay Kumar Plant Biotechnology Lab, Division of Research and Development, Lovely Professional University, Phagwara, Punjab, India
Department of Biotechnology, Lovely Faculty of Technology and Sciences, Lovely Professional University, Phagwara, Punjab, India

Rebekah Eileen Lanning Department of Biomedical Engineering and Informatics, Luddy School of Informatics, Computing, and Engineering, Indiana University Indianapolis, Indianapolis, IN, USA

Guangze Li State Key Laboratory of Crop Stress Resistance and High-Efficiency Production, College of Agronomy, Northwest A&F University, Yangling, Shaanxi, China

Felipe López-Hernández Corporación Colombiana de Investigación Agropecuaria – AGROSAVIA, Rionegro, Colombia

Yan Lu College of Horticulture and Plant Protection, Inner Mongolia Agricultural University, Hohhot, China

Mónica Meijón Department of Organisms and Systems Biology, University of Oviedo, Oviedo, Spain
Biotechnology Institute of Asturias (IUBA), University of Oviedo, Oviedo, Spain

David Morales Department of Organisms and Systems Biology, University of Oviedo, Oviedo, Spain
Biotechnology Institute of Asturias (IUBA), University of Oviedo, Oviedo, Spain

Aarifa Nabi College of Biology, State Key Laboratory of Chemo/Biosensing and Chemometrics, and Hunan Province Key Laboratory of Plant Functional Genomics and Developmental Regulation, Hunan University, Changsha, People's Republic of China

R. Aswati Nair Department of Biochemistry and Molecular Biology, Central University of Kerala, Kasaragod, India

M. J. Nishanth Department of Biotechnology, School of Life Sciences, St Joseph's University, Bengaluru, Karnataka, India

S. B. Parvathy Devi Department of Genomic Science, Central University of Kerala, Kasaragod, India

Jesús Pascual Biotechnology Institute of Asturias (IUBA), University of Oviedo, Oviedo, Spain
Department of Functional Biology, University of Oviedo, Oviedo, Spain

Fabio Pasin Center for Biological Research "Margarita Salas" (CIB), Spanish National Research Council (CSIC), Madrid, Spain

Padmesh Pandaram Pillai Department of Genomic Science, Central University of Kerala, Kasaragod, India

Charlotte Poschenrieder Plant Physiology Laboratory, Bioscience Faculty, Universitat Autònoma de Barcelona, Barcelona, Spain

Siya Pradeep Department of Genomic Science, Central University of Kerala, Kasaragod, India

Muhammad Qadir College of Biology, State Key Laboratory of Chemo/Biosensing and Chemometrics, and Hunan Province Key Laboratory of Plant Functional Genomics and Developmental Regulation, Hunan University, Changsha, People's Republic of China

Chao Qin State Key Laboratory of Genome and Multi-omics Technologies, BGI Research, Shenzhen, China
BGI Research, Beijing, China

Saikat Sena Plant Biotechnology Lab, Division of Research and Development, Lovely Professional University, Phagwara, Punjab, India
Department of Biotechnology, Lovely Faculty of Technology and Sciences, Lovely Professional University, Phagwara, Punjab, India

Atsushi Shimada Plant Epigenetics Unit, OIST, Okinawa, Japan

Yashika Sood Department of Biotechnology, Lovely Faculty of Technology and Sciences, Lovely Professional University, Phagwara, Punjab, India

Swaraj Madan Thorat Department of Biomedical Engineering and Informatics, Luddy School of Informatics, Computing, and Engineering, Indiana University Indianapolis, Indianapolis, IN, USA

Prasanth Kumar Thuthika Department of Biomedical Engineering and Informatics, Luddy School of Informatics, Computing, and Engineering, Indiana University Indianapolis, Indianapolis, IN, USA

Peng Tian State Key Laboratory of Crop Stress Resistance and High-Efficiency Production, College of Agronomy, Northwest A&F University, Yangling, Shaanxi, China

Luis Valledor Department of Organisms and Systems Biology, University of Oviedo, Oviedo, Spain
Biotechnology Institute of Asturias (IUBA), University of Oviedo, Oviedo, Spain

Haijuan Wang College of Horticulture and Plant Protection, Inner Mongolia Agricultural University, Hohhot, China

Long Wang College of Biology, State Key Laboratory of Chemo/Biosensing and Chemometrics, and Hunan Province Key Laboratory of Plant Functional Genomics and Developmental Regulation, Hunan University, Changsha, People's Republic of China
National Center of Technology Innovation for Saline-Alkali Tolerant Rice, Gerater by Area Institute For Innovation, Hunan University, Changsha, People's Republic of China
Interdisciplinary and Intelligent Seed Industry Equipment Research Department, Yue Lushan Laboratory, Changsha, People's Republic of China

Keke Xia State Key Laboratory of Genome and Multi-omics Technologies, BGI Research, Shenzhen, China
BGI Research, Beijing, China

Jianying Yue College of Horticulture and Plant Protection, Inner Mongolia Agricultural University, Hohhot, China
College of Agriculture, Inner Mongolia Agricultural University, Hohhot, China

Mingmin Zhao College of Horticulture and Plant Protection, Inner Mongolia Agricultural University, Hohhot, China

Chapter 1
Plant Transcriptomics and Epitranscriptomics: An Introduction

Parvathy Devi S. B, Siya Pradeep, R. Aswati Nair, and Padmesh Pandaram Pillai

Abstract Transcriptomics and epitranscriptomics have revolutionized our understanding of RNA biology in plants, providing detailed insights into gene expression, regulatory mechanisms and adaptive responses to environmental signals. Recent advances in sequencing technologies, including next-generation and third-generation sequencing, have enabled high-resolution transcriptome analysis, revealing the complexity of coding and noncoding RNAs, alternative splicing and gene regulatory networks. Epitranscriptomics further expands this landscape by uncovering post-transcriptional RNA modifications, such as N6-methyladenosine (m6A), 5-methylcytosine (m5C) and pseudouridine (Ψ), which influence RNA stability, translation and stress adaptation. The integration of these approaches with computational tools and other "omics" technologies is driving innovations in plant science, with promising applications in crop improvement, stress resilience and medicinal plant research. This introductory chapter explores the evolution of transcriptomic and epitranscriptomic technologies, current methodologies, their impact on modern plant biology and future challenges.

Keywords Transcriptomics · Epitranscriptomics · Next-generation sequencing (NGS) · m6A · m5C · Pseudouridine · Gene regulation · Computational biology

Abbreviations

2′-O-Me	2′-O-methylation
A-to-I	Adenosine-to-Inosine
ac4C	N4-acetylcytidine

Parvathy Devi S. B · S. Pradeep · P. P. Pillai (✉)
Department of Genomic Science, Central University of Kerala, Kasaragod, India
e-mail: padmeshpillai@cukerala.ac.in

R. A. Nair
Department of Biochemistry and Molecular Biology, Central University of Kerala, Kasaragod, India

L. M. Vaschetto (ed.), *Plant Transcriptomics and Epitranscriptomics*, Concepts and Strategies in Plant Sciences, https://doi.org/10.1007/978-981-95-5183-5_1

Aza-IP	5-Azacytidine-mediated RNA ImmunoPrecipitation
bp	Base pairs
CAGE	Cap analysis of gene expression
cDNA	Complementary DNA
circRNAs	circular RNAs
CMCT	Carbodiimide treatment
CRISPR	Clustered regularly interspaced short palindromic repeats
DNA	Deoxyribonucleic acid
dNTP	Deoxynucleotide triphosphates
DRS	Direct RNA sequencing
ESTs	Expressed sequence tags
ESI-TOF-MS	Electrospray ionization time-of-flight mass spectrometry
FLC	*FLOWERING LOCUS C*
GEO	Gene Expression Omnibus
GO	Gene ontology
GSEA	Gene set enrichment analysis
HAMR	High-throughput annotation of modified ribonucleotides
HPLC	High-performance liquid chromatography
hm5C	5-Hydroxymethylcytosine
iCLIP	Individual-nucleotide resolution UV-crosslinking and ImmunoPrecipitation
LC-MS	Liquid chromatography-mass spectrometry
LC-MS/MS	Liquid chromatography-tandem mass spectrometry
m/z	mass-to-charge ratio
MeRIP	Methylated RNA immunoprecipitation
m1A	N1-methyladenosine
m3C	3-Methylcytidine
m5C	5-methylcytosine
m6A	N6-methyladenosine
m7G	N7-methylguanosine
miRNAs	microRNAs
mRNA	messenger RNA
MS	Mass spectrometry
ncm5U	5-carbamoylmethyluridine
ncRNAs/lncRNAs	noncoding RNAs/long noncoding RNAs
NGS	Next-generation sequencing
nm5U	5-aminomethyluridine
ONT	Oxford nanopore technologies
PacBio SMRT sequencing	Pacific Biosciences Single-Molecule Real-Time sequencing
PAR-CLIP	PhotoActivatable-Ribonucleoside-Enhanced CrossLinking and ImmunoPrecipitation

PCR	Polymerase chain reaction
PTC	Peptidyl Transferase Centre
QC	Quality control
RBPs	RNA-binding proteins
RNA	Ribonucleic acid
RNases	Ribonucleases
RNA-Seq	RNA sequencing
RNAMDB	RNA modification database
rRNA	ribosomal RNA
RT	Reverse transcription
RT-qPCR	Real-time quantitative PCR
SAGE	Serial analysis of gene expression
scRNA-Seq	Single-cell RNA sequencing
SNP	Single-nucleotide polymorphism
SOLiD	Sequencing by oligonucleotide ligation and detection
SRA	Sequence Read Archive
snoRNAs	small nucleolar RNAs
snRNAs	small nuclear RNAs
SSRs	Simple sequence repeats
SVM	Support vector machines
TGS	Third-generation sequencing
tRNAs	transfer RNAs
UMIs	Unique molecular identifiers
UTRs	Untranslated regions
Ψ	Pseudouridine

1.1 Introduction

The study of plant transcriptomics and epitranscriptomics has revolutionized our understanding of metabolic pathways, stress responses and gene regulation in plants. Transcriptomics, which examines the complete set of RNA molecules in a cell or tissue, provides insight into gene expression patterns and functional genomics. Recent developments in RNA sequencing (RNA-Seq) have further transformed the field by enabling accurate quantification of gene expression and facilitating the discovery of novel transcripts. Consequently, the integration of transcriptomics with other "omics" approaches has significantly advanced our research in agriculture, biotechnology, and medicinal plant science.

Epitranscriptomics, on the other hand, refers to chemical modifications in RNA and adds an additional layer of regulatory control over gene expression. Modifications such as N6-methyladenosine (m6A), 5-methylcytosine (m5C), pseudouridine (Ψ) and N1-methyladenosine (m1A) significantly influence RNA stability, translation and degradation. These modifications play a crucial role in plant growth, development and stress adaptation, offering promising applications in crop improvement. Further, recent advances in sequencing technologies and mass spectrometry (MS) have enhanced our ability to study the epitranscriptome, uncovering its functional significance in plants.

This chapter explores key techniques in transcriptomics and epitranscriptomics, highlighting their methodologies, applications and potential in plant research. A deeper understanding of these molecular processes is expected to lead to innovations in plant science and sustainable agriculture.

1.1.1 Plant Transcriptomics

The transcriptome refers to the complete set of RNA molecules—including messenger RNA (mRNA) and noncoding RNAs (ncRNAs)—transcribed from a cell, tissue or organism at a specific developmental stage or physiological condition (Tyagi et al. 2022). Plant transcriptomics employs high-throughput sequencing technologies, such as RNA sequencing (RNA-Seq), to enable precise and comprehensive transcriptome analysis. RNA-Seq allows accurate quantification of gene expression, discovery of novel transcripts, detection of alternative splicing events and identification of ncRNAs, thereby enriching our holistic understanding of transcriptional regulation in plants.

Comparative transcriptomic analysis, using multiple samples under defined conditions, enables the identification of signalling pathways and regulatory genes involved in stress tolerance. These insights contribute to crop improvement programs through genetic modification and informed breeding strategies, ultimately facilitating the development of enhanced plant varieties. Additionally, transcriptomics aids in the discovery of functional genes involved in metabolic pathways, particularly those responsible for the biosynthesis of medicinally important secondary metabolites, thereby advancing applications in synthetic biology.

When transcriptomics is integrated with other "omics" layers, such as genomics, proteomics, and metabolomics, it provides a systems-level understanding of plant physiology and stress responses, thereby driving innovation in agriculture and biotechnology.

1.1.1.1 Evolution of Transcriptomics

In the 1980s, low-throughput Sanger sequencing was widely used to generate expressed sequence tags (ESTs)—short sequences representing individual RNA transcripts. ESTs enabled gene identification without the need to sequence entire genomes, making them particularly valuable for studying environmental samples and mixed-species transcriptomes. Owing to their versatility, ESTs were also utilized in analyses of genetic variability at the species level. The term "transcriptome" was first introduced in the 1990s (Velculescu et al. 1997) and subsequent technological advances have since transformed transcriptomics into a mainstream research discipline.

In 1995, serial analysis of gene expression (SAGE) was developed, employing Sanger sequencing to analyse concatenated random transcript fragments, with quantification performed by matching these fragments to known genes. Further, the advent of microarray technology in the mid-1990s enabled the simultaneous analysis of thousands of transcripts through hybridization with complementary probes. The subsequent emergence of next-generation sequencing (NGS) technologies in 2005 marked a significant turning point in transcriptomics, generating vast amounts of sequencing data, facilitating genome-wide transcript analysis and enabling the discovery of genetic markers such as simple sequence repeats (SSRs) and single-nucleotide polymorphisms (Wang et al. 2015).

RNA sequencing (RNA-Seq), which sequences complementary DNA (cDNA) to quantify transcript abundance, rapidly replaced microarrays due to its superior resolution, precision, and ability to detect novel genes and fusion transcripts. By the 2010s, single-cell RNA sequencing (scRNA-Seq) had further advanced the field by enabling gene expression analysis at the level of individual cells (Seyfferth et al. 2021). More recently, emerging techniques such as spatial transcriptomics and long-read sequencing have provided deeper insights into RNA biology, bridging gaps between gene expression and functional genomics (Burgess 2019).

The study of individual RNA transcripts began well before transcriptomics emerged as a defined area of interest. In the 1970s, Walter Fiers and his team pioneered RNA sequencing by decoding the complete transcriptome of the MS2 bacteriophage. Initially, the inherent instability of RNA—due to its single-stranded nature and susceptibility to degradation—posed a significant challenge to transcriptome sequencing. The discovery of reverse transcriptase allowed scientists to convert the unstable messenger RNA (mRNA) and noncoding RNA (ncRNA) into stable complementary DNA (cDNA). Around the same time, Frederick Sanger developed DNA sequencing technology, which later became pivotal in the advancement of transcriptomics.

1.1.1.2 Transcriptomics Technologies

In the rapidly evolving field of molecular biology, comprehensive analysis of gene expression through transcriptomics enables the detection of broad, coordinated patterns that targeted assays may fail to detect or overlook. Transcriptomics technologies encompass all techniques used to study an organism's transcriptome. Advancements in these technologies have continuously reshaped the field, rendering earlier methods obsolete. Techniques such as EST libraries, microarrays, and RNA-Seq have facilitated high-throughput analysis, enabling the identification of thousands of differentially expressed genes. These tools have revolutionized research by improving our ability to assess cellular responses, identify molecular markers, analyse functional gene expression and explore biosynthetic pathways during plant development.

1.1.1.3 RNA Isolation

The first step in any transcriptomic technique is the isolation of high-quality RNA, which is essential for the success of downstream processes. Despite biological diversity across species, RNA extraction methods generally follow a standard workflow: mechanical disruption of cells or tissues in the presence of liquid nitrogen, inhibition of RNases using chaotropic agents, separation of RNA from other biomolecules such as DNA and concentration of RNA through precipitation or elution (Chomczynski and Sacchi 2006). DNase treatment is commonly applied to eliminate residual DNA and prevent interference in downstream applications. Since total RNA consists of approximately 98% ribosomal RNA (rRNA), enrichment of mRNA is typically performed using either poly-A affinity selection or rRNA depletion with sequence-specific probes (Zhao et al. 2014). As RNA degradation can significantly affect the accuracy of experimental results, additional precautionary measures are necessary to ensure RNA quality. To preserve RNA integrity, tissues are usually snap-frozen prior to extraction and RNase exposure is minimized during processing.

1.1.1.4 Serial and Cap Analysis of Gene Expression (SAGE/CAGE)

Serial analysis of gene expression (SAGE) was the first sequencing-based technique developed to simultaneously assess transcript abundance. It is based on the assumption that a short DNA sequence (tag) of 9–11 base pairs (bp), extracted from a defined location within a transcript, contains sufficient information to specifically identify that transcript, provided the position of the tag within the transcript is known. Transcript abundance is quantified by counting the frequency of these tags. These cDNA tags are concatenated to form longer fragments exceeding 500 bp, which are then sequenced. Following deconvolution, the tags are mapped to a reference genome to identify the corresponding genes. In the absence of a reference genome, tags that are differentially expressed under specific conditions can still serve as condition-specific molecular markers.

Unlike microarrays, which rely on known sequences to detect transcripts, SAGE can identify novel and low-abundance transcripts. It quantifies transcript levels by tag counting and allows straightforward comparison across samples. However, short tags may align to multiple genes with similar coding sequences, making annotation occasionally challenging. Therefore, access to reliable EST or genome database is essential for accurate transcript identification.

Cap analysis of gene expression (CAGE) is a modification of SAGE that sequences tags from the 5′ end of mRNA transcripts, enabling the identification of transcription start sites when aligned to a reference genome. This is particularly useful for full-length cDNA cloning and promoter analysis. Although SAGE and CAGE offer more comprehensive gene expression information than single EST sequencing, their sample preparation and analysis workflow are considerably more labour-intensive, which limits their practical utility.

1.1.1.5 Microarray

Microarrays are high-throughput techniques used to simultaneously assess the expression levels of thousands of genes by hybridizing fluorescently labelled cDNA or RNA transcripts to immobilized short nucleotide oligomers, known as 'probes', on a solid medium. Unlike SAGE, transcript abundance in microarrays is determined by quantifying hybridization signal intensity.

During gene expression, transcription produces multiple mRNA copies, which are translated into proteins. Therefore, assessing mRNA levels provides insights into gene expression patterns and facilitates the understanding of genetic regulation. However, mRNA is inherently unstable and is typically converted into cDNA for stabilization. The resulting cDNA is then labelled with Cy3 and Cy5 fluorochromes for detection. Microarrays operate on the principle of hybridization between the labelled target sequences and complementary probes fixed on the array surface. These probes are synthesized in situ by sequential nucleotide addition until the desired length is achieved. The amplified and purified cDNA is deposited at defined locations using specialized delivery systems to ensure precision and accuracy. Transcript abundance is inferred from fluorescence intensity at each probe site.

cDNA synthesis enables rapid transcriptome profiling across diverse biological samples, making microarrays well-suited for gene function analysis. Key parameters for microarray design and evaluation include probe density, throughput, reproducibility, data quality, cost, and ease of prototyping, all of which are largely relied on the availability of known gene sequences (Gao et al. 2004). Therefore, the major limitation of microarray is the requirement for prior genomic knowledge, such as an annotated genome or EST library, to design specific probes. Consequently, microarrays can only detect known genes. Moreover, since transcript abundance is inferred indirectly through hybridization signals, the technique is susceptible to false positives. Additionally, microarrays have limited capability in detecting transcripts arising from alternative splicing events.

1.1.1.6 Transcriptome Sequencing (RNA-Seq)

RNA sequencing (RNA-Seq) combines high-throughput sequencing with computational analysis to quantify the transcripts present in an RNA extract. The next-generation sequencing (NGS)-based method reveals expressed sequences in specific tissues at defined time points. The typical workflow includes RNA extraction, enrichment, fragmentation, amplification and sequencing. However, RNA-Seq requires the conversion of RNA into cDNA prior to sequencing, thereby making it compatible with existing genomic sequencing platforms.

During library preparation, cDNA is amplified using polymerase chain reaction (PCR) and spike-in controls are incorporated to assess data quality and facilitate normalization. Additionally, unique molecular identifiers are employed to quantify RNA molecules more accurately and to correct for amplification bias. The sequencing can be performed using single-end reads, which are faster and more cost-effective, or paired-end reads, which provide greater accuracy for gene annotation. Strand-specific protocols preserve transcriptional directionality, thereby improving the resolution of gene prediction and isoform detection.

Following the development of RNA-Seq, various NGS technologies have emerged, each offering unique methodologies and advantages. Platforms such as Roche 454, Illumina, and SOLiD have significantly advanced genomic and transcriptomic research. The evolution of sequencing technologies has progressed through three generations: first-generation sequencing enabled clonal DNA sequencing; second-generation sequencing increased throughput through massively parallel reactions; and third-generation sequencing allows direct sequencing of single DNA molecules without the need for amplification.

NGS technologies revolutionized molecular biology by providing a fast, cost-effective and high-throughput method for sequencing genomes and transcriptomes (Sathyanarayana et al. 2017; Tyagi et al. 2022). Unlike traditional Sanger sequencing, which is slow, labour-intensive and expensive, NGS can simultaneously sequence millions of nucleic acid molecules, greatly accelerating genome assembly and gene discovery. NGS has profoundly impacted research in plant genomics, transcriptomics, and genetic marker development, facilitating studies on gene expression, phylogenetics and responses to environmental stress. While NGS offers considerable advantages, it generates vast volumes of data that require advanced computational analysis for accurate interpretation (Zhao et al. 2014).

Introduced in 2005, Roche 454 sequencing is a high-throughput platform based on pyrosequencing that offers relatively long read lengths and short run times. It determines DNA sequences by measuring light intensity generated during nucleotide incorporation, including within homopolymer repeats. Unlike hybridization-based methods, this technology does not rely on probes. However, it is prone to insertion and deletion errors during sequencing (Margulies et al. 2005).

Illumina's NGS platform is widely used for large-scale genomic and transcriptomic studies, particularly in nonmodel plant and animal species. Solexa (now Illumina) sequencing, based on the sequencing-by-synthesis method, randomly attaches DNA fragments to a flow cell and undergoes bridge amplification to generate millions of distinct clusters. Fluorescently labelled deoxynucleotide triphosphates are incorporated during the extension process and the emitted fluorescence is captured to determine the nucleotide sequence. This platform is well-suited for genome-wide studies, including those lacking a reference genome. However, its main limitations include relatively short read lengths and the computational challenges associated with de novo assembly. The sequencing workflow involves multiple imaging cycles, where accurate identification of flow cell is essential for precise base calling.

SOLiD (Sequencing by Oligonucleotide Ligation and Detection) sequencing, introduced in 2007, is a massively parallel method using magnetic beads to facilitate cost-effective DNA library preparation. It is based on the sequencing-by-ligation principle, in which fluorescently labelled oligonucleotides are sequentially ligated to a DNA template, providing high base-calling accuracy. However, sequencing errors can propagate across cycles and complicate data decoding, making error correction more challenging (Guo et al. 2021).

The emergence of RNA sequencing (RNA-Seq) has further advanced transcriptomic studies by enabling deep sequencing of RNA for functional genomics and molecular marker identification, greatly benefiting plant research and breeding programs. RNA-Seq facilitates the investigation of gene expression under different conditions, the discovery of novel genes, comparative expression analysis across samples and the detection of sequence variations. Importantly, it provides insights into alternative splicing events that generate diverse mRNA transcripts and protein variants, thereby enhancing our understanding of plant cellular function and metabolic pathways.

RNA-Seq has evolved in parallel with advancements in sequencing technology. Its early applications began in 2006 using 454 sequencing (Lowe et al. 2017), while widespread adoption followed the release of Solexa/Illumina platforms in 2008, which enabled large-scale transcriptome analysis. Consequently, RNA-Seq has been extensively applied to model plants such as maize, *Arabidopsis* and rice, as well as to a wide range of nonmodel species.

Unlike microarrays, which have limited sensitivity and struggle to detect low-abundance transcripts, RNA-Seq can identify subtle changes in gene expression and discover novel genes in the absence of prior genome annotations. Currently, RNA-Seq is the dominant transcriptomic approach, having largely replaced microarrays due to its higher sensitivity, a broader dynamic range, and superior capacity to detect gene isoforms.

Advancements in RNA-Seq have significantly enhanced transcriptome characterization across various cell types, supporting genomic studies, EST sequencing, and molecular marker development. This technique has been successfully applied to marker development in numerous plant species, for instance, underutilized legumes like faba bean (*Vicia faba*) (Kaur et al. 2012), rice bean (*Vigna umbellata*) (Chen et al. 2016), grass pea (*Lathyrus sativus*), and Bambara groundnut (*Vigna subterranea*) (Chapman 2015). Additionally, RNA-Seq has proven useful for genomic studies in woody and medicinal plants, which remain genetically undercharacterized (Guo et al. 2021; Sathyanarayana et al. 2017).

RNA-Seq generates millions of short reads from cDNA fragments derived from sample RNA, facilitating transcriptome analyses such as quantification, differential expression, gene annotation, and de novo transcript assembly in the absence of a reference genome. Typically, reads are mapped to a reference genome, enabling the identification of isoforms. For species lacking a reference genome, RNA-Seq supports de novo assembly. This approach is particularly valuable for nonmodel plants, allowing the discovery of unique sequences and expression patterns.

1.1.1.7 Third-Generation Sequencing

Third-generation sequencing (TGS) overcomes several limitations of NGS, which requires PCR amplification and is prone to base mismatches, template switching, and

GC bias. In contrast, TGS provides long-read sequencing, making it especially useful for transcriptome sequencing, structural variation analysis, epitranscriptomics, and methylation studies. Although early TGS technologies had lower accuracy than NGS, continuous advancements have improved its reliability and hence TGS is now widely used in biological research (Xie et al. 2023).

The primary TGS technologies include Pacific Biosciences Single-Molecule Real-Time (PacBio SMRT) sequencing, and Oxford Nanopore Technologies (ONT) sequencing. PacBio SMRT sequencing employs a synthesis-based principle and does not require amplification, which reduces background noise and enables self-correction. ONT sequences DNA or RNA by passing molecules through a protein nanopore and detecting base sequences via changes in electrical signals (Van der Verren et al. 2020; Xie et al. 2023). Although ONT offers long reads and high throughput at a relatively low cost, it has relatively a high error rate (5–20%). Notably, nanopore technology has shown potential for peptide sequencing and is therefore not limited to DNA and RNA sequencing.

1.1.1.8 Single-Cell RNA Sequencing Technology (scRNA-Seq)

Single-cell RNA sequencing (scRNA-Seq) analyses gene expression at the individual cell level and enables the identification of cell subtypes and expression heterogeneity. Since its emergence in 2009, scRNA-Seq has provided high-resolution insights into cellular diversity, revolutionizing biological research by offering enhanced spatial and temporal resolution. In plants, scRNA-Seq has uncovered novel cell types and interactions and has facilitated the tracking of stem cell differentiation (Macosko et al. 2015; Bawa et al. 2022).

For example, in maize, scRNA-Seq has been used to map inflorescence development and genetic redundancy, aiding crop improvement (Xu et al. 2021a). In rice roots, the technique has been employed to reconstruct cell differentiation trajectories and uncover key regulatory mechanisms. A separate study on bundle sheath differentiation identified unique cellular traits relevant to bioengineering applications (Bezrutczyk et al. 2021). As scRNA-Seq continues to be applied across a wider range of crop species, it holds considerable promise for advancing agricultural research and breeding.

1.2 Transcriptome Data Analysis

RNA transcript data can be obtained through sequencing individual transcripts—such as ESTs and RNA-Seq—or by hybridizing transcripts to probe arrays, as in microarrays. Transcriptome reconstruction from raw sequencing reads requires specialized assembly tools that operate using either a reference genome, de novo assembly, or a combination of both approaches (Martin and Wang 2011).

In plants, NGS-based transcriptome analysis presents unique challenges, as plant RNA is highly susceptible to degradation by ribonucleases (RNases) and interference from secondary metabolites such as phenolics and polysaccharides. As a result, assessing RNA quality is especially critical for successful sequencing in plants—often more so than in animal systems. Furthermore, the continual evolution of sequencing platforms can influence experimental outcomes in unforeseen ways. Therefore, understanding the impact of both technical and biological variables is essential for the effective application of NGS technologies in plant molecular biology.

RNA-Seq enables both gene- and transcript-level expression analysis under varying physiological or environmental conditions, supported by a wide array of bioinformatics tools that utilize diverse algorithms and statistical models. A simplified illustration of the RNA-Seq workflow is presented in Fig. 1.1. Broadly, this framework supports customizable pipelines for transcriptome assembly and expression quantification, allowing researchers to tailor RNA-Seq workflow for specific objectives in plant systems.

The routine RNA-Seq workflow typically involves the following steps: sample preparation and assembly, preprocessing raw data, read alignment, transcriptome reconstruction, expression quantification, differential expression analysis and gene function annotation (Yang and Kim 2015). RNA-Seq generates vast amounts of raw

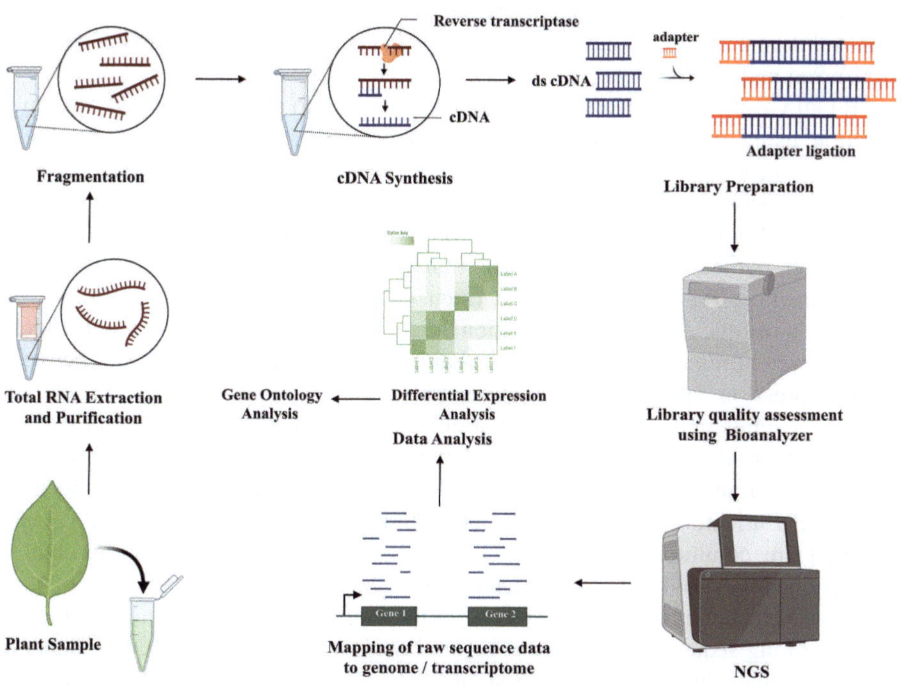

Fig. 1.1 The workflow illustrating the major events during RNA sequencing. (Created with BioRender.com)

sequence reads that must be processed using bioinformatics tools tailored to the experimental design.

The initial step in RNA-Seq data analysis involves quality control (QC) of raw sequencing reads to assess overall quality and identify potential contaminants or sequencing artefacts. This is followed by read alignment, where reads are mapped to a reference genome or transcriptome. Additional QC steps are often required at this stage, particularly when working with degraded or low-quality RNA samples. In cases where reference genome is unavailable, de novo assembly is employed as an alternative strategy for transcriptome reconstruction.

Transcriptome assembly enables the identification of all expressed transcripts based on mapped read data, followed by accurate estimation of transcript abundance. Raw sequencing data are typically stored in compressed FASTQ format, while processed count data are substantially small in size. Public repositories such as the Sequence Read Archive (SRA) and Gene Expression Omnibus (GEO) support data storage and sharing, thereby promoting reanalysis and reproducibility.

Further, differential expression analysis is conducted using specialized statistical tools to compare gene expression levels across different conditions or treatments, offering valuable insights into underlying biological variations.

Quantifying gene expression using an independent method is essential after identifying key genes and pathways regulating a biological system. Real-time quantitative PCR (RT-qPCR) is widely used for this purpose due to its high sensitivity, specificity, reproducibility, and accuracy—making it the gold standard for validating transcriptomic data obtained from large-scale studies such as RNA-Seq and microarrays.

RT-qPCR involves reverse transcription of RNA into cDNA, followed by amplification using gene-specific primers to quantify expression levels. It measures relative gene expression against internal reference (housekeeping) genes and generates expression profiles comparable to those from RNA-Seq, albeit limited to amplicons shorter than 300 base pairs. Primer specificity is critical to ensure accurate validation, particularly when analysing transcript isoforms.

In practice, RT-qPCR validation often demonstrates strong correlations with RNA-Seq results. Additionally, functional validation of key genes—through knockdown, overexpression or rescue experiments—further establishes the link between gene expression patterns and phenotypic traits.

The accurate and precise interpretation of biological information from differentially expressed genes still remains a major challenge in transcriptome studies, despite the transformative potential of RNA-Seq. Bioinformatics tools such as Gene Set Enrichment Analysis and Gene Ontology (GO)-based functional categorization are commonly employed to identify overrepresented ontologies and to classify genes accordingly (Subramanian et al. 2005).

Several technical and computational limitations persist, including difficulties in aligning long sequencing reads and accurately inferring expression level. The core computational challenge lies in assigning short sequencing reads to the correct transcripts and reliably quantifying gene expression. Benchmarking studies have often relied on simulated or long RNA-based datasets, which limits their applicability of total RNA analyses—particularly those involving small RNAs.

RNA-Seq data frequently include reads that map to multiple genes or transcript isoforms, especially in de novo assemblies where reference genome is unavailable. Experimental design parameters—such as read number, read length, and the use of single-end versus paired-end sequencing—significantly affect data quality and interpretability.

Given the high-throughput nature of transcriptomics studies, substantial computational resources are required to extract biologically meaningful insights. Moreover, RNA-Seq generates large datasets that often hold value beyond the scope of original study. Data deposited in public repositories such as the Sequence Read Archive (SRA) and Gene Expression Omnibus (GEO) promotes transparency, reproducibility and global accessibility.

1.3 Plant Epitranscriptomics

It is well established that plant development is regulated by both intrinsic genetic programs and external environmental cues, with gene expression controlled by complex, multilayered networks. RNA plays a pivotal role in this regulation, not only as a structural (rRNA), messenger (mRNA), and translator (tRNA) molecule involved in protein translation, but also as a regulator of gene expression, including small nuclear RNA (snRNAs) and other classes of regulatory RNAs.

The remarkable process of splicing increases transcript diversity, while post-transcriptional modifications further enhance the structural and functional complexity of the RNA molecule. The first RNA modification was reported in 1957 (Davis and Allen 1957; Cayir et al. 2020) and subsequent research has significantly expanded our understanding of these modifications and their roles in diverse biological processes. Occurring after transcription, these chemical modifications influence RNA structure, stability, metabolism and function, thereby contributing to transcriptomic diversity.

The study of RNA species and their co- and post-transcriptional modifications is collectively referred to as the *epitranscriptome*—a rapidly evolving field that has gained considerable attention in recent years. Epitranscriptomics encompasses the chemical and covalent modifications of ribonucleotides within RNA transcripts, emerging as dynamic regulator of gene expression. Parallel to epigenetic mechanisms, plants also employ epitranscriptomic modifications to sense, respond and to adapt under environmental stresses. While epigenetics primarily acts on the DNA level, epitranscriptomics regulates gene expression post-transcriptionally, at the RNA level.

Recent advances in mass spectrometry and high-throughput sequencing technologies have enabled detection and characterization of a wide range of RNA modifications, significantly accelerating progress in this field. To date, more than 170 distinct RNA modifications have been identified across organisms (Dhingra et al. 2023). Among the most prominent are N6-methyladenosine (m6A), 5-methylcytosine (m5C), inosine, pseudouridine (Ψ) and N1-methyladenosine (m1A) (Meyer et al. 2012; Dominissini et al. 2016; Morena et al. 2018). These modifications influence key aspects of RNA metabolism, including splicing, stability, translation and degradation.

Among them, m6A is the most extensively studied, present in 70–100% of mRNA transcripts with frequency and distribution varying across species and organelles. It can occur at multiple sites within a single transcript and functional studies on mutants deficient in m6A machinery have highlighted its essential role in plant development and stress responses. Similarly, m5C modifications in *Arabidopsis* and rice have been associated with developmental regulation and responses to oxidative stress and heat (Chmielowska-Bąk et al. 2023).

The plant epitranscriptomic machinery plays crucial roles in regulating physiological processes, development, and environmental adaptation. As a rapidly advancing field at the intersection of molecular biology, bioinformatics and epigenetics, epitranscriptomics holds significant promise for agricultural applications—particularly in improving crop resilience, productivity, and stress tolerance. Ongoing research aims to further unravel how specific RNA modifications could influence gene expression at both molecular and organismal levels.

1.3.1 RNA Modifications

Although chemical modifications of DNA and RNA have been known for over half a century, their genome-wide characterization began only in the late 1990s (Xie et al. 2023). Over the past decade, significant advancements have deepened our understanding of these modifications and their role in regulating transcript stability, translation efficiency, and intracellular localization within cells and plant organs. These modifications range from simple nucleotide alterations to complex structural changes that occur across all major classes of RNA.

Internal m6A modifications in mRNAs and m7G (N7-methylguanosine) in the 5′ cap were identified in 1970s. Later, several studies reported that post-transcriptional addition or deletion of nucleotides could change protein sequence and/or its length and these changes were collectively termed ‘RNA editing’ (Dhingra et al. 2023). Technological advancements have since enabled the detection of widespread RNA modifications across diverse organisms, including plants (Sharma et al. 2023).

Given that the structural modifications often impact distinct functional consequences; considerable research has focused on unravelling their diverse biological roles in plants. Key enzymatic players involved in “writing,” “reading,” and “erasing” these modifications have been identified, which offers critical insights into their molecular and physiological relevance. The epitranscriptome spans both coding RNAs (mRNAs) and various classes of noncoding RNAs, as discussed in the following sections.

1.3.1.1 Coding RNA (mRNA)

Among the key modifications, N7-methylguanosine (m7G) forms the cap structure at the 5′ end of eukaryotic mRNAs, playing a vital role in transcript stability, nuclear export, and efficient translation. In addition to this cap function, internal m7G modifications have also been identified within mRNAs. The 3′ end mRNA a poly (A) tail—comprising approximately 150–200 adenylate residues—and a conserved AAUAAA signal, which together enhance transcript stability, promote nuclear export and facilitate translation (Crick 1979; Zhao et al. 1999; Bawankar et al. 2010).

mRNA undergoes splicing, a post-transcriptional mechanism that removes introns from pre-mRNA and contributes significantly to transcriptome and proteome diversity. Furthermore, a variety of chemically modified nucleosides have been identified in mature, cytoplasmic and mitochondrial mRNAs, contributing to a highly heterogeneous mRNA population. The mRNA epitranscriptome spans untranslated regions (UTRs), exons and introns, where modifications such as m6A, m1A, m5C, m3C (3-methylcytosine), ψ and ac4C (N4-acetylcytidine) influence mRNA metabolism and gene expression. These modifications are dynamically regulated by specific enzymes categorized as writer, reader and eraser proteins, which collectively fine-tune gene expression at the post-transcriptional level.

Amongst, N6-methyladenosine (m6A) is the most abundant internal modification in plant mRNAs, involving methylation at the nitrogen-6 position of adenosine (Morena et al. 2018). Discovered in the 1970s, transcriptome-wide mapping has since revealed its roles in regulating mRNA stability, splicing and translation. m6A is often enriched near stop codons and within the 3′ UTR, where it contributes to plant growth development and stress responses. Notable m6A is the first RNA modification for which a complete set of regulatory components, including writers, erasers, and readers, has been identified (Cayir et al. 2020; Ramakrishnan et al. 2022).

Another prevalent modification, 5-methylcytosine (m5C), has been detected in plant coding sequences. It typically occurs two nucleotides downstream of the start codon or just upstream of the stop codon, suggesting a conserved positional preference (Cui et al. 2017). Although its precise function remains under investigation, m5C has been implicated in mRNA mobility across graft junctions. Similarly, 5-hydroxymethylcytosine (hm5C), predominantly located in exons, is believed to enhance translation efficiency, although its exact mechanism remains unclear.

Other notable RNA modifications include 3-methylcytosine (m3C), identified through computational predictions and RNA immunoprecipitation, although its biological significance remains unclear. N1-methyladenosine (m1A), though relatively rare, is present in both nuclear-encoded mRNA and mitochondrial transcripts and may influence translation efficiency. Pseudouridine (Ψ), the most abundant RNA modification, has been detected in plant mRNAs and is implicated in RNA stability, translation initiation, and ribosome interaction (Ontiveros et al. 2019). Adenosine-to-inosine (A-to-I) editing alters base-pairing properties and can subsequently affect mRNA structure, fate, and function. Additionally, 2′-O-methylation

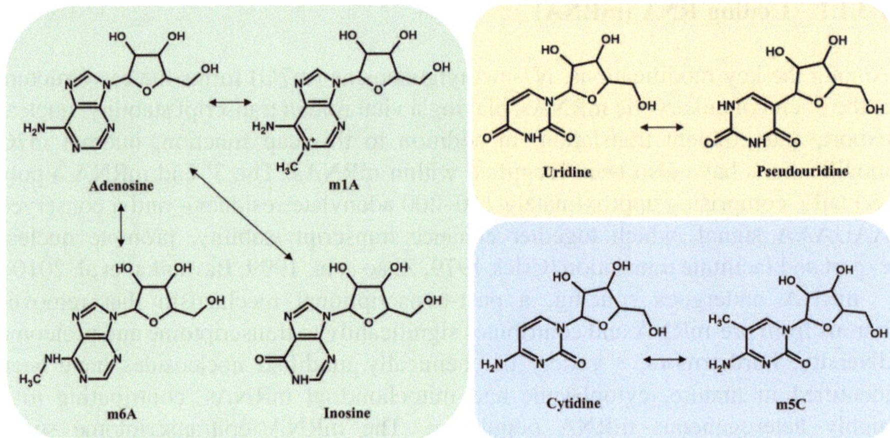

Fig. 1.2 Predominant RNA modifications occurring at selected nitrogen bases

(2′-O-Me) guided by small nucleolar RNAs (snoRNAs), may contribute to self/non-self RNA discrimination, enhance RNA stability, and regulate translation.

Collectively, these modifications play critical roles in regulating mRNA metabolism, thereby shaping plant growth, development, and responses to environmental cues. Representative examples of RNA base modifications are illustrated in Fig. 1.2.

1.3.1.2 Noncoding RNA (ncRNA)

The epitranscriptome encompasses a diverse array of post-transcriptional modifications across all RNA classes, including transfer RNA (tRNA), ribosomal RNA (rRNA), and regulatory noncoding RNAs (ncRNAs) such as long noncoding RNAs (lncRNAs), microRNAs (miRNAs), and circular RNAs (circRNAs). While modifications in tRNA and rRNA are extensively characterized and known to play critical roles in translation and ribosome function, our understanding of modifications in other noncoding RNA species, as well as their biological significance, remains incomplete and forms subject of ongoing studies.

Ribosomal RNA (rRNA) undergoes numerous post-transcriptional modifications during ribosome biogenesis, primarily catalysed by writer enzymes or directed by antisense small nucleolar RNAs (snoRNAs). Among the most prevalent rRNA modifications are 2′-O-methylation (2′-O-Me) and uridine-to-pseudouridine (Ψ) isomerization. These epitranscriptomic modifications are essential for stabilizing rRNA secondary structures, promoting correct folding and enhancing ribosomal catalytic activity. Notably, rRNA modifications are enriched in functionally critical regions of ribosome, such as the peptidyl transferase centre, where they fine-tune ribosomal architecture and facilitate accurate processing. 2′-O-Me modifications contribute to structural stability by shielding the phosphate backbone from hydrolysis and influencing the ribose conformation, thereby supporting the formation of complex secondary and tertiary structures vital for ribosome function. Pseudouridine (Ψ) modifications are predominantly localized within domain V of the large ribosomal subunit, a key region for peptidyl transferase activity. Although the precise mechanistic role of Ψ remains under active investigation, their absence has been correlated with defective translation and reduced cellular growth. While Ψ is also present in the small subunit, it has not been detected in its functionally essential domains. Collectively, these rRNA modifications are indispensable for proper ribosomal assembly, structural integrity and translation efficiency, underscoring their fundamental role in gene expression.

Transfer RNAs (tRNAs) are small, noncoding RNAs typically ranging from 70 to 100 nucleotides in length, and constitute the most abundant RNA class in the cells. They play essential roles in translation by accurately decoding mRNA codons and delivering the corresponding amino acids to the ribosome for protein synthesis. Remarkably, tRNAs are among the most heavily modified RNA species, with an estimated 10–20% of their nucleotides undergoing post-transcriptional chemical modifications. These modifications are critical for maintaining tRNA structural stability, improving aminoacylation efficiency and ensuring high fidelity codon-anticodon pairing during translation. In *Arabidopsis thaliana*, conserved modification sites have been identified in key structural regions of tRNA, including the D stem, anticodon loop and TΨC-loop. Interestingly, certain stress-responsive modifications such as 5-aminomethyluridine (nm5U) and 5-carbamoylmethyluridine (ncm5U) have been observed to increase under salt stress, implicating them in plant stress adaptation mechanisms. Overall, tRNA modifications are highly conserved across eukaryotes and are subject to dynamic regulation in responses to developmental cues and environmental stimuli, underscoring their fundamental role in plant growth, translation and stress resilience.

Epitranscriptomic modifications are increasingly being recognized for their pivotal role in regulating RNA functionality; however, their specific contribution to lncRNAs remain relatively underexplored. Recent findings have demonstrated that the antisense lncRNA *COOLAIR*, transcribed from the 3′ end of the *FLOWERING LOCUS C* (*FLC*) gene, requires m6A methylation to effectively silence *FLC* expression, thereby promoting flowering in plants (Kramer et al. 2018; Xu et al. 2021b; Sharma et al. 2023). Similarly, chemical modifications have been identified in mature miRNAs, which are key mediators of RNA silencing (Cai et al. 2025). Among these, m6A modification has been reported as a positive regulator of miRNA biogenesis (Sharma et al. 2023). In addition, 2′-O-Me at the 3′ end of mature miRNA protects them from degradation (Ontiveros et al. 2019). Small nuclear RNAs (snRNAs), the integral components of the spliceosome, also undergo extensive post-transcriptional modifications, notably Ψ and 2′-O-Me, especially in U2 snRNA. These modifications are essential for spliceosome assembly and enhance splicing accuracy, ultimately impacting gene expression (Wu et al. 2021). In circular RNAs (circRNAs), m6A modifications have been identified through immunoprecipitation-based assays and computational prediction. Studies in plant species such as moso bamboo and petunia have reported both m6A and m1A sites in

Table 1.1 A brief summary of major RNA modifications and their functional effect on RNA species

RNA modification	Chemical structure	RNA type	Function
N7-Methylguanosine (m7G)	Methylation at N7 position of guanosine	mRNA	5′ cap structure for mRNA stability, translation efficiency
Poly(A) tail	Adenylate residues (150–200)	mRNA	Enhances mRNA stability, nuclear export, and translation
N6-Methyladenosine (m6A)	Methylation at N6 position of adenosine	mRNA, lncRNA, circRNA	Influences mRNA structure, stability, translation, stress response regulates RNA processing in crcRNA
N1-Methyladenosine (m1A)	Methylation at N1 of adenosine	mRNA, mitochondrial RNA	Affects translation efficiency
A-to-I editing	Deamination of adenosine to inosine	mRNA	Alters base-pairing, affects RNA function
5-Methylcytosine (m5C)	Methylation at C5 of cytosine	mRNA	Affects mRNA mobility, stability, translation
5-Hydroxymethylcytosine (hm5C)	Hydroxylation of m5C	mRNA	Enhances translation efficiency
3-Methylcytosine (m3C)	Methylation at N3 of cytosine	mRNA	Unknown function
Pseudouridine (Ψ)	Isomerization of uridine	mRNA, rRNA, snRNA	RNA stability, translation initiation, ribosome interaction ribosome catalytic activity
2′-O-methylation (2′-O-Me)	Methylation at 2′-OH of ribose	mRNA, rRNA, miRNA, snRNA	Increases RNA stability, self/non-self RNA discrimination

circRNAs, often localized near exon splice sites, suggesting a potential role in RNA processing, stability and possibly translational regulation (Sharma et al. 2023).

RNA modifications are involved in regulating molecular and physiological processes; however, their effects can vary depending on biological context, and sometimes they may even exhibit contradictory roles within the same cell (Table 1.1). To comprehensively understand these modifications, it is essential to investigate the roles of RNA-binding proteins (RBPs) that mediate their addition, recognition and removal (i.e., writers, readers and erasers). These site-specific enzymatic modifications include methylation, deamination, thiolation, glycosylation, and isomerization, forming an intricate epitranscriptomic regulatory network. Remarkably, this system has been conserved across all domains of life following similar epigenetic mechanisms, but with more sophistication.

Despite its high conservation level in eukaryotic systems, RNA modifications remain largely uncharted, with much still to be discovered. While current research focuses on mapping these epitranscriptomic marks, it is equally important to investigate their functional roles, particularly in plants, to fully appreciate their biological significance.

1.3.1.3 Techniques for Studying RNA Modifications

The study of RNA modifications progressed slowly over the past five decades due to the lack of precise detection methods. However, advancements in molecular biology, NGS and bioinformatics have revolutionized the field, enabling transcriptome-wide identification of RNA modifications (Kladwang et al. 2020). These technological breakthroughs have significantly advanced epitranscriptome research, leading to the cataloguing of over 170 distinct RNA modifications (Dhingra et al. 2023).

Although detection of epitranscriptomic marks remains challenging, various NGS and mass spectrometry-based approaches, often combined with chemical, antibody or enzymatic methods, have been developed to identify and quantify RNA modifications. Detection strategies generally fall into three categories: immunoprecipitation-based methods, reverse transcription-based methods and mass spectrometry. The first two is frequently coupled with high-throughput sequencing. Rapid and continuous advancements in these techniques are driving deeper insights into RNA biology.

1.3.1.4 Immunoprecipitation-Based RNA Enrichment

Methylated RNA immunoprecipitation (MeRIP) is a widely used technique for detecting modified nucleotides such as m6A, m5C, hm5C and m1A. This method relies on antibodies that selectively bind to specific methylated bases, enabling transcript enrichment for downstream analysis. However, MeRIP may introduce precipitation bias, affecting the accuracy of modification site localization. To enhance precision, it is often combined with techniques such as Individual-nucleotide resolution UV-crosslinking and ImmunoPrecipitation or PhotoActivatable-Ribonucleoside-Enhanced CrossLinking and ImmunoPrecipitation. These methods improve enrichment efficiency by using ultraviolet light to crosslink RNA at reactive bases prior to antibody selection, thereby increasing the specificity of detected modifications. Furthermore, alternative enrichment strategies such as 5-Azacytidine-mediated RNA ImmunoPrecipitation (Aza-IP) and biotin-based clickable chemistry targets modified bases directly. These approaches introduce selective chemical modifications that induce reverse transcription abortion at modification sites, allowing precise mapping without the need for specific antibodies. Nevertheless, antibody-based methods are restricted to the availability of validated antibodies.

1.3.1.5 Reverse Transcription (RT)-Based Detection

Reverse transcription (RT)-based approaches identify modifications by detecting disruptions during cDNA synthesis (Xie et al. 2023). Certain modifications such as m1A induce mis-incorporation or premature termination during primer extension, resulting in characteristic signatures in sequencing data. Similarly, inosine modifications from adenosine-to-inosine (A-to-I) by RNA editing generate distinctive RT signatures. However, some modifications are RT-insensitive and require prior chemical treatment to become detectable. For instance, bisulphite conversion selectively deaminates cytosine to uracil, enabling discrimination of m5C and hm5C by analysing retained cytosine residues. Similarly, pseudouridine (Ψ) detection is enhanced by carbodiimide treatment, which causes RT stalling at Ψ sites, leading to reduced sequencing coverage at those positions. RT-based methods offer high-resolution, single-nucleotide mapping of RNA modifications, making them valuable tools for epitranscriptomic studies.

1.3.1.6 Mass Spectrometry-Based Detection

Mass spectrometry (MS) is a powerful technique for identifying and quantifying RNA modifications with high specificity (Yan et al. 2021). Unlike immunoprecipitation or RT-based methods, MS can detect broad range of modifications by analysing the mass-to-charge ratio (m/z) of modified nucleosides, enabling identification of both known and novel modifications. Prior to MS analysis, RNA is enzymatically digested into nucleosides or standardized fragments. High-performance liquid chromatography is often employed to separate modified nucleosides before analysis with techniques such as liquid chromatography-mass spectrometry (LC-MS), tandem MS (LC-MS/MS), or electrospray ionization time-of-flight MS (ESI-TOF-MS) (Stephenson et al. 2022). Additionally, two-dimensional hydrophobic end-labelling strategies can enhance resolution and improve detection of modifications within complex RNA populations. The purity of the RNA sample is critical for ensuring accurate analysis.

The analysis of transcriptome-wide patterns in a context-dependent manner remains challenging. To address this, high-throughput profiling technologies that combine short-read sequencing with antibody-based, chemical and enzymatic methods, and Nanopore long-read direct RNA sequencing, have been developed (Shen et al. 2023). Additionally, specialized sequencing approaches have emerged for detecting specific base modifications. However, systematic analysis is hindered by the low abundance of many RNA modifications and technical constraints associated with simultaneous detection of multiple modifications.

Third-generation sequencing (TGS) has revolutionized epitranscriptomic studies by directly analysing native RNA molecules, thereby eliminating PCR biases and reducing the need for repeated library preparations. This technology enables the identification of multiple RNA modifications in a single run and provides full-length transcript data, including details of 5′ cap and 3′ poly(A) tail, which are essential for studying alternative transcription initiation, cleavage and polyadenylation. Both PacBio Single-Molecule Real-Time (SMRT) sequencing and Oxford Nanopore Technologies (ONT) generate full-length transcripts, but their capabilities in detecting base modifications differ. PacBio SMRT HiFi sequencing detects DNA methylations via polymerase kinetics but requires additional treatments for comprehensive RNA modification analysis (Xie et al. 2023). In contrast, ONT provides both cDNA sequencing and direct RNA sequencing (DRS), allowing direct detection of base modifications without additional processing, making it a more versatile tool for epitranscriptomic research.

1.3.1.7 Computational Methods for Detection and Estimation of RNA Modifications

Advancements in mass spectrometry (MS) and sequencing technologies over the past decade have significantly expanded transcriptomic and epitranscriptomic datasets. Effective interpretation of these large-scale data depends on continuous development of computational tools. Notably, in silico methods, including support vector machines and other machine learning algorithms, have demonstrated high accuracy in predicting post-transcriptional modification sites. To complement the experimental approaches, various computational techniques have emerged to identify and quantify RNA modifications systematically. For such a purpose, specialized databases have also been developed, providing repositories for epitranscriptomic information accessible to the scientific community.

RNA modification databases serve as essential resources for understanding the diversity and functions of chemical modifications in RNA molecules. The RNA Modification Database, established in 1997, remains as a key reference, cataloguing approximately 128 RNA modifications predominantly found in tRNAs, rRNAs, and mRNAs. It also provides valuable tools for MS-based RNA modification analysis. MODOMICS, a database of RNA modifications and related information, is the most comprehensive repository of RNA modification pathways, currently listing over 300 modifications along with functionally characterized enzymes involved in their biosynthesis (available at https://iimcb.genesilico.pl/modomics/). This database integrates LC-MS data, structural files and detailed biosynthetic pathway information. Several other specialized databases focus on distinct aspects of RNA modifications. For instance, RMBase compiles high-throughput sequencing data to explore post-transcriptional modifications and their associations with diseases, RNA-binding proteins, and miRNAs; PRMdb provides detailed annotations of RNA modifications; while PED and REDIdb focus specifically on RNA editing events, especially in plants.

Computational approaches have significantly advanced RNA modification studies. High-throughput annotation of modified ribonucleotides predicts global RNA modifications by analysing characteristic sequencing errors introduced by reverse transcriptase enzymes, successfully identifying novel m3C sites in *Arabidopsis*. Similarly, PEA-m5C, a machine-learning model, predicts m5C modifications transcriptome-wide based on sequence features. Together, these databases and computational tools continue to deepen our understanding of RNA modifications and their diverse biological roles.

1.4 Functional Role of Transcriptomics and Epitranscriptomics in Plants

In brief, both transcriptomics and epitranscriptomics play crucial roles in advancing plant research and agriculture. Transcriptomics, through the analysis of gene expression patterns, regulatory networks and molecular processes, helps uncover the genetic basis of important plant traits. This knowledge facilitates the development of improved crop varieties with enhanced productivity, including stress tolerance and nutritional value, thereby supporting sustainable agriculture. The applications of transcriptomics in plants are extensive, including hormone signalling, responses to abiotic and biotic stresses, developmental gene expression, carbon fixation and secondary metabolite biosynthesis, among others.

One of the key applications of transcriptomics is gene discovery, which enables the identification of novel genes involved in specific biological processes. This is essential for understanding plant traits and guiding crop improvement strategies. Additionally, gene expression profiling allows researchers to examine how genes are regulated across different tissues, developmental stages and environmental conditions, providing insights into plant growth and stress responses. Transcriptomic data also play a crucial role in gene function annotation by enabling comparisons with existing genomic databases and linking them to specific phenotypes. For instance, in *Arabidopsis*, genes associated with metal hyper accumulation have been associated with metal uptake, tolerance and homeostasis (Verbruggen et al. 2009). Moreover, integration of RNA-Seq across tissues has enhanced gene function annotation in species like cucumber (Li et al. 2011).

Transcriptomics contributes to systems biology by integrating data from multiple omics fields, such as proteomics and metabolomics (Chen 2025). This holistic approach facilitates mapping of regulatory networks and predicting the effects of genetic modifications on plant phenotypes. The analysis of gene expression also plays a significant role in plant biotechnology, particularly in genetic engineering. It is employed to assess the impact of gene editing or insertion on overall transcript levels, thereby ensuring the safety of genetically modified crops for human consumption and environmental sustainability (Ricroch et al. 2011).

In medicinal plant research, transcriptome sequencing has been instrumental in functional gene discovery, molecular marker development, identification of secondary metabolite pathway and understanding developmental mechanisms. These advancements provide valuable insights for crop breeding and enhance the production of bioactive compounds. Traditionally, SSR markers (microsatellites) have been widely used due to their high polymorphism, simple detection, codominance and broad applications in genetic mapping, diversity analysis and parentage identification (Guo et al. 2021; Tyagi et al. 2022). A notable example is the de novo transcriptome assembly of two *Menispermum* species, which generated 53,712 and 78,921 unigenes, respectively. From these datasets, researchers identified 521 polymorphic EST-SSRs, providing a valuable resource for marker development and population genetics studies (Hina et al. 2020).

Transcriptomics enables the identification of genes and pathways involved in responses to environmental stresses through a nontargeted approach that uncovers novel transcriptional networks. For example, studies in *Candida albicans*, biofilm formation revealed a coregulated set of genes essential for biofilm maintenance (Lowe et al. 2017). Additionally, transcriptomic profiling has provided insights into mechanisms underlying drug resistance (Lowe et al. 2017). These findings offer valuable foundations for future functional genomic research and metabolite engineering.

RNA modifications, including nucleotide methylation and pseudouridylation, have been observed across multiple plant species. Recent studies have provided new insights into the functional roles of these RNA modifications in plants, along with the identification of the enzymes responsible for addition, removal, and recognition of these modifications.

In maize, global m6A methylation generally exhibits a negative correlation with translation, except when located at the 5′ end of transcripts, where it enhances translation efficiency. Studies on cold stress indicate that m6A-enriched transcripts are translated more actively, linking this modification to increased gene expression (Govindan et al. 2022). To further elucidate the functional roles of RNA modifications in plants, integrative approaches combining genetic manipulation with proteomics are essential.

The influence of RNA modifications on RNA stability is complex and not yet fully understood. Studies on the widely studied m6A modification indicate that its depletion in plants leads to reduced stability of target mRNAs, suggesting a stabilizing role (Sharma et al. 2023). Similarly, other modifications such as m1A in *Petunia* and m5C in *Arabidopsis* have also been associated with increased RNA stability (David et al. 2017; Yang et al. 2020).

RNA modifications influence transcript stability and translation by altering secondary structure and base pairing. They help stabilize tRNA and rRNA by modifying thermodynamic properties in specific regions, which is critical for proper RNA folding and function. Recent research has also highlighted the essential role of mRNA modifications in RNA transport. For example, m5C has been shown to be crucial for the transport of specific methylated mRNAs across graft junctions via the phloem (Sharma et al. 2023).

RNA modifications are involved in all aspects of plant development, influencing both vegetative and floral growth, as well as stress responses (Qin et al. 2021; Yu et al. 2021). While numerous reviews address their types, detection methods, and roles in growth, studies specifically focusing on their function in plant stress responses remain limited. The presence, alteration or absence of specific RNA marks across eukaryotic species can lead to severe consequences, including embryonic lethality and organ-specific abnormalities.

RNA modifications also play a significant role in plant stress responses by shaping the epitranscriptome. They influence transcript stability and function, thereby impacting gene expression regulation. For example, m6A and m5C methylations are crucial for fine-tuning gene expression during stress. In *Arabidopsis*, m6A profiling revealed widespread methylation changes in transcripts under salt stress (Sharma et al. 2023). Similarly, cold stress increased m6A enrichment in responsive transcripts, enhancing both their abundance and translation (Govindan et al. 2022).

1.4.1 Transcriptomics and Epitranscriptomics: Challenges and Future Directions

Transcriptomics has significantly advanced our understanding of gene expression, enabling the study of plant biology at an unprecedented scale. RNA sequencing (RNA-Seq), a high-throughput next-generation sequencing (NGS) technology, has become an essential tool for such activities as functional gene mining, molecular marker development, secondary metabolite pathway identification, and overall plant developmental studies. The rapid advancement of sequencing technologies has made transcriptomics more accessible and allowed integration with other omics approaches like proteomics and metabolomics, for a comprehensive understanding of cellular processes.

Although significant progresses have already been achieved, challenges such as short read lengths and the dependence for reference genomes remain to be fully addressed. Though TGS technologies offer longer read lengths and enable complete transcript sequencing, they also exhibit relatively high error rates. Therefore, an effective combination of TGS and NGS approaches may improve accuracy and enhance productivity of transcriptome research. Coupled with decreasing costs and increasing accuracy, sequencing technologies will continue to propel transcriptomics forward.

Transcriptomics provides insights into gene expression, regulatory networks, and molecular processes, thereby advancing plant and agricultural research. Current studies focus on both sequenced and nonsequenced species, revealing genes linked to commercially valuable traits (Agarwal et al. 2014). This has led to the production of crop varieties with improved productivity, nutritional quality and stress tolerance. Evidently, emerging frontiers like single-cell transcriptomics, spatial transcriptomics, comparative transcriptomics, and long-read sequencing technologies are reshaping plant transcriptomics. Single-cell transcriptomics for plants is making it possible for researchers to study cell fate, regulatory networks and developmental processes, while spatial transcriptomics helps uncover cellular interactions, tissue-specific gene expression patterns and the spatial organization of gene expression. Comparative transcriptomics will help explain how gene expression evolves, how plants adapt and how regulatory networks are conserved or diverge across species. Progress in long-read sequencing and data analysis will advance our understanding of gene structure, alternative splicing and transcript diversity in plants (Daszkowska-Golec et al. 2023). Transcriptomics, combined with high-throughput functional genomics (clustered regularly interspaced short palindromic repeats (CRISPR)-based editing) and computational approaches such as machine learning, is accelerating functional annotation (Pan et al. 2023). Given the vast amount of data generated, the next priority should be the experimental validation of predicted pathways and networks to clarify the roles of genes across plant species.

Over the past decade, research on plant epitranscriptomics has gained considerable momentum. RNA modifications and editing are now recognized as key players in gene regulation. Dynamic reprogramming of the epitranscriptome makes it possible to regulate growth, development, and stress responses without altering genome integrity (Dhingra et al. 2023). Early studies in epitranscriptome editing already highlight its practical value. For example, Shen et al. (2023) demonstrated that introducing human m6A demethylase into rice and potato was associated with enhanced yield and drought tolerance. Still, most studies are restricted to m6A modification, and even this area is at a nascent stage (Dhingra et al. 2023). Thus, exploring the functions of other RNA modifications and different RNA editing types is a critical research area. RNA modifications are often context-dependent, so high-resolution analysis at cellular and tissue levels will likely be essential, together with the development of various detection methods using Nanopore–DRS technology and refined bioinformatic tools.

Challenges in plant transcriptomics include the lack of single-nucleotide resolution maps in crops, limited understanding of the biological outcomes, the need for optimized editing strategies, and the high error rates and false positives in detecting and mapping them. Although new approaches like CRISPR-based editing are promising, they require further refinement to deliver reliable and high-resolution datasets. Recent progress with CRISPR/Cas9-mediated regulation of RNA modification proteins and targeted epitranscriptomic alterations using dCas13-fused proteins, as reported by Sharma et al. (2023), shows that precise manipulation of the epitranscriptome is possible. Detailed knowledge of expression patterns, substrates, and functions of writers, readers, and erasers of RNA modification pathways in crops is particularly essential for understanding stress adaptation. In future, combining multiomics approaches may provide a more comprehensive view of how RNA modifications interact with epigenetic and metabolic pathways. Thus, consistent research on transcriptomics and epitranscriptomics will ensure food security by producing stress-resilient, high-yielding crops.

1.5 Conclusion

Plant transcriptomics and epitranscriptomics have revolutionized our understanding of gene expression, regulatory mechanisms and plant adaptation to environmental stresses. High-throughput sequencing technologies, particularly RNA-Seq, have enabled precise transcriptome analysis, facilitating gene discovery, metabolic pathway elucidation, and crop improvement strategies. Advances in single-cell RNA sequencing and third-generation sequencing continue to refine our understanding of plant biology at unprecedented resolution.

Epitranscriptomics adds an additional layer of gene regulation by influencing RNA stability, translation and stress responses. Modifications such as m6A and m5C play a critical role in plant development and environmental adaptation. Emerging techniques, including mass spectrometry and nanopore sequencing, are rapidly enhancing the detection and functional characterization of these modifications.

Against the significant achievements made, the challenges still remain, particularly in achieving single-nucleotide resolution mapping of RNA modifications and understanding their dynamic regulation. Therefore, future research integrating transcriptomics with multiomics approaches will be crucial to unlock the full potential of RNA-based regulation in plant science. However, advancements hold great promise for improving crop resilience, productivity and sustainability in the face of global agricultural challenges.

Conflict of Interest Disclosure The authors declare that they have no conflict of interest relevant to the content of this chapter.

References

Agarwal P, Parida SK, Mahto A (2014) Expanding frontiers in plant transcriptomics in aid of functional genomics and molecular breeding. Biotechnol J 9:1480–1492. https://doi.org/10.1002/biot.201400063

Bawa G, Liu Z, Yu X et al (2022) Single-cell RNA sequencing for plant research: insights and possible benefits. Int J Mol Sci 23:4497. https://doi.org/10.3390/ijms23094497

Bawankar P, Shaw PJ, Sardana R et al (2010) 5′ and 3′ end modifications of spliceosomal RNAs in plasmodium falciparum. Mol Biol Rep 37:2125–2133. https://doi.org/10.1007/s11033-009-9682-4

Bezrutczyk M, Zöllner NR, Kruse CPS et al (2021) Evidence for phloem loading via the abaxial bundle sheath cells in maize leaves. Plant Cell 33:531–547. https://doi.org/10.1093/plcell/koaa055

Burgess DJ (2019) Spatial transcriptomics coming of age. Nat Rev Genet 20:317–317. https://doi.org/10.1038/s41576-019-0129-z

Cai J, Shen L, Kang H et al (2025) RNA modifications in plant adaptation to abiotic stresses. Plant Commun 6. https://doi.org/10.1016/j.xplc.2024.101229

Cayir A, Byun H-M, Barrow TM (2020) Environmental epitranscriptomics. Environ Res 189:109885. https://doi.org/10.1016/j.envres.2020.109885

Chapman MA (2015) Transcriptome sequencing and marker development for four underutilized legumes. Appl Plant Sci 3:apps.1400111. https://doi.org/10.3732/apps.1400111

Chen J-T (ed) (2025) Biotechnology, multiple omics, and precision breeding in medicinal plants. CRC Press, Boca Raton. https://doi.org/10.1201/9781003475491

Chen H, Chen X, Tian J et al (2016) Development of gene-based SSR markers in rice bean (Vigna umbellata L.) based on transcriptome data. PLoS One 11:e0151040. https://doi.org/10.1371/journal.pone.0151040

Chmielowska-Bąk J, Searle IR, Wakai TN et al (2023) The role of epigenetic and epitranscriptomic modifications in plants exposed to non-essential metals. Front Plant Sci 14. https://doi.org/10.3389/fpls.2023.1278185

Chomczynski P, Sacchi N (2006) The single-step method of RNA isolation by acid guanidinium thiocyanate-phenol-chloroform extraction: twenty-something years on. Nat Protoc 1:581–585. https://doi.org/10.1038/nprot.2006.83

Crick F (1979) Split genes and RNA splicing. Science 204:264–271. https://doi.org/10.1126/science.373120

Cui X, Liang Z, Shen L, Zhang Q et al (2017) 5-Methylcytosine RNA methylation in Arabidopsis Thaliana. Mol Plant 10:1387–1399. https://doi.org/10.1016/j.molp.2017.09.013

Daszkowska-Golec A, Mascher M, Zhang R (2023) Editorial: applications of long-read sequencing in plant genomics and transcriptomics. Front Plant Sci 14:1141429. https://doi.org/10.3389/fpls.2023.1141429

David R, Burgess A, Parker B et al (2017) Transcriptome-wide mapping of RNA 5-Methylcytosine in Arabidopsis mRNAs and noncoding RNAs. Plant Cell 29:445–460. https://doi.org/10.1105/tpc.16.00751

Davis FF, Allen FW (1957) Ribonucleic acids from yeast which contain a fifth nucleotide. J Biol Chem 227:907–915

Dhingra Y, Gupta S, Gupta V et al (2023) The emerging role of epitranscriptome in shaping stress responses in plants. Plant Cell Rep 42:1531–1555. https://doi.org/10.1007/s00299-023-03046-1

Dominissini D, Nachtergaele S, Moshitch-Moshkovitz S et al (2016) The dynamic N(1)-methyladenosine methylome in eukaryotic messenger RNA. Nature 530:441–446. https://doi.org/10.1038/nature16998

Gao X, Gulari E, Zhou X (2004) In situ synthesis of oligonucleotide microarrays. Biopolymers 73:579–596. https://doi.org/10.1002/bip.20005

Govindan G, Sharma B, Li Y-F et al (2022) mRNA N6 -methyladenosine is critical for cold tolerance in Arabidopsis. Plant J 111:1052–1068. https://doi.org/10.1111/tpj.15872

Guo J, Huang Z, Sun J et al (2021) Research progress and future development trends in medicinal plant transcriptomics. Front Plant Sci 12. https://doi.org/10.3389/fpls.2021.691838

Hina F, Yisilam G, Wang S et al (2020) De novo transcriptome assembly, gene annotation and SSR marker development in the moon seed genus Menispermum (Menispermaceae). Front Genet 11. https://doi.org/10.3389/fgene.2020.00380

Kaur S, Pembleton LW, Cogan NOI et al (2012) Transcriptome sequencing of field pea and faba bean for discovery and validation of SSR genetic markers. BMC Genomics 13:104. https://doi.org/10.1186/1471-2164-13-104

Kladwang W, Topkar VV, Liu B et al (2020) Anomalous reverse transcription through chemical modifications in polyadenosine stretches. Biochemistry 59:2154–2170. https://doi.org/10.1021/acs.biochem.0c00020

Kramer MC, Anderson SJ, Gregory BD (2018) The nucleotides they are a-changin': function of RNA binding proteins in post-transcriptional messenger RNA editing and modification in Arabidopsis. Curr Opin Plant Biol 45:88–95. https://doi.org/10.1016/j.pbi.2018.05.010

Li Z, Zhang Z, Yan P et al (2011) RNA-Seq improves annotation of protein-coding genes in the cucumber genome. BMC Genomics 12:540. https://doi.org/10.1186/1471-2164-12-540

Lowe R, Shirley N, Bleackley M et al (2017) Transcriptomics technologies. PLoS Comput Biol 13:e1005457. https://doi.org/10.1371/journal.pcbi.1005457

Macosko EZ, Basu A, Satija R et al (2015) Highly parallel genome-wide expression profiling of individual cells using nanoliter droplets. Cell 161:1202–1214. https://doi.org/10.1016/j.cell.2015.05.002

Margulies M, Egholm M, Altman WE et al (2005) Genome sequencing in microfabricated high-density picolitre reactors. Nature 437:376–380. https://doi.org/10.1038/nature03959

Martin JA, Wang Z (2011) Next-generation transcriptome assembly. Nat Rev Genet 12:671–682. https://doi.org/10.1038/nrg3068

Meyer KD, Saletore Y, Zumbo P et al (2012) Comprehensive analysis of mRNA methylation reveals enrichment in 3′ UTRs and near stop codons. Cell 149:1635–1646. https://doi.org/10.1016/j.cell.2012.05.003

Morena F, Argentati C, Bazzucchi M et al (2018) Above the epitranscriptome: RNA modifications and stem cell identity. Genes 9:329. https://doi.org/10.3390/genes9070329

Ontiveros RJ, Stoute J, Liu KF (2019) The chemical diversity of RNA modifications. Biochem J 476:1227–1245. https://doi.org/10.1042/BCJ20180445

Pan C, Li G, Bandyopadhyay A (2023) Guide RNA library-based CRISPR screens in plants: opportunities and challenges. Curr Opin Biotechnol 79:102883. https://doi.org/10.1016/j.copbio.2022.102883

Qin T, Zhao P, Sun J, Zhao Y et al (2021) Research progress of PPR proteins in RNA editing, stress response, plant growth and development. Front Genet 12. https://doi.org/10.3389/fgene.2021.765580

Ramakrishnan M, Rajan KS, Mullasseri S et al (2022) The plant epitranscriptome: revisiting pseudouridine and 2′-O-methyl RNA modifications. Plant Biotechnol J 20:1241–1256. https://doi.org/10.1111/pbi.13829

Ricroch AE, Bergé JB, Kuntz M (2011) Evaluation of genetically engineered crops using transcriptomic, proteomic, and metabolomic profiling techniques. Plant Physiol 155:1752–1761. https://doi.org/10.1104/pp.111.173609

Sathyanarayana N, Pittala RK, Tripathi PK et al (2017) Transcriptomic resources for the medicinal legume Mucuna pruriens: de novo transcriptome assembly, annotation, identification and validation of EST-SSR markers. BMC Genomics 18:409. https://doi.org/10.1186/s12864-017-3780-9

Seyfferth C, Renema J, Wendrich JR et al (2021) Advances and opportunities in single-cell transcriptomics for plant research. Annu Rev Plant Biol 72:847–866. https://doi.org/10.1146/annurev-arplant-081720-010120

Sharma B, Prall W, Bhatia G et al (2023) The diversity and functions of plant RNA modifications: what we know and where we go from here. Annu Rev Plant Biol 74:53–85. https://doi.org/10.1146/annurev-arplant-071122-085813

Shen L, Ma J, Li P et al (2023) Recent advances in the plant epitranscriptome. Genome Biol 24:43. https://doi.org/10.1186/s13059-023-02872-6

Stephenson W, Razaghi R, Busan S et al (2022) Direct detection of RNA modifications and structure using single-molecule nanopore sequencing. Cell Genomics 2:100097. https://doi.org/10.1016/j.xgen.2022.100097

Subramanian A, Tamayo P, Mootha VK et al (2005) Gene set enrichment analysis: a knowledge-based approach for interpreting genome-wide expression profiles. Proc Natl Acad Sci USA 102:15545–15550. https://doi.org/10.1073/pnas.0506580102

Tyagi P, Singh D, Mathur S et al (2022) Upcoming progress of transcriptomics studies on plants: an overview. Front Plant Sci 13:1030890. https://doi.org/10.3389/fpls.2022.1030890

Van der Verren SE, Van Gerven N, Jonckheere W et al (2020) A dual-constriction biological nanopore resolves homonucleotide sequences with high fidelity. Nat Biotechnol 38:1415–1420. https://doi.org/10.1038/s41587-020-0570-8

Velculescu VE, Zhang L, Zhou W et al (1997) Characterization of the yeast transcriptome. Cell 88:243–251. https://doi.org/10.1016/s0092-8674(00)81845-0

Verbruggen N, Hermans C, Schat H (2009) Molecular mechanisms of metal hyperaccumulation in plants. New Phytol 181:759–776. https://doi.org/10.1111/j.1469-8137.2008.02748.x

Wang X, Shi W, Rinehart T (2015) Transcriptomes that confer to plant defense against powdery mildew disease in Lagerstroemia indica. Int J Genomics 2015:e528395. https://doi.org/10.1155/2015/528395

Wu S, Wang Y, Wang J et al (2021) Profiling of RNA ribose methylation in Arabidopsis thaliana. Nucleic Acids Res 49:4104–4119. https://doi.org/10.1093/nar/gkab196

Xie Y, Chan L-Y, Cheung M-Y et al (2023) Current technical advancements in plant epitranscriptomic studies. Plant Genome 16:e20316. https://doi.org/10.1002/tpg2.20316

Xu X, Crow M, Rice BR et al (2021a) Single-cell RNA sequencing of developing maize ears facilitates functional analysis and trait candidate gene discovery. Dev Cell 56:557–568.e6. https://doi.org/10.1016/j.devcel.2020.12.015

Xu C, Wu Z, Duan H-C et al (2021b) R-loop resolution promotes co-transcriptional chromatin silencing. Nat Commun 12:1790. https://doi.org/10.1038/s41467-021-22083-6

Yan T-M, Pan Y, Yu M-L et al (2021) Full-range profiling of tRNA modifications using LC–MS/MS at single-base resolution through a site-specific cleavage strategy. Anal Chem 93:1423–1432. https://doi.org/10.1021/acs.analchem.0c03307

Yang IS, Kim S (2015) Analysis of whole transcriptome sequencing data: workflow and software. Genomics Inform 13:119–125. https://doi.org/10.5808/GI.2015.13.4.119

Yang W, Meng J, Liu J et al (2020) The N1-methyladenosine methylome of petunia mRNA. Plant Physiol 183:1710–1724. https://doi.org/10.1104/pp.20.00382

Yu X, Sharma B, Gregory BD (2021) The impact of epitranscriptomic marks on post-transcriptional regulation in plants. Brief Funct Genomics 20:113–124. https://doi.org/10.1093/bfgp/elaa021

Zhao J, Hyman L, Moore C (1999) Formation of mRNA 3′ ends in eukaryotes: mechanism, regulation, and interrelationships with other steps in mRNA synthesis. Microbiol Mol Biol Rev 63:405–445. https://doi.org/10.1128/MMBR.63.2.405-445.1999

Zhao W, He X, Hoadley KA et al (2014) Comparison of RNA-Seq by poly (A) capture, ribosomal RNA depletion, and DNA microarray for expression profiling. BMC Genomics 15:419. https://doi.org/10.1186/1471-2164-15-419

Chapter 2
RNA Sequencing Platforms and Bioinformatics Tools

Pelayo G. de Lena, David Morales, Mónica Escandón, Mónica Meijón, Luis Valledor, María Jesús Cañal, and Jesús Pascual

Abstract Transcriptomics refers to the comprehensive analysis of ribonucleic acid (RNA) molecules present in a cell or organism at a specific time and under defined conditions. Positioned, at a molecular level, between genomics and proteomics, transcriptomic studies offer a dynamic view of gene regulation and expression. RNA sequencing (RNA-seq) encompasses a range of technologies developed to identify and quantify diverse RNA species, playing a pivotal role in advancing plant research. These technologies have enabled deep insights into gene expression patterns, physiological states, developmental processes, and responses to environmental stimuli. The evolution of RNA-seq technologies is inseparable from the development of bioinformatic tools designed to manage and analyse the vast amounts of data these technologies produce. In this chapter, we provide an overview of the main RNA-seq technologies from a historical perspective, along with the key

Pelayo G. de Lena and David Morales contributed equally to this work and are listed in alphabetical order by surname.

P. G. de Lena
Department of Organisms and Systems Biology, University of Oviedo, Oviedo, Spain

Health Research Institute of Asturias (ISPA), Oviedo, Spain

Foundation for Biomedical Research and Innovation in Asturias (FINBA), Oviedo, Spain
e-mail: gonzalezrpelayo@uniovi.es

D. Morales · M. Escandón · M. Meijón · L. Valledor · M. J. Cañal
Department of Organisms and Systems Biology, University of Oviedo, Oviedo, Spain

Biotechnology Institute of Asturias (IUBA), University of Oviedo, Oviedo, Spain
e-mail: moralesdavid@uniovi.es; escandonmonica@uniovi.es; meijonmonica@uniovi.es; valledorluis@uniovi.es

J. Pascual (✉)
Biotechnology Institute of Asturias (IUBA), University of Oviedo, Oviedo, Spain

Department of Functional Biology, University of Oviedo, Oviedo, Spain
e-mail: pascualjesus@uniovi.es

L. M. Vaschetto (ed.), *Plant Transcriptomics and Epitranscriptomics*, Concepts and Strategies in Plant Sciences, https://doi.org/10.1007/978-981-95-5183-5_2

bioinformatics tools used throughout the typical RNA-seq analysis workflow. We also highlight specialized tools tailored to specific analyses, including differential gene expression, alternative splicing, small RNA characterization, and single-cell RNA-seq.

Keywords Transcriptomics · RNA sequencing · Sequencing technologies · Bioinformatic tools · Data analysis pipelines

Abbreviations

APS	Adenosine 5′-phosphosulfate
ATP	Adenosine triphosphate
bp	Base pair
CCS	Circular consensus sequencing
cDNA	Complementary DNA
CPM	Counts per million
CTAB	Cetyltrimethylammonium bromide
DAS	Differential alternative splicing
DGE	Differential gene expression
dNTP	Deoxynucleotide triphosphate
ddNTP	Dideoxynucleotide triphosphate
DTU	Differential transcript usage
FDR	False discovery rate
FLNC	Full-length nonconcatemer
FPKM	Fragments per kilobase of transcript per million mapped reads
GNN	Graph neural network
GO	Gene ontology
GSEA	Gene set enrichment analysis
HiFi	High fidelity
KEGG	Kyoto Encyclopedia of Genes and Genomes
lncRNA	Long noncoding RNA
miRNA	Micro RNA
mRNA	Messenger RNA
ncRNA	Noncoding RNA
NGS	Next-generation sequencing
Oligo-dT	Oligo-deoxythymidine
ONT	Oxford Nanopore Technologies
PacBio	Pacific Biosciences
PCA	Principal component analysis
PCR	Polymerase chain reaction
piRNA	Piwi-interacting RNA
PPi	Pyrophosphate
PTP	PicoTiterPlate

qPCR	Quantitative polymerase chain reaction
QC	Quality control
RNA-seq	RNA sequencing
RIN	RNA integrity number
RLE	Relative Log Expression
rRNA	Ribosomal RNA
RPKM	Reads per kilobase of transcript per million mapped reads
SBX	Sequencing by expansion
scRNA-seq	Single-cell RNA sequencing
siRNA	Small interfering RNA
SMRT	Single-molecule real-time sequencing
sncRNA	Small noncoding RNA
snRNA	Small nuclear RNA
TMM	Trimmed mean of M-values
tRNA	Transfer RNA
TPM	Transcripts per million
t-SNE	t-Distributed stochastic neighbour embedding
UMAP	Uniform manifold approximation and projection
UMI	Unique molecular identifier
ZMW	Zero-mode waveguide

2.1 Introduction

The transcriptome is the set of different ribonucleic acids (RNAs) that are present in a cell, tissue, organ or organism at a given moment or time under specific conditions. It includes various kinds of RNAs, such as messenger RNA (mRNA), transfer RNA (tRNA), ribosomal RNA (rRNA), small nucleolar RNA (snoRNA), long noncoding RNA (lncRNA), small noncoding RNA (sncRNA), or micro RNA (miRNA), each one playing distinct roles within the cell (Santos et al. 2014). RNA sequencing (RNA-seq) refers to the technologies used to obtain the specific sequence of nucleotides that integrate a set of RNA molecules as well as their quantities. While the term "RNA-seq" was popularized around 2008, most notably by the seminal paper of Mortazavi et al. (2008), earlier studies had already begun to explore next-generation sequencing (NGS) approaches for transcriptome analysis. Noteworthy, the development of RNA-seq technologies has come hand in hand with an array of bioinformatic tools specifically devoted to the output data from each of them and addressing their inherent limitations. In this chapter, we will discuss the importance of RNA-seq in plant sciences, set the basics of a typical transcriptomics workflow, differentiate the various options for sample sequencing and display the array of bioinformatics tools most commonly used for the downstream processing of the data.

2.1.1 *RNA-seq in Plant Research*

RNA-seq provides information about an organism's transcriptome, offering an overview of the gene expression under certain conditions. The study of expressed or transcriptionally active genes gives an insight into the physiological state of the plant, environmental and stress responses, state of development or pathway regulation (Guo et al. 2021; Reed et al. 2022). More precisely, transcriptomics enables evaluation of differential gene expression (DGE) profiles between experimental conditions, identifying candidate genes with potentially relevant roles in diverse pathways and responses (C.-Q. Zhou et al. 2021). It also facilitates advanced transcriptomics analysis, such as the characterization of alternative splicing events (Roces et al. 2022), noncoding RNAs (ncRNAs) and regulation of gene expression (T. Yu et al. 2019), or epitranscriptomic modifications (Hu et al. 2022) among others.

DGE analysis is the most widespread use of RNA-seq, followed by splicing and ncRNA studies (Stark et al. 2019). These RNA-seq approaches are of great use due to their throughput and cost-effectiveness (Trapnell et al. 2012; Wang et al. 2009). However, in the last years several trends have been rising in RNA-seq that allow for a deeper understanding of the transcriptomic level. Single-cell RNA-seq (scRNA-seq) enables the analysis of gene expression at the cellular level, revealing cellular diversity and cell populations (Kolodziejczyk et al. 2015). On the other hand, spatial RNA-seq provides transcriptomic information with additional data about spatial distribution of samples (Li and Wang 2021). Other techniques such as nascent RNA-seq capture recently synthesized RNAs with no further modifications (Wissink et al. 2019). Transcriptomics is the omic layer between genomics and proteomics, standing out for being prone to several regulatory processes. Transcription does not necessarily involve complete gene function execution but represents the first step of gene activation, standing as a middle point from genotype to phenotype. These are aspects of great scientific interest and form the basis of numerous studies worldwide (Cai et al. 2022a; Xu et al. 2021).

Overall, transcriptomics plays an important role in the general understanding of plant dynamics, having applications in crop science and production (Yu et al. 2021), biotic and abiotic stress responses (Zandalinas et al. 2021), biodiversity and evolution (Wu et al. 2022), or development (Cai et al. 2022b). It is important to note that although the focus of this work highlights transcriptomics approaches, the integration with data from other omic layers such as proteomics or metabolomics (discussed in later chapters) is of special relevance for the characterization and overall comprehension of plant biology (Dwivedi et al. 2023; Escandón et al. 2024; Shu et al. 2022).

2.2 Experimental Design and Wet-Lab Workflow of a Typical RNA-seq Experiment in Plants

A typical RNA-seq experiment workflow has several consecutive steps that may vary from one case to another. In this section, we outline the most common and widespread workflow concerning experimental design and the wet-lab procedures

Fig. 2.1 Scheme of a common RNA-seq experiment workflow. Numbers indicate the chapter sections in which the different steps of the workflow are described in detail

performed prior to actual RNA-seq itself and the subsequent data analysis (Fig. 2.1), covered in detail in Sects. 2.3 and 2.4, respectively. It is important to note that we refer to rather general or common aspects, but there might be specific variations in the case of each experiment.

2.2.1 *Experimental Design*

When planning an RNA-seq experiment, meticulous attention to experimental design is paramount. In plants, complex transcriptomes, polyploidy, and tissue-specific gene expression patterns further complicate experimental setups, making the careful consideration of design parameters essential for obtaining robust and reproducible results (Conesa et al. 2016). Furthermore, plant transcriptomes vary significantly throughout the day due to circadian rhythms (Hayes et al. 2010). Sampling points must therefore occur at a constant time between conditions and replicates to reduce possible circadian-based biases.

Another critical component of experimental design is the determination of the appropriate number of biological replicates. The optimal number depends on a balance between intra-group variation, the anticipated rates of false positives and negatives, logistical constraints, and available resources. Inadequate replication can lead to diminished statistical power and unreliable conclusions, while excessive replication may not be cost-effective. Therefore, an accurate estimation of sample size and statistical power is vital to the success of the research. Although it is challenging to establish general standards, previous studies have shown that four replicates can be adequate for most cases, though six is usually ideal (Stark et al. 2019).

2.2.2 *RNA Extraction and Library Construction*

The next critical step in an RNA-seq experiment workflow is the extraction of total RNA from the sample. RNA is a very labile molecule, very prone to degradation, so biological samples should preferably be preserved at −80 °C until extraction, although the ideal is to isolate RNA sooner rather than later and store the purified molecule at −80 °C instead. Obtaining RNA from plant tissues poses additional challenges due to the high concentration of polysaccharides and phenolic compounds present. Polysaccharides usually precipitate alongside polynucleotide chains, and phenolic compounds tend to bind to nucleotides, reducing the extraction yield (White et al. 2008). A plethora of methods have been described to optimize the quantity and quality of RNA yields in several plant species, including the cetyltrimethylammonium bromide, silica columns or Trizol methods to name a few (S. Chang et al. 1993; Inglis et al. 2018; Liyanage et al. 2021; Sánchez et al. 2016; Valledor et al. 2014; Xiao et al. 2012). There are also numerous commercial kits available for multiple purposes in this regard (Tiwari and Vis 2024), although in many cases they are not suitable for plant material as they are very rarely developed for this type of samples. Additionally, it is important to note that each species and tissue has their limitations, so preliminary tests and optimizations are generally recommended.

Once extracted, RNA should be quantified and tested for potential degradation. Usually, RNA-seq techniques require medium to high quantities of RNA (1–10 μg) and high integrity levels, although specific requirements vary from one sequencing method to another. RNA integrity number (RIN) is a widely used parameter developed by Agilent which quantifies RNA integrity on a scale from 1 ("completely degraded") to 10 ("completely intact") (Gallego Romero et al. 2014). While alternative methods for integrity evaluation are on the rise (S. Li et al. 2024), RIN still remains a standard nowadays. In plants, it is based on the proportion of 25S and 18S rRNA. For most plant RNA-seq applications, a RIN score of at least 8 is often preferred, which is generally high for plant samples. Nevertheless, RINs of 6–7 are usually acceptable (even as low as 4 for some sequencing platforms) (Gallego Romero et al. 2014). At this stage, samples should be DNA-free, which can be achieved with a DNase treatment.

Following steps vary depending on the focus of the study. Commonly, mRNA analysis is the objective of projects as it provides information about gene expression as well as splicing, among other processes. However, rRNA is the most abundant RNA in the cells (up to 80–90%, depending on tissue), and it gives minimal information about gene expression or regulation. Therefore, several procedures can be carried out to enrich the sample in the desired mRNA. Taking advantage of mRNA poly-A 3′ ends, methods applying binding to oligo-deoxythymidines (oligo-dT) of 18–20 nucleotides are used to capture these transcripts. Chromatography with oligo-dT columns or magnetic beads with oligo-dT are broadly used to this effect at a reduced cost. Although widely used, enriched poly-A fractions are not suitable for all purposes. Transcripts lacking poly-A tails such as histone mRNA, prokaryotic mRNA or non-coding RNAs cannot be captured through this approach. Alternatives for depleting rRNA in the samples are required in these cases, such as probes or oligos binding specifically to rRNA, though costs rise significantly. For studies focusing on sRNAs, capillary electrophoresis can be an effective method to eliminate other RNA fractions and DNA alike. There are also some protocols that achieve the separation of large and small RNA during the extraction process (Valledor et al. 2014).

Once rRNA fraction has been depleted, retrotranscription to cDNA is usually required, except with direct sequencing methods later explained. Standard methods such as oligo-dT or random hexamers are normally used. cDNA libraries are the input of most of the sequencing platforms. The specific downstream workflow differs from one platform to another. Usually, libraries are fragmented by sonication or restriction endonucleases and specific adapters for various uses are ligated to 5′ and 3′ ends. Next, sequencing is performed in the platform of choice and raw sequencing data are obtained.

Another relevant option to consider is the possibility to multiplex the samples using barcoding. The technique is based on adding additional adapters with different and known sequences (named barcodes) to different samples or treatments. This allows mixing them in the same run for the sequencer, as they will be differentiated in the subsequent bioinformatics analysis. Multiplexing by barcoding reduces the number of samples to sequence, reducing the cost significantly. The downside is a lower number of reads per sample, so low-abundance transcripts may not get detected. Depending on the species' genome complexity, the common desirable read depth is between 10 and 30 million reads per sample to ensure a complete transcriptome coverage, although one million is enough to cover 50% of the more highly expressed transcriptome (Stark et al. 2019).

2.3 RNA Sequencing Platforms

2.3.1 First-Generation Sequencing

The ability to reliably decode nucleotide sequences has always been of great interest in life sciences. In the 1970s, the first DNA sequencing methods were starting to emerge. The most successful approach was the chain-termination method, popularly known as Sanger sequencing after its developer Frederick Sanger. Sanger sequencing was the main sequencing method for 40 years, and it is still of great use until this day. This first generation of sequencing technology enabled the sequencing of the first genomes, including the model plant species *Arabidopsis thaliana*, which was completed in the year 2000 (The Arabidopsis Genome Initiative 2000).

Sanger sequencing is based on the use of both normal deoxynucleotides (dNTPs) and chain-terminating dideoxynucleotides (ddNTPs). ddNTPs lack the hydroxyl group at the 3′ position of the ribose, which prevents further elongation of the DNA strand once incorporated by a DNA polymerase. Because ddNTPs are incorporated randomly, this produces DNA fragments of all varying lengths corresponding to the amplified sequence. The sequence is determined using fluorescently labelled ddNTPs (each with a distinct fluorophore), a modern improvement over earlier radioactive labelling. This allows all four ddNTPs to be included in a single reaction and analysed via capillary electrophoresis.

The modern method achieves reads of up to 750–1000 base pairs (bp), it is reliable and cheap and still has several uses, such as vector cloning or mutant construction confirmations. However, it is not suitable for RNA-seq and transcriptomic studies, as it is limited to one sequence read per run. This inherent limitation made early genome sequencing projects extremely time consuming and economically inefficient. To put it into perspective, the first human genome (Human Genome Project) was carried out in 13 years and cost an estimate of 3 billion dollars. Nowadays, one human genome can be sequenced in less than a week for less than 1000 dollars.

2.3.2 Next-Generation Sequencing

In the 2000s, new approaches to sequencing were rising in popularity. Since then, a plethora of sequencing methods have been developed, having completely shifted the standards from the classic first-generation approach.

All NGS methods rely on massive parallelization, generating thousands to millions of reads simultaneously, coupled with computational analysis (Satam et al. 2023). The jump from generating one sequence per experiment to high-throughput of short-sequence reads, paired with rapid computational development in the early 2000s was essential to the rise of NGS as the new standard for sequencing. Before, NGS studies were limited by techniques, costs, and time. Nowadays, the complexity of gene expression and its regulation is available at a reasonable cost with the right computational power and the proper knowledge. As an example, projects like the One Thousand Plant Transcriptomes (One Thousand Plant Transcriptomes Initiative 2019) aim to universalize the transcriptomics approach to non-model plant species, something that was completely unthinkable just 20 years ago.

NGS encompasses two distinct technology generations: second- and third-generation sequencing. Second-generation sequencing introduced NGS itself, bringing massive parallelization and computing. The main detection systems for basecalling (i.e. the process of assigning nucleotides to a signal) were either optical or ionic. These methods produce short reads, they all involve RNA retrotranscription into cDNA and a PCR reaction or library amplification step, and sequencing is based on DNA synthesis by a DNA polymerase, so-called sequencing by synthesis.

Third-generation sequencing introduced the single molecule real-time sequencing (SMRT), which allows for the reading of unique nucleic acid molecules instead of the output of a library amplified by PCR. This technology also allows the generation of longer reads, which pose some advantages compared to the shorter reads in second-generation methods, while maintaining parallelization. Longer reads are more reliable for transcript assembly, especially for nonmodel species without a reference genome and for repetitive regions. They also provide greater accuracy when it comes to splicing isoform detection and reconstruction (García-Campa et al. 2023). Third-generation sequencing also allows for direct RNA-seq, eliminating the need to retrotranscribe RNA to cDNA, which has opened new frontiers in transcriptomics research, especially in the field of epitranscriptomics as noncanonical bases carrying modifications are not lost in retrotranscription and PCR reactions, and therefore can be sequenced.

2.3.3 Second-Generation Technologies

2.3.3.1 Roche 454 (Pyrosequencing)

Pyrosequencing was the first available NGS method, launched in 2005 by Life Sciences (Goodwin et al. 2016). The technology takes advantage of the pyrophosphate (PPi) released as a by-product during nucleotide incorporation by a DNA polymerase during sequencing. The liberated PPi serves as substrate, along with APS, for ATP production via an ATP sulfurylase. The generated ATP is subsequently used by a luciferase to catalyse the oxidation of luciferin, producing a detectable light signal. Therefore, each nucleotide incorporation event is coupled to a pulse of light, which is recorded by a detector. As dNTPs are added sequentially in a known order, each recorded light pulse is automatically associated with the addition of a known nucleotide, enabling basecalling based on signal timing and intensity.

A typical pyrosequencing workflow begins with the construction of a cDNA library, which is fragmented to an average length of 300–800 bp. Adapters are then ligated to both 5′ and 3′ ends of the fragments. The resulting library is heat denatured and bound to microspheres (28 μm in diameter) coated with oligonucleotides complementary to the adapter sequences. These microspheres are soaked in an aqueous solution containing the cDNA and the reagents for PCR reaction. By high-speed rotation, the mixture is emulsified with mineral oil, forming a microemulsion in which droplets, called microreactors, ideally contain a single bead and one cDNA fragment, along with all necessary PCR components. Within these droplets, clonal amplification, known as droplet PCR, takes place. After amplification, the emulsion is broken, and the beads, now bearing millions of identical copies of a single cDNA fragment, are loaded into a PicoTiterPlate, a photosensitive glass slide containing up to 1.6 million microwells (each 44 μm in diameter), designed to hold only one bead per well. During sequencing, dNTPs are added sequentially across the plate. DNA polymerase incorporates nucleotides where complementary bases are present, releasing PPi. This PPi is used in the previously described coupled reaction with ATP sulfurylase and luciferase, generating light pulses that are detected by optical sensors. The intensity of the light is proportional to the number of nucleotides incorporated. Between nucleotide additions, apyrase degrades any unincorporated dNTPs and excess ATP, preparing the system for the next cycle.

This method is fast and yields 600 bp reads on average (300–1000 bp). However, it has issues detecting homopolymers (sequences where one base is repeated several times) of more than 7 or 8 nucleotides. Its cost is high and around one million reads are obtained per run. Although it set a benchmark for NGS, its usage has dropped in recent times for RNA-seq due to its lower coverage at a higher prize compared to other methods. It has applications in directed approaches still, such as bisulphite sequencing or species identification by DNA barcoding (metabarcoding).

2.3.3.2 Solexa-Illumina

Solexa-Illumina, most commonly just Illumina, was released in 2006. The main innovation introduced by Illumina is the use of modified dNTPs known as reversible terminators. Similar to Sanger sequencing, these nucleotides block DNA strand elongation when incorporated by a DNA polymerase. However, unlike in Sanger sequencing, this termination is reversible. Reversible terminators fall into two main categories, distinguished by the position of the reversible blocking group. In 3'-O-blocked reversible terminators, the blocking group is attached to the oxygen atom of the 3′-hydroxyl (3'-OH) group on the ribose. In contrast, 3′-unblocked reversible terminators have the reversible blocking group attached to the base, along with a fluorescent reporter that also serves as part of the reversible terminator (Goodwin et al. 2016). In both types, the fluorescent reporter is cleavable, allowing it to be removed after each cycle of primer extension and termination, thus enabling successive rounds of sequencing. Regarding library construction workflow, RNAs are first sonicated, followed by cDNA synthesis, and repair and adenylation of 5′ and 3′ ends, respectively. Then, specific adaptors are ligated to both ends. Samples are then denatured and passed through a flowcell with oligonucleotides complementary to the adapters covalently fixed to the surface. cDNA fragments are captured at random positions within the flowcell and library amplification is performed before sequencing. This amplification consists of several steps. First, an initial DNA strand is synthesized, anchoring the molecule to the flow cell surface. Then, a so-called bridge PCR reaction is carried out, using surface-bound oligonucleotides as primers to the reaction. During this process, the DNA bends over to hybridize with a nearby oligonucleotide, forming a bridge structure. Repeated cycles of bridge PCR result in the formation of clusters of identical cDNA molecules localized in close proximity. These dense clusters are needed to amplify the signal during sequencing, ensuring sufficient sensitivity for accurate basecalling. For sequencing, dNTPs marked with different fluorochromes are added. These dNTPs temporarily block DNA elongation when incorporated. However, and contrary to what happened in Sanger sequencing, once fluorescence has been detected, blocking is chemically reversed, allowing for the next nucleotide to be added. Library fragments are commonly sequenced from both ends, yielding so-called paired-end reads. Paired-end reads provide more robust data, as reads are confirmed by duplicate to different degrees depending on the overlapping between the reads. They allow for a better assemble of contigs (both with a reference genome and from a de novo approach) and improved splicing isoform analysis (Magar et al. 2022). However, they are more expensive and require more downstream computational treatment. In contrast, single-end reads are more cost-effective and allow for a higher number of reads per sample, but they offer reduced reliability and information content compared to paired-end reads.

The reversible chain terminators method is precise, cost-efficient and capable of generating a high number of reads per run, ranging from ten million to 2.5 billion (Goodwin et al. 2016). In addition, it deals well with homopolymers and paired-end reads are simple to generate. On the other hand, each individual read is usually

150 bp long, which is shorter than other approaches and can limit assembly capabilities. Also, operational costs of the equipment are high.

There are several models of Illumina sequencers (MiniSeq, HiSeq, NovaSeq, NextSeq), each one with specific characteristics. Different models yield various throughputs, reads per run, read length, price and applications (Goodwin et al. 2016; Illumina Inc. 2025). The choice of one over another should be considered depending on the goal of the project and the available budget.

Illumina is arguably the most popular sequencing method to this date. More than 95% of the short-read RNA-seq data have been generated with an Illumina platform (Stark et al. 2019), positioning itself as the main method of choice for short sequencing. Illumina offers a remarkable relationship between reads per run and price, making it a demanded option to this date.

2.3.3.3 Ion Torrent

Ion Torrent sequencing platforms were introduced to the market in 2010. They are a second-generation NGS method that uses ionic nucleotide identification. This ionic detection was the major innovation introduced by this sequencing technology. When a nucleotide is added to a DNA strand, it liberates a proton to the media, which causes a change in both electrical conductivity as well as in the pH level. These variables are measured by a metal oxide semiconductor, as well as an ion-sensitive field effect transistor (Goodwin et al. 2016). As in previous platforms' workflow, the first step is the fragmentation of the cDNA library and the attachment of specific adaptors to 5′ and 3′ ends. As pyrosequencing, Ion Torrent utilizes the droplet PCR technology showcased above. In this case, microspheres are placed in a silicon-based chip with wells allowing only one microbead in as well. dNTPs are applied sequentially, elongating a DNA strand only when the base complementarity allows it. When a nucleotide is added, the liberated proton is detected by a change in the electric conductivity and a decrease in the pH by the current-sensitive components detailed above.

There is an array of Ion Torrent platforms available. They are easier to use, fast, and have a moderate price, costing an estimated 2–10 times more per Gb than Illumina (Goodwin et al. 2016). This technology shares many characteristics with the pyrosequencing approach, using both the droplet PCR and the sequential addition of nucleotides. Like in pyrosequencing, homopolymers are a source of inaccuracies in basecalling. The pH drops 0.02 units per proton liberated, but this relationship is not linear, thus yielding errors with homopolymers of longer than 6–8 nucleotides. Reads range from 300 to 400 bp and up to 80 million reads are yielded per run. Ion Torrent has also fallen in popularity due to its lower coverage and higher price compared to Illumina. However, it is still useful for directed approaches. In comparison, pyrosequencing is less used than Ion Torrent nowadays, due to a more abrupt loss of yield and cost-efficiency compared to other methods (Goodwin et al. 2016).

2.3.4 *Third-Generation Technologies*

2.3.4.1 Pacific Biosciences (PacBio)

PacBio was introduced in 2011. It utilizes the single molecule real-time (SMRT) technology, yielding higher read lengths. However, unlike other third-generation methods, it still uses a DNA polymerase as in second-generation approaches (Stark et al. 2019).

PacBio has several signature features. The main one is the use of sequencing cavities known as zero-mode waveguides (ZMW). ZMWs have sub-wavelength size (typically 100 nm in diameter and 100–150 nm deep) allowing for high-sensitivity fluorescence detection in an extremely small volume (in the order of zeptoliters-10^{-21}). This ultralow volume is critical for single-molecule resolution at the core of SMRT sequencing, as it restricts the excitation of fluorescent signals to a very small region near the bottom of the well, where a DNA polymerase is immobilized (Iizuka et al. 2022). In this small volume, only fluorescently labelled nucleotides that are actively being incorporated by the polymerase can be detected, reducing background fluorescence from free-floating nucleotides. The ZMW structure thus allows real-time detection of nucleotide incorporation events, one base at a time, by measuring the characteristic fluorescence of each labelled base before cleavage of the fluorescent tag.

Another innovation brought by PacBio sequencing is related precisely to fluorescent tags. They are linked to the terminal phosphate group of dNTPs instead of to the bases, as in second-generation NGS technologies. This difference couples dye cleavage to nucleotide incorporation. Thus, incorporated labelled nucleotides emit a brief burst of fluorescence while in the active site of the DNA polymerase, long enough to be detected, and then flows away as soon as it is incorporated, avoiding interference with further extension and allowing continuous, real-time sequencing without blocking DNA synthesis unlike reversible terminators previously described (Iizuka et al. 2022).

Previous feature makes fluorescence detection and basecalling challenging due to DNA polymerase high synthesis rate, leading to high error rates. To solve this, PacBio has developed circular consensus sequencing, a sequencing mode based on sequencing circularized DNA molecules instead of linear fragments. These circularized DNA molecules are called SMRTbell templates and are created by ligating hairpin adapters to both ends of a double-stranded DNA or cDNA fragment. DNA polymerase continuously sequences around the circular template, reading the same fragment multiple times in both forward and reverse directions. Each pass generates a subread and, by aligning multiple subreads of the same fragment, it is produced a high-accuracy consensus sequence, also known as a HiFi read, reaching accuracy over 99%.

Despite being a third-generation sequencing, it still uses optical detection of nucleotides and an amplification reaction. Several platforms using this technology are available, such as Revio, Vega or Onso systems, with different features and prices.

PacBio read lengths range from 10,000 to 25,000 bp, representing a substantial increase over second-generation methods. Although not as reliable as Nanopore technology (described below), epigenetic modifications can be inferred from delays in nucleotide incorporation (Zhao et al. 2019), although in RNA-seq applications this is not possible as it does not perform a direct sequencing approach. The main downsides of PacBio are its elevated cost, lower throughput and higher error rates in comparison to other technologies.

2.3.4.2 Oxford Nanopore Technologies: Nanopore Sequencing

The other main third-generation NGS technology available in the market in addition to PacBio was developed by Oxford Nanopore Technologies (ONT) and has been commercialized since 2015. Like PacBio, ONT is a SMRT, long-read sequencing technology. However, ONT's approach is radically different (Zhao et al. 2019).

ONT involves a flow cell containing thousands of nanopores, each one allowing an electrical current to flow, and through which single nucleic acid molecules are translocated by the action of an adapter-ligated motor protein that controls the speed of the process. When a nucleotide chain traverses these nanopores, the current flow is disturbed differently based on which nucleotides pass through it and whether it contain any modifications or not. Pore diameter ranges between 1.3 and 2.4 nm and membrane plates can be as thin as 1.4 nm. Pores can either be biological or solid-state. Biological nanopores are made from membrane proteins, typically alpha haemolysin variants, whereas solid-state nanopores are basically tiny holes in a plate. Both are used nowadays, although solid-state nanopores may have greater reusability in the long term (MacKenzie and Argyropoulos 2023). Membrane plates are made from an array of materials such as graphene, glass capillaries or MoS_2, but the most widespread ones are made from silicon (MacKenzie and Argyropoulos 2023). The basecalling is performed with neural networks and machine learning algorithms, translating current flow disruption to nucleotide sequence.

ONT has introduced several ground-breaking technical advances to sequencing methods. Firstly, it has introduced an approach that does not involve the use of an amplification reaction, thus removing potential amplification bias and DNA polymerase misreads. The ability to read libraries themselves and not PCR products is called direct sequencing. This is particularly relevant in RNA-seq, as RNA transcripts can be sequenced directly, without the need to generate a cDNA library by retrotranscription. By doing so, ONT also allows to detect modified bases as they disrupt electric current differently from canonical bases, opening the way for epitranscriptomic studies.

There are several platforms using ONT technology (MinION, GridION, PromethION), having different pore numbers, throughputs, applications, and expenses (Oxford Nanopore Technologies plc. 2025).

To sum up, ONT sequencing is fast, simple, and can be portable. It yields very long reads typically ranging from 1000 to 30,000 bp (up to even millions of bases), and it can detect epitranscriptomic modifications as RNA can be directly sequenced. On the other hand, its price ranges from moderate to high, it can give errors with homopolymers, and its accuracy is reduced compared to other methods. However, important improvements have been made through the development of improved sequencing chemistries, more reliable basecalling algorithms, and basecalling error-correction tools. The typical accuracy ranges from 87% to 98% (MacKenzie and Argyropoulos 2023). Also, reads per run range from 0.5 to one million and libraries must be minimally degraded.

2.3.5 General Considerations on Platform Choice

When selecting a sequencing platform, it is important to carefully consider several factors, as the chosen method will impact downstream applications and results. Therefore, thorough evaluation should be conducted in advance to ensure the best outcome for the project. The hypothesis to be tested and the objectives of the experiment are crucial to this end. Table 2.1 provides a summary of the most important aspects of the main sequencing platforms, serving as a reference for making an informed choice. Additionally, some common considerations are outlined below.

The main objective of the project is fundamental for platform choice. Is the main purpose of the research a simpler initial approach to DGE between conditions or is splicing isoform, regulatory analysis and ncRNAs more desirable? Simple DGE and splicing analysis are usually performed using Illumina as the sequencing technology of choice, especially in well-characterized species with a reference genome available. sRNA analyses do not benefit from long-read approaches either, unless the focus is on epitranscriptomic modifications. However, for non-model species without a reference genome, third-generation technologies might be preferable for better de novo approaches. Short-read approaches in these cases are also possible, especially if the budget is narrowing the study to some extent. On the other hand, coverage is also important, as transcripts with lower expression values may go below the detection threshold. Species with more complex genomes may need higher coverage yielding technologies such as Illumina to ensure full detection of transcripts or lower represented splicing isoforms. PacBio approaches are more suitable for repetitive regions, for example, concerning the expression of transposable elements. Furthermore, if epitranscriptomic modifications are of particular interest for the project, nanopore sequencing should be the technology of choice, as no other sequencing method yields direct information in this matter. If budget permits, hybrid approaches (e.g., PacBio or ONT for isoform discovery + Illumina for quantification) can leverage the strengths of different technologies. However, each project has unique goals and constraints, so selecting the right strategy is critical for success.

Table 2.1 Brief summary of NGS technologies' features and general advantages and disadvantages

Platform	Avg. read length	Avg. reads per run	Avg. cost	Detection of modified bases	Error rates (%)	General advantages	General disadvantages
454	600 bp	1 M	High	No	~1	Fast, medium-sized reads	Homopolymer errors, low cost-efficiency
Illumina	150 bp	10 M–2.5 B	Low	No	~0.1–1	Cost-efficient with good coverage	Reduced read length
Ion torrent	300 bp	80 M	Medium	No	~1–2	Easier to use and maintain	Homopolymer errors
PacBio	10 k–25 kbp	0.5 M	High	Yes	~0.1–1[a]	Long reads with high basecalling accuracy	Low reads per run at higher costs
ONT	1 k–30 k bp	0.5 M–1 M	Medium-high	Yes	~1–5[b]	Long reads at a moderate prize. Direct sequencing and non-canonical base detection	Low basecalling accuracy and reduced reads per run. Homopolymer errors

[a] For HiFi mode
[b] For sequencing with chemistry kits 14 and R10.4.1 flow cells

2.4 Data Analysis and Bioinformatic Tools

All sequencing platforms displayed above can yield data reaching up to several hundreds of gigabytes of information. Therefore, bioinformatic tools are required for their analysis and the mining of relevant biological information. In fact, RNA-seq comes hand-in-hand with a sometimes overwhelming variety of data analysis tools. In general, no single analysis workflow can be used in all cases provided the wide and increasing variety of different applications and analysis scenarios deriving from both the biological sphere as well as the sequencing platforms or technologies currently available. In this section, we will cover major steps and approaches in RNA-seq data analysis, highlighting the main tools available for each of them (Fig. 2.2) while providing some guidance on better choices for specific purposes. We will also cover some specific analysis, highlighting the main challenges in each case.

2.4.1 Raw Data Quality Control (QC)

Regardless of the experimental aim, the first step in any RNA-seq analysis is quality control (QC) of the raw reads. Subsequent steps may include additional QC checkpoints, but this section focuses on initial assessments of the unprocessed sequencing output.

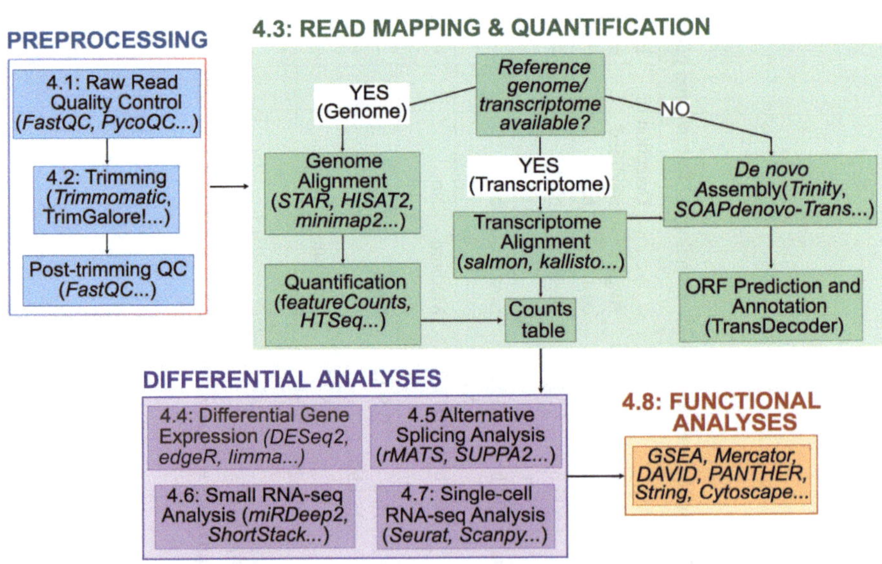

Fig. 2.2 Scheme of a typical RNA-seq data analysis workflow including main analyses and main tools covered in chapter Sect. 2.4. Numbers indicate the chapter sections in which the different steps of the workflow are described in detail

RNA-seq data can harbour quality issues that distort results and lead to erroneous conclusions. Thus, raw reads must undergo rigorous QC before downstream analysis. Common metrics include total read count, per-base sequence quality, GC content, adapter contamination, overrepresented or duplicated sequences, and read length distribution. These parameters help detect sequencing errors, amplification biases, or contaminants.

2.4.1.1 Tools for Short-Read Data QC

For short-read data, *FastQC* (Andrews 2010) remains a standard, offering rapid, visual summaries of quality metrics (e.g., base-level quality, adapter content). It is lightweight (<2 GB RAM/sample) and multiplatform, but lacks batch-level integration. *MultiQC* (Ewels et al. 2016) addresses this by aggregating *FastQC* outputs into interactive cross-sample reports, aiding outlier detection in large datasets. Noteworthy, *MultiQC* can handle the output of a wide collection of tools including some not exclusively related to QC.

Fastp (Chen et al. 2018) streamlines preprocessing by combining adapter trimming, quality filtering, and QC in a single multithreaded step. It processes millions of reads in minutes but offers limited visualizations compared to *FastQC*. Another tool, *AfterQC* (Chen et al. 2017), extends functionality with real-time error correction and rRNA contamination detection, but it is optimized for Illumina data.

For biologically contextualized QC, *RSeQC* (Wang et al. 2012) evaluates strand specificity and read distribution across genomic features (e.g., exons, introns), critical for complex plant transcriptomes with polyploidy or alternative splicing. However, it requires aligned BAM files, so it does not work strictly with raw data.

2.4.1.2 Long-Read and Single-Cell Data QC

Long-read technologies (ONT, PacBio) present unique challenges regarding variable lengths and higher error rates. *pycoQC* (Leger and Leonardi 2019) provides interactive visualizations of yield and read quality but requires structured metadata and > 4 GB RAM for large datasets. *NanoPlot* (De Coster et al. 2018) complements *pycoQC* with command-line plotting of read-length distributions and quality scores, although it demands Python proficiency. *NanoPlot* is compatible with both ONT and PacBio data and can handle large datasets efficiently. Moreover, it is frequently integrated into *Snakemake* (Köster and Rahmann 2012) or *Nextflow* (Di Tommaso et al. 2017) pipelines to automate QC across multiple runs.

For scRNA-seq, QC involves additional layers of complexity. These include detecting the expression of mitochondrial markers for cell stress or death, ambient RNA contamination, and multiplets or doublets, i.e., more than one cell captured and sequenced together. *Seurat* is a common choice (Pereira et al. 2021) as it offers integrated filtering and normalization, but it is resource-intensive (≥32 GB RAM) and requires R expertise.

2.4.1.3 Identification of Batch Effects

Batch effects represent a critical but frequently overlooked aspect of quality control in transcriptomic studies. These systematic technical variations arising from differences in sample processing, library preparation, or sequencing runs, can obscure true biological signals and inflate false discovery rates, particularly in plant research where environmental variability and multibatch experimental designs are common. Several computational approaches have been developed to address this challenge. *ComBat* (Zhang et al. 2020) employs an empirical Bayes framework to adjust for known batch variables, proving particularly effective when batch effects are not confounded with biological factors. When batch information is incomplete, the *sva* package (Leek et al. 2012) estimates surrogate variables representing hidden technical variation, while *RUVSeq* (Risso et al. 2014) utilizes control genes or replicates to remove unwanted variation through factor analysis. For simpler linear adjustments, *limma's removeBatchEffect* function (Ritchie et al. 2015) provides a straightforward solution by modelling batch as a covariate. In scRNA-seq workflows, where batch effects often dominate dimensionality reduction outputs, tools like *Harmony* (Korsunsky et al. 2019) align shared biological structures across batches, while *scVI* (Lopez et al. 2018) employs variational autoencoders to model both biological and technical variation. Regardless of the method employed, visual inspection before and after correction remains essential, as does careful experimental design to ensure batch correction preserves meaningful biological signal while minimizing technical artifacts.

2.4.2 *Trimming*

Trimming is a crucial preprocessing step in RNA-seq analysis alongside QC. Trimming aims at removing sequencing adapters, low-quality bases, and other artifacts that could compromise alignment and downstream quantification. For short-read Illumina data, commonly used tools include *Trim Galore!* (Krueger 2015), which combines *Cutadapt* (Martin 2011) for adapter removal with *FastQC* for QC, providing an automated, user-friendly solution optimized for typical Illumina read lengths. *Trimmomatic* (Bolger et al. 2014), another widely adopted tool, offers greater flexibility, supporting sliding-window trimming, quality score thresholds, and head-cropping, making it suitable for complex trimming strategies in large-scale studies. However, it demands more parameter tuning and computational resources. For long-read technologies (ONT, PacBio), specialized tools address platform-specific challenges such as variable read lengths and higher error rates. ONT workflows often employ *NanoFilt* (De Coster et al. 2018) for quality- and length-based filtering, while *Porechop* (Wick 2017) handles adapter trimming.

PacBio data benefits from *Lima*[1] for demultiplexing and trimming of barcoded reads, and *IsoSeq3*[2] for processing full-length non-concatemer reads, i.e., single full-length cDNA reads, during isoform reconstruction.

Post-trimming, it is advisable to check the resulting files with QC tools to verify the removal of artifacts and assess the impact of filtering on data quality.

2.4.3 Read Mapping and Counting

Following initial QC and trimming, read mapping and counting form the core steps of RNA-seq analysis, with some tools performing both tasks simultaneously. Mapping determines each read's genomic or transcriptomic origin, while counting quantifies reads per gene or transcript to enable downstream differential quantitative analysis. The choice of the mapping strategy depends on experimental goals, species characteristics, and available genomic resources.

A fundamental decision regarding mapping involves choosing between genome or transcriptome alignment. Genome mapping enables the discovery of novel transcripts and splicing events but increases computational demands and analytical complexity. Transcriptome alignment offers streamlined gene expression quantification but limits analysis to annotated features, potentially missing novel isoforms. This trade-off between discovery power and analytical efficiency is particularly relevant in plant studies, where alternative splicing and limited genome annotations pose unique challenges.

Mapping tools fall into two categories: traditional aligners and pseudo-aligners. Aligners, like *STAR* (Dobin et al. 2013), *HISAT2* (Kim et al. 2015), and *minimap2* (Li 2018), perform base-level alignment, enabling detection of novel transcripts, splicing variants, and genomic variants, crucial for studies of transcriptome complexity. However, their computational intensity can be prohibitive for large plant genomes. In contrast, pseudoaligners such as *kallisto* (Bray et al. 2016) and *salmon* (Patro et al. 2017) use k-mer matching for rapid transcript quantification, ideal for expression studies in well-annotated species but unsuitable for novel isoform discovery.

When working with specific RNA biotypes or sequencing technologies, the choice of tool must also reflect data type. For instance, *sRNAMapper* (Zytnicki and Gaspin 2022) is specialized for short RNA species (e.g., miRNAs, siRNAs), whereas *minimap2* is currently one of the most robust and flexible aligners for long-read RNA-seq, providing accurate alignments across full-length transcripts.

For species lacking a reference genome, de novo transcriptome assembly is a necessary preliminary step. Assemblers such as *Trinity* (Grabherr et al. 2011) reconstruct transcripts from read overlaps, typically using a fixed *k*-mer size of 25.

[1] https://github.com/PacificBiosciences/barcoding.

[2] https://github.com/PacificBiosciences/IsoSeq.

The choice of *k*-mer is a fundamental parameter in de novo assembly, as it determines the minimum overlap required between reads to construct contigs. Short *k*-mers, as used in *Trinity*, increase sensitivity to detect low-abundance transcripts but often lead to fragmented assemblies and misassemblies due to repetitive elements. In contrast, assemblers like *SOAPdenovo-Trans* (Xie et al. 2014) and *Oases* (Schulz et al. 2012) allow users to adjust the *k*-mer size, enabling optimization for specific dataset characteristics, such as read length and expected transcriptome complexity. These tools can achieve higher contiguity by leveraging longer overlaps, though they may miss rare or highly variable isoforms. Post-processing tools, such as *CD-HIT-EST* (W. Li and Godzik 2006) and *TransDecoder* (Haas et al. 2013), are often required to cluster redundant contigs and predict coding regions. More recently, long-read sequencing technologies have transformed de novo transcriptome reconstruction. Tools like *RNA-Bloom2* (Nip et al. 2023) enable full-length transcript assembly without relying on *k*-mers or short-read overlaps, offering a powerful solution for resolving alternative splicing events and complex isoform structures in plant transcriptomes.

2.4.3.1 Read Counting

Quantifying gene and transcript expression is a fundamental step in RNA-Seq workflows, with significant implications for downstream analyses. Several widely used tools offer distinct methodological approaches, computational efficiencies, and analytical resolutions, making them suitable for different experimental needs. Noteworthy, some of the tools perform read counting exclusively, taking mapping files (SAM/BAM) as an input, while others encapsulate both mapping and counting, performing both.

HTSeq (Anders et al. 2015) and *featureCounts* (Liao et al. 2014) employ rule-based counting strategies, requiring aligned reads (SAM/BAM mapping files) and gene annotations (GTF/GFF) to assign reads to genomic features. *HTSeq* specializes in gene-level quantification, whereas *featureCounts* extends this to any annotated feature, including exons and custom genomic regions. In contrast, *StringTie* (Pertea et al. 2015) performs transcript assembly before quantification, reconstructing full-length isoforms and estimating their abundances, making it particularly valuable for identifying novel transcript isoforms. *Salmon* (Patro et al. 2017) adopts a fundamentally different approach, using lightweight pseudomapping to quantify transcripts directly from FASTQ files without prior alignment, though it can also process aligned data. This probabilistic model significantly reduces computational overhead while maintaining high accuracy. A key distinction between these tools lies in transcript assembly. *HTSeq* and *featureCounts* rely entirely on existing annotations, whereas *StringTie* actively reconstructs transcript structures, enabling isoform discovery. *Salmon*, while not performing assembly, quantifies known or user-provided transcripts at high resolution, making it ideal for studies prioritizing transcript-level accuracy.

Computational efficiency varies substantially between tools. *HTSeq* is single-threaded and moderately paced, while *StringTie*, though efficient in data handling, incurs overhead from assembly. *featureCounts* excels in speed due to multi-threading and optimized memory usage. *Salmon* is the fastest, processing millions of reads in minutes, even on standard workstations. Ease of use also differs. *HTSeq* and *StringTie* require command-line proficiency, whereas *featureCounts* and *salmon* offer simpler interfaces with extensive documentation. Normalization approaches further distinguish these tools. *HTSeq* and *featureCounts* output raw counts requiring external normalization, while *StringTie* and *salmon* provide built-in normalization, facilitating immediate comparative analyses. Details regarding normalization must be considered in relation to the input requirements of tools used for downstream analysis.

2.4.4 Differential Gene Expression Analysis

Determining DGE is one of the most common applications of RNA-seq, enabling the comparison of transcript abundance between experimental conditions to identify genes involved in specific biological processes or pathways. Accurate quantification, however, depends critically on the normalization method used to account for differences in sequencing depth, transcript length, and compositional biases. Early approaches, such as reads per kilobase of transcript per million mapped reads (RPKM) and fragments per kilobase of transcript per million mapped reads (FPKM), aimed to normalize for gene length and library size, but have been shown to yield inconsistent results across samples (Wagne et al. 2012). Transcripts per million (TPM) improves interpretability by ensuring that normalized transcript counts sum to a constant, facilitating comparisons across samples (Li et al. 2010). More robust between-sample normalization approaches have been developed to support rigorous differential expression testing. The trimmed mean of M-values (TMM) method implemented in *edgeR* (Robinson et al. 2010), one of the most widely adopted R packages in RNA-seq analysis (R Core Team 2024), adjusts for library composition by excluding extreme fold changes and highly variable genes. *DESeq2* (Love et al. 2014), another referential R package in the field, employs geometric mean-based size factors to reduce the influence of highly expressed transcripts, while its relative log expression (RLE) approach scales counts relative to a pseudo-reference sample. Upper quartile normalization, also available in *edgeR* and other packages such as *EDASeq* (Risso et al. 2011), uses the 75th percentile of expression values to minimize the impact of low-abundance genes. While quantile normalization can align expression distributions across samples, its assumption of globally similar expression profiles makes it less suitable for studies involving strong biological perturbations.

The choice of differential expression analysis tools depends on experimental design and dataset characteristics. The three most widely adopted packages, *edgeR*, *DESeq2*, and *limma-voom* (Law et al. 2014), employ distinct statistical frameworks each with particular strengths. *edgeR* utilizes a negative binomial model with empirical Bayes dispersion estimation and excels in complex experimental designs, though it requires familiarity with generalized linear models. *DESeq2* similarly employs a negative binomial approach but also incorporates additional shrinkage estimators for dispersion that improve performance with low-count genes. The *limma-voom* pipeline transforms count data to logCPM (counts per million) values with precision weights, offering computational efficiency for large datasets while maintaining statistical power. All three tools incorporate features to handle technical noise. *DESeq2* and *edgeR* implement independent filtering and variance stabilization, while *limma-voom's* precision weighting models the mean-variance relationship. For studies requiring batch effect correction, these tools can incorporate covariates in their statistical models or interface with specialized packages previously mentioned, such as *sva* (Leek et al. 2012) or *RUVSeq* (Risso et al. 2014) (see Sect. 2.4.1 for further details).

Despite the availability of well-validated methods, no single approach for DGE analysis consistently outperforms others across all RNA-seq datasets and experimental designs. Comparative studies have shown that tool selection can meaningfully influence results, particularly for genes with low expression or borderline significance (Soneson and Delorenzi 2013; Williamson et al. 2017). This underscores the importance of careful method selection based on study design characteristics and the value of interpreting results with appropriate caution. When analysing non-model species or complex plant transcriptomes, additional considerations arise regarding the completeness of genome annotations and the prevalence of alternative splicing, factors that may favour certain tools or require specialized analytical approaches (Conesa et al. 2016). Ultimately, rigorous experimental design coupled with appropriate bioinformatic methods provides the foundation for reliable differential expression analysis in RNA-seq studies.

In RNA-seq differential expression analysis, testing thousands of genes increases the chance of false positives. To control the false discovery rate, methods like the Benjamini-Hochberg correction (Benjamini and Hochberg 1995) are applied, yielding *q-values*. A common significance threshold is $q < 0.05$, though stricter or more tolerant cutoffs may be used depending on the context. Tools such as *DESeq2*, *edgeR*, and *limma* report *q-values* by default, which should be used instead of raw *p-value*s to assess significance.

2.4.5 Tools for Differential Alternative Splicing Analysis

Alternative splicing has gained increasing recognition as a fundamental mechanism in the regulation of gene expression, contributing to transcriptomic complexity and proteome diversity in biological processes such as stress response and memory. This growing interest has been facilitated by advances in RNA-seq technologies, particularly the advent of long-read sequencing, which enables more comprehensive detection of splicing variants (Liu et al. 2017).

Analysing alternative splicing using RNA-seq presents several challenges. Accurate detection of splicing events is complicated by the inherent complexity of splicing patterns, the low abundance of many isoforms, and in the case of short-read sequencing data, the difficulty of reconstructing full-length transcripts and unambiguously assigning reads to specific isoforms (Tognacca et al. 2023). A critical step in alternative splicing analysis is the precise identification and quantification of splicing events before differential analysis can be performed. This process begins with preprocessed RNA-seq reads (quality-controlled and trimmed) and involves mapping splice junctions, exon-intron boundaries, and isoform structures to characterize splicing variation.

There are several strategies for detecting splicing events. The most straightforward approach relies on a well-annotated reference genome or transcriptome. In non-model organisms, where such annotations are often lacking, researchers may use de novo transcriptome assembly tools (e.g., *Trinity* [Grabherr et al. 2011], *RNA-Bloom2* [Nip et al. 2023]), though this introduces challenges such as transcript redundancy and fragmentation. Regardless of the approach, accurate read alignment is, as stated above, essential, requiring splice-aware aligners (e.g., *STAR* [Dobin et al. 2013], *HISAT2* [Kim et al. 2015], *minimap2* [Li 2018]) capable of mapping reads across exon-exon junctions.

The downstream workflow consists of two main tasks: transcript reconstruction and quantification, and differential splicing analysis. For isoform reconstruction, tools like *FLAIR* (Tang et al. 2020), *StringTie* (Pertea et al. 2015), and *Cufflinks* (Trapnell et al. 2012) assemble full-length transcripts, either guided by reference annotations or de novo, enabling transcript discovery. Long-read technologies are particularly powerful in this regard, as they can capture entire isoforms in single reads. Short reads spanning over splice junctions are limited by their length, making alternative splicing detection possible, but not as accurate as with long reads (Sahlin and Mäkinen 2021). Quantification of isoform expression relies on tools such as *kallisto* (Bray et al. 2016), *salmon* (Patro et al. 2017), and *featureCounts* (Liao et al. 2014), which estimate transcript-level abundances or assign reads to exonic features, enabling differential transcript usage (DTU) analysis. Alternatively, some methods focus on splice junctions or exon bins rather than full isoforms, allowing differential alternative splicing (DAS) analysis without complete transcript reconstruction. These methods rely on the assessment of splicing events and their quantification as percent spliced-in (Ψ) values. *rMATS* (Shen et al. 2014), *DEXSeq* (Anders et al. 2012), *SUPPA2* (Trincado et al. 2018), *MAJIQ* (Vaquero-Garcia et al. 2016) and *MISO* (Katz et al. 2010) are examples of commonly used DAS tools. They have become prominent for their ability to model diverse splicing patterns, accommodate biological replicates and handle complex transcriptome data (Jiang et al. 2023).

2.4.6 Tools for sRNA-seq Analysis

sRNA-seq analysis demands specialized bioinformatics approaches to accommodate the unique features of sRNAs, including their short length, non-polyadenylated ends, and diverse biogenesis pathways. Unlike mRNA-seq, sRNA-seq specifically profiles regulatory molecules like miRNAs, siRNAs, and piRNAs that play critical roles in gene regulation, stress responses, and genome defence (Yu et al. 2019).

The standard sRNA-seq workflow parallels conventional RNA-seq pipelines, encompassing QC, trimming, read mapping, annotation, quantification, and differential expression analysis, with the potential addition of novel sRNAs prediction. While general QC and trimming tools (see Sects. 2.4.1 and 2.4.2) remain applicable, parameter adjustments must account for shorter read lengths.

Alignment typically employs traditional mappers like *Bowtie2* (Langmead et al. 2009) or *STAR* (Dobin et al. 2013), which efficiently handle short reads and tolerate mismatches inherent to sRNA datasets. Ungapped alignments are generally preferred to maintain precision when distinguishing closely related sRNA family members. Specialized tools have also emerged. They can be divided in two types: gene-centric and cluster-centric. Gene-centric tools focus on the annotation and quantification to discrete genomic loci, either regular genes or sRNA loci, particularly miRNAs. *miRDeep2* (Friedländer et al. 2012) remains widely used for miRNA quantification and discovery, particularly in non-model organisms.

Cluster-centric tools, in turn, analyse sRNAs as genomic regions or clusters, identifying contiguous genomic regions producing sRNAs, what allows sRNAs discovery and transposon analysis. *ShortStack* (Axtell 2013) is a widely used tool representing this approach. Other tools within this family, like *proTRAC* (Rosenkranz and Zischler 2012) or *sPARTA* (Kakrana et al. 2014), are devoted to specific sRNA classes.

2.4.7 Tools for scRNA-seq Analysis

The analysis of scRNA-seq data presents unique computational challenges due to high technical variability, sparse expression matrices (i.e. "dropouts"), and batch effects. These challenges necessitate specialized analytical workflows that differ substantially from those used in bulk RNA-seq.

The scRNA-seq pipeline begins as always with QC and preprocessing of raw data. This step involves filtering out low-quality cells based on metrics such as the number of detected genes per cell, total unique molecular identifiers (i.e., short random sequences added during library preparation) counting, and the proportion of mitochondrial gene reads, commonly used as an indicator of cellular stress or apoptosis. Tools such as *Seurat* (Pereira et al. 2021), *Scater* (McCarthy et al. 2017), and *CellRanger* (10x Genomics 2025) provide integrated solutions for assessing these parameters. Additionally, specialized tools like *Scrublet* (Wolock et al. 2019) and *DoubletFinder* (McGinnis et al. 2019) detect and remove doublets (droplets containing multiple cells), which can otherwise lead to artefactual clustering and biased downstream analyses.

Following QC, normalization is performed to account for technical variation in sequencing depth and capture efficiency across cells. Traditional methods like CPM are inadequate for scRNA-seq due to data sparsity and zero inflation. Instead, more robust approaches such as RLE, quantile normalization, and deconvolution-based methods (e.g. *scran* [Lun et al. 2016]) are preferred. *SCTransform* (Hafemeister and Satija 2019), implemented in *Seurat*, employs regularized negative binomial regression to stabilize variance and improve downstream clustering and differential expression analysis.

The next step involves feature selection and dimensionality reduction. Highly variable genes are identified to focus on the most biologically informative features. These genes serve as input for principal component analysis, which captures major sources of variation. Non-linear dimensionality reduction techniques, such as *t-SNE* (Van Maaten and Hinton 2008) and *UMAP* (McInnes et al. 2020), are then applied to visualize cell populations in a low-dimensional space. Other tools, like *zinbwave* (Risso et al. 2018), are designed to further enhance dimensionality reduction by explicitly modelling zero-inflated distributions common in scRNA-seq data.

Clustering is performed to group cells into transcriptionally distinct populations, which may correspond to different cell types or states. Graph-based clustering algorithms (e.g., Louvain, Leiden), implemented in frameworks like *Seurat*, *Scanpy* (Wolf et al. 2018), and *Monocle3* (Cao et al. 2019), enable high-resolution clustering with adjustable granularity. For cell-type annotation, reference-based methods such as *SingleR* (Aran et al. 2019), and pretrained classifiers like *CellTypist* (Domínguez Conde et al. 2022) or *Azimuth* (Hao et al. 2021) facilitate label transfer from well-annotated datasets.

DGE analysis identifies marker genes that distinguish cell populations or experimental conditions. Due to the unique distribution of scRNA-seq data (e.g., bimodality, excess zeros), specialized statistical models are required. *MAST* (Finak et al. 2015), which uses hurdle models to separately model gene detection rates and expression levels, is a widely adopted approach. Nonparametric methods like the Wilcoxon rank-sum test (available in *Seurat* and *Scanpy*) offer computational efficiency but may lack sensitivity in low-coverage datasets. While bulk RNA-seq tools (e.g., *DESeq2* [Love et al. 2014], *limma-voom* [Law et al. 2014], *edgeR* [Robinson et al. 2010]) can be adapted, their assumptions are less suited to single-cell data and require careful parameter tuning.

For studies investigating dynamic processes (e.g., differentiation, cell state transitions), trajectory inference methods reconstruct pseudotemporal ordering of cells. *Monocle3* (Cao et al. 2019) integrates clustering and trajectory analysis using reversed graph embedding, while *Slingshot* (Street et al. 2018) and *PAGA* (partition-based graph abstraction) (Wolf et al. 2019) model branching events. *Velocyto* (La Manno et al. 2018) extends this by leveraging unspliced mRNA reads to infer RNA velocity, a method that uses both splice and unspliced mRNA counts to infer the direction and speed of gene expression, and thereby predict future cell states.

2.4.8 Tools for Functional Analysis

Once differential analyses have been completed and lists of statistically significant features have been generated, a common next step is often to interpret these results in a biological context through functional enrichment analysis. This approach identifies overrepresented biological processes, molecular functions, cellular components, and pathways among differential features, helping uncover key mechanisms underlying experimental conditions or phenotypes. Commonly used tools include *gene set enrichment analysis* (Subramanian et al. 2005), which evaluates coordinated shifts in predefined gene sets without relying on arbitrary expression thresholds, and platforms like *DAVID* (Huang et al. 2009), *PANTHER* (Mi et al. 2019) and *Cytoscape* (Shannon et al. 2003) that provide interfaces for functional clustering, gene ontology term enrichment, and interaction network visualization. These tools typically integrate with curated databases such as the *Gene Ontology (GO)*, *KEGG* for pathway analysis (Kanehisa and Goto 2000), and *STRING* for protein-protein interaction networks (Szklarczyk et al. 2025), enabling comprehensive pathway reconstruction and mapping of differential genes onto biological cascades.

While these approaches yield robust biological insights in well-annotated model organisms, functional interpretation becomes more challenging in non-model or orphan species due to incomplete annotation. For such cases, specialized tools like *Mercator* and *MapMan* ontology (Thimm et al. 2004) enable functional categorization of transcripts by mapping them onto conserved metabolic bins or plant-specific ontologies based on sequence similarity. Additionally, the integration of machine learning into functional analysis pipelines has gained momentum in recent years. Supervised and unsupervised learning algorithms are increasingly being used to classify genes into functional groups, predict gene functions based on coexpression patterns or sequence features, and detect latent structures within high-dimensional expression data (Yan and Wang 2022). Particularly, combining expression data with network-based models, such as graph neural networks, enables the identification of regulatory modules and candidate biomarkers, especially in complex pathways and poorly annotated genomes (Chang et al. 2024).

2.5 Future Directions, Main Challenges and Emerging Trends in RNA-seq in Plant Sciences

2.5.1 Methodological Challenges in Plant RNA-seq Approaches

Plant RNA-seq analysis encounters unique challenges stemming from both genomic complexity and physiological characteristics. Polyploidy, affecting approximately 17% of modern crops, including wheat and sugarcane (Meyer et al. 2014; Weeks 2017), introduces analytical complications due to increased genome size, abundant repetitive regions and difficulties in distinguishing expression patterns among homologous genes. This is particularly evident in allopolyploids like wheat, where unequal expression of homologs has been well-documented (Ramírez-González et al. 2018). The high repetitive content in plant genomes (up to 85% in cereals and gymnosperms [Kong et al. 2023]) presents additional challenges, as conventional approaches often discard reads mapping to these regions, despite their potential biological significance. Recent studies highlight the regulatory importance of transposable elements, particularly under stress conditions (Ramakrishnan et al. 2022), emphasizing the need for improved analytical approaches.

The limited availability of reference genomes for non-model species remains another significant constraint, underscoring the importance of data sharing through public repositories and standardized pipelines. In addition, temporal factors introduce further complexity, as circadian rhythms can alter gene expression, e.g., by up to 25% in photosynthetic tissues (Hayes et al. 2010; Hudson 2010), necessitating careful experimental design to control for these effects (Hotta 2021). All these factors can influence plant RNA-seq experiments and results to some extent. Therefore, being aware of their existence and planning accordingly will be critical for reducing biases in future works.

2.5.2 The Future of RNA-seq: Latest Breakthroughs, Trends, and New Tool Development

Technological innovations continue to transform plant RNA-seq capabilities. Future techniques are poised to overcome current limitations in read length, error rates, and throughput, especially concerning SMRT technologies. The emerging sequencing by expansion, improving ONT sequencing capabilities is a good example (Kokoris et al. 2025). On the other hand, recent advances in spatial and temporal dimensions in transcriptomics and single-cell approaches enable unprecedented resolution in studying gene expression dynamics. Altogether, these advances allow us to foresee a near future in which we will be able to capture full-length transcripts, nascent, unprocessed RNAs, and complex isoforms with greater accuracy, thereby deepening our understanding of plant molecular biology and its regulation, and to do it in experimental systems allowing greater resolution at different levels. scRNA-seq approaches, on the rise, are a clear example of this trend, deepening the understanding of rare cell populations and their role in plant physiology.

As it has been highlighted and vertebrated throughout the chapter, NGS advances require bioinformatic developments to take place in parallel. In this field, the integration of artificial intelligence represents the last frontier and promises to revolutionize data analysis with tools like *scVelo* (Bergen et al. 2020) and *DeepVelo* (Chen et al. 2022), enabling precise characterization of transcriptional dynamics. Automated systems such as *AutoBA* (Zhou et al. 2023) have demonstrated how large language models can streamline complex RNA-seq workflows, representing a paradigm shift toward adaptive, learning-based analytical frameworks to come.

2.5.3 *Frontiers of Knowledge in Plant Sciences*

At the frontier of plant science, multiomics integration is revealing new layers of complexity in gene regulation, development, and stress adaptation. The integration of genomic, transcriptomic, proteomic, epigenomic, and metabolomic data is enabling researchers to construct comprehensive models that link genotype and phenotype in unprecedented detail. This integrative approach is not only deepening our fundamental understanding of plant biology but also paving the way for practical applications in crop improvement, resilience to climate change, and sustainable agriculture. Future research will likely focus on leveraging these insights to engineer plants with enhanced productivity and stress tolerance, while also exploring the evolutionary mechanisms that drive biodiversity. As multiomics integration methodologies continue to mature, they promise to unlock new frontiers in plant science, offering transformative insights into the molecular foundations of plant life.

Acknowledgement Our research group was funded by SEKUENS (IDE/2024/000701) and the Spanish Ministry of Science and Innovation (PID2019-107107GB-I00 and PID2020-113896GB-I00). PGL was supported by the Spanish Ministry of Science and Innovation (PRE2019-091395), DM by the FPU programme (FPU21/02839), and ME by a Ramón y Cajal contract (RYC2023-043866-I).

Competing Interests The authors have no conflicts of interest to declare that are relevant to the content of this chapter.

References

10x Genomics (2025). https://www.10genomics.com/. Accessed Mar 2025

Anders S, Reyes A, Huber W (2012) Detecting differential usage of exons from RNA-Seq data. Nat Preced 1–1. https://doi.org/10.1038/npre.2012.6837.2

Anders S, Pyl PT, Huber W (2015) HTSeq—a Python framework to work with high-throughput sequencing data. Bioinformatics 31(2):166–169. https://doi.org/10.1093/bioinformatics/btu638

Andrews S (2010) FastQC: a quality control tool for high throughput sequence data [Online]. Available from: https://www.bioinformatics.babraham.ac.uk/projects/fastqc/

Aran D, Looney AP, Liu L et al (2019) Reference-based analysis of lung single-cell sequencing reveals a transitional profibrotic macrophage. Nat Immunol 20(2):163–172. https://doi.org/10.1038/s41590-018-0276-y

Axtell MJ (2013) ShortStack: comprehensive annotation and quantification of small RNA genes. RNA 19(6):740–751. https://doi.org/10.1261/rna.035279.112

Benjamini Y, Hochberg Y (1995) Controlling the false discovery rate: a practical and powerful approach to multiple testing. J R Stat Soc Ser B Methodol 57(1):289–300. https://doi.org/10.1111/j.2517-6161.1995.tb02031.x

Bergen V, Lange M, Peidli S et al (2020) Generalizing RNA velocity to transient cell states through dynamical modeling. Nat Biotechnol 38(12):1408–1414. https://doi.org/10.1038/s41587-020-0591-3

Bolger AM, Lohse M, Usadel B (2014) Trimmomatic: a flexible trimmer for Illumina sequence data. Bioinformatics 30(15):2114–2120. https://doi.org/10.1093/bioinformatics/btu170

Bray NL, Pimentel H, Melsted P et al (2016) Near-optimal probabilistic RNA-seq quantification. Nat Biotechnol 34(5):525–527. https://doi.org/10.1038/nbt.3519

Cai J, Lv L, Zeng X et al (2022a) Integrative analysis of metabolomics and transcriptomics reveals molecular mechanisms of Anthocyanin metabolism in the Zikui Tea Plant (*Camellia sinensis* cv. Zikui). Int J Mol Sci 23(9). https://doi.org/10.3390/ijms23094780

Cai Z, Cai Z, Huang J et al (2022b) Transcriptomic analysis of tuberous root in two sweet potato varieties reveals the important genes and regulatory pathways in tuberous root development. BMC Genomics 23(1):473. https://doi.org/10.1186/s12864-022-08670-x

Cao J, Spielmann M, Qiu X et al (2019) The single-cell transcriptional landscape of mammalian organogenesis. Nature 566(7745):496–502. https://doi.org/10.1038/s41586-019-0969-x

Chang S, Puryear J, Cairney J (1993) A simple and efficient method for isolating RNA from pine trees. Plant Mol Biol Report 11(2):113–116. https://doi.org/10.1007/BF02670468

Chang L, Jin X, Rao Y et al (2024) Predicting abiotic stress-responsive miRNA in plants based on multi-source features fusion and graph neural network. Plant Methods 20(1):33. https://doi.org/10.1186/s13007-024-01158-7

Chen S, Huang T, Zhou Y et al (2017) AfterQC: automatic filtering, trimming, error removing and quality control for fastq data. BMC Bioinform 18(3):80. https://doi.org/10.1186/s12859-017-1469-3

Chen S, Zhou Y, Chen Y et al (2018) fastp: an ultra-fast all-in-one FASTQ preprocessor. Bioinformatics 34(17):884–890. https://doi.org/10.1093/bioinformatics/bty560

Chen Z, King WC, Hwang A et al (2022) DeepVelo: single-cell transcriptomic deep velocity field learning with neural ordinary differential equations. Sci Adv 8(48):3745. https://doi.org/10.1126/sciadv.abq3745

Conesa A, Madrigal P, Tarazona S et al (2016) A survey of best practices for RNA-seq data analysis. Genome Biol 17(1):13. https://doi.org/10.1186/s13059-016-0881-8

De Coster W, D'Hert S, Schultz DT et al (2018) NanoPack: visualizing and processing long-read sequencing data. Bioinformatics 34(15):2666–2669. https://doi.org/10.1093/bioinformatics/bty149

Di Tommaso P, Chatzou M, Floden EW et al (2017) Nextflow enables reproducible computational workflows. Nat Biotechnol 35(4):316–319. https://doi.org/10.1038/nbt.3820

Dobin A, Davis CA, Schlesinger F et al (2013) STAR: ultrafast universal RNA-seq aligner. Bioinformatics 29(1):15–21. https://doi.org/10.1093/bioinformatics/bts635

Domínguez Conde C, Xu C, Jarvis LB et al (2022) Cross-tissue immune cell analysis reveals tissue-specific features in humans. Science 376(6594):5197. https://doi.org/10.1126/science.abl5197

Dwivedi AK, Singh V, Anwar K et al (2023) Integrated transcriptome, proteome and metabolome analyses revealed secondary metabolites and auxiliary carbohydrate metabolism augmenting drought tolerance in rice. Plant Physiol Biochem 201:107849. https://doi.org/10.1016/j.plaphy.2023.107849

Escandón M, Valledor L, Lamelas L et al (2024) Multiomics analyses reveal the central role of the nucleolus and its machinery during heat stress acclimation in *Pinus radiata*. J Exp Bot 75(8):2558–2573. https://doi.org/10.1093/jxb/erae033

Ewels P, Magnusson M, Lundin S et al (2016) MultiQC: summarize analysis results for multiple tools and samples in a single report. Bioinformatics 32(19):3047–3048. https://doi.org/10.1093/bioinformatics/btw354

Finak G, McDavid A, Yajima M et al (2015) MAST: a flexible statistical framework for assessing transcriptional changes and characterizing heterogeneity in single-cell RNA sequencing data. Genome Biol 16(1):278. https://doi.org/10.1186/s13059-015-0844-5

Friedländer MR, Mackowiak SD, Li N et al (2012) miRDeep2 accurately identifies known and hundreds of novel microRNA genes in seven animal clades. Nucleic Acids Res 40(1):37–52. https://doi.org/10.1093/nar/gkr688

Gallego Romero I, Pai AA, Tung J et al (2014) RNA-seq: impact of RNA degradation on transcript quantification. BMC Biol 12(1):42. https://doi.org/10.1186/1741-7007-12-42

García-Campa L, Valledor L, Pascual J (2023) The integration of data from different long-read sequencing platforms enhances proteoform characterization in *Arabidopsis*. Plants 12(3). https://doi.org/10.3390/plants12030511

Goodwin S, McPherson JD, McCombie WR (2016) Coming of age: ten years of next-generation sequencing technologies. Nat Rev Genet 17(6):333–351. https://doi.org/10.1038/nrg.2016.49

Grabherr MG, Haas BJ, Yassour M et al (2011) Trinity: reconstructing a full-length transcriptome without a genome from RNA-Seq data. Nat Biotechnol 29(7):644–652. https://doi.org/10.1038/nbt.1883

Guo Q, Li X, Niu L et al (2021) Transcription-associated metabolomic adjustments in maize occur during combined drought and cold stress. Plant Physiol 186(1):677–695. https://doi.org/10.1093/plphys/kiab050

Haas BJ, Papanicolaou A, Yassour M et al (2013) De novo transcript sequence reconstruction from RNA-seq using the trinity platform for reference generation and analysis. Nat Protoc 8(8): 1494–1512. https://doi.org/10.1038/nprot.2013.084

Hafemeister C, Satija R (2019) Normalization and variance stabilization of single-cell RNA-seq data using regularized negative binomial regression. Genome Biol 20(1):296. https://doi.org/10.1186/s13059-019-1874-1

Hao Y, Hao S, Andersen-Nissen E et al (2021) Integrated analysis of multimodal single-cell data. Cell 184(13):3573–3587. https://doi.org/10.1016/j.cell.2021.04.048

Hayes KR, Beatty M, Meng X et al (2010) Maize global transcriptomics reveals pervasive leaf diurnal rhythms but rhythms in developing ears are largely limited to the core oscillator. PLoS ONE 5(9):12887. https://doi.org/10.1371/journal.pone.0012887

Hotta CT (2021) From crops to shops: how agriculture can use circadian clocks. J Exp Bot 72(22): 7668–7679. https://doi.org/10.1093/jxb/erab371

Hu J, Cai J, Xu T et al (2022) Epitranscriptomic mRNA modifications governing plant stress responses: underlying mechanism and potential application. Plant Biotechnol J 20(12): 2245–2257. https://doi.org/10.1111/pbi.13913

Huang DW, Sherman BT, Zheng X et al (2009) Extracting biological meaning from large gene lists with DAVID. Curr Protoc Bioinformatics 27(1):13.11.1–13.11.13. https://doi.org/10.1002/0471250953.bi1311s27

Hudson KA (2010) The circadian clock-controlled transcriptome of developing soybean seeds. Plant Genome 3(1). https://doi.org/10.3835/plantgenome2009.08.0025

Iizuka R, Yamazaki H, Uemura S (2022) Zero-mode waveguides and nanopore-based sequencing technologies accelerate single-molecule studies. Biophys Physicobiol 19. https://doi.org/10.2142/biophysico.bppb-v19.0032

Illumina Inc (2025). https://www.illumina.com/. Accessed Mar 2025

Inglis PW, Pappas MDCR, Resende LV et al (2018) Fast and inexpensive protocols for consistent extraction of high quality DNA and RNA from challenging plant and fungal samples for high-throughput SNP genotyping and sequencing applications. PLoS ONE 13(10). https://doi.org/10.1371/journal.pone.0206085

Jiang M, Zhang S, Yin H et al (2023) A comprehensive benchmarking of differential splicing tools for RNA-seq analysis at the event level. Brief Bioinform 24(3). https://doi.org/10.1093/bib/bbad121

Kakrana A, Hammond R, Patel P et al (2014) sPARTA: a parallelized pipeline for integrated analysis of plant miRNA and cleaved mRNA data sets, including new miRNA target-identification software. Nucleic Acids Res 42(18). https://doi.org/10.1093/nar/gku693

Kanehisa M, Goto S (2000) KEGG: kyoto encyclopedia of genes and genomes. Nucleic Acids Res 28(1):27–30. https://doi.org/10.1093/nar/28.1.27

Katz Y, Wang ET, Airoldi EM et al (2010) Analysis and design of RNA sequencing experiments for identifying isoform regulation. Nat Methods 7(12):1009–1015. https://doi.org/10.1038/nmeth.1528

Kim D, Langmead B, Salzberg SL (2015) HISAT: a fast spliced aligner with low memory requirements. Nat Methods 12(4):357–360. https://doi.org/10.1038/nmeth.3317

Kokoris M, McRuer R, Nabavi M et al (2025) Sequencing by expansion (SBX)—a novel, high-throughput single-molecule sequencing technology. bioRxiv. https://doi.org/10.1101/2025.02.19.639056
Kolodziejczyk AA, Kim JK, Svensson V et al (2015) The technology and biology of single-cell RNA sequencing. Mol Cell 58(4):610–620. https://doi.org/10.1016/j.molcel.2015.04.005
Kong W, Wang Y, Zhang S et al (2023) Recent advances in assembly of complex plant genomes. Genom Proteom Bioinforms 21(3):427–439. https://doi.org/10.1016/j.gpb.2023.04.004
Korsunsky I, Millard N, Fan J et al (2019) Fast, sensitive and accurate integration of single-cell data with Harmony. Nat Methods 16(12):1289–1296. https://doi.org/10.1038/s41592-019-0619-0
Köster J, Rahmann S (2012) Snakemake—A scalable bioinformatics workflow engine. Bioinformatics 28(19):2520–2522. https://doi.org/10.1093/bioinformatics/bts480
GitHub repository: https://github.com/FelixKrueger/TrimGalore
La Manno G, Soldatov R, Zeisel A et al (2018) RNA velocity of single cells. Nature 560(7719): 494–498. https://doi.org/10.1038/s41586-018-0414-6
Langmead B, Trapnell C, Pop M et al (2009) Ultrafast and memory-efficient alignment of short DNA sequences to the human genome. Genome Biol 10(3):25. https://doi.org/10.1186/gb-2009-10-3-r25
Law CW, Che Y, Shi W et al (2014) voom: precision weights unlock linear model analysis tools for RNA-seq read counts. Genome Biol 15(2):29. https://doi.org/10.1186/gb-2014-15-2-r29
Leek JT, Johnson WE, Parker HS et al (2012) The sva package for removing batch effects and other unwanted variation in high-throughput experiments. Bioinformatics 28(6):882–883. https://doi.org/10.1093/bioinformatics/bts034
Leger A, Leonardi T (2019) pycoQC, interactive quality control for Oxford Nanopore Sequencing. J Open Source Softw 4(34):1236. https://doi.org/10.21105/joss.01236
Li H (2018) Minimap2: pairwise alignment for nucleotide sequences. Bioinformatics 34(18): 3094–3100. https://doi.org/10.1093/bioinformatics/bty191
Li W, Godzik A (2006) Cd-hit: a fast program for clustering and comparing large sets of protein or nucleotide sequences. Bioinformatics 22(13):1658–1659. https://doi.org/10.1093/bioinformatics/btl158
Li X, Wang C-Y (2021) From bulk, single-cell to spatial RNA sequencing. Int J Oral Sci 13(1):1–6. https://doi.org/10.1038/s41368-021-00146-0
Li B, Ruotti V, Stewart RM et al (2010) RNA-Seq gene expression estimation with read mapping uncertainty. Bioinformatics 26(4):493–500. https://doi.org/10.1093/bioinformatics/btp692
Li S, Liu J, Zhao M et al (2024) RNA quality score evaluation: a preliminary study of RNA integrity number (RIN) and RNA integrity and quality number (RNA IQ). Forensic Sci Int 357:111976. https://doi.org/10.1016/j.forsciint.2024.111976
Liao Y, Smyth GK, Shi W (2014) featureCounts: an efficient general purpose program for assigning sequence reads to genomic features. Bioinformatics 30(7):923–930. https://doi.org/10.1093/bioinformatics/btt656
Liu X, Mei W, Soltis PS et al (2017) Detecting alternatively spliced transcript isoforms from single-molecule long-read sequences without a reference genome. Mol Ecol Resour 17(6):1243–1256. https://doi.org/10.1111/1755-0998.12670
Liyanage NMN, Chandrasekara BCHWM, Bandaranayake PCG (2021) A CTAB protocol for obtaining high-quality total RNA from cinnamon (*Cinnamomum zeylanicum* Blume). 3 Biotech 11(4):201. https://doi.org/10.1007/s13205-021-02756-1
Lopez R, Regier J, Cole MB et al (2018) Deep generative modeling for single-cell transcriptomics. Nat Methods 15(12):1053–1058. https://doi.org/10.1038/s41592-018-0229-2
Love MI, Huber W, Anders S (2014) Moderated estimation of fold change and dispersion for RNA-seq data with DESeq2. Genome Biol 15(12):550. https://doi.org/10.1186/s13059-014-0550-8
Lun ATL, McCarthy DJ, Marioni JC (2016) A step-by-step workflow for low-level analysis of single-cell RNA-seq data with bioconductor. F1000Research 5:2122. https://doi.org/10.12688/f1000research.9501.2

MacKenzie M, Argyropoulos C (2023) An introduction to Nanopore Sequencing: past, present, and future considerations. Micromachines 14(2). https://doi.org/10.3390/mi14020459

Magar ND, Shah P, Harish K et al (2022) Gene expression and transcriptome sequencing: basics, analysis, advances. In: Gene expression. IntechOpen. https://doi.org/10.5772/intechopen.105929

Martin M (2011) Cutadapt removes adapter sequences from high-throughput sequencing reads. EMBnetJ 17(1). https://doi.org/10.14806/ej.17.1.200

McCarthy DJ, Campbell KR, Lun ATL et al (2017) Scater: pre-processing, quality control, normalization and visualization of single-cell RNA-seq data in R. Bioinformatics 33(8): 1179–1186. https://doi.org/10.1093/bioinformatics/btw777

McGinnis CS, Murrow LM, Gartner ZJ (2019) DoubletFinder: doublet detection in single-cell RNA sequencing data using artificial nearest neighbors. Cell Syst 8(4):329–337. https://doi.org/10.1016/j.cels.2019.03.003

McInnes L, Healy J, Melville J (2020) UMAP: uniform manifold approximation and projection for dimension reduction. https://doi.org/10.48550/arXiv.1802.03426

Meyer N, Gregor T, Meierott L et al (2014) Diploidy suggests hybrid origin and sexuality in *Sorbus* subgen. *Tormaria* from Thuringia, Central Germany. Plant Syst Evol 300(10):2169–2175. https://doi.org/10.1007/s00606-014-1043-7

Mi H, Muruganujan A, Huang X et al (2019) Protocol update for large-scale genome and gene function analysis with the PANTHER classification system (v.14.0). Nat Protoc 14(3):703–721. https://doi.org/10.1038/s41596-019-0128-8

Mortazavi A, Williams BA, McCue K et al (2008) Mapping and quantifying mammalian transcriptomes by RNA-Seq. Nat Methods 5(7):621–628. https://doi.org/10.1038/nmeth.1226

Nip KM, Hafezqorani S, Gagalova KK et al (2023) Reference-free assembly of long-read transcriptome sequencing data with RNA-Bloom2. Nat Commun 14(1):2940. https://doi.org/10.1038/s41467-023-38553-y

One Thousand Plant Transcriptomes Initiative (2019) One thousand plant transcriptomes and the phylogenomics of green plants. Nature 574(7780):679–685. https://doi.org/10.1038/s41586-019-1693-2

Oxford Nanopore Technologies plc (2025). https://nanoporetech.com/. Accessed Mar 2025

Patro R, Duggal G, Love MI et al (2017) Salmon provides fast and bias-aware quantification of transcript expression. Nat Methods 14(4):417–419. https://doi.org/10.1038/nmeth.4197

Pereira WJ, Almeida FM, Conde D et al (2021) Asc-Seurat: analytical single-cell Seurat-based web application. BMC Bioinform 22(1):556. https://doi.org/10.1186/s12859-021-04472-2

Pertea M, Pertea GM, Antonescu CM et al (2015) StringTie enables improved reconstruction of a transcriptome from RNA-seq reads. Nat Biotechnol 33(3):290–295. https://doi.org/10.1038/nbt.3122

R Core Team (2024) R: a language and environment for statistical computing. R foundation for statistical computing. https://www.R-project.org/

Ramakrishnan M, Satish L, Sharma A et al (2022) Transposable elements in plants: recent advancements, tools and prospects. Plant Mol Biol Report 40(4):628–645. https://doi.org/10.1007/s11105-022-01342-w

Ramírez-González RH, Borrill P, Lang D et al (2018) The transcriptional landscape of polyploid wheat. Science 361(6403). https://doi.org/10.1126/science.aar6089

Reed RC, Bradford KJ, Khanday I (2022) Seed germination and vigor: ensuring crop sustainability in a changing climate. Heredity 128(6):450–459. https://doi.org/10.1038/s41437-022-00497-2

Risso D, Schwartz K, Sherlock G et al (2011) GC-content normalization for RNA-Seq data. BMC Bioinform 12(1):480. https://doi.org/10.1186/1471-2105-12-480

Risso D, Ngai J, Speed TP et al (2014) Normalization of RNA-seq data using factor analysis of control genes or samples. Nat Biotechnol 32(9):896–902. https://doi.org/10.1038/nbt.2931

Risso D, Perraudeau F, Gribkova S et al (2018) A general and flexible method for signal extraction from single-cell RNA-seq data. Nat Commun 9(1):284. https://doi.org/10.1038/s41467-017-02554-5

Ritchie ME, Phipson B, Wu D et al (2015) Limma powers differential expression analyses for RNA-sequencing and microarray studies. Nucleic Acids Res 43(7). https://doi.org/10.1093/nar/gkv007

Robinson MD, McCarthy DJ, Smyth GK (2010) edgeR: a bioconductor package for differential expression analysis of digital gene expression data. Bioinformatics 26(1):139–140. https://doi.org/10.1093/bioinformatics/btp616

Roces V, Lamelas L, Valledor L et al (2022) Integrative analysis in *Pinus* revealed long-term heat stress splicing memory. Plant J 112(4):998–1013. https://doi.org/10.1111/tpj.15990

Rosenkranz D, Zischler H (2012) proTRAC – a software for probabilistic piRNA cluster detection, visualization and analysis. BMC Bioinform 13(1):5. https://doi.org/10.1186/1471-2105-13-5

Sahlin K, Mäkinen V (2021) Accurate spliced alignment of long RNA sequencing reads. Bioinformatics 37(24):4643–4651. https://doi.org/10.1093/bioinformatics/btab540

Sánchez C, Villacreses J, Blanc N et al (2016) High quality RNA extraction from Maqui berry for its application in next-generation sequencing. Springerplus 5(1):1243. https://doi.org/10.1186/s40064-016-2906-x

Santos CA, Blanck DV, De Freitas PD (2014) RNA-seq as a powerful tool for penaeid shrimp genetic progress. Front Genet 5. https://doi.org/10.3389/fgene.2014.00298

Satam H, Joshi K, Mangrolia U et al (2023) Next-generation sequencing technology: current trends and advancements. Biology 12(7). https://doi.org/10.3390/biology12070997

Schulz M, Zerbino DR, Vingron M et al (2012) Oases: robust de novo RNA-seq assembly across the dynamic range of expression levels. Bioinformatics 28(8):1086–1092. https://doi.org/10.1093/bioinformatics/bts094

Shannon P, Markiel A, Ozier O et al (2003) Cytoscape: a software environment for integrated models of biomolecular interaction networks. Genome Res 13(11):2498–2504. https://doi.org/10.1101/gr.1239303

Shen S, Park JW, Lu Z et al (2014) rMATS: robust and flexible detection of differential alternative splicing from replicate RNA-Seq data. Proc Natl Acad Sci 111(51):5593–5601. https://doi.org/10.1073/pnas.1419161111

Shu J, Ma X, Ma H et al (2022) Transcriptomic, proteomic, metabolomic, and functional genomic approaches of *Brassica napus* L. during salt stress. PLoS ONE 17(3). https://doi.org/10.1371/journal.pone.0262587

Soneson C, Delorenzi M (2013) A comparison of methods for differential expression analysis of RNA-seq data. BMC Bioinform 14(1):91. https://doi.org/10.1186/1471-2105-14-91

Stark R, Grzelak M, Hadfield J (2019) RNA sequencing: the teenage years. Nat Rev Genet 20(11):631–656. https://doi.org/10.1038/s41576-019-0150-2

Street K, Risso D, Fletcher RB et al (2018) Slingshot: cell lineage and pseudotime inference for single-cell transcriptomics. BMC Genomics 19(1):477. https://doi.org/10.1186/s12864-018-4772-0

Subramanian A, Tamayo P, Mootha VK et al (2005) Gene set enrichment analysis: a knowledge-based approach for interpreting genome-wide expression profiles. Proc Natl Acad Sci 102(43):15545–15,550. https://doi.org/10.1073/pnas.0506580102

Szklarczyk D, Nastou K, Koutrouli M et al (2025) The STRING database in 2025: protein networks with directionality of regulation. Nucleic Acids Res 53(1):730–737. https://doi.org/10.1093/nar/gkae1113

Tang AD, Soulette CM, van Baren MJ et al (2020) Full-length transcript characterization of SF3B1 mutation in chronic lymphocytic leukemia reveals downregulation of retained introns. Nat Commun 11(1):1438. https://doi.org/10.1038/s41467-020-15171-6

The Arabidopsis Genome Initiative (2000) Analysis of the genome sequence of the flowering plant Arabidopsis thaliana. Nature 408(6814):796–815. https://doi.org/10.1038/35048692

Thimm O, Bläsing O, Gibon Y et al (2004) mapman: a user-driven tool to display genomics data sets onto diagrams of metabolic pathways and other biological processes. Plant J 37(6):914–939. https://doi.org/10.1111/j.1365-313X.2004.02016.x

Tiwari S, Vis M (2024) Fine-tuning RNA extraction protocols for polysaccharide-rich freshwater red algae (Rhodophyta). microPubl Biol. https://doi.org/10.17912/micropub.biology.001180
Tognacca RS, Rodríguez FS, Aballay FE et al (2023) Alternative splicing in plants: current knowledge and future directions for assessing the biological relevance of splice variants. J Exp Bot 74(7):2251–2272. https://doi.org/10.1093/jxb/erac431
Trapnell C, Roberts A, Goff L et al (2012) Differential gene and transcript expression analysis of RNA-seq experiments with TopHat and Cufflinks. Nat Protoc 7(3):562–578. https://doi.org/10.1038/nprot.2012.016
Trincado JL, Entizne JC, Hysenaj G et al (2018) SUPPA2: fast, accurate, and uncertainty-aware differential splicing analysis across multiple conditions. Genome Biol 19(1):40. https://doi.org/10.1186/s13059-018-1417-1
Valledor L, Escandón M, Meijón M et al (2014) A universal protocol for the combined isolation of metabolites, DNA, long RNAs, small RNAs, and proteins from plants and microorganisms. Plant J 79(1):173–180. https://doi.org/10.1111/tpj.12546
Van der Maaten L, Hinton G (2008) Visualizing data using t-SNE. J Mach Learn Res 9(11): 2579-2605.
Vaquero-Garcia J, Barrera A, Gazzara MR et al (2016) A new view of transcriptome complexity and regulation through the lens of local splicing variations. eLife 5. https://doi.org/10.7554/eLife.11752
Wagne GP, Kin K, Lynch VJ (2012) Measurement of mRNA abundance using RNA-seq data: RPKM measure is inconsistent among samples. Theor Biosci 131(4):281–285. https://doi.org/10.1007/s12064-012-0162-3
Wang Z, Gerstein M, Snyder M (2009) RNA-Seq: a revolutionary tool for transcriptomics. Nat Rev Genet 10(1):57–63. https://doi.org/10.1038/nrg2484
Wang L, Wang S, Li W (2012) RSeQC: quality control of RNA-seq experiments. Bioinformatics 28(16):2184–2185. https://doi.org/10.1093/bioinformatics/bts356
Weeks DP (2017) gene editing in polyploid crops: wheat, camelina, canola, potato, cotton, peanut, sugar cane, and citrus. Prog Mol Biol Transl Sci 149:65–80. https://doi.org/10.1016/bs.pmbts.2017.05.002
White EJ, Venter M, Hiten NF et al (2008) Modified Cetyltrimethylammonium bromide method improves robustness and versatility: the benchmark for plant RNA extraction. Biotechnol J 3(11):1424–1428. https://doi.org/10.1002/biot.200800207
Wick R (2017) Porechop: adapter trimmer for Oxford Nanopore reads. https://github.com/rrwick/Porechop/. Accessed Mar 2025
Williamson L, Saponaro M, Boeing S et al (2017) UV irradiation induces a Non-coding RNA that functionally opposes the protein encoded by the same gene. Cell 168(5):843–855. https://doi.org/10.1016/j.cell.2017.01.019
Wissink EM, Vihervaara A, Tippens ND et al (2019) Nascent RNA analyses: tracking transcription and its regulation. Nat Rev Genet 20(12):705–723. https://doi.org/10.1038/s41576-019-0159-6
Wolf FA, Angerer P, Theis FJ (2018) SCANPY: large-scale single-cell gene expression data analysis. Genome Biol 19(1):15. https://doi.org/10.1186/s13059-017-1382-0
Wolf FA, Hamey FK, Plass M et al (2019) PAGA: graph abstraction reconciles clustering with trajectory inference through a topology preserving map of single cells. Genome Biol 20(1):59. https://doi.org/10.1186/s13059-019-1663-x
Wolock SL, Lopez R, Klein AM (2019) Scrublet: computational identification of cell doublets in single-cell transcriptomic data. Cell Syst 8(4):281–291. https://doi.org/10.1016/j.cels.2018.11.005
Wu Q, Tong W, Zhao H et al (2022) Comparative transcriptomic analysis unveils the deep phylogeny and secondary metabolite evolution of 116 Camellia plants. Plant J 111(2): 406–421. https://doi.org/10.1111/tpj.15799
Xiao Y, Yang Y, Cao H et al (2012) Efficient isolation of high quality RNA from tropical palms for RNA-seq analysis. Plant Omics 5(6):584–589. https://doi.org/10.3316/informit.017946107336522

Xie Y, Wu G, Tang J et al (2014) SOAPdenovo-trans: de novo transcriptome assembly with short RNA-Seq reads. Bioinformatics 30(12):1660–1666. https://doi.org/10.1093/bioinformatics/btu077

Xu X, Legay S, Sergeant K et al (2021) Molecular insights into plant desiccation tolerance: transcriptomics, proteomics and targeted metabolite profiling in *Craterostigma plantagineum*. Plant J 107(2):377–398. https://doi.org/10.1111/tpj.15294

Yan J, Wang X (2022) Unsupervised and semi-supervised learning: the next frontier in machine learning for plant systems biology. Plant J 111(6):1527–1538. https://doi.org/10.1111/tpj.15905

Yu T, Tzeng DTW, Li R et al (2019) Genome-wide identification of long non-coding RNA targets of the tomato MADS box transcription factor RIN and function analysis. Ann Bot 123(3): 469–482. https://doi.org/10.1093/aob/mcy178

Yu P, He X, Baer M et al (2021) Plant flavones enrich rhizosphere Oxalobacteraceae to improve maize performance under nitrogen deprivation. Nat Plants 7(4):481–499. https://doi.org/10.1038/s41477-021-00897-y

Zandalinas SI, Sengupta S, Fritschi FB et al (2021) The impact of multifactorial stress combination on plant growth and survival. New Phytol 230(3):1034–1048. https://doi.org/10.1111/nph.17232

Zhang Y, Parmigiani G, Johnson WE (2020) ComBat-seq: batch effect adjustment for RNA-seq count data. NAR Genom Bioinform 2(3). https://doi.org/10.1093/nargab/lqaa078

Zhao L, Zhang H, Kohnen MV et al (2019) Analysis of transcriptome and epitranscriptome in plants using PacBio Iso-Seq and nanopore-based direct RNA sequencing. Front Genet 10. https://doi.org/10.3389/fgene.2019.00253

Zhou C-Q, Lu C-H, Mai L et al (2021) Response of rice (*Oryza sativa* L.) roots to nanoplastic treatment at seedling stage. J Hazard Mater 401:123412. https://doi.org/10.1016/j.jhazmat.2020.123412

Zhou J, Zhang B, Chen X et al (2023) Automated bioinformatics analysis via AutoBA. arXiv. https://doi.org/10.48550/arXiv.2309.03242

Zytnicki M, Gaspin C (2022) srnaMapper: an optimal mapping tool for sRNA-Seq reads. BMC Bioinform 23(1):495. https://doi.org/10.1186/s12859-022-05048-4

Chapter 3
Transcriptomic Insights into Plant Adaptation

Glòria Escolà, Antoni Garcia-Molina, Maria Almira-Casellas, Charlotte Poschenrieder, and Silvia Busoms

Abstract The ability of plants to adapt to rapidly changing environments depends on gene regulatory mechanisms that control stress perception, signaling, and short- and long-term responses. Transcriptomics has become a key tool for elucidating these adaptive processes, enabling researchers to capture the dynamic gene reprogramming of plants under abiotic and biotic stresses. This chapter provides an in-depth overview of transcriptomic technologies and their application in plant stress research. We trace the evolution of transcriptomic techniques, from Northern blotting and RT-qPCR to RNA-sequencing, single-cell RNA-seq, and spatial transcriptomics, and discuss how these tools have advanced our understanding of plant adaptive mechanisms. Particular focus is given to the identification of differentially expressed genes, transcription factors, alternative splicing events, and gene regulatory networks (GRNs) that mediate plant responses to drought, salinity, temperature extremes, flooding, nutrient deficiencies, herbivory, and pathogens. We further highlight how comparative transcriptomics across species and populations reveals both conserved and divergent molecular strategies for adaptation, with special attention to the evolutionary plasticity of GRNs. The chapter also explores the impact of beneficial plant-associated microbiomes, including rhizobacteria and mycorrhizal fungi, on host transcriptomic landscapes, emphasizing

G. Escolà · M. Almira-Casellas · C. Poschenrieder
Plant Physiology Laboratory, Bioscience Faculty, Universitat Autònoma de Barcelona, Barcelona, Spain
e-mail: gloria.escola@uab.cat; MariaJose.Almira@uab.cat

A. Garcia-Molina
Centre for Research in Agricultural Genomics (CRAG), CSIC-IRTA-UAB-UB, Edifici CRAG, Barcelona, Spain
e-mail: antoni.garcia@cragenomica.es

S. Busoms (✉)
Plant Physiology Laboratory, Bioscience Faculty, Universitat Autònoma de Barcelona, Barcelona, Spain

Plant Physiology Laboratory, Department of Environmental Sciences, Universitat de Girona, Girona, Spain
e-mail: Silvia.busoms@udg.edu

L. M. Vaschetto (ed.), *Plant Transcriptomics and Epitranscriptomics*, Concepts and Strategies in Plant Sciences, https://doi.org/10.1007/978-981-95-5183-5_3

the potential of microbial interactions to reprogram plant gene expression and enhance stress tolerance. Finally, we discuss emerging trends in integrating transcriptomic data with multiomics approaches and machine learning, which offer powerful avenues for predictive modeling of gene function and stress-responsive networks.

Keywords Plant stress responses · RNA sequencing · Gene expression · Abiotic and biotic stress adaptations · Gene regulatory networks · Comparative transcriptomics · Plant-microbiome interactions

Abbreviations

ABA	Abscisic acid
AMF	Arbuscular mycorrhizal fungi
AS	Alternative splicing
ATAC-seq	Assay for transposase-accessible chromatin using sequencing
cDNA	Complementary DNA
CV2	Coefficient of variation squared
DEGs	Differentially expressed genes
ELD	Expression level dominance
GA	Gibberellin
GRNs	Gene regulatory networks
HSP	Heat shock protein
HSFs	Heat shock factors
HVGs	Highly variable genes
JA	Jasmonic acid
miRNA	MicroRNA
mRNA	Messenger RNA
NUpE	Nitrogen uptake efficiency
NUE	Nitrogen use efficiency
ONT	Oxford Nanopore Technologies
PacBio	Pacific Biosciences
PAE	Phosphorus acquisition efficiency
PUE	Phosphorus use efficiency
PAP	Purple acid phosphatase
PHR1	Phosphate starvation response 1
PR	Pathogenesis related
RNA-seq	RNA sequencing
ROS	Reactive oxygen species
RT-qPCR	Reverse transcription quantitative polymerase chain reaction
SAGE	Serial analysis of gene expression
scRNA-seq	Single-cell RNA sequencing
SHAP	Shapley additive explanations
SOC	Seed oil content
SPX	SYG1/PHO81/XPR1 domain-containing proteins

Stereo-seq	Spatially enhanced resolution omics-sequencing
TCA	Tricarboxylic acid (cycle)
TFs	Transcription factors
WGAN-GP	Wasserstein generative adversarial network with gradient penalty

3.1 Introduction

Plants inhabit dynamic environments and must continuously adapt to various abiotic and biotic stresses to survive and persist. Transcriptomics, the study of the RNA transcripts, has become a powerful tool for resolving the complex regulatory networks that underlie plant responses to environmental stimuli. In recent decades, transcriptomic techniques have evolved from traditional methods like Northern blotting and RT-qPCR to high-throughput approaches, which include RNA-sequencing (RNA-seq), single-cell RNA-seq (scRNA-seq), and spatial and landscape transcriptomics. These innovations have allowed scientists to explore stress-responsive pathways with unprecedented resolution, discovering novel genes, isoforms, and regulatory elements.

An integrative framework for studying plant responses to environmental stress combines natural variation prospection, controlled experiments, and advanced genomic and transcriptomic tools. By collecting plants, seeds, and associated microbiomes from contrasting habitats, researchers can explore how plants adapt to diverse environmental conditions. Transcriptomic techniques are then applied to capture the dynamic regulation of gene expression under both abiotic and biotic stressors, revealing processes from early signaling to long-term acclimation (Kollist et al. 2019; Javaid et al. 2022). This chapter reviews how such technologies have contributed to the identification of differentially expressed genes (DEGs), transcription factors (TFs), alternative splicing events, and gene regulatory networks (GRNs) that underpin plant adaptation. We highlight transcriptomic insights into plant tolerance to drought, salinity, extreme temperatures, nutrient limitations, flooding, herbivory, and pathogen attack, while also examining the role of hormone signaling and beneficial microbial interactions in shaping transcriptomic responses. Furthermore, the integration of transcriptomic datasets with machine learning and multiomics approaches is presented as a powerful strategy to improve gene function prediction and uncover regulatory mechanisms. Together, these approaches form a comprehensive research pipeline to dissect plant stress responses and guide the development of resilient crops (Fig. 3.1).

3.2 Transcriptomic Techniques from Past to Present

Since the discovery of RNA in 1961, when François Jacob and Jacques Monod proposed the concept of messenger RNA (mRNA), techniques for quantifying and sequencing RNA have become fundamental tools in biological sciences. Methods

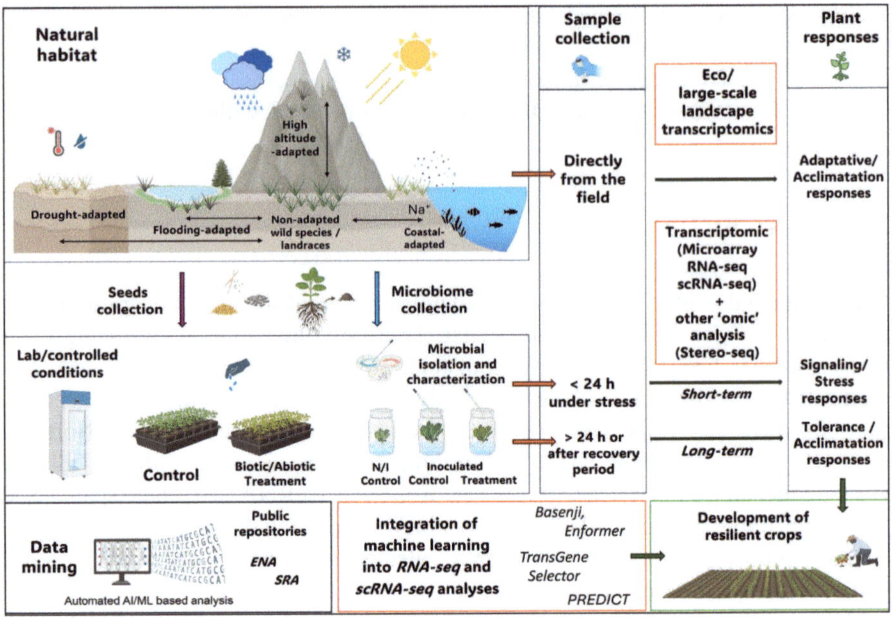

Fig. 3.1 Overview of transcriptomic strategies to study plant responses to environmental stress. This scheme summarizes the experimental and analytical workflow covered in the chapter. It highlights natural variation in plant adaptation across habitats, the collection of seeds and microbiomes for controlled experiments, and the application of transcriptomic techniques. Downstream analyses capture both short- and long-term plant responses, from stress signaling to acclimation. Integration of transcriptomic datasets with machine learning tools enhances predictive insights into gene function and regulation, supporting the development of resilient crops

for detecting and quantifying RNA marked the first steps toward the development of transcriptomic techniques. In 1977, James Alwine, David Kemp, and George Stark introduced the Northern Blot technique, which still remains a useful molecular biology approach to quantify the RNA expression or validate the results obtained through other techniques (Alwine et al. 1977; Moustafa and Cross 2016). This method is based on separating the RNA fragments according to their size using electrophoresis, followed by a hybridization of these fragments with specific target probes; the final exposure to X-ray allows the quantification (Bhattacharya et al. 2020). The first projects using this technique analyzed the expression levels of heat shock RNAs under arsenite and cadmium treatments (Edelman et al. 1988) and the nontissue-specific overexpression of a novel rice gene under water-stress conditions (Mundy and Chua 1988). Thus, Northern Blot represented a significant upgrade for gaining new insights about how plants regulate gene expression under stress conditions. Nowadays, it is recognized as a useful approach to detect small RNAs such as microRNAs (miRNAs) and short-interfering RNAs (siRNAs). Through the migration of small RNAs during gel electrophoresis, it accurately displays the size of small RNAs, allowing the simultaneous detection of both size and quantity (De la Rosa and Reyes 2019).

The first report of *in vitro* synthesis of cDNA was published in 1972 (Verma et al. 1972), providing the base for the development of the quantitative reverse transcription PCR (RT-qPCR). This technique is based on the first conversion of the RNA template to cDNA, followed by PCR amplification, where the cDNA copies produced are measured in real-time. Plant research progressively adopted RT-qPCR during the 1990s and 2000s. The first studies started to identify plant gene expression profiles under specific situations, such as the p50C gene expression. Under osmotic stress, the expression of this membrane-associated protein differed between adapted and nonadapted tobacco cells and was dependent on the age of the tissues, indicating the impact of the developmental stage and the environment on the regulation of these genes (Chang et al. 1996). Already in the 2000s, Kinoshita et al. (2001) quantified the different expression of two mRNA variants (*HI-5a* and *HI-5b)* from a single *HI-5* gene, coming from an alternative RNA splicing. The study observed that *HI-5b* was constitutively expressed across all the conditions, while *HI-5a* was increased after heat stress exposure. Still today, RT-qPCR is highly used, especially when the aim is the analysis of a few genes or the validation of results from other techniques.

Although the discovery of the RT-qPCR marked significant progress due to its higher sensitivity compared to Northern Blot, both methodologies are limited to analyses of a small number of genes (Bhattacharya et al. 2020). These limitations were overcome in 1995 by two new approaches, serial analysis of gene expression (SAGE) and microarrays, both allowing the detection of a significantly larger number of genes simultaneously, now reaching thousands (Hu and Polyak 1995; Schena et al. 1995). The SAGE methodologies involve isolating short tags of cDNA from the RNA sample that uniquely identify each transcript, which are then linked together and cloned into a vector. Via bioinformatic tools, they are identified, and their frequency is counted (Boheler and Stern 2003). This untargeted technology allows the discovery of new genes since no prior knowledge of them is required; however, it demands a high amount of initial RNA, and the short tags may not always be unique (van Ruissen et al. 2005).

The first study of microarrays presented the gene expression analysis of a total of 45 cDNAs of *Arabidopsis thaliana*, using three additional cDNAs from other organisms as controls (Schena et al. 1995). This pioneering work also discussed the possibility of developing arrays to detect 20,000 cDNA. In fact, by 1998, microarray technologies were able to measure the gene expression of 1400 cDNA of *A. thaliana*. In 2005, Ma et al. (2005) successfully compared the whole transcriptome of rice and *A. thaliana*, highlighting the progressive improvement of the technique. In the first analysis, the labeling step was used to differentiate between two distinct sample groups, such as shoot versus root tissues, or control versus treatment conditions, employing different fluorescence dyes (Aharoni and Vorst 2001; Kathiresan et al. 2006). This strategy allowed more detailed studies, such as the analysis of the influence of biotic and abiotic stress factors on molecular pathways regulating gene expression (Taji et al. 2004; Satoh et al. 2014; Li et al. 2015). Over the years, microarray databases have been significantly expanded, incorporating new data from different conditions and species. More recently, large-scale integrative analysis, combining diverse publicly available data resources, has been performed mainly focused on differentially expressed core genes under multiple stress factors in *Arabidopsis* (Pathania and Kumar 2022; Sham et al. 2015, 2019; Shen et al. 2017), rice (Azad et al. 2024; De Abreu and Frei 2016), or wheat (Nemati et al. 2023; Shokri-Gharelo et al. 2024), among others.

Despite the high contribution of the microarray to deciphering plant responses to stress, its low sensitivity and the challenge to design a standard array led to the development of more advanced methodologies overcoming the disadvantages of the former techniques. In this context, RNA-seq emerged in 2006 with the work of Bainbridge et al. (2006) and in 2008 when it was officially named RNA-seq. First transcriptome analyses were performed with *A. thaliana, Mus musculus, Saccharomyces cerevisae,* and *Schizosaccharomyces pombe* (Lister et al. 2008; Mortazavi et al. 2008; Nagalakshmi et al. 2008; Wilhelm et al. 2008). Now, RNA-seq has become the principal and standard approach for transcriptomic sequencing. Despite microarrays and RNA-seq results can be reasonably comparable for average-abundance gene expression, RNA-seq offers higher sensitivity and more accuracy, allowing the detection of a high number of DEGs and different molecular pathways responses (Rao et al. 2019; Zogopoulos et al. 2022).

RNA-seq technology comprises two alternative methodologies: short-read and long-read approaches. Short-read techniques work with libraries of 200 to 900 bp cDNA fragments, with Illumina and Ion Torrent being the most used platforms for sequencing. Short-read sequencing is robust and widely used, and most of the computational tools have been specifically developed for its analysis (Hu et al. 2021; Stark et al. 2019). However, as this methodology consists of the fragmentation of the mRNA into small pieces, it is not possible to sequence full-length transcriptomic isoforms (Su et al. 2023). In contrast, long-read sequencing allows the detection of alternative splicing (AS) events and the transcription initiation sites, because it can reach up to 50 kb DNA fragments (Daszkowska-Golec et al. 2023; Su et al. 2023). Pacific Biosciences (PacBio) and Oxford Nanopore (ONT) are the most common platforms for this methodology. In recent years, long-read sequencing has allowed the generation of reliable transcriptome assemblies, which are fundamental for an accurate differential gene expression analysis (Feng et al. 2019; Hoang et al. 2017; Xie et al. 2020). For instance, the full characterization of the transcriptome of the allotetraploid *Brassica napus* was challenging. Combining Illumina and PacBio isoform sequencing (Iso-Seq), Yao et al. (2020) sequenced the transcriptome and identified a total of 558 unannotated genes and 128,967 novel isoforms. Moreover, as long-read sequencing allows the identification of AS events, it is possible to determine the role of these AS in the modulation of plant stress responses. As an example, a combined analysis of PacBio and Illumina RNA-seq revealed that poplar trees adapt to cold stress through pre-mRNA alternative splicing (Yang et al. 2021). The higher tolerance of some populations may be attributed to their more diverse isoforms and the quicker reaction in front of the stress.

The significant improvement in transcriptomic research has also led to the development of even more specialized approaches, such as scRNA-seq, revealing the heterogeneity of gene expression at the single-cell level. Tang et al. (2009)

introduced scRNA-seq in 2009, and since then, it has been widely adopted in plant research, with studies in *Arabidopsis* and various crop species (see Bawa et al. 2022), creating single-cell maps and identifying developmental trajectories (e.g., Shahan et al. 2022).

This progress in sequencing methodologies, particularly microarrays and RNA-Seq, was accompanied by the development of computational tools designed to efficiently process and analyze the huge amount of data obtained in each run. Bioinformatic pipelines are essential to characterize and to correctly interpret the results. Optimal outcomes are only achieved when the most suitable method is selected based on a clear understanding of the conceptual differences among all these approaches. Furthermore, the integration of multiomics data with transcriptome results enhances biological interpretation, obtaining a more comprehensive analysis of the mechanistic details and complex biological systems. With ongoing advancements, several approaches and methods are being developed to integrate the multiomic data sets, especially transcriptomics and metabolomics, offering new insights into biological regulation (Depuydt et al. 2023).

Understanding gene evolution and function has traditionally relied on comparative genomic analyses focused on tracking orthogroup dynamics such as gene family gains, losses, expansions, or contractions. While informative, these approaches alone often fall short in elucidating the roles of genes lacking clear sequence similarity to characterized homologues. To overcome this, Julca et al. (2021) combined comparative genomics with gene expression atlases across a phylogenetically broad set of land plants. This integrative strategy allowed for the identification of organ-, tissue-, and cell-specific genes, leading to novel and missing components involved in organ development and gametogenesis. Large-scale comparative transcriptomics, thus, is also a useful tool to integrate and analyze multiple gene expression datasets and to identify both shared and distinct molecular pathways that organisms utilize to sense and respond to various stress conditions. To identify genes whose expression levels are influenced by various stress conditions, researchers have focused on extracting "highly variable genes" (HVGs) using normalized gene expression data, such as Robust Multi-array Average (RMA)-normalized values for microarray datasets or DESeq2-normalized counts for RNA-seq or scRNA-seq datasets. As an example, Sriden and Charoensawan (2022) assessed variability applying two statistical measures: standard deviation (SD) to capture the spread of expression values, and Analysis Of Variance (ANOVA)-derived *p*-values to test for statistically significant differences across conditions. Genes were classified as HVGs if they exhibited both relatively high SDs and low ANOVA *p*-values compared to other genes in the genome. Given the variability in data types, tissue sources, and transcriptomic platforms, the thresholds for SD and *p*-values were determined empirically for each dataset by identifying inflection points on scree plots. Alternatively, Wu et al. (2024b) used a dropout-based feature selection to identify HVGs in the gene expression atlas of maize and rice by fitting the function between the coefficient of variation squared (CV^2) and mean expression.

Altogether, these methodological and computational advances (summarized in Table 3.1) are refining our ability to analyze transcriptomic data across diverse

Table 3.1 Overview of transcriptomic techniques used in plant stress research

Technique	Description/Key notes
Northern blot	For RNA expression quantification and validation; useful for miRNA and siRNA detection.
RT-qPCR	Quantifies cDNA from RNA; used for validating gene expression of a few genes and splicing variants.
SAGE (serial analysis of gene expression)	Untargeted RNA quantification using short cDNA tags; allows new gene discovery.
Microarrays	Measure expression of thousands of genes; used for stress studies and DEGs detection.
RNA-seq (short-read)	Illumina, Ion Torrent; highly sensitive and accurate for gene expression analysis.
RNA-seq (long-read)	PacBio, Oxford Nanopore; detects full-length isoforms and AS events.
Iso-seq (PacBio isoform sequencing)	Used for transcriptome assembly and splicing analysis.
Single-cell RNA-seq (scRNA-seq)	Reveals gene expression at single-cell resolution.
Spatial transcriptomics (Stereo-seq)	Integrates gene expression data with spatial tissue context, enabling high-resolution mapping without protoplasting.
Machine learning and deep learning methods	Used to predict stress signatures and identify biomarkers (e.g., *TransGeneSelector*, *PREDICT*).

This table summarizes the main transcriptomic methodologies discussed in the chapter, highlighting their key features, applications, and technical advantages for studying plant responses to stress

conditions and plant species, and they also enable a more integrated, systems-level understanding of gene function, evolution, and plant adaptive responses.

3.3 Transcriptomic Insights into Biotic and Abiotic Stress Tolerance

Adverse climatic factors and low soil fertility bear major constraints to crop production worldwide. Fortunately, there are major differences in plants' tolerance to these abiotic stress factors. The identification of the underlying genetic mechanisms can boost specific breeding efforts to improve crop yields. Transcriptomic analysis enables comprehensive profiling of genome-wide gene expression in response to stress, facilitating the elucidation of complex regulatory networks underlying plant stress adaptability and tolerance. Meta-analysis of comparative transcriptomic data of plants exposed to different biotic and abiotic stress factors revealed that both adverse conditions induce some common pathways (Biniaz et al. 2022). Also, plants exposed to different abiotic stresses like chilling, osmotic stress, and salinity share DEGs related to tolerance (Buti et al. 2019). Common abiotic stress-activated pathways related to tolerance typically include reactive oxygen species (ROS) signaling, stress hormone signaling (mainly abscisic acid (ABA), jasmonic acid

(JA), and ethylene), ROS scavenging and antioxidants like flavonoids, and stomatal regulation (Raza et al. 2021; Sachdev et al. 2021; Parwez et al. 2022).

High levels of antioxidant production are essential to tolerate all environmental stress factors. However, improving crop yields under stress conditions additionally requires specific mechanisms to avoid, tolerate, and/or escape from the prevailing stress situations in the target cropping area. A major challenge is the elucidation of the genetic bases of these stress-specific mechanisms. Here, we will provide an overview of transcriptomic research regarding abiotic stress situations currently aggravated by climate change, such as extreme temperatures, drought, and flood, as well as the main crop-limiting soil-derived stress factors, including salinity, nitrogen, phosphate, and iron deficiency.

3.3.1 Transcriptomic Responses to Selected Climate-Change Related Stress Factors

3.3.1.1 Heat Stress

Heat stress rapidly disrupts plant and seed development by impairing storage biosynthesis, cell growth, and key physiological processes within hours. After reviewing recent data on physiological, hormonal, and transcriptional responses to heat stress, Balfagón et al. (2020) highlighted the importance of photodamage avoidance, abscisic and JA signaling, and the upregulation of genes involved in oxidation–reduction processes, photosynthesis, and protein metabolism as key mechanisms for plant acclimation to high temperatures. For example, transcriptome analysis on barley caryopses revealed distinct early responses involving heat shock proteins (HSPs), ROS scavenging, and compatible solute production (Mangelsen et al. 2011). Increased cell division, activation of heat shock proteins (HSP), dissimilation, and antioxidant transcript amounts were correlated with successful resistance to high-temperature stress in the cool-season forage *Festuca arundinace* (Hu et al. 2014).

Through comparative transcriptomics and leveraging publicly available datasets, Sriden and Charoensawan (2022) systematically gathered and reanalyzed 259 transcriptomic profiles derived from 139 independent *A. thaliana* experiments conducted in shoot, root, and seedling tissues exposed to a broad spectrum of temperature conditions, including freezing, cold, suboptimal and elevated ambient temperatures, as well as heat shock. Despite notable transcriptomic differences across tissues, they successfully identified distinct gene sets exhibiting strong transcriptional responses to specific temperature regimes, with some specific responses conserved across tissues and others showing tissue specificity. By comparing the highly variable heat shock factors (HSFs) and HSP genes identified in high-temperature-associated clusters, the authors found three HSP genes (*HSP18.5*, *HSP23.5*, and *HSP90.1*) consistently classified as HVGs in all four analyses. This uniform activation across tissues highlights their potential as universal biomarkers for plant responses to elevated temperatures. Moreover, they identified *SAP10* and *SAUR32* with specific functions in roots under high temperatures and the genes *DNAJ, FKBP65, HOP3, SGT1a, MBF1C, SR30, SR45A,* and *AT5G12110* up- or downregulated in all the high temperature clusters.

Notably, canonical temperature-responsive genes, such as members of the HSP family, were upregulated not only under heat stress but also under cold conditions, suggesting broader roles in temperature response. By integrating DNA-binding specificity data for key temperature-regulating TFs such as *HSFA1a*, *PIF4*, and C-repeat binding factor (CBF), Sriden and Charoensawan (2022) delineated distinct binding patterns correlated with the differential expression of target genes across temperature stresses.

3.3.1.2 Chilling and Freezing Stress

Chilling and freezing stresses are major abiotic challenges that endanger plant growth, development, and survival, especially in temperate and alpine regions. To cope with these low-temperature conditions, plants have evolved a range of adaptive mechanisms, including transcriptomic reprogramming. Thousands of DEGs, miRNAs, alternative splicing genes, and many TFs have been identified as involved in cold stress (e.g., Calixto et al. 2018; Waititu et al. 2021; Kidokoro et al. 2022; Guo et al. 2023). However, researchers have established cold-responsive genes (CORs) and cold-responsive regulatory networks for several plant species. The most well-studied TFs involved in cold response are the C-repeat binding factor (CBF) genes, which are rapidly and transiently induced upon exposure to cold stress. These TFs enhance freezing tolerance by binding to specific cis-elements (CCGAC) in the promoters of CORs (Stockinger et al. 1997). Activation of CBF-regulated CORs such as *COR47*, *KIN1*, and *RD29A* has been shown to significantly improve plant freezing tolerance (Jaglo-Ottosen et al. 1998; Novillo et al. 2007).

In *A. thaliana*, *HSFA6B* and *HSFC1* have been identified as “low-temperature” highly variable genes (HVGs) in shoot and root transcriptomes, with notably stronger transcriptional responses observed in roots (Sriden and Charoensawan 2022). Interestingly, in the *cbf123* triple mutant exposed to cold (4 °C for 1 h), *HSFC1* expression was still transiently upregulated, whereas classical CBF-dependent cold-responsive genes exhibited minimal transcriptional changes. This suggests that *HSFC1* may function through a CBF-independent cold response pathway (Zhao et al. 2016). In parallel, *HSFA6B* has been shown to directly bind to heat shock elements (HSEs) in the promoter region of *DREB2A*, a key TF involved in abiotic stress responses, implying that *HSFA6B* may participate in cold stress responses via *DREB2A*-mediated pathways (Wang and Zhu 2023). In rice, similar regulatory components appear to function in response to cold and water stress. For instance, *OsHSFA2a* and *OsHSFA3* are upregulated under chilling conditions, while *ICE* (inducer of CBF expression), a MYC-type basic helix-loop-helix TF, plays a pivotal role by enhancing CBF transcription (Wang et al. 2022a). Moreover, a phylotranscriptomic analysis performed in cold-treated seedlings from five representative eudicots identified 35 high-confidence conserved cold-responsive TF orthogroups (CoCoFos) that included the cold-responsive regulators CBFs, HSFC1, ZAT6/10, and CZF1, among others (Wang et al. 2023).

The cold responsiveness of plants is also regulated by the circadian clock system. Some circadian clock-related TFs, such as *CCA1*, *LHY*, and *RVEs*, have been proven to be involved in cold tolerance by directly regulating the expression of downstream CORs (Guo et al. 2023). As expected, hormone signaling is also key in cold stress early responses. For instance, protein–protein interaction analyses revealed that ABA signaling via protein phosphatase 2Cs was the most important signal transduction pathway associated with adaptation to cold stress in sugarcane (Huang et al. 2021). The study of global hormone-regulated transcript expression patterns in *Tetrastigma hemsleyanum* suggested that gibberellin (GA) and ABA could work antagonistically to balance the cold stress responses (Peng et al. 2021), while transcriptome analysis in cold-tolerant and cold-sensitive common bean cultivars revealed considerably greater expression of ABA, GA, and JA in the tolerant cultivar (Yang et al. 2023).

3.3.1.3 Drought Stress

Freezing and drought stress have similar effects in plant responses, as both can lead to cellular dehydration, oxidative stress, and the activation of overlapping signaling pathways and stress-responsive genes. Thus, ABA signaling and genes involved in the circadian clock are also implicated in day and night drought-responsive pathways (Yi et al. 2022). More specifically, the up and down regulation of a wide range of TFs (*NAC, AP2/ERF, DREB, bZIP, WRKY, bHLH*), circadian rhythm-plant players (such as *PRR7*), proline synthesis regulation genes (*PRODH*, *OAT*, *ALDH18A1*, *P5CS*, *P4HA*), and carbon metabolism and photosynthesis hub genes (*GBSS1, PORA, SGR1*) have been identified in several drought-adapted species (e.g., Capelari et al. 2021; Villanueva et al. 2023; He et al. 2025).

To explore such a complex and multifaceted stressor, systems biology approaches and functional genomics have been employed across both model and crop plants. These strategies involve constructing gene coexpression networks, prioritizing key TFs, and applying dynamic mathematical models alongside computational simulations to capture regulatory dynamics. For example, in *A. thaliana*, Kumar et al. (2023) identified distinct temporal transcriptional signatures during drought stress and uncovered numerous novel TFs that could serve as valuable tools for genetic engineering. Expanding beyond model systems, functional genomics has been increasingly applied to wild accessions and crop landraces that have evolved traits enabling them to withstand harsh agro-climatic conditions such as drought. In barley, Marok et al. (2021) synthesized findings from multiple studies to highlight core drought-tolerance regulators, including TFs, heat shock proteins, late embryogenesis proteins, and key components of ABA signaling, proline biosynthesis, and redox regulation. Similarly, in soybean, Shahriari et al. (2022) performed protein-protein interaction network analyses and genome-wide surveys of cis-acting regulatory elements, identifying hub genes and major TFs involved in drought-responsive pathways. In *Populus*, Tahmasevi et al. (2023) integrated meta-analysis and machine learning to pinpoint key water stress-responsive genes. Their coexpression analysis revealed *NATA1*-like acetyltransferase and cytochrome P450 genes as pivotal players, along with prominent roles for MYB (v-myb avian myeloblastosis viral oncogene homolog) and MAPK (Mitogen-Activated Protein Kinase) TFs. Notably, *ARF2*-like and *PYL4*-like genes were proposed as promising candidates for improving drought tolerance through breeding programs. Together, these examples underscore the power of integrative functional genomics in uncovering complex gene regulations driving plant adaptation to drought stress.

3.3.1.4 Flooding Stress

Due to climate change, flooding events are increasingly challenging crop performance even under arid and semi-arid climates. Flooding causes severe metabolic and growth reduction in sensitive plants primarily due to hypoxia and indirectly due to the formation of inorganic and organic toxicants in the waterlogged soil. The capacity to form aerenchyma and functional adventitious roots (AR) is key under waterlogging conditions as it allows efficient oxygen diffusion from nonsubmerged plant parts to the submerged organs. Survival to complete submergence in terrestrial plants needs either stimulation of enhanced elongation to reach the above water level or starvation tolerance. Several transcriptomic studies have contributed to the identification of the genetic basis underlying these mechanisms. Comparative transcriptomics of five Brassicaceae species differing in tolerance to complete submergence identified close to 17,000 orthogroups. Tolerance in *Rorripa sp.* was associated with strong downregulation of cell cycle-related genes, while in tolerant *Cardamine sp.,* the global transcriptomic response was small. In all cases, tolerance was related to starvation responses associated not only with cell cycle restriction but also with autophagy, senescence, photosynthesis, and fermentation (van Veen et al. 2025).

Building on these findings, further transcriptomic investigations have deepened our understanding of species-specific tolerance mechanisms. In wheat, tolerance to waterlogging supported by ethylene and ROS-induced aerenchyma formation is related to regulation through *P2/ERF* and *LOB/AS2* TFs (Gao et al. 2025). Upregulation of genes from the lignin biosynthesis pathway underlies the formation of barriers to radial oxygen loss in the roots (Kim et al. 2024). Furthermore, mitigation of energy deficit in tolerant barley and quinoa is related to genes responsible for higher sugar concentrations and more efficient alcohol metabolism (Wang et al. 2024a; Guo et al. 2022). Comparative transcriptomics in cucumber differing in (AR) formation in response to waterlogging identified *CsPrx73* as a key gene in this process. Moreover, it was shown that the *CsERF7-3* TF binds to the ATCTA-box motif in the promoter of this gene, initiating its transcription (Pan et al. 2024). Together, these transcriptomic insights provide a valuable framework for dissecting and improving flooding tolerance mechanisms in diverse crop species.

3.3.2 Transcriptomic Responses to Selected Soil-Derived Stress Factors

3.3.2.1 Salinity

Salinity is one of the major factors limiting crop productivity, affecting at least 20% of the irrigated land worldwide, whereas all major crops are glycophytes (sensitive to NaCl levels above 50 mM). When assessing plant tolerance and adaptation to salinity, it is essential to distinguish between neutral (mainly NaCl and Na_2SO_4 ions) and alkaline (NaCl and $NaHCO_3$/Na_2CO_3 ions) salinity types due to their differential impact on plant crops. In fact, Na^+ exclusion mechanisms are insufficient for crop productivity under alkalinity (Pérez-Martín et al. 2022; Chen et al. 2024). At the transcriptomic level, salinity induces differential expression of genes involved in plant signaling pathways related to general stress responses, ion homeostasis, phytohormone signaling, and cellular structural modifications, and this alteration is exacerbated under alkaline salinity (Fang et al. 2021; Almira-Casellas et al. 2024). Gene expression changes encompass protein kinases and phosphatases, reactive oxygen species (ROS) scavengers, regulators of osmolyte and antioxidant production, as well as a wide array of TFs (Wang et al. 2023; Jamil et al. 2011; Krasensky and Jonak 2012). ABA signaling plays a crucial part in such regulatory changes, as salinity (like drought and low temperatures) imposes osmotic stress, disrupting plant water relations. Plant acclimation responses to salinity through differential expression of genes include the regulation of ABA receptor (PYL/PYR), serine/threonine-protein kinases (SnRKs), and ABA-responsive genes from the Abscisic Acid Insensitive (ABI) family (Zhang et al. 2024). Upregulation of ABA biosynthesis genes and TFs such as 9-cis-epoxycarotenoid dioxygenase (NCED), zeaxanthin epoxidase (ZEP), and ABA 8′ -hydroxylase (ABA 8′ -OH) also mitigates the effects of salt stress (Martínez-Andújar et al. 2020). The target regulation of other hormone-related genes, such as *JAZ* and *GH3* genes, has also been identified in salt and salt-alkaline-tolerant species (Almira-Casellas et al. 2024; Wang et al. 2024c).

Comparative transcriptome analyses of salt-sensitive and salt-tolerant species or among crop varieties with contrasting salt sensitivity have contributed significantly to elucidating salt-responsive molecular signatures of halophytes and glycophytes. Several studies provide evidence that the upregulation of genes from the Salt Overly Sensitive (SOS) pathway is a salt tolerance response (Kumar et al. 2009; Sathee et al. 2022). In halophytes like *Thellungiella halophila,* differential protein interaction, modification and redistribution processes, rather than TF regulation, result advantageous under salinity (Gong et al. 2005). *Puccinellia tenuiflora* outperformed other glycophytic gramineous relatives through an enhanced expression of ion transport regulation genes (mostly a KUP/HAK/KT K^+ transporter, the HKT1;3 gene, two voltage-gated potassium channel genes, two NHXs, and a V-H+-ATPase), flavonoid and sucrose biosynthesis genes, and a rate-controlling gene for the pentose phosphate pathway (the glucose-6-phosphate dehydrogenase) (Guo et al. 2021). A differential transcriptome reconfiguration leading to a tolerant genotype specific response under salinity has been further addressed in other major crops like alfalfa, with an attenuation in the ROS (including *PODs, GSTs, APXs, GRXs*, and AOXs) and Ca^{2+} (mainly Calcium-Dependent Protein Kinase (CDPKs) and CBL-Interacting Protein Kinase (CIPKs)) signaling pathways (Lei et al. 2018), or in maize, with the upregulation of genes homologous to WRKY TFs in tomato (WRY3) and cotton (WRKY41) reducing oxidative stress, increasing proline content, decreasing Na^+ and enhancing K^+ and Ca^{2+} contents (Wang et al. 2019). The transcriptomic studies of crop wild relatives such as *Brassica Gossypium purpurascens* or *Brassica fruticulosa* also reveal potential genetic targets of recently evolved salinity adaptations (Peng et al. 2023; Busoms et al. 2024)

3.3.2.2 Nitrogen Deficiency

Nitrogen (N) deficiency is a major constraint for crop yield, but there are major differences among plants' ability to take up and efficiently use the absorbed soil N. Nitrogen uptake efficiency (NUpE) and N use efficiency are the key traits in breeding programs (Xu and Takahashi 2020). The genetic bases of these characteristics have been explored by transcriptomics along with physiological and metabolomic approaches. Results are strongly influenced by the stress level, which depends on the stress exposure time and the genotype. Under long-term N-deprivation, many genes related to distress, growth inhibition, photosynthesis decline, and senescence are activated. Genes related to N uptake and metabolism may initially have been upregulated, but expression can decline after long-term N-deficiency (Yang et al. 2020). The most interesting results are being obtained comparing the gene expression profiles between N-efficient and N-inefficient varieties under milder stress conditions. As an example, experiments with potato revealed higher expression of genes related to N-acquisition and N and C metabolism in the efficient cultivar. Enrichment of the nicotinate-nicotinamide metabolism reflects the importance for N-efficiency of enhanced redox capacity and energy metabolism. Specially upregulated in the efficient cultivar were the high-affinity nitrate transporter *StNRT2.7*, the nitrate reductase *StNR*, and the two genes involved in NH_4 incorporation to glutamate, *StNADH-GOGAT* and *StGDH*. Enhanced N assimilation in efficient cultivars can further be accompanied by upregulation of photosynthesis-related genes, the tricarboxylic acid cycle and the C metabolism, reinforcing the view of a close coordination between N and C metabolism (Zhang et al. 2021; Wang et al., 2022b; Xie et al. 2024). Especially interesting is also the enhancement of photorespiration in N-efficient wheat (Wang et al., 2022b), which may provide enhanced N uptake and improve yield under N-limiting conditions as observed in rice (Chen et al. 2025). Moreover, phytohormone metabolism and TFs *WRKY28, TFY11*, and *NAMB2* were differently expressed in efficient wheat lines (Zhang et al. 2021). It is well established that nitrate activates transcriptional networks that have a strong influence on root architecture, enhancing N-uptake (Vidal et al. 2015).

Stress due to N-deficiency seldom comes alone, and transcriptomic research is increasingly addressing combinatory stress effects to elucidate tolerance mechanisms. Recently, the transcriptomic responses of *A. thaliana* to N deficiency and drought, individually and in combination (N/W), have been evaluated using the standard accession Col-0 and different natural populations (Xue et al. 2024). Mild drought diminished N deficiency effects, probably due to reduced growth. Although the transcriptomic response to combined stress was accession dependent. However, a subset of DEGs specifically responding to the stress combination could be identified. Specifically upregulated by combined N/W stress were *PIN8, FUSCA3, APL4, E25, FUT8,* and *EIL* while downregulated were *SAUR12 BGL2, PBP1, RPL24, FUT4, RSW10, GLR2.5, GLR2.9 CPK31,* and *MPK11*. Under the combined N/W stress, genes involved in the regulation of auxin and carbohydrate metabolism were also outstanding. Especially interesting is the downregulation of the glutamate receptor GLR2.5, which plays a crucial role in root architecture (Gong et al. 2025), a key trait in the response to both N-deficiency and drought.

3.3.2.3 Phosphate Deficiency

Phosphate (P) deficiency is an important growth-limiting factor of crop plants, especially on alkaline and acid soils where inorganic phosphate tends to form sparingly soluble P-compounds with calcium, iron, aluminum, and/or manganese (Ducousso-Détrez et al. 2022). Even a short deficiency can cause a huge alteration in gene expression, as shown in leaves of soybean plants with 533 DEGs after 24 h exposure to P deprivation. Among the upregulated P_i sensing genes, five genes coding for SYG1/PHO81/XPR1 domain-containing proteins (SPX) domain-containing proteins were identified (Zeng et al. 2018). Inositol phosphates can bind to these SPX domains, acting as a cellular P-status sensor and regulating SPX interaction with PHR1, a key regulator of the phosphate starvation response (Collins et al. 2024). Transcriptomics comparing rice landraces differing in P-efficiency revealed specific upregulation of Os*SPX2* in P-efficient races (Prodhan et al. 2022).

Besides quick sensing of deficient P supply, plant P-efficiency comprises both P-acquisition efficiency (PAE) and P-use efficiency (PUE) (Vandamme et al. 2016). PAE requires the ability to mobilize soil P through root exudates and efficient uptake mechanisms. PUE, among others, depends on the ability to mobilize P from internal storage sites and efficient transport mechanisms. Exudation of organic acids, especially citrate, favors solubilization of inorganic soil P, while phosphatases mobilize P from external and internal organic sources. In fact, comparative transcriptomics with two contrasting cultivars of *Stylosanthes guianesis* revealed DEGs related to Pi uptake and transport, and Pi mobilization, like purple acid phosphatase (PAP), and multidrug and toxin extrusion (MATE) involved in organic acid exudation. DEGs also included TFs and Pi signaling-related genes, like *PHOSPHATE STARVATION RESPONSE 1-like* (*PHR1*), *WRKY,* and the *SYG1/PHO81/XPR1* (SPX) domain (Chen et al. 2021).

Comparative transcriptomics in soya isogenic lines spotted upregulation of 22 phosphatase genes in the efficient line, along with the enhancement of the transporter genes *Glyma08G171100* and *Glyma07G163300* in roots and leaves, respectively (Zhang et al. 2017). Transcriptomic studies on P efficiency can be especially intricate in nodulated legumes. Nodulation has a high P demand. Comparison of chickpea-N-fixing bacterial associations differing in nodulation efficiency under P-deficiency revealed in the inefficient association a much higher number of DEGs, among others, associated with nodulation, Pi remobilization, key metabolic pathways, ion transport, TFs, and signaling. This more intense differential regulation represents rather a higher stress level in the inefficient association than a more efficient adaptation (Zhang et al. 2017). Moreover, the influence of P starvation on gene regulation within the nodules must specifically be addressed. As an example, more than 2000 DEGs were found in soybean nodules under P deficiency, including eight P transporters (PTs), 16 purple acid phosphatases (Xue et al. 2018). GMPT7, a P transporter located in the nodules, has been identified as a key player in symbiotic N_2 fixation (Chen et al. 2019). Phosphate deficiency can further alter gene expression through the regulation of alternative transcription start sites, producing different transcript isoforms. P-deficient *A. thaliana*, among others, highly expressed ALT_{ECT4}, a short isoform of the m^{6}A reader gene *ECT4* strongly involved in root growth regulation (Reis et al. 2024).

3.3.2.4 Iron Deficiency

Iron (Fe) deficiency affects crop yield on alkaline soils where the high pH limits Fe solubility and bicarbonate interferes with plants' Fe acquisition and transport. Iron acquisition in plants is controlled by bHLH FER-like Fe deficiency-induced transcription factor (FIT), both in strategy I dicots and strategy II monocot grasses. Transcriptomics comparing *fit* knock-out mutant and FIT overexpressing lines revealed 448 FIT-regulated genes in *A. thaliana* (Mai et al. 2016). In Fe-deficient wheat, a monocot with strategy II for Fe acquisition, almost 6000 and 2600 DEGs were found in leaves and roots, respectively (Wang et al. 2020). Of special interest is the differential gene expression in the methionine salvage pathway and the biosynthesis of phytosiderophores, the key Fe-binding compounds in strategy II species (Wang et al. 2020). Transcriptomic analysis of 18 soybean genotypes after exposure to Fe deficiency for only 60 min revealed 9141 and 2848 DEGs in leaves and roots, respectively, that were specific for iron-efficient genotypes (Kohlhase et al. 2021). However, only 123 DEGs were shared by 3 efficient genotypes, and only 1 DEG was common to 4 efficient genotypes. These common DEGs were mainly related to stress signaling and tolerance. The unique DEG common to 4 efficient genotypes was the downregulated *Glyma.13G155200*. This gene is homologous to *AtTIL*, a temperature-induced lipocalin. Its overexpression in *A. thaliana* promotes the delay of flowering and senescence.

Tolerance to bicarbonate (Bic) is a key factor for Fe efficiency. Transcriptomics revealed a quick upregulation of Fe homeostasis genes after only 3 h exposure in Bic sensitive Arabidopsis plants, while in the tolerant ones, signal transduction, GS, sulfur acquisition, and cell cycle precede the induction of Fe homeostasis mechanisms, yielding an efficient response to bicarbonate stress (Pérez-Martín et al. 2021).

Later, the combination with other genomic techniques allowed the researchers to discover 8 candidate genes (*AHH*, *CSLD2*, *DAO1*, *GDH2*, *FBN6*, *INV-E*, *JAZ10*, and *TBL19*) involved in bicarbonate stress responses and soil carbonate tolerance (Busoms et al. 2023). On the other hand, Fe deficiency is a major nutritional disorder in fruit trees cultivated in calcareous soils. The key factors for differential Fe deficiency tolerance have been investigated in peach and apple rootstocks and genes related to specific Fe transporters, photosynthetic recovery, and carbon fixation have been identified (Sun et al. 2022; Li et al. 2025).

3.3.3 Transcriptomic Responses to Plant Pathogens and Herbivory

Recent advances in plant transcriptomics have exposed adaptive gene expression programs activated in response to biotic stressors, including herbivores and pathogens, revealing how plants perceive, signal, and activate tailored defenses shaped by both the nature of the attacker and the environmental context.

In plants, insect attacks represent a complex stimulus that can be conceptually divided into three phases: pest recognition, signal transduction, and defense activation. Distinguishing between wound-induced and herbivore-specific responses is crucial. The existing transcriptome studies provide valuable insights into gene regulation in attacked plants, but they also underscore the need for synchronized experimental designs, as most research to date has focused on individual herbivore attacks. Heidel-Fischer et al. (2014) and Santamaria et al. (2021) contributed concise overviews of plant transcriptomic responses to multiple herbivore interactions. More recent studies focus on investigating gene expression changes in plants following herbivore attack emphasize how plants tailor their defense strategies based on the type of herbivore, with distinct responses between chewing vs. piercing-sucking insects and aboveground vs. belowground feeders (Leclerc et al. 2024; Ye et al. 2021). Many of these studies highlight the role of phytohormones in defense, JA for chewing herbivores, and salicylic acid for piercing-sucking herbivores. Additionally, ABA and other signaling molecules are involved in modulating these responses. Herbivore-induced responses extend beyond the direct site of attack, affecting systemic signaling, root responses, and secondary metabolism. For example, Montesinos et al. (2024) examined how *A. thaliana* responded to herbivores with different feeding strategies, comparing the chewing caterpillar *Pieris brassicae* and the piercing-sucking mite *Tetranychus urticae*. While both pests triggered transcriptomic changes, *P. brassicae* induced a short-term, stress-related response involving JA and ABA regulation, whereas *T. urticae* elicited a prolonged but lower-intensity response, primarily affecting SA signaling. Notably, these responses were detected in noninfested tissues, affecting root hormonal signaling, flavonoid regulation, and cell wall reorganization.

Several transcriptomic studies identified pathways such as "the phenylpropanoid pathway," "flavonoid biosynthesis," and "secondary metabolite production" as key components of plant resistance to herbivory (e.g., Wang et al. 2024b). Thus, the combination of transcriptomic and metabolomic techniques is highly valuable for the study of plant defense mechanisms and has been used in several plant species and predators or pathogens (see Ye et al. 2021). Recently, Lou et al. (2025) investigated maize defense responses to *Spodoptera exigua* infestation and discovered that the upregulation of biosynthetic pathways, amino acid metabolism, and specific TFs is linked with a strong modulation of phenylpropanoid and benzoxazinoid biosynthesis after 24 hours of infestation. Kariñho Betancourt et al. (2024) revealed a species-specific regulation of specialized metabolism in *Datura*, showing that JA signaling and associated TFs drive metabolic plasticity in response to herbivory. Regarding pathogen defense, Ding et al. (2021) investigated transcriptional and metabolic responses in *Brachypodium distachyon* roots challenged with *Fusarium graminearum* (Fg). They identified shared gene expression patterns between infected roots and spikes, particularly associated with the mycotoxin deoxynivalenol. Overexpression of *BdMYB78* enhanced root resistance, and roots were able to recognize Fg before physical contact by modulating gene expression and metabolite secretion. The integration of these omic techniques has also been used to study postharvest infections. For example, Cheng et al. (2023) investigated gene expression and metabolite changes in peach fruit in response to *Monilinia fructicola* infection. Transcriptomic analysis revealed that most DEGs were upregulated within 12 hours, while metabolomic analysis identified 357 differential metabolites involved in various metabolic pathways. Peach defense responses were mediated by JA at 12 hours and SA at 48 hours, highlighting a dynamic molecular framework for systemic resistance to this pathogen.

Ongoing environmental changes are increasingly reshaping plant–pathogen interactions, highlighting the profound influence of abiotic stresses on biotic responses across molecular, cellular, and physiological levels. Environmental conditions directly modulate factors such as plant architecture, tissue organization, and both primary and secondary metabolism, ultimately impacting plants' susceptibility or resistance to pathogens. Furthermore, abiotic stress can alter the timing and intensity of plant defense mechanisms, as well as pathogen virulence traits (Zarattini et al. 2021). Recent studies have provided significant insights into how abiotic factors such as drought, heat, and nutrient limitation intersect with biotic challenges like herbivory and fungal infection to shape plant immune responses. For example, Bihani et al. (2024) applied a spatial transcriptomics approach to explore the transcriptional responses of cotton under combined insect infestation and drought stress. Their analysis uncovered a distinct separation between broadly responsive "stress-general" genes, such as *WRKY28*, *WRKY40*, *WRKY53*, *ERF4*, and *ERF5*, and "stress-specific" genes like *GH3.1*, *ACS1*, and *CYP74A*, which were selectively activated depending on the type of insect infestation. Their results suggest that stress-general genes could serve as key targets for improving crop resilience to multiple simultaneous stresses. Zhong et al. (2024) demonstrated how transient high temperatures modulate herbivory-induced defenses in *Solanum tuberosum*.

Heat stress enhanced the growth of *Phthorimaea operculella* larvae and significantly altered the potato defense transcriptome. The expression profile shifted from an evolutionary hourglass pattern to a "vase" pattern, characterized by suppression of both early and late-stage defense gene expression, alongside an accelerated intermediate response. Additionally, high temperatures disrupted jasmonate signaling and reduced the accumulation of defense compounds. These findings, supported by experiments in JA-signaling modified plants, confirmed that elevated temperatures compromise JA-mediated defense mechanisms.

Nutrient availability can also modulate biotic interactions through mechanisms of nutritional immunity; however, the nature of the immune response is species-dependent, as seen in a comparative study examining *A. thaliana* and rice responses exposed to fungal pathogens under different Fe nutritional regimes (Garcia-Molina et al. 2025). Despite their contrasting strategies to handle Fe, the GRNs of both species showed remarkable overlap in regulatory control, with approximately 50% of equivalent TFs. This indicates a conserved core of transcriptional regulators that are co-opted across species to coordinate defense responses under nutrient-limited conditions.

Taken together, these studies underscore the complexity of abiotic and biotic stress integration in plants (Table 3.2). The use of advanced transcriptomic and regulatory network analyses can help to understand how plants modulate gene expression to fine-tune their responses under combined stresses. Such insights are essential for developing crops with broad-spectrum resistance suited to fluctuating and extreme environmental conditions.

3.4 Comparative Transcriptomics Reveal Conserved and Divergent Adaptive Strategies Among Plant Species

The dynamicity of ecological systems involves that plants are continuously exposed to changing environments and they explore the colonization of new habitats. This scenario poses a recurrent arm-race evolutive impasse for plant acclimation responses: the implementation of programs to withstand hostile environments is essential to ensure survival, but paradoxically, the associated energetic costs can trade off plant vegetative growth and reproduction (Bostock et al. 2014). Then, the adaptation of plants to the niches they occupy is the consequence of having evolved plasticity to transiently readjust physiology and metabolism to face recurrent daily and seasonal environmental fluctuations, as well as frequent adversities, without compromising fitness. Systemically, the transitions from the basal to the distinct acclimation states obey sophisticated reconfigurations in the transcriptome landscape orchestrated by gene regulatory networks. GRNs are defined as the hierarchical entities recapitulating the interplays between TFs and cognate target genes that mediate adequate spatiotemporal transcriptome changes (Minelli and Valero-Gracia 2022). The prevalent role of GRNs in the achievement of acclimation states underlies that the architecture they adopt entangles the range of permissibility of the plant

Table 3.2 Transcription factors (TFs) and stress-responsive genes involved in plant responses to abiotic and biotic stresses

TF/Gene/DEG	Stress type	Notes
HSP18.5, HSP23.5, HSP90.1	Heat	Universal heat stress markers, highly variable genes (HVGs)
HSFA1a, HSFA6B, HSFC1	Heat, cold	TFs involved in temperature regulation
CBFs, COR47, KIN1, RD29A	Cold	CBFs activate CORs; key for freezing tolerance
DREB2A	Cold, drought	Mediated by HSFA6B; key abiotic stress TF
OsHSFA2a, OsHSFA3, ICE	Cold	Rice TFs responsive to cold stress
CCA1, LHY, RVEs	Cold, drought	Circadian clock TFs involved in cold tolerance
PRR7	Drought	Circadian-regulated drought-responsive gene
NAC, AP2/ERF, DREB, bZIP, WRKY, bHLH	Drought	General drought-responsive TFs
PRODH, OAT, ALDH18A1, P5CS, P4HA	Drought	Proline biosynthesis genes
GBSS1, PORA, SGR1	Drought	Photosynthesis and carbon metabolism genes
MYB, MAPK, ARF2-like, PYL4-like	Drought	Coexpression in *Populus*
P2/ERF, LOB/AS2	Flooding	Ethylene-regulated TFs in wheat under waterlogging
CsERF7-3, CsPrx73	Flooding	TF and gene regulating adventitious root formation in cucumber
PYL/PYR, SnRKs, ABI, NCED, ZEP, ABA 8′-OH	Salinity	ABA signaling components for salinity tolerance
Methionine biosynthesis genes	Salinity (microbe-induced)	Lead to ethylene production for stress relief
MAPKs, Ca2+ signalling genes	Drought, Salinity (microbial)	Enhanced via PGPR or AMF inoculation
StNRT2.7, StNR, StNADH-GOGAT, StGDH	Nitrogen deficiency	Genes for N acquisition and assimilation in potato
WRKY28, TFY11, NAMB2	Nitrogen deficiency	Differentially expressed in N-efficient wheat
PIN8, FUSCA3, APL4, FUT8, EIL	Combined N-deficiency + drought	Specifically upregulated under combined stress
SPX-domain genes, PHR1, WRKY	Phosphate deficiency	Involved in Pi sensing and starvation response
PAP, MATE, PTs, ALTECT4	Phosphate deficiency	Mobilization and transport of P
FIT, FER-like, CDPKs, PP2Cs	Iron deficiency	TFs and kinases in Fe homeostasis and signaling

(continued)

Table 3.2 (continued)

TF/Gene/DEG	Stress type	Notes
BdMYB78	Pathogen	Overexpression enhances root resistance
GH3.1, ACS1, CYP74A	Biotic stress (herbivory)	Stress-specific genes under insect attack
WRKY28, WRKY40, WRKY53, ERF4, ERF5	Combined biotic and abiotic stress	Stress-general genes

This table lists key TFs, DEGs, and HVGs identified through transcriptomic techniques, organized by the type of abiotic or biotic stress they are associated with

to overcome evolutionary pressures. This concept attracted researchers' interest to interrogate the evolutionary aspects of plant adaptation from the perspective of GRNs.

Daisuke Uranos' group contributed seminal works that investigated evolutionary mechanisms for plant adaptation to adverse environments by addressing cross-kingdom comparisons of GRNs reconstructed from time-series transcriptome responses between distant species. Wu et al. (2021) examined the similarity in dynamics and the biological relevance of transcriptome changes between *Marchantia polymorpha* (early terrestrial plant) and *A. thaliana* (flowering plant) in response to salt stress and unveiled that the GRNs operating in both species were governed by a conserved core of WRKYs, namely Ath*WRKY6/25/26/33/46/71* and Mp*WRKY1/3/6/12/13/14*. However, topological analysis revealed that, compared to the single low-scale *M. polymorpha* GRN, the Arabidopsis network has been expanded and structured as two disentangled regulatory modules, distinguishing early from late responsive features. Nevertheless, the core of *M. polymorpha* WRKYs is subspecialized into a first subgroup, which is prevalent during early time-points connected to transcripts related to ABA response, and a second subgroup operating as a late-responsive module targeting general stress and tissue development transcripts. In addition, predictive models highlighted high divergence in the putative cis-regulatory elements (pCREs) of target transcripts recognized by WRKYs in each species. The apparent coevolution of conserved WRKYs and the pCREs in targeted transcripts from *M. polymorpha* to Arabidopsis sets the basis for an adaptive response to preserve the regulatory components while refining the selectivity for responsive features. As a result, this model explains the transition towards a specialized two-phase response in higher plants consisting of an initial program to counteract osmotic stress, followed by a second round of events to cope with Na^+ toxicity under high salinity.

Dong et al. (2024) used a similar approach to address adaptive mechanisms in terrestrialization based on cross-comparison of nutritional responses between *Klebsormidium nitens* (algae before terrestrial colonization) and *M. polymorpha.* In this case, the reconstruction of GRNs operating under nitrate, phosphate, and sucrose starvation, as a key challenge for terrestrial colonization, revealed profound divergences in the regulatory networks between species. *M. polymorpha* GRNs showed larger scales and were commonly governed by three master TFs, namely

Mp*CSD,* Mp*HD9,* and Mp*ERF20*. The common TFs operating the GRNs in *K. nitens* failed to attract direct orthologues for *M. polymorpha,* and only one equivalence was found for Mp*HD9*. However, the Mp*HD9* was connected to cytokinin (CK) signaling components, including the regulator of CK responses Mp*CSD*, whereas the orthologue in *K. nitens* was rather involved in cell division and carbon metabolism. Importantly, CKs are fundamental signaling hormones to initiate acclimation to nutrient scarcity in land plants. Thus, the dissection of discrete GRNs modules in both species allowed us to propose a co-optation event to expand the connections of an ancestral regulator to recruit components for novel functions to overcome the constraints in the terrestrial habitats.

The models for GRN evolution unveiled adaptive strategies to face the abrupt transition from aquatic to terrestrial environments, consisting of the sudden diversification of TFs dedicated to expanding counteracting responses to adversities in the novel niches. Multiple cross-comparisons of heat shock (HS) TFs revealed major conservation in the HSFA subfamily compared to HSFB, being the principal structural divergences entailed within the DNA-binding domains (Wu et al. 2022). HSFB diversified prior to territorialization, whereas HSFC separated recently from HSFA. Indeed, Mp*HSFA1* sustains the canonical heat stress response by triggering target genes whose promoters encompass the heat shock element (HSE) motif, while Mp*HSFB1* activates complementary targets operating distinct biological processes containing other stress-related pCREs. Afterwards, approximately half of the HSF in green algae duplicated in at least two subclasses in land plants, which subsequently underwent further specialized functions during heat stress response.

Predictive models on temporal transcriptome profiles have also shed light on differential responsiveness among species due to a particular adaptation case such as domestication. The wild-type tomato *Solanum pennellii* implements extensive and persistent transcriptome reconfigurations after wounding, especially to repress genes, compared to the domesticated *S. lycopersicum* (Liu et al. 2018). Moreover, orthologue gene pairs displayed differential expression trends, highlighting profound divergences in responsiveness between species. Inference of pCREs in wounding-responsive transcripts revealed the existence of hundreds of TF binding sites that could be putatively activated by the same families of TFs, although more than half of the pCREs were exclusively enriched in one of the two species. Further sequence analysis correlated the degree of sequence preservation and the expression pattern between orthologs, suggesting a slow evolution of TF affinity for targets during the domestication process but more profound refining capacity by divergences in the binding preference for the DNA domain (Liu et al. 2018).

Besides these studies focused on GRNs, other research has focused on the investigation of expression changes between diploid ancestors and polyploid descendants to understand subgenome functional divergence. Polyploidy is a key feature of all angiosperm genomes and remains an important driver of speciation. Allopolyploidy involves genomic merger and doubling, leading to various non-Mendelian interactions such as sequence elimination, epigenetic modifications, gene activation, and homoeologous exchanges. Variations in duplicate gene expression are equally diverse, ranging from expression conservation similar to diploid progenitors (expression dominance) to the silencing of one homoeolog or to novel patterns of up- and downregulation (bias or unequal homoeolog expression) (Grover et al. 2012). Thus, the assessment of homoeolog expression bias (HEB) and expression level dominance is commonly used for understanding the reassignment of biological functions in polyploids. For instance, HEB in tetraploid coffee and cotton species revealed important mechanisms underlying temperature, salt, and other environmental stress responses (Combes et al. 2013; Dong et al. 2022; Wu et al. 2024a). Moreover, prior to transcriptome analysis, subgenome phasing and homoeology inference are essential to ensure an accurate and complete interpretation of gene expression patterns. Aufiero et al. (2024) reviewed the most challenging aspects of allopolyploid transcriptome analysis and suggested high-throughput RNA-sequencing (bulk RNA-seq) as the best method for studying allopolyploid gene expression. This approach targets mRNA and/or noncoding RNA libraries and effectively captures gene expression from both parental genomes, facilitating a comprehensive analysis of their combined profiles (e.g., Zhang et al. 2023).

On the other hand, time-series eco-transcriptomics allows the study of perennial and evergreen species that are expected to have evolved specialized strategies for persistence under highly variable environments. By capturing gene expression dynamics over multiple time points and across different seasons, this approach enables the identification of transcriptional patterns associated with acclimation, phenological transitions, and stress memory. It also facilitates the detection of key regulatory genes and networks that mediate responses to recurring abiotic and biotic stresses, providing insights into long-term adaptation mechanisms and the temporal coordination of growth, defense, and metabolic processes in natural settings. For instance, a long-term transcriptomic study of *Arabidopsis halleri subsp. gemmifera* revealed that both temperature and day length jointly regulate seasonal and diurnal gene expression in natural habitats. Over two years, thousands of genes showed seasonal and diurnal oscillations, with temperature emerging as the dominant factor in shaping seasonal expression dynamics (Nagano et al. 2019). The study of the transcriptome dynamics in an evergreen shrub from high elevation environments revealed that transcriptome changes were higher during the cold de-acclimation than during the acclimatation phase. They observed that de-acclimation was not just a reversal of cold acclimation but also involved specific changes implicated in growth and development (Rathore et al. 2022). These studies highlight the importance of investigating transcriptome remodeling and reconfiguration in plants adapted to environmental variability.

3.5 The Role of Beneficial Microorganisms in Shifting Plant Transcriptomic Responses

Beneficial microorganisms, such as plant growth-promoting rhizobacteria (PGPR) and mycorrhizal fungi, play a key role in modulating plant transcriptomic responses under both normal and stress conditions. By interacting with plant roots, they can trigger systemic changes in gene expression that improve stress tolerance, nutrient uptake, and growth. In particular, bacterial interactions are known to regulate the expression of a wide range of plant genes simultaneously, which can be involved in hormone signaling, defense pathways, and antioxidant systems, among others, demonstrating the broad influence of the bacteria on the plant transcriptome. Even in nonstress conditions, bacterial colonization has been shown to induce the expression of genes related to auxins and CK biosynthesis (Hanifah et al. 2023), as well as genes involved in nitrate acquisition in *Zea mays* in interaction with two diazotrophic bacterial strains (Hardoim et al. 2020). Under stress conditions, this transcriptomic reprogramming may become more pronounced. Recent work by Feng et al. (2024) demonstrated that the interaction between *Bacillus subtilis* and tomato plants under salinity stress activates genes involved in ROS scavenging, enhancing the plant antioxidant system. This effect was primarily led by an induction of plant methionine biosynthesis, a precursor of ethylene synthesis. Therefore, beyond improving redox homeostasis, the increased production of methionine upregulated the ethylene pathways, resulting in an alleviation of salinity stress. Kruasuwan et al. (2023) observed that suberin biosynthesis was exclusively induced in rice inoculated with *Streptomyces* sp., indicating the impact of the bacteria on the regulation of these molecules known to control root-to-shoot Na^+ translocation. In addition to salinity, other abiotic stresses presented similar microbial modulation. For example, *B. megaterium* inoculation improved drought tolerance, increasing gene expression of proteins involved in Ca^{2+} signaling transduction, enhancing glycine betaine and proline to maintain osmoregulation, and upregulating MAPKs involved in stress tolerance (Morcillo et al. 2021).

Similarly, fungal members of plants' microbiomes can also be fundamental for plant adaptation. The positive effects of Arbuscular Mycorrhizal Fungi (AMF) on stress alleviation are well established. For instance, Puccio et al. (2023) reported a general downregulation of stress-related genes in plants inoculated with AMF, indicating a lower impact of salinity stress. Additionally, higher levels of Ca^{2+} signaling pathways and MAPK expression were observed, suggesting an induction of cellular signaling responses by the fungus interaction. Moreover, AMF have also been shown as inductors of genes involved in pigment biosynthesis, photosynthesis, and sugar transport under stress conditions (Li et al. 2024). Beyond AMF, other fungal species are also known to improve abiotic stress tolerance via transcriptomic responses modulation, such as *Trichoderma longibrachiatum* in tomato plants under drought conditions (De Palma et al. 2021), or *Piriformospora indica* in *A. thaliana* under freezing stress (Jiang et al. 2021).

Microbial interactions also influence transcriptome responses that contribute to biotic stress adaptations. Several bacterial and fungal species demonstrated an impact on overall plant gene expression in response to pathogen infection. For instance, Scot pine seedlings infected with *Heterobasidion annosum* showed lower enhanced expression of pathogen-related (PR) genes in plants inoculated with the beneficial fungus *Suillus luteus* (Wen et al. 2025). Otherwise, ion transporter genes were upregulated, indicating a better nutrient supply, as well as hormone signaling was altered to limit pathogen growth. Similarly, Oukala et al. (2021) reviewed how bacterial endophytes participate in biotic stress tolerance. They reported that an initial distinct transcriptomic response is observed depending on whether plants interact with beneficial or pathogenic bacteria. Moreover, beneficial bacteria may trigger the induction of PR activity and defense-related enzymes, eventually increasing plant resistance to pathogens.

In this context, bacterial and fungal presence can reshape the plant's transcriptomic landscape, becoming essential for shifting plant performance from stress-sensitive to a more robust adaptive state.

3.6 Future Directions

Transcriptomic technologies, such as sequence-based methods, RNA sequencing, massively parallel signature sequencing, and Cap analysis of gene expression, are widely utilized in transcriptomics. Advancements in these technologies have significantly enhanced sensitivity and affordability, leading to a rapid increase in publicly available gene expression data. Raw sequencing data from platforms such as Illumina, 454, IonTorrent, Complete Genomics, PacBio, and Oxford Nanopore (ONT) are deposited in primary nucleotide sequence databases like the Sequence Read Archive or the European Nucleotide Archive. By 2024, over 300,000 RNA-seq experiments covering hundreds of Viridiplantae species had been made publicly accessible. While originally generated for diverse research purposes, these datasets can be repurposed for large-scale comparative analyses (Julca et al. 2021). Pipelines like the one developed by Wulf et al. (2024), which provide guidelines for experiment selection, efficient mapping of data onto reference transcriptomes, and the inspection of global transcript abundances and patterns, offer valuable tools for exploring gene function across broad phylogenetic groups or even the entire kingdom.

As exposed in this chapter, single-cell RNA sequencing is particularly valuable for investigating cellular heterogeneity in response to stress. As single-cell transcriptome analysis advances, both the diversity of measurable parameters per cell and the total number of identifiable cells are expected to expand significantly. Thus, there is a growing demand to integrate single-cell data with spatial information. Spatially enhanced-resolution omics-sequencing (Stereo-seq) is an advanced technology that integrates scRNA-seq with spatial transcriptomics to provide unbiased, high-resolution spatial maps of gene expression without requiring protoplasting. It offers broad tissue coverage, deep sequencing, and precise subcellular resolution, enabling detailed analyses of plant development and stress responses. By capturing spatial and temporal gene activity, Stereo-seq helps uncover new regulatory elements and cell-type-specific dynamics in organs like roots, leaves, flowers, and fruits (see Bawa et al. 2024).

The integration of machine learning into RNA-seq and scRNA-seq analyses is particularly promising, offering powerful tools for identifying new markers (see Brendel et al. 2022). Machine learning, a cutting-edge advancement in artificial intelligence, has proven valuable in extracting meaningful and concise features from noisy, heterogeneous, and high-dimensional scRNA-seq data. For example, Panahi (2024) applied machine learning to analyze transcriptomic data from 515 RNA-seq profiles across 18 independent studies of barley, covering 11 abiotic and three biotic stresses. Among different machine learning models, the C4.5 algorithm showed the highest accuracy in predicting stress signatures. After rigorous data preprocessing and feature selection, 400 core stress-responsive genes were prioritized, and key genes, including *bHLH119* and *DRIP2*, were identified as potential biomarkers.

Additionally, several models to predict the expression level of a given genomic sequence (Basenji, ExPecto, Enformer) have been created and thoroughly reviewed by Hwang et al. (2024). As an example *in planta*, Tan et al. (2022) applied eQTL mapping using genomic data from 505 *Brassica napus* accessions and 583 transcriptomes across two seed developmental stages. To refine the identification of regulatory elements and gene–TF relationships, the authors employed XGBoost and Basenji models trained on 856 RNA-seq and 59 ATAC-seq datasets from the cultivar ZS11. This integrative approach led to the prediction of 141 regulatory hotspots and the identification of key TFs, such as *NAC13* and *SCL31*, involved in seed oil content regulation.

Improving machine learning models for regulatory biology depends on high-quality training and benchmarking datasets. To achieve this, public databases must prioritize metadata accuracy and quality control, which can be facilitated by standardized analysis pipelines tailored to each functional genomics assay and sequencing platform. Huang et al. (2025) have developed the *TransGeneSelector*, a novel deep learning framework that enables efficient gene mining from small-scale transcriptomic datasets. By integrating a Wasserstein Generative Adversarial Network with Gradient Penalty for synthetic sample generation and a Transformer-based architecture for classification, *TransGeneSelector* effectively captures both global gene regulatory interactions and context-specific biological processes. Applied to *A. thaliana*, the model demonstrated superior performance in predicting seed germination and heat stress responses compared to traditional approaches such as random forest and support vector machines. Analysis using Shapley Additive Explanations and gene regulatory network reconstruction revealed that *TransGeneSelector* reliably identified upstream regulatory genes enriched in key pathways relevant to stress adaptation. In parallel, Thalimaraw et al. (2024) developed *PREDICT*, a user-friendly web tool designed to identify precise cis-regulatory elements and associated TFs across plant species under diverse abiotic stresses. It leverages a rapid *k-mer* pipeline and integrates GRN analysis to enhance sequence-to-expression prediction. By pinpointing stress-specific regulatory motifs and TFs, *PREDICT* provides a powerful resource for exploring plant gene regulation and advancing applications in synthetic biology.

Overall, these advances in transcriptomic technologies, spatial resolution, and integrative machine learning approaches are reshaping our ability to decode gene regulation and plant adaptation. This chapter highlights the potential of advanced transcriptomic techniques to uncover regulatory mechanisms in plant stress responses, to identify key regulatory elements and TFs, and to accelerate the development of resilient crops.

Acknowledgments Special thanks to Xinyuan Huang and Lvi Yant for their valuable advice on transcriptomic analysis in plants. Refined and formatted with the assistance of artificial intelligence (AI) tools developed by OpenAI, ChatGPT (*GPT-4*), 2025. Every section was carefully reviewed, revised, and approved by the authors to ensure that the text is appropriate in terms of content and scientific accuracy, maintaining the integrity of the original material.

Conflict of Interest Disclosure All figures presented in this work are original and created by the authors using publicly available features in *BioRender.com*. The authors declare no conflicts of interest related to this study.

References

Aharoni A, Vorst O (2001) DNA microarrays for functional plant genomics. Plant Mol Biol 48:99–118

Almira-Casellas M, Busoms S, Pérez-Martín L et al (2024) At the core of salinity: Divergent transcriptomic responses to neutral and alkaline salinity in Arabidopsis thaliana. Environ Exp Bot 228:105982

Alwine JC, Kemp J, Stark GR (1977) Method for detection of specific RNAs in agarose gels by transfer to diazobenzyloxymethyl-paper and hybridization with DNA probes. Proc Natl Acad Sci USA 74(12):5350–5354

Aufiero G, Fruggiero C, D'Angelo D et al (2024) Homoeologs in Allopolyploids: navigating redundancy as both an evolutionary opportunity and a technical challenge—A transcriptomics perspective. Gene 15(8):977

Azad M, Tohidfar M, Ghanbari Moheb Seraj R et al (2024) Identification of responsive genes to multiple abiotic stresses in rice (*Oryza sativa*): a meta-analysis of transcriptomics data. Sci Rep 14:5463

Bainbridge MN, Warren RL, Hirst M et al (2006) Analysis of the prostate cancer cell line LNCaP transcriptome using a sequencing-by-synthesis approach. BMC Genomics 7:1–11

Balfagón D, Zandalinas SI et al (2020) High temperatures modify plant responses to abiotic stress conditions. Physiol Plant 170(3):335–344

Bawa G, Liu Z, Yu X et al (2022) Single-cell RNA sequencing for plant research: insights and possible benefits. Int J Mol Sci 23(9):4497

Bawa G, Liu Z, Yu X et al (2024) Introducing single cell stereo- sequencing technology to transform the plant transcriptome landscape. Trends Plant Sci 2S(2):249–265

Bhattacharya A, Chaturvedi G, Singhal K et al (2020) Experimental toolkit to study RNA level regulation. In: RNA-based regulation in human health and disease. Academic, pp 371–396

Bihani D, Khuman A, Chaudhary B (2024) Comparative transcriptomics reveals novel spatial gene expression profiles in cotton (*Gossypium hirsutum* L) under herbivory and drought stress. J Plant Growth Regul 43(11):4018–4037

Biniaz Y, Tahmasebi A, Afsharifar A et al (2022) Meta-analysis of common and differential transcriptomic responses to biotic and abiotic stresses in *Arabidopsis thaliana*. Plants 11(4):502

Boheler KR, Stern MD (2003) The new role of SAGE in gene discovery. Trends Biotechnol 21(2): 55–57

Bostock RM, Pye MF, Roubtsova TV (2014) Predisposition in plant disease: exploiting the nexus in abiotic and biotic stress perception and response. Annu Rev Phytopathol 52(1):517–549

Brendel M, Su C, Bai Z et al (2022) Application of deep learning on single-cell RNA sequencing data analysis: a review. Genom Proteom Bioinform 20(5):814–835

Busoms S, Pérez-Martín L et al (2023) Combined genomics to discover genes associated with tolerance to soil carbonate. Plant Cell Environ 46(12):3986–3998

Busoms S, da Silva AC, Escolà G et al (2024) Local cryptic diversity in salinity adaptation mechanisms in the wild outcrossing Brassica fruticulosa. Proc Natl Acad Sci USA 121(40): e2407821121

Buti M, Baldoni E, Formentin E et al (2019) A meta- analysis of comparative transcriptomic data reveals a set of key genes involved in the tolerance to abiotic stresses in rice. Int J Mol Sci 20(22):5662

Calixto C, Guo W, James AB et al (2018) Rapid and dynamic alternative splicing impacts the Arabidopsis cold response transcriptome. Plant Cell 30:1424–1444

Capelari EF, Dos Anjos L et al (2021) Transcriptional profiling and physiological responses reveal new insights into drought tolerance in a semiarid adapted species A*nacardium occidentale*. Plant Biol 23(6):1074–1085

Chang PFL, Damsz B, Kononowicz AK et al (1996) Alterations in cell membrane structure and expression of a membrane-associated protein after adaptation to osmotic stress. Physiol Plant 98(3):505–516

Chen L, Qin L, Zhou L et al (2019) A nodule-localized phosphate transporter Gm PT 7 plays an important role in enhancing symbiotic N2 fixation and yield in soybean. New Phyt 221(4): 2013–2025

Chen Z, Song J, Li X et al (2021) Physiological responses and transcriptomic changes reveal the mechanisms underlying adaptation of *Stylosanthes guianensis* to phosphorus deficiency. BMC Plant Biol 21:1–15

Chen X, Zhao C, Yun P et al (2024) Climate-resilient crops: lessons from xerophytes. Plant J 117(6):1815–1835

Chen G, Li Y, Jin K et al (2025) Synthetic photorespiratory bypass improves rice productivity by enhancing photosynthesis and nitrogen uptake. Plant Cell 37(1):koaf015

Cheng C, Yan CY, Qi C et al (2023) Metabolome and transcriptome analysis of postharvest peach fruit in response to fungal pathogen *Monilinia fructicola* infection. LWT 173:114301

Collins E, Shou H, Mao C et al (2024) Dynamic interactions between SPX proteins, the ubiquitination machinery, and signalling molecules for stress adaptation at a whole-plant level. Biochem J 481(5):363–385

Combes MC, Dereeper A, Severac D et al (2013) Contribution of subgenomes to the transcriptome and their intertwined regulation in the allopolyploid coffea arabica grown at contrasted temperatures. New Phyt 200(1):251–260

Daszkowska-Golec A, Mascher M, Zhang R (2023) Editorial: applications of long-read sequencing in plant genomics and transcriptomics. Front Plant Sci 14:1141429

De Abreu Neto JB, Frei M (2016) Microarray meta-analysis focused on the response of genes involved in redox homeostasis to diverse abiotic stresses in rice. Front Plant Sci 6: 1260

De la Rosa C, Reyes JL (2019) Northern blot analysis of microRNAs and other small RNAs in plants. In: de Folter S (ed) Plant microRNAs. Springer, pp 121–129

De Palma M, Salzano M, Villano C et al (2021) Transcriptome modulation by the beneficial fungus Trichoderma longibrachiatum drives water stress response and recovery in tomato. Environ Exp Bot 188:104588

Depuydt T, De Rybel B, Vandepoele K (2023) Charting plant gene functions in the multi-omics and single-cell era. Trends Plant Sci 28(3):283–296

Ding Y, Gardiner DM, Powell JJ et al (2021) Adaptive defence and sensing responses of host plant roots to fungal pathogen attack revealed by transcriptome and metabolome analyses. Plant Cell Environ 44(12):3756–3774

Dong Y, Hu G, Grover CE et al (2022) Parental legacy versus regulatory innovation in salt stress responsiveness of allopolyploid cotton (Gossypium) species. Plant J 111(3):872–887

Dong Y, Krishnamoorthi S, Tan GZ H, Poh ZY, Urano D (2024) Co-option of plant gene regulatory network in nutrient responses during terrestrialization. Nat plants 10(12): 1955–1968

Ducousso-Détrez A, Fontaine J et al (2022) Diversity of phosphate chemical forms in soils and their contributions on soil microbial community structure changes. Microorganisms 10:609

Edelman L, Czarnecka E, Key JL (1988) Induction and accumulation of heat shock-specific Poly(A +) RNAs and proteins in soybean seedlings during arsenite and cadmium treatments. Plant Physiol 86(4):1048–1056

Fang S, Hou X, Liang X (2021) Response mechanisms of plants under saline-alkali stress. Front Plant Sci 12:667458

Feng S, Xu M, Liu F et al (2019) Reconstruction of the full-length transcriptome atlas using PacBio Iso-Seq provides insight into the alternative splicing in *Gossypium australe*. BMC Plant Biol 19: 365

Feng L, Li Q, Zhou D et al (2024) B subtilis CNBG-PGPR-1 induces methionine to regulate ethylene pathway and ROS scavenging for improving salt tolerance of tomato. Plant J 117(1): 193–211

Gao H, Chen M, Jin N et al (2025) A comprehensive analytical method 'Regulatome revealed a novel pathway for aerenchyma formation under waterlogging in wheat. Physiol Plant 177(2): e70157

Garcia-Molina A, Martín-Cardoso H, Ribaya M et al (2025) Transcriptome-derived networks reconstruct distinct immune strategies to counteract fungal infection under different iron availability between Arabidopsis and rice. Plant Stress 15:100756

Gong Q, Li P, Ma S et al (2005) Salinity stress adaptation competence in the extremophile *Thellungiella halophila* in comparison with its relative *Arabidopsis thaliana*. Plant J 44(5): 826–839

Gong Y-Y, Wu C-Z, Wu Y-S et al (2025) A glutamate receptor-like gene AtGLR2.5 with its unusual splice variant has a role in mediating glutamate-elicited changes in Arabidopsis root architecture. Plant Cell Environ 48(5):3778–3792

Grover CE, Gallagher JP, Szadkowski EP et al (2012) Homoeolog expression bias and expression level dominance in allopolyploids. New Phyt 196(4):966–971

Guo R, Zhao L, Zhang K et al (2021) Comparative genomics and transcriptomics of the extreme halophyte *Puccinellia tenuiflora* provides insights into salinity tolerance differentiation between halophytes and glycophytes. Front Plant Sci 12:649001

Guo Y, Wang Q, Zhang H et al (2022) Responses to waterlogging stress in quinoa seedlings based on metabolomic and transcriptomic analysis. Environ Exp Bot 203:105044

Guo L, Xu Z, Wang S et al (2023) Integrative multi- omics analysis of three early diverged rosid species reveals an ancient hierarchical cold-responsive regulatory network. Physiol Plant 175(2):e13892

Hanifah NAS, Ghadamgahi F, Ghosh S et al (2023) Comparative transcriptome profiling provides insights into the growth promotion activity of *Pseudomonas fluorescens* strain SLU99 in tomato and potato plants. Front Plant Sci 14:1141692

Hardoim PR, de Carvalho TL, Ballesteros HG et al (2020) Genome-wide transcriptome profiling provides insights into the responses of maize (*Zea mays* L) to diazotrophic bacteria. Plant Soil 451:121–143

He C, Liu W, Jin M et al (2025) Analysis of full length transcriptome and resistance characteristics of Atraphaxis bracteata under drought. Sci Rep 15(1):807

Heidel-Fischer HM, Musser RO, Vogel H (2014) Plant transcriptomic responses to herbivory. Annu Plant Rev Insect-Plant Interact 47:155–196

Hoang NV, Furtado A, Mason PJ et al (2017) A survey of the complex transcriptome from the highly polyploid sugarcane genome using full-length isoform sequencing and de novo assembly from short read sequencing. BMC Genomics 18(1):395

Hu M, Polyak K (1995) Serial analysis of gene expression. Nat Protoc 270:484–486

Hu T, Sun X, Zhang X et al (2014) An RNA sequencing transcriptome analysis of the high-temperature stressed tall fescue reveals novel insights into plant thermotolerance. BMC Genomics 15:1–13

Hu T, Chitnis N, Monos D et al (2021) Next-generation sequencing technologies: an overview. Hum Immunol 82:801–811

Huang X, Liang Y, Zhang B et al (2021) Integration of transcriptional and post-transcriptional analysis revealed the early response mechanism of sugarcane to cold stress. Front Genet 11: 581993

Huang K, Tian J, Sun L et al (2025) TransGeneSelector: using a transformer approach to mine key genes from small transcriptomic datasets in plant responses to various environments. BMC Genomics 2C(1):259

Hwang H, Jeon H, Yeo N et al (2024) Big data and deep learning for RNA biology. Exp Mol Med 5C(6):1293–1321

Jaglo-Ottosen KR, Gilmour SJ, Zarka DG et al (1998) Arabidopsis CBF1 overexpression induces COR genes and enhances freezing tolerance. Science 280:104–106

Jamil A, Riaz S, Ashraf M et al (2011) Gene expression profiling of plants under salt stress. Crit Rev Plant Sci 30(5):435–458

Javaid MH, Khan AR, Salam A et al (2022) Exploring the adaptive responses of plants to abiotic stresses using transcriptome data. Agriculture 12(2):211

Jiang W, Pan R, Buitrago S et al (2021) Transcriptome analysis of Arabidopsis reveals freezing-tolerance related genes induced by root endophytic fungus *Piriformospora indica*. Physiol Mol Biol Plants 27:189–201

Julca I, Ferrari C, Flores-Tornero M et al (2021) Comparative transcriptomic analysis reveals conserved programmes underpinning organogenesis and reproduction in land plants. Nat Plants 7(8):1143–1159

Kariñho Betancourt E, Calderón Cortés N et al (2024) Comparative transcriptome profiling reveals distinct regulatory responses of secondary defensive metabolism in Datura species (Solanaceae) under plant development and herbivory-mediated stress. Ecol Evol 14(7):e11496

Kathiresan A, Lafitte HR, Chen J et al (2006) Gene expression microarrays and their application in drought stress research. Field Crop Res 97:101–110

Kidokoro S, Shinozaki K, Yamaguchi-Shinozaki K (2022) Transcriptional regulatory network of plant cold-stress responses. Trends Plant Sci 27:922–935

Kim YC, Chen M, Kim GW et al (2024) Comparative transcriptomic analysis in roots of a domestic wheat cultivar (*Triticum aestivum* L cv. Keumkang) seedlings under waterlogging stress. J Plant Biol C7(1):35–48

Kinoshita S, Kaneko G, Lee H et al (2001) A novel heat stress-responsive gene in the marine diatom *Chaetoceros compressum* encoding two types of transcripts, a trypsin-like protease and its related protein, by alternative RNA splicing. Eur J Biochem 268:4599–4609

Kohlhase DR, McCabe CE et al (2021) Comparing early transcriptomic responses of 18 soybean (Glycine max) genotypes to iron stress. Int J Mol Sci 22:11643

Kollist H, Zandalinas SI, Sengupta S et al (2019) Rapid responses to abiotic stress: priming the landscape for the signal transduction network. Trends Plant Sci 24(1):25–37

Krasensky J, Jonak C (2012) Drought, salt, and temperature stress-induced metabolic rearrangements and regulatory networks. J Exp Bot 63(4):1593–1608

Kruasuwan W, Lohmaneeratana K, Munnoch JT et al (2023) Transcriptome landscapes of salt-susceptible rice cultivar IR29 associated with a plant growth promoting endophytic Streptomyces. Rice 16(1):6

Kumar G, Purty RS, Sharma MP et al (2009) Physiological responses among Brassica species under salinity stress show strong correlation with transcript abundance for SOS pathway-related genes. J Plant Physiol 166(5):507–520

Kumar N, Mishra BK, Liu J et al (2023) Network biology analyses and dynamic modeling of gene regulatory networks under drought stress reveal major transcriptional regulators in Arabidopsis. Int J Mol Sci 24(8):7349

Leclerc L, Nguyen TH, Duval P et al (2024) Early transcriptomic responses of rice leaves to herbivory by *Spodoptera frugiperda*. Sci Rep 14(1):2836

Lei Y, Xu Y, Hettenhausen C et al (2018) Comparative analysis of alfalfa (*Medicago sativa* L) leaf transcriptomes reveals genotype-specific salt tolerance mechanisms. BMC Plant Biol 18:1–14

Li B, Ning L, Zhang J et al (2015) Transcriptional profiling of Petunia seedlings reveals candidate regulators of the cold stress response. Front Plant Sci 6:118

Li W, Wu H, Hua J et al (2024) Arbuscular mycorrhizal fungi enhanced resistance to low-temperature weak-light stress in snapdragon (*Antirrhinum majus* L) through physiological and transcriptomic responses. Front Plant Sci 15:1330032

Li Y, Chen P, Zeng F et al (2025) Transcriptome and metabolome analysis reveal the mechanisms of iron absorption differences in apple rootstocks under alkaline condition. Physiol Plant 177(1): e70134

Lister R, O'Malley RC, Tonti-Filippini J et al (2008) Highly integrated single-base resolution maps of the Epigenome in Arabidopsis. Cell 133(3):523–536

Liu MJ, Sugimoto K, Uygun S et al (2018) Regulatory divergence in wound-responsive gene expression between domesticated and wild tomato. Plant Cell 30(7):1445–1460

Luo Q, Duan F, Song W (2025) Transcriptomics integrated with metabolomics reveals the defense response of insect-resistant *Zea mays* infested with *Spodoptera exigua*. Heliyon 11(4):e42565

Ma L, Chen C, Liu X et al (2005) A microarray analysis of the rice transcriptome and its comparison to Arabidopsis. Genome Res 15(9):1274–1283

Mai HJ, Pateyron S, Bauer P (2016) Iron homeostasis in *Arabidopsis thaliana*: transcriptomic analyses reveal novel FIT-regulated genes, iron deficiency marker genes and functional gene networks. BMC Plant Biol 16(1):221

Mangelsen E, Kilian J, Harter K et al (2011) Transcriptome analysis of high-temperature stress in developing barley caryopses: early stress responses and effects on storage compound biosynthesis. Mol Plant 4(1):97–115

Marok MA, Marok-Alim D, Rey P (2021) Contribution of functional genomics to identify the genetic basis of water-deficit tolerance in barley and the related molecular mechanisms. J Agron Crop Sci 207(6):913–935

Martínez-Andújar C, Martínez-Pérez A, Ferrández-Ayela A et al (2020) Impact of overexpression of 9-cis-epoxycarotenoid dioxygenase on growth and gene expression under salinity stress. Plant Sci 2S5:110268

Minelli A, Valero-Gracia A (2022) Spatially and temporally distributed complexity—A refreshed framework for the study of GRN evolution. Cells 11(11):1790

Montesinos Á, Sacristán S, del Prado-Polonio P et al (2024) Contrasting plant transcriptome responses between a pierce-sucking and a chewing herbivore go beyond the infestation site. BMC Plant Biol 24(1):120

Morcillo RJ, Vílchez JI, Zhang S et al (2021) Plant transcriptome reprograming and bacterial extracellular metabolites underlying tomato drought resistance triggered by a beneficial soil bacteria. Metabolites 11(6):369

Mortazavi A, Williams BA, McCue K et al (2008) Mapping and quantifying mammalian transcriptomes by RNA-Seq. Nat Methods 5(7):621–628

Moustafa K, Cross JM (2016) Genetic approaches to study plant responses to environmental stresses: an overview. Biology 5(2):1–18

Mundy J, Chua NH (1988) Abscisic acid and water-stress induce the expression of a novel rice gene. EMBO J 7(8):2279–2286

Nagalakshmi U, Wang Z, Waern K et al (2008) The transcriptional landscape of the yeast genome defined by RNA sequencing. Science 320(5881):1344 1349

Nagano AJ, Kawagoe T, Sugisaka J et al (2019) Annual transcriptome dynamics in natural environments reveals plant seasonal adaptation. Nat Plants 5(1):74–83

Nemati M, Zare N, Hedayat-Evrigh N et al (2023) identification of key gene network modules and hub genes associated with wheat response to biotic stress using combined microarray meta-analysis and WGCN analysis. Mol Biotechnol 65:453–465

Novillo F, Medina JC, Salinas J (2007) Arabidopsis CBF1 and CBF3 have a different function than CBF2 in cold acclimation and define different gene classes in the CBF regulon. Proc Natl Acad Sci USA 104:21002–21007

Oukala N, Aissat K, Pastor V (2021) Bacterial endophytes: the hidden actor in plant immune responses against biotic stress. Plants 10(5):1012

Pan J, Song J, Sohail H et al (2024) RNA-seq-based comparative transcriptome analysis reveals the role of CsPrx73 in waterlogging-triggered adventitious root formation in cucumber. Hortic Res 11(4):uhae062

Panahi B (2024) Transcriptome signature for multiple biotic and abiotic stress in barley (*Hordeum vulgare* L) identifies using machine learning approach. Curr Plant Biol 40:100416

Parwez R, Aftab T, Gill SS et al (2022) Abscisic acid signaling and crosstalk with phytohormones in regulation of environmental stress responses. Environ Exp Bot 1SS:104885

Pathania S, Kumar M (2022) An integrative computational approach to predict stress-specific candidate and shared genes in multiple plant stresses. Plant Gene 30:100356

Peng X, Li M, Wu H et al (2021) Co-regulation role of endogenous hormones and transcriptomics profiling under cold stress in *Tetrastigma hemsleyanum*. J Plant Growth Regul 40:1992–2006

Peng Z, Rehman A, Li X et al (2023) Comprehensive evaluation and transcriptome analysis reveal the salt tolerance mechanism in semi-wild cotton (*Gossypium purpurascens*). Int J Mol Sci 24(16):12853

Pérez-Martín L, Busoms S, Tolrà R et al (2021) Transcriptomics reveals fast changes in salicylate and jasmonate signaling pathways in shoots of carbonate-tolerant *Arabidopsis thaliana* under bicarbonate exposure. Int J Mol Sci 22(3):1226

Pérez-Martín L, Busoms S, Amira MJ et al (2022) Evolution of salt tolerance in *Arabidopsis thaliana* on siliceous soils does not confer tolerance to saline calcareous soils. Plant Soil 47C(1): 455–475

Prodhan MA, Pariaska-Tanaka J, Ueda Y et al (2022) Comparative transcriptome analysis reveals a rapid response to phosphorus deficiency in a phosphorus-efficient rice genotype. Sci Rep 12: 9460

Puccio G, Ingraffia R, Mercati F et al (2023) Transcriptome changes induced by Arbuscular mycorrhizal symbiosis in leaves of durum wheat (*Triticum durum* Desf) promote higher salt tolerance. Sci Rep 13(1):116

Rao MS, Van Vleet TR, Ciurlionis R et al (2019) Comparison of RNA-Seq and microarray gene expression platforms for the toxicogenomic evaluation of liver from short-term rat toxicity studies. Front Genet 9: 636

Rathore N, Kumar P, Mehta N et al (2022) Time-series RNA-Seq transcriptome profiling reveals novel insights about cold acclimation and de-acclimation processes in an evergreen shrub of high altitude. Sci Rep 12(1):15553

Raza A, Charagh S, Zahid Z et al (2021) Jasmonic acid: a key frontier in conferring abiotic stress tolerance in plants. Plant Cell Rep 40(8):1513–1541

Reis RS, Clúa J, Jaskolowski A et al (2024) Phosphate deficiency alters transcript isoforms via alternative transcription start sites. Plant J 120:218–233

Sachdev S, Ansari SA, Ansari MI et al (2021) Abiotic stress and reactive oxygen species: generation, signaling, and defense mechanisms. Antioxidants 10(2):277

Santamaria ME, Garcia A, Arnaiz A et al (2021) Comparative transcriptomics reveals hidden issues in the plant response to arthropod herbivores. J Integr Plant Biol 63(2):312–326

Sathee L, Sairam RK, Chinnusamy V et al (2022) Upregulation of genes encoding plastidic isoforms of antioxidant enzymes and osmolyte synthesis impart tissue tolerance to salinity stress in bread wheat. Physiol Mol Biol Plants 28(9):1639–1655

Satoh K, Saji S, Ito S et al (2014) Gene response in rice plants treated with continuous fog influenced by pH, was similar to that treated with biotic stress. Rice 7:1–11

Schena M, Shalon D, Davis RW et al (1995) Quantitative monitoring of gene expression patterns with a complementary DNA microarray. Science 270:467–470

Shahan R, Hsu CW, Nolan TM et al (2022) A single-cell Arabidopsis root atlas reveals developmental trajectories in wild-type and cell identity mutants. Dev Cell 57(4):543–560

Shahriari AG, Soltani Z, Tahmasebi A et al (2022) Integrative system biology analysis of transcriptomic responses to drought stress in soybean (*Glycine max* L.). Gene 13(10):1732

Sham A, Moustafa K, Al-Ameri S et al (2015) Identification of Arabidopsis candidate genes in response to biotic and abiotic stresses using comparative microarrays. PLoS ONE 10(5): e0125666

Sham A, Al-Ashram H, Whitley K et al (2019) Metatranscriptomic analysis of multiple environmental stresses identifies RAP2.4 gene associated with Arabidopsis immunity to *Botrytis cinerea*. Sci Rep 9(1):1–15

Shen P, Hour A, Liu LD (2017) Microarray meta-analysis to explore abiotic stress-specific gene expression patterns in Arabidopsis. Bot Stud 58(22)

Shokri-Gharelo R, Derakhti-Dizaji M, Dadashi D et al (2024) Bioinformatics and meta-analysis of expression data to investigate transcriptomic response of wheat root to abiotic stresses. Biosystems 237:105165

Sriden N, Charoensawan V (2022) Large-scale comparative transcriptomic analysis of temperature-responsive genes in Arabidopsis thaliana. Plant Mol Biol 110(4):425–443

Stark R, Grzelak M, Hadfield J (2019) RNA sequencing: the teenage years. Nat Rev Genet 20:631–656

Stockinger EJ, Gilmour SJ, Thomashow MF (1997) *Arabidopsis thaliana* CBF1 encodes an AP2 domain-containing transcriptional activator that binds to the C- repeat/DRE, a cis-acting DNA regulatory element that stimulates transcription in response to low temperature and water deficit. Proc Natl Acad Sci USA 94:1035–1040

Su T, Hollas MAR, Fellers RT et al (2023) Identification of splice variants and isoforms in transcriptomics and proteomics. Annu Rev Biomed Data Sci 6:357–376

Sun S, Li J, Song H et al (2022) Comparative transcriptome and physiological analyses reveal key factors in the tolerance of peach rootstocks to iron deficiency chlorosis. 3 Biotech 12(1):38

Tahmasebi A, Niazi A, Akrami S (2023) Integration of meta-analysis, machine learning and systems biology approach for investigating the transcriptomic response to drought stress in *Populus* species. Sci Rep 13(1):847

Taji T, Seki M, Satou M et al (2004) Comparative genomics in salt tolerance between Arabidopsis and Arabidopsis-related halophyte salt cress using Arabidopsis microarray. Plant Physiol 135(3):1697–1709

Tan Z, Peng Y, Xiong Y et al (2022) Comprehensive transcriptional variability analysis reveals gene networks regulating seed oil content of Brassica napus. Genome Biol 23(1):233

Tang F, Barbacioru C, Wang Y et al (2009) mRNA-Seq whole- transcriptome analysis of a single cell. Nat Methods 6(5):377–382

Thalimaraw L, Henry WX, Liu MJ et al (2024) PREDICT: advancing accurate gene expression prediction and motif identification in plant stress responses. bioRxiv 2024 Mar 31:2024–03

van Ruissen F, Ruijter JM, Schaaf GJ et al (2005) Evaluation of the similarity of gene expression data estimated with SAGE and Affymetrix GeneChips. BMC Genomics 6: 91

van Veen H, Triozzi PM, Loreti E (2025) Metabolic strategies in hypoxic plants. Plant Physiol 197(1):kiae564

Vandamme E, Rose T, Saito K et al (2016) Integration of P acquisition efficiency P utilization efficiency and low grain P concentrations into P- efficient rice genotypes for specific target environments. Nutr Cycl Agroecosyst 104:413–427

Verma IM, Temple GF, Fan H et al (1972) In vitro synthesis of DNA complementary to rabbit Reticulocyte 10S RNA. Nat New Biol 235:226–229

Vidal EA, Álvarez JM, Moyano TC et al (2015) Transcriptional networks in the nitrate response of *Arabidopsis thaliana*. Curr Opin Plant Biol 27:125–132

Villanueva G, Vilanova S, Plazas M et al (2023) Transcriptome profiles of eggplant (*Solanum melongena*) and its wild relative *S. dasyphyllum* under different levels of osmotic stress provide insights into response mechanisms to drought. Curr Plant Biol 33:100276

Waititu JK, Cai Q, Sunn Y et al (2021) Transcriptome profiling of maize (*Zea mays* L) leaves reveals key cold-responsive genes, transcription factors, and metabolic pathways regulating cold stress tolerance at the seedling stage. Gene 12(10):1638

Wang Q, Zhu Z (2023) Transcription factors in the regulation of plant heat responses. Crit Rev Plant Sci 42(6):385–398
Wang M, Wang Y, Zhang Y et al (2019) Comparative transcriptome analysis of salt-sensitive and salt-tolerant maize reveals potential mechanisms to enhance salt resistance. Genes Genom 41: 781–801
Wang M, Gong JC, Bhullar NK (2020) Iron deficiency triggered transcriptome changes in bread wheat. Comput Struct Biotechnol J 18:2709–2722
Wang X, Song Q, Liu Y et al (2022a) The network centered on ICEs play roles in plant cold tolerance, growth and development. Planta 255(4):81
Wang H, Ma Q, Shan F et al (2022b) Transcriptional regulation mechanism of wheat varieties with different nitrogen use efficiencies in response to nitrogen deficiency stress. BMC Genomics 23: 727
Wang S, Shen Y, Deng D et al (2023) Orthogroup and phylotranscriptomic analyses identify transcription factors involved in the plant cold response: a case study of Arabidopsis BBX29. Plant Commun 4: 6
Wang F, Zhou Z, Liu X et al (2024a) Transcriptome and metabolome analyses reveal molecular insights into waterlogging tolerance in Barley. BMC Plant Biol 24(1):385
Wang M, Wang Y, Li X et al (2024b) Integration of metabolomics and transcriptomics reveals the regulation mechanism of the phenylpropanoid biosynthesis pathway in insect resistance traits in Solanum habrochaites. Hortic Res 11(2):uhad277
Wang D, Xu M, Xu TY et al (2024c) Specific physiological responses to alkaline carbonate stress in rice (*Oryza sativa*) seedlings: organic acid metabolism and hormone signalling. Funct Plant Biol 51:10
Wen Z, Manninen MJ, Asiegbu FO (2025) Beneficial mutualistic fungus *Suillus luteus* provided excellent buffering insurance in Scots pine defense responses under pathogen challenge at transcriptome level. BMC Plant Biol 25(1):12
Wilhelm BT, Marguerat S, Watt S et al (2008) Dynamic repertoire of a eukaryotic transcriptome surveyed at single-nucleotide resolution. Nature 453(26):1239–1243
Wu TY, Goh H, Azodi CB et al (2021) Evolutionarily conserved hierarchical gene regulatory networks for plant salt stress response. Nat Plants 7(6):787–799
Wu TY, Hoh KL, Boonyaves K et al (2022) Diversification of heat shock transcription factors expanded thermal stress responses during early plant evolution. Plant Cell 34(10):3557–3576
Wu Y, Sun R, Huan T et al (2024a) An insight into the gene expression evolution in Gossypium species based on the leaf transcriptomes. BMC Genomics 25(1):179
Wu Z, Sun Y, Zhao X et al (2024b) Phenotype prediction in plants is improved by integrating large-scale transcriptomic datasets. NAR Genom Bioinform 6(4):lqae184
Wulf D, Bräutigam A, Eisenhut M (2024) Using publicly available RNA-seq Data for expression analysis of genes of interest. In: Photorespiration: methods and protocols. Springer, New York, pp 241–250
Xie L, Teng K, Tan P et al (2020) PacBio single- molecule long-read sequencing shed new light on the transcripts and splice isoforms of the perennial ryegrass. Mol Gen Genomics 295:475–489
Xie R, Jin X, Fang J et al (2024) Exploring the molecular landscape of nitrogen use efficiency in potato (*Solanum tuberosum* L) under low nitrogen stress: a transcriptomic and metabolomic approach. Agronomy 14:2000
Xu G, Takahashi H (2020) Improving nitrogen use efficiency: from cells to plant systems. J Exp Bot 71(15):4359–4364
Xue Y, Zhuang Q, Zhu S et al (2018) Genome wide transcriptome analysis reveals complex regulatory mechanisms underlying phosphate homeostasis in soybean nodules. Int J Mol Sci 19(10):2924
Xue Z, Ferrand M, Gilbault E et al (2024) Natural variation in response to combined water and nitrogen deficiencies in Arabidopsis. Plant Cell 36(9):3378–3398

Yang S, Li S, Wag S et al (2020) Dynamic transcriptome analysis indicates extensive and discrepant transcriptomic reprogramming of two rapeseed genotypes with contrasting NUE in response to nitrogen deficiency. Plant Soil 456:369–390

Yang J, Lv W, Shao L et al (2021) PacBio and Illumina RNA sequencing identify alternative splicing events in response to cold stress in two poplar species. Front Plant Sci 12: 737004

Yang X, Liu C, Li M et al (2023) Integrated transcriptomics and metabolomics analysis reveals key regulatory network that response to cold stress in common Bean (*Phaseolus vulgaris* L). BMC Plant Biol 23(1):85

Yao S, Liang F, Gill RA et al (2020) A global survey of the transcriptome of allopolyploid *Brassica napus* based on single- molecule long-read isoform sequencing and Illumina-based RNA sequencing data. Plant J 103:843–857

Ye J, Zhang L, Zhang X et al (2021) Plant defense networks against insect-borne pathogens. Trends Plant Sci 2C(3):272–287

Yi F, Huo M, Li J et al (2022) Time-series transcriptomics reveals a drought-responsive temporal network and crosstalk between drought stress and the circadian clock in foxtail millet. Plant J 110(4):1213–1228

Zarattini M, Farjad M, Launay A et al (2021) Every cloud has a silver lining: how abiotic stresses affect gene expression in plant-pathogen interactions. J Exp Bot 72(4):1020–1033

Zeng H, Zhang X, Zhang X et al (2018) Early transcriptomic response to phosphate deprivation in soybean leaves as revealed by RNA- sequencing. Int J Mol Sci 19:2145

Zhang D, Zhang H, Chu S et al (2017) Integrating QTL mapping and transcriptomics identifies candidate genes underlying QTLs associated with soybean tolerance to low-phosphorus stress. Plant Mol Biol 93:137–150

Zhang X, Ma Q, Li F et al (2021) Transcriptome analysis reveals different responsive patterns to nitrogen deficiency in two wheat near-isogenic lines contrasting for nitrogen use efficiency. Biology 10:1126

Zhang L, He C, Lai Y et al (2023) Asymmetric gene expression and cell-type-specific regulatory networks in the root of bread wheat revealed by single-cell multiomics analysis. Genome Biol 24(1):65

Zhang Z, Xia Z, Zhou C et al (2024) Insights into salinity tolerance in wheat. Gene 15(5):573

Zhao C, Zhang Z, Xie S et al (2016) Mutational evidence for the critical role of CBF transcription factors in cold acclimation in Arabidopsis. Plant Physiol 171(4):2744–2759

Zhong J, Zhang J, Zhang Y et al (2024) Heat stress reprograms herbivory-induced defense responses in potato plants. BMC Plant Biol 24(1):677

Zogopoulos VL, Saxami G, Malatras A et al (2022) Approaches in gene coexpression analysis in Eukaryotes. Biology 11:1019

Chapter 4
Transcriptomic Reprogramming in Plants: Insights into Gene Expression Modulation and Stress Adaptation

Saikat Sena, Yashika Sood, and Vijay Kumar

Abstract Plants are continuously dealing with a dynamic range of environmental cues, including drought, salt, temperature fluctuations, and pathogen attacks, prompting immediate and coordinated molecular responses. Transcriptomic reprogramming is the mechanism that allows plants to modify their gene expression profiles in response to internal and external stimuli, and it is necessary for regulating these adaptive responses. This chapter explores the mechanisms that regulate transcriptome reprogramming in plants, focusing on essential players such as transcription factors, noncoding RNAs, epigenetic modifications, and hormonal interactions. We address the role of these regulatory networks in controlling plant growth, development, and adaptation to stress, especially in response to both abiotic and biotic stresses. Additionally, we discuss the applications of transcriptome reprogramming knowledge in plant biotechnology, highlighting its potential for developing crops with enhanced stress tolerance via genetic engineering and synthetic biology methodologies.

Keywords Abiotic stress · Biotic stress · Plant Biotechnology · Transcriptomic reprogramming

S. Sena · V. Kumar (✉)
Plant Biotechnology Lab, Division of Research and Development, Lovely Professional University, Phagwara, Punjab, India

Department of Biotechnology, Lovely Faculty of Technology and Sciences, Lovely Professional University, Phagwara, Punjab, India
e-mail: vijay.24374@lpu.co.in; vijay.srm23@gmail.com

Y. Sood
Department of Biotechnology, Lovely Faculty of Technology and Sciences, Lovely Professional University, Phagwara, Punjab, India

L. M. Vaschetto (ed.), *Plant Transcriptomics and Epitranscriptomics*, Concepts and Strategies in Plant Sciences, https://doi.org/10.1007/978-981-95-5183-5_4

Abbreviations

4CLs	4-coumarate-coA ligases
ABA	Abscisic acid
ABREs	Abscisic acid-responsive elements
ARF	Auxin-responsive factor
CAT	Catalase
CCR	Cinnamoyl-coA reductase
COMTs	Caffeic acid 3-O-methyltransferases
CSE	Caeffeoyl shikimate esterase
ETI	Effector-triggered immunity
HSTs	Shikimate O-hydroxycinnamoyltransferases
JA	Jasmonate
LEAs	Late embryogenesis-abundant proteins
NPR1	Non-expressor of pathogenesis-related genes 1
PALs	Phenylalanine ammonia-lyase
PAMPs	Pathogen-associated molecular patterns
PODs	Peroxidases
PRRs	Pattern recognition receptors
PTI	Pattern-triggered immunity
SA	Salicylic acid
TFs	Transcription factors
uORFs	Upstream open reading frames

4.1 Introduction

Plants, as stationary organisms, continuously face a variety of environmental challenges, such as drought, salinity, fluctuating temperatures, and pathogen attacks. To cope with these forces, plants initiate rapid and coordinated molecular responses at the cellular level (Verma et al. 2016). Central to this process is transcriptomic reprogramming: a dynamic mechanism where plants alter their gene expression profiles in response to both external and internal stimuli. This adaptation involves the integration of numerous regulatory elements including transcription factors, noncoding RNAs, epigenetic modifications, and hormonal signals that collectively regulate plant growth, stress tolerance, and development (Padilla et al. 2023). This includes a complex interaction of regulatory elements, including transcription factors (TFs), such as auxin-responsive factor (ARF), v-myb avian myeloblastosis viral oncogene homolog (MYB), and NAC (NAM/ATAF1/2/CUC2) families; noncoding RNAs (ncRNAs); epigenetic modifications (e.g., DNA methylation, histone acetylation, and methylation); and hormonal signals like jasmonate (JA) and abscisic acid (ABA) (Praveen et al. 2023). Together, these factors modulate gene expression at both transcriptional and translational levels, controlling plant development, growth, and stress tolerance. Transcriptomic reprogramming in plants is a multifaceted and dynamic process that enables these organisms to adapt to a wide range of environmental stimuli (Table 4.1) (Liu et al. 2024a).

Plants enhance the concentration of endogenous ABA in response to abiotic stress, a hormone crucial for cellular tolerance to osmotic stress. This hormone initiates various physiological processes, including the closing of stomata, modulations of the root system, soil microbial community organization, transcription, post-transcription gene expression activation, and metabolic adjustments (EL Sabagh et al. 2022). Consequently, ABA promotes the expression of genes encoding enzymes responsible for synthesizing osmoprotectants, including trehalose, late embryogenesis-abundant proteins (LEAs), and dehydrins, which collectively enhance cellular protection under stress (Magwanga et al. 2018). ABA biosynthesis is reported to be regulated by the enzyme NCED (9-cis epoxy carotenoid dioxygenase) (Huang et al. 2018). Recent studies on transcriptional regulatory networks reveal that under osmotic stress conditions, gene expression is regulated through an ABA-dependent pathway by the ABA-responsive element binding proteins (AREB/ABFs) (Fidler et al. 2022). These elements, known as abscisic acid-responsive elements (ABREs), are *cis*-regulatory elements that mediate ABA-dependent gene expression. The NAC regulon is implicated in responses to osmotic stress (Nakashima et al. 2014). The NAC (CUC2, ATAF1–2, and NAM) family of plant-specific TFs constitutes a significant and extensive group of TFs in plants, regulating genes related to abiotic stress tolerance. Epigenetic modifications such as DNA methylation lead to transcriptional silencing of specific genes during stress responses or developmental transitions (Yuan et al. 2019).

Recent advances in high-throughput technologies, such as RNA-sequencing and single-cell transcriptomics, have significantly expanded our understanding of plant transcriptomes. These tools enable the identification of genes and regulatory pathways responsive to environmental stresses, which is essential for crop improvement. Transcriptome studies in *Arabidopsis* and many other plants have revealed the existence of several pathways involving both ABA-dependent and independent responses to environmental stressors. This implies that complex regulatory mechanisms control stress tolerance or sensitivity at the transcriptional level (Lv et al. 2024).

Developing crop varieties with enhanced resilience to abiotic stresses such as salinity, extreme temperatures, and drought, as well as biotic stresses like pathogen attacks, requires a comprehensive understanding of transcriptomic reprogramming. This knowledge is essential for promoting sustainable agriculture amid global climate change and can be harnessed through advanced plant biotechnology approaches, including genetic engineering and synthetic biology (Abdulraheem et al. 2024). This chapter will briefly discuss the mechanisms that regulate transcriptome reprogramming in plants, focusing on essential players such as TFs, ncRNAs, epigenetic modifications, and hormonal interactions. Here, we address the role of these regulatory networks in controlling plant growth, development, and adaptation to stress, especially in response to both biotic and abiotic stresses, while also discussing the applications of transcriptome reprogramming knowledge in plant

Table 4.1 Major transcription factors involved in plant stress responses across various species and stress types

Stress response	Plant species	Transcription factor	Genes involved	Functions	Reference
Abiotic stress response					
Drought	*Arabidopsis thaliana*	MYB94 and MYB96	*myb94*, *myb96*	Key regulators of cuticular wax biosynthesis in response to drought	Lee et al. (2016)
	Triticum aestivum	TaAREB3	*RD29A*, *RD29B*, *COR15A*, and *COR47*	Overexpression of TaAREB3 in *Arabidopsis* increased sensitivity to ABA	Wang et al. (2016)
	Arabidopsis thaliana	NGATHA1 (NGA1)	*NCED3*	Encodes 9-cis-epoxycarotenoid dioxygenase for ABA biosynthesis	Sato et al. (2018)
	Vitis vinifera L.	VhMYB2	*NCED3*, *SnRK2.6*, *CAT1*, and *P5CS*	ABA signal transduction, proline synthesis, ROS scavenging	Ren et al. (2023)
	Oryza sativa	OsNAC5	*OsCCR10*	Production of drought-induced lignin	Bang et al. (2022)
	Setaria italica	R2R3-MYB	*SiMYB56*	Regulate lignin biosynthesis and ABA signaling pathway	Xu et al. (2020)
Salt	*Glycine max*	NAC	*GmNAC06*	Gene promotes the accumulation of proline and glycine betaine, which counteracts ROS damage, and regulates Na+/K+ ratios in hairy roots	Li et al. (2021)
	Arabidopsis	R2R3-MYB AtMYB20	*AtMYB20*	Downregulates the expression of type 2C serine/threonine protein phosphatases	Cui et al. (2013)
	Oryza sativa	OsWRKY54	*OsWRKY54*	Regulates the expression of genes related to salt tolerance, such as *OsHKT1;5* and *OsNHX4*.	Huang et al. (2022)
	Vitis vinifera L.	VhMYB2	*SOS1* and *NHX1*	Facilitating the function of *AtNHX1* and *AtSOS1/2/3* to regulate intracellular Na^+ equilibrium	Ren et al. (2023)
Heat	*Arabidopsis thaliana*	MYB30	*ANN1* and *ANN4*	Bound to the promoters of *ANN1* and *ANN4* repressing their expression therefore regulating Ca^{2+} signaling	Liao et al. (2017)

Cold	*Japonica cultivar*	bZIP73[Jap]	*OsINV4*, *OsMST7* and *OsMST8*	Activation of bZIP71-bZIP73Jap-qLTG3-1Nip-sugar transport pathway	Liu et al. (2019)
	Arabidopsis thaliana	MYB15	*CBF*	Key component of a complex TF network that regulates the expression of *CBFs*	Agarwal et al. (2006)
	Malus domestica	MdMYB308L	*MdCBF2*, *MdDFR*	Increase cold resistance and anthocyanin levels in apple trees	An et al. (2020)
Biotic stress response					
Pathogen attack	*Solanum lycopersicum*	WRKY	*SlWRKY8*	Expression was upregulated upon *Pseudomonas syringae* pv. tomato DC3000 infestation	Gao et al. (2020)
	Malus x domestica	WRKY	*WRKY40*, *WRKY6*, *WRKY18* and *WRKY33*	WRKY genes adjust plant defenses to pathogens, influencing their virulence	Balan et al. (2018)
Pest resistance	*Gossypium hirsutum*	WRKY	*WRKY40*, *GhMPK3*	Involved in MPK-WRKY-JA and ET signaling pathways against whitefly (*Bemisia tabaci*)	Li et al. (2016)
	Cucurbita moschata Duch.		*HSF*, *MLO3*, *bHLH87*, *ERF014*, *WRKY21 and SGT1*	Involved in regulating defense response against *cucurbit* powdery mildew caused by *Podosphaera xanthii*	Guo et al. (2018)
	Hemerocallis citrina	MYB, WRKY, bHLH, and AP2/ERF		Response against *Thrips palmi* triggers the production of antioxidants, JA, SA, and metabolites for plant defense	Sun et al. (2024)

biotechnology, highlighting its possibility for cultivating crops with improved tolerance to stress via genetic engineering and synthetic biology methodologies.

4.2 Mechanisms of Transcriptomic Reprogramming in Plants

Transcriptomic reprogramming in plants is a dynamic and complex mechanism that allows these organisms to adjust to various environmental stressors, including both biotic and abiotic stresses. This reprogramming involves significant alterations in gene expression at both transcriptional and translational levels, allowing plants to effectively modulate their growth, defense mechanisms, and metabolic pathways (Abdulraheem et al. 2024). At the transcriptional level, plants utilize a diverse array of transcription factors to regulate gene expression. The ARF family, for example, plays a critical role in auxin signaling, which governs key developmental processes such as cell elongation, root formation, and fruit development (Liu et al. 2024b). Similarly, the MYB and NAC families of TFs are involved in stress responses, regulating genes associated with secondary metabolite production and stress tolerance (Shao et al. 2015). Epigenetic modifications, including histone acetylation, DNA methylation, and histone methylation, are also critical. These alterations affect gene accessibility and chromatin structure (Abdulraheem et al. 2024). DNA methylation can lead to transcriptional silencing of specific genes during stress responses or developmental transitions (Qiao et al. 2025). Additionally, histone modifications can either promote or inhibit transcription depending on the nature of the modification. The interplay between these epigenetic marks is essential for establishing a memory of stress exposure, enabling plants to respond more rapidly to subsequent stresses (Nunez-Vazquez et al. 2022).

In response to pathogens, plants activate immune responses combining effector-triggered immunity and pattern-triggered immunity (PTI). Pattern recognition receptors (PRRs) detect conserved microbial features called pathogen-associated molecular patterns (PAMPs), triggering PTI. It also initiates a variety of signaling processes that activate certain TFs including NPR1 (Non-expressor of Pathogenesis Related genes 1), which orchestrates the expression of defense-related genes. Concurrently, translational regulation is mainly important for improving the defense capacity of the plant. For example, the GCN2-eIF2α-TBF1 module is integral to this process; GCN2 senses amino acid starvation or stress signals and phosphorylates eIF2α, leading to selective translation of mRNAs involved in stress responses (Cai et al. 2024). Moreover, during PTI, translational reprogramming is mediated by upstream open reading frames that modulate the efficiency of translation initiation for specific mRNAs. This mechanism allows plants to prioritize the synthesis of proteins essential for defense while downregulating nonessential processes. The R-motif present in certain mRNAs enhances their translation under stress conditions by promoting ribosome loading. Processing bodies (P-bodies) also play a significant role in this context by regulating mRNA stability and translation efficiency; they sequester mRNAs for degradation or storage during unfavorable conditions (Son and Park 2023). Beyond immune responses, plants also utilize miRNAs and siRNAs as post-transcriptional regulators that fine-tune gene expression during various developmental stages and stress responses. miRNAs are short RNA molecules that bind to target mRNAs to promote their degradation or inhibit their translation. For example, miR393 targets auxin receptors like TIR1 to modulate auxin signaling pathways during stress conditions. Similarly, siRNAs (small interfering RNA) can silence transposable elements or viral RNA, contributing to genome stability and defense against viral infections (Ding et al. 2024).

Another crucial element of transcriptome reprogramming in plants is the JA signaling pathway. As a plant hormone, JA controls many different physiological processes, including activation of defense against pathogens, herbivores, and growth inhibition. Upon herbivore attack or pathogen infection, JA biosynthesis is triggered, leading to the activation of JA-responsive genes involved in secondary metabolite production. These metabolites often serve as deterrents against herbivory or pathogens (Li et al. 2022). Furthermore, the integration of various signaling pathways allows plants to achieve a coordinated response that balances growth and defense strategies. For example, crosstalk between the JA pathway and salicylic acid (SA) signaling influences the outcome of plant immune responses; while JA primarily regulates defenses against herbivores and necrotrophic pathogens, SA is crucial for resistance against biotrophic pathogens. This intricate network ensures that plants can allocate resources efficiently based on prevailing environmental conditions (Aerts et al. 2021). Another layer of complexity in transcriptomic reprogramming arises from environmental factors such as light quality and temperature fluctuations. Photoreceptors such as phytochromes and cryptochromes detect changes in light conditions and initiate signaling cascades that alter gene expression associated with photomorphogenesis, including modifications in leaf expansion, stem elongation, and flowering time. Temperature fluctuations can also induce epigenetic changes that affect gene expression patterns related to thermal stress tolerance (Xu et al. 2025). Moreover, recent studies have highlighted the role of lncRNAs (long non-coding RNAs) in regulating gene expression during various physiological processes in plants. lncRNAs can interact with chromatin-modifying complexes or TFs to influence gene expression patterns without coding for proteins themselves. Their involvement in stress responses suggests they may serve as important regulatory elements in transcriptomic reprogramming (Yadav et al. 2024).

4.3 Stress-Responsive Transcriptomic Reprogramming

4.3.1 Abiotic Stress Response

Plants cannot escape adverse environmental conditions; therefore, they cope with stress by regulating the expression of tolerance-related genes, which in turn elicit

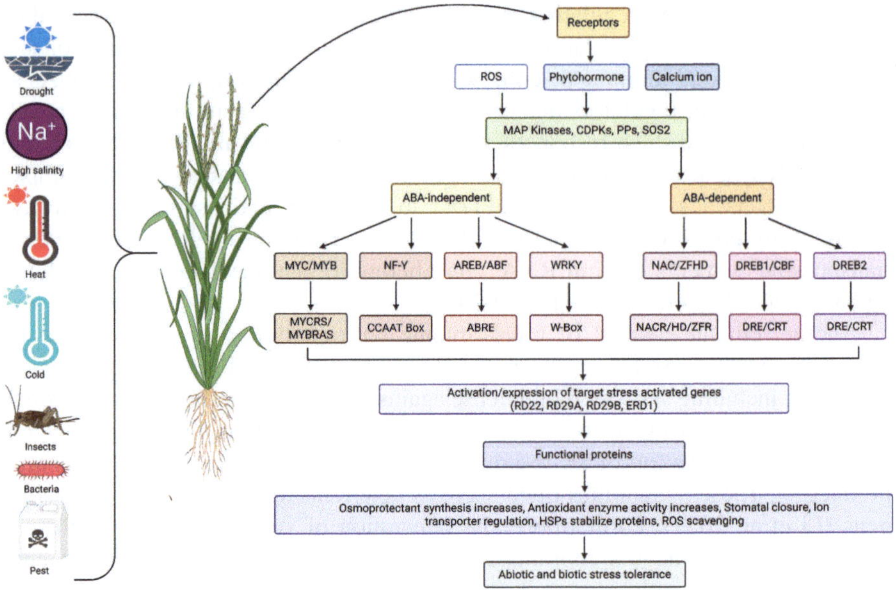

Fig. 4.1 The diagram summarizes plant abiotic/biotic stress tolerance mechanisms, whereby stress signals activate ABA-dependent/independent pathways via transcription factors to increase stress-responsive gene expression and promote tolerance by osmoprotection, antioxidant activity, and stomatal regulation

specific physiological and cellular responses (Fig. 4.1) (Nguyen et al. 2022). Among the various factors of abiotic stress are salinity, drought, and temperature.

4.3.1.1 Drought Stress

Drought stress has a notable influence on plant growth and their functioning, which contributes to extensive transcriptomic reprogramming in order for plant survival. This reprogramming encompasses a multitude of molecular mechanisms, which consist of alterations in gene expression, epigenetic modifications, and the regulation of microRNAs. The genes showing differential expression under drought stress include those encoding essential enzymes involved in hormone biosynthesis, proteins associated with osmotic regulation and cellular protection, and various signaling proteins such as kinases, phosphatases, and TFs. Under drought conditions, plants increase endogenous ABA levels, a hormone pivotal for osmotic stress tolerance. ABA induces physiological adjustments such as stomatal closure, root system modulation, microbial community organization, and activation of transcriptional and post-transcriptional gene expression changes (Bharath et al. 2021). ABA promotes the expression of various genes encoding enzymes for producing osmoprotectants, such as trehalose, LEAs, dehydrins, and other proteins for protection (Nguyen et al. 2022). The enzyme NCED regulates ABA biosynthesis, with the transcription factor NGATHA1 (NGA1) of *A. thaliana*, activating *NCED3* expression during drought, and its accumulation increasing under stress (Sato et al. 2018). ABA-dependent physical interactions between its key component, PYR/PYL/RCAR (Pyrabactin resistance/pyrabactin resistance-like/regulatory component of ABA receptors), ABA receptors, Clade A protein phosphatases type 2C (AtPP2Cs), sucrose nonfermenting (SNF) 1-related kinases 2s (AtSnRK2s), and ABA-responsive TFs affect the ABA signaling pathway in guard cells (Wang et al. 2013; Bai et al. 2019; Rehman et al. 2021). The PYL/RCAR complex positively controls the AREB/ABF SnRK2 pathway, thereby releasing SnRK2s. Significantly important in both stress responses and seed maturation, SnRK2s also control downstream TFs, including FBH3/AKS1, AREB/ABFs, and SNS1. Several studies have shown AREB/ABFs as important transcriptional activators controlling the expression of genes linked to ABA signaling. As part of the plant's reaction to drought stress, phosphorylated AREB/ABF TFs, including ABF1, ABF3, AREB1, and AREB2 bind to the promoter sites of certain target genes, hence increasing their expression. Under ABA treatment, TFs AREB1/ABF2, ABF3, and AREB/ABF4 activate drought tolerance in vegetative cells. It has been shown that overexpressing AREB1 improves drought tolerance in *Arabidopsis* and soybean (Joshi et al. 2016). Overexpression of wheat TF TaAREB3 has been shown to increase ABA sensitivity and drought tolerance by directly binding to the promoters of genes such as *COR47*, *COR15A*, *RD29A*, and *RD29B*, activating their expression (Wang et al. 2016). ABA also regulates the expression of CPK6, which activates calcium-dependent protein kinases (CDPKs), MAPKs, and CIPKs under drought stress. These protein kinases phosphorylate ABA-related TFs and AREBs, enhancing their transcriptional activity. This signaling cascade is essential for the adaptive response of plants to drought stress (Zhang et al. 2020). In two different studies on rice plants, overexpression of *SiMYB56* and *OsCCR10* led to drought-induced lignin production, thereby enhancing drought tolerance (Xu et al. 2020; Bang et al. 2022). ABRE motifs in the promoters of drought-responsive genes (*PtrNAC006*, *PtrNAC120*, *PtrNAC007*) in *Populus trichocarpa* help to increase acetylation at lysine residue 9 of histone H3 (H3K9ac), hence activating these genes (Li et al. 2019). Overexpression of maize *ZmWRKY40* in transgenic *Arabidopsis thaliana* improved drought tolerance by modulating the expression of abiotic stress-related genes and decreasing ROS levels through enhanced peroxidase (POD) and catalase (CAT) activities under drought stress (Wang et al. 2018). The study by Ren et al. (2023) identified VhMYB2, a TF from the 'Beta' grape rootstock, as key to enhancing resistance to salinity and drought stress. *VhMYB2* promotes the downstream-drought stressed-related genes *P5CS*, *NCED3*, *CAT1*, and *SnRK2.6*, therefore improving drought tolerance. Qian et al. (2021) reported transcriptomic changes enhance sucrose, starch, and carotenoid biosynthesis in drought stress in *Polygonatum kingianum* tubers. Key target genes involved in enhanced lignin production in *P. kingianum* tubers include *CSE, HSTs*, *COMTs*, *CCRs*, *4CLs*, *PALs*, and *PODs*.

DREB TFs, part of the AP2/EREBP family, contain a single AP2/EREBP domain, around 60 amino acids long, with conserved elements YRG and RAYD. This domain enables DREB proteins to selectively bind to the dehydration-responsive element/C-repeat (DRE/CRT), with the core sequence 5′-CCGAC-3′ and therefore regulate the expression of genes involved in abiotic stress (Wu et al. 2022). Wang et al. (2021a) studied the genes *DREB2*, *DREB6* and *Wdreb2* in wheat AK58 and their expression under osmotic stress. They found that these genes have distinct roles in the stress response with tissue-specific expression patterns in leaves. Promoter methylation of these genes is stress-responsive, with *Wdreb2* exhibiting lower methylation under osmotic stress, which negatively correlates with its expression.

In drought priming-induced stress tolerance, jasmonic acid acts upstream of ABA. Drought priming enhances drought tolerance in wheat by activating downstream stress-responsive genes, including LOX, AOS, OPR3, NCED, and AOC, which are involved in the biosynthesis and signaling pathways of ABA and JA (Wang et al. 2021b). The ABA-mediated stress responses and their corresponding signaling pathways affect the expression of MYB TFs. In *Arabidopsis thaliana*, the ABA-regulated drought stress response involves several MYB transcription factors, including *AtMYB44*, *AtMYB96*, *AtMYB2*, *AtMYB15*, and *AtMYB77* (Baldoni et al. 2015). Overexpression of MYB12 and MYB75 in transgenic plants significantly enhances the accumulation of flavonoids with strong antioxidant properties, thereby improving tolerance to abiotic stresses, such as oxidative stress and drought. MYB94 and MYB96 are key regulators of cuticular wax biosynthesis under drought conditions, as they activate the expression of genes associated with wax production. MYB96 overexpression increases cuticular wax synthesis, therefore increasing *Arabidopsis* and *Camelina sativa* drought tolerance (Lee et al. 2016; Wang et al. 2021c).

4.3.1.2 Salt Stress

Salt stress can cause significant damage to cell wall structure and development. High salinity, prevalent in both natural and agricultural environments, induces osmotic and ionic stress, which limits water absorption by cells and results in cytoplasmic and organelle toxicity. Salinity toxicity in plants primarily leads to reduced photosynthetic efficiency, oxidative damage caused by reactive oxygen species (ROS), alterations in cell wall composition, and the suppression of cell growth and division. This stress also activates genes involved in cell wall degradation, which may weaken and damage the cell wall, especially in sensitive cultivars (Islam et al. 2024). Salt stress upregulates several genes related to cell wall degradation, including *CSLC9* (cellulose synthase), chitinase family proteins (CHIT2, CHIT3), and OsMan04 (Endo-Beta-Mannanase), which contribute to the breakdown of cell wall components like mannan polymers. This gene activation suggests a role in salt-induced softening or damage to the cell wall, particularly in salt-sensitive cultivars (Feng et al. 2018). A high concentration of sodium ions in the soil controls salinity by inducing ion toxicity and osmotic stress in plants (Acosta-Motos et al. 2017). In response to increased salinity stress, plants activate various transcription factors (TFs) gene families, including NAC, WRKY, bZIP, bHLH, MYB, ERFs, MYCs (JA-responsive factors), and AP2/ERF, such as DREB, which belongs to the

AP2/EREBP family. These TFs are expressed at different levels and regulate distinct sets of stress-responsive genes (Mareri et al. 2022; Bhat et al. 2024). Salinity tolerance has been demonstrated to be improved in *Arabidopsis*, tomato, and rice by overexpression of *AtNHX1* and its orthologs. In tomato, *AtNHX1* enhances vacuolar K^+ accumulation and facilitates K^+ transport from root to shoot, contributing to improved intracellular K^+/Na^+ ratios and thereby reducing Na^+-induced stress (Barragán et al. 2012).

Plant responses to salt stress are substantially mediated by WRKY TFs. *Arabidopsis* tolerance to salt stress is positively regulated by *WRKY74* and *WRKY8*. In tomatoes, overexpression of *SlWRKY8* enhances plant tolerance to salt stresses (Gao et al. 2020). *OsWRKY54* is a key regulatory factor of salt tolerance in rice, controlling the expression of essential salt tolerance genes, which include *OsNHX4 and OsHKT15* (Huang et al. 2024). Under salinity stress, the tomato WRKY transcription factor SlWRKY57 functions as a negative regulator by directly repressing the transcription of salt-responsive genes *SlRD29B* and *SlDREB2*, as well as the ion homeostasis gene *SlSOS1* (Ma et al. 2023). Birhanu et al. (2020) explored the role of WRKY12 and WRKY13 as negative regulators of salt stress in transgenic tomatoes. They showed that RNAi silencing of these genes improved salt tolerance, upregulating *SOS1* in WRKY13_RNAi lines, potentially reducing Na^+ accumulation, and increasing APX expression in both WRKY12 and WRKY13_RNAi lines, indicating their involvement in salinity tolerance mechanisms.

MYB TFs contribute to the plant salt stress response by regulating the expression of specific downstream target genes associated with ion homeostasis, osmotic protection, and stress signaling. The R2R3-MYB transcription factor AtMYB20 enhances salt tolerance by downregulating the expression of type 2C serine/threonine protein phosphatases (PP2Cs), which are negative regulators of the ABA signaling pathway (Cui et al. 2013). Overexpression of AtMYB44 contributes to the salt stress response by regulating the expression of PP2Cs, key negative regulators of the ABA signaling pathway. Overexpression of AtMYB44 in *Arabidopsis thaliana* significantly reduced the salinity-induced expression of PP2C genes, including AtPP2CA, HAB1, ABI2, ABI1, and HAB2 (Wang et al. 2021c). MdSIMYB1 in apple and TaMYB56-B/MYB TFs in wheat play a role in regulating salinity stress (Wang et al. 2014). *VhMYB2* improves salt tolerance by activating salt-related genes, which include *SOS1* and *NHX1* via the *SOS2-SOS3* pathway (Ren et al. 2023). Abscisic acid, stress, and ripening-induced (ASR) genes are associated with the response to abiotic stressors. ABA induces the *OsASR1* gene in rice, and salt stress is the primary factor in the stress response. Overexpression of *OsASR1* in transgenic rice enhances water regulation under salt stress conditions by promoting osmolyte accumulation, stomatal closure, and reduced transpiration. The salt-tolerant rice landrace Pokkali shows higher levels of OsASR1 proteins, thereby contributing to salt tolerance since *OsASR1* acts as an activator of the ABA signaling pathway (Park et al. 2019). GhPYL9-5D and GhPYR1-3A positively regulate salt stress tolerance in cotton by influencing the expression of stress-associated genes (Wang et al. 2023c). The overexpression of *GmNAC06* in *Glycine max* (soybean), a NAC TF gene, elevates glycine and proline levels, contributing to the mitigation of

ROS accumulation and thus improving the plant's tolerance to salinity stress (Li et al. 2021). Gan et al. (2024) reported the use of high-throughput RNA-sequencing to explore the transcriptome and alternative splicing changes in tomato roots under salt stress. More than 700 transcription factors, including members of the MYB, bHLH, AP2/ERF, and WRKY families, have been identified as potential regulators of tomato root responses to salt stress. The R1-MYB type TF gene, *SlARS1*, which influences ABA-mediated stomatal conductance under salt stress (Campos et al. 2016), was significantly upregulated under salt stress. NaCl-induced cytosolic Ca^{2+} is sensed by calcium sensors such as CBL/CIPK complexes, calmodulin (CaM), calcium-dependent protein kinases (CDPKs), and calmodulin-like proteins (CMLs), with CAMTA6 (calmodulin-binding transcription activator 6) playing a key role in regulating salt-responsive genes during germination in *Arabidopsis* (Shkolnik et al. 2019). In *Hordeum vulgare*, HvCaM1 and HvCAMTA4 regulate salt tolerance by modulating *HvHKT1* expression (Shen et al. 2020). HbERF6 enhances salt tolerance in *Arabidopsis* by inducing *AtCIPK24* (*SOS2*) expression (Jiang et al. 2022).

4.3.1.3 Temperature Stress (Heat and Cold)

Plants have evolved adaptable strategies to reach thermotolerance in order to survive high temperatures. Heat stress can activate multiple signaling pathways, leading to extensive reprogramming that includes reactive oxygen species (ROS) production, upregulation of heat stress transcription factors (HSFs) and heat shock proteins (HSPs), activation of mitogen-activated and Ca^{2+}-dependent protein kinase cascades, and accumulation of osmolytes, all of which influence membrane fluidity and cellular morphology (Glushkevich et al. 2022). HSPs function as molecular chaperons and act to protect the denaturation of intracellular proteins, maintaining their stability and functioning by facilitating proper protein folding (Mikołajczak et al. 2023). Under nonstress conditions, HSPs bind to HSF1 TF, maintaining the monomeric form inactive in the cytosol. When plants are exposed to heat stress, the accumulation of misfolded and aggregated proteins prompts the dissociation of HSF1 from its chaperones. This release allows HSF1 to trimerize, translocate to the nucleus, and bind to its target sequences, initiating the expression of heat-inducible genes that encode molecular chaperones (Pessa et al. 2024). Disruption of protein interaction under heat stress leads to the activation of downstream target genes such as *DREB2A*, *HSFA7B*, *HSFA7A*, *MBF1C*, and *HSFA2*. HSFA1s are a master regulator of HS-induced gene expression due to their ability to interact with other TFs, thus amplifying and coordinating the heat stress response. Activation of HSFA1s stimulates the expression of *DREB2A*, which in turn directly activates *HSFA3*, a heat-inducible gene involved in enhancing thermotolerance and heat stress resistance (Ohama et al. 2016). *VOZ1* (*VASCULAR PLANT ONE ZINC FINGER PROTEIN*), a member of the NAC family, is also an inhibitor of *DREB2C*, which prevent *HSFA3* downstream effects (Song et al. 2018). Under heat stress, *DREB2A* is activated by *JUB1* (*JUNGBRUNNEN1*), *MBF1c* (*MULTIPROTEIN BINDING*

FACTOR 1c) and *HsfA1s*. Under nonstress conditions, *DREB2A* expression is inhibited by *GRF7* (*GROWTH REGULATING FACTOR 7*) (Ding et al. 2020). In *Arabidopsis thaliana*, transcriptional activation of the heat-responsive genes *HSFA3* and *UVH6* (*UV-HYPERSENSITIVE 6*) is mediated by the histone acetyltransferase (HAT) GCN5 (Hu et al. 2015).

OsTT1 (*Thermo-tolerance 1*), identified in African rice (*Oryza glaberrima*), has a role in the degradation of ubiquitinated, denatured proteins by encoding the α2 subunit of the 26S proteasome, enhancing cellular protection against heat (Li et al. 2015). MYB30 maintains oxidative and thermal stress responses by modulating calcium signaling via the repression of *ANN* (*ANNEXIN*) genes, which control cytosolic calcium levels. MYB30 directly binds to the promoter sequences of ANN1 and ANN4, inhibiting their expression. In plants with a MYB30 mutation, this suppression is lost, resulting in an abnormally large increase in cytosolic calcium in response to H_2O_2 and heat stress (Liao et al. 2017).

Plant development is substantially affected by low temperatures, which affect metabolic processes and gene expression. Plants promptly activate *CBFs* (*C-REPEAT BINDING FACTORS*) TFs in response to cold stress, which results in the expression of downstream *COR* (*COLD RESPONSIVE*) genes (Liu et al. 2018). MYB TFs significantly contribute to cold tolerance by controlling CBF genes. Arabidopsis MYB14 and MYB15 negatively regulate cold tolerance by inhibiting the expression of CBF genes. In response to low temperatures, AtMYB15 has been identified as a critical regulator of *CBF* genes. Cold stress induces the expression of MYB15, which subsequently represses *CBF* genes by binding to their promoters. The signal transduction pathway activated by cold stress is primarily regulated by CBF/DREB1 TFs. *CBF/DREB1* gene activation relies on the upstream regulator *ICE1* (*INDUCER OF CBF EXPRESSION 1*). *Myb15* mutants show increased freezing tolerance, while overexpression reduces it. MYB15 also interacts with *ICE1* to modulate *CBF* expression, playing multiple roles in cold stress response (Agarwal et al. 2006). Cold stress triggers MYB96, which works as an upstream regulator, which enhances cold tolerance by regulating *CBF* expression through HHP (HEPTAHELICAL PROTEIN). MYB96 directly activates *HHP1–3*, which subsequently activates ICE1 and downregulates *CBF* expression (Lee and Seo 2015).

MYB TFs exhibit diverse functions in plant growth and development by interacting with bHLH TFs. By interaction with MdbHLH33, MdMYB308L functions as a positive regulator for increased tolerance to cold and anthocyanin accumulation in apple trees, increasing MdbHLH33 binding affinity to the *MdCBF2* and *MdDFR* promoters. Apple cold tolerance is negatively regulated by MdbHLH4 via *MdCBF1/3* inhibition, *MdICE1L* binding interference, *MdCAX3L-2* upregulation, and *MdICE1L* degradation during cold stress (Yang et al. 2023). The composition of MYB-bHLH protein complexes is a key factor in plant growth and development regulation (An et al. 2020).

The japonica variant of the *bZIP73* gene (*bZIP73*^{Jap}) enhances cold stress tolerance in rice seedlings and reproductive stages by utilizing a specific pathway involving bZIP71, qLTG3-1Nip, and sugar transport (Liu et al. 2019). Also in rice,

MYB factors have been shown to play a role in cold and freezing stress responses. Plants exhibit improved cold tolerance when *MYB3R-2*, *MYBS3*, *MYB2*, and *MYB4* are overexpressed (Vannini et al. 2004). The increase in cytoplasmic Ca^{2+} levels directly or indirectly influences the expression of certain cold-related genes, thereby modulating cold tolerance. In the endoplasmic reticulum (ER), OsCRT3 promotes Ca2+ release and increases cytosolic calcium levels in response to cold stress (Guo et al. 2023). Plant-specific TFs under the TIFY [TIF (F/Y)XG] protein family regulate plant stress responses. A study by Wang et al. (2023b) exploring the roles of *CaTIFY* genes in *Capsicum annuum* L. revealed that *CaTIFY7* and *CaTIFY10b* are induced by cold stress. *CaTIFY7* was highly expressed in young leaves, and *CaTIFY10b* in roots. *Arabidopsis thaliana* possesses nine genes of the AREB/ABF/ABI5 subfamily, and, among them, ABF1 is known to be responsive to cold and low-temperature stress (Sharma et al. 2011).

4.3.2 Biotic Stress Response

Plants are susceptible to attacks by insects and pathogens and rely on diverse defense mechanisms to counter these threats (Fig. 4.1). Key signaling molecules, including SA, jasmonic acid (JA), and ethylene (ET), are involved in plant responses to both microbial pathogens and herbivores. After attack, *Arabidopsis thaliana* exhibits distinct kinetics in the production of ET, SA, and JA, varying in both quantity and timing. This triggers transcriptional reprogramming, marked by the overexpression of stress-related genes. Comparative analysis of transcriptional profiles reveals a substantial overlap in gene expression changes triggered by various attack patterns, whether induced by pathogens or insect herbivores.

4.3.2.1 Pathogen Attack

To defend against viruses and other pathogens, plants rely on PRRs (pattern recognition receptors) at the plasma membrane, which recognize PAMPs (pathogen-associated molecular patterns) and subsequently activate defense signaling pathways, ultimately leading to PAMP-triggered immunity (PTI) (Niehl and Heinlein 2019). A recently discovered mechanism suggests that potato plants are highly vulnerable to PVY (potato virus Y) due to an overabundance of PARylated proteins in infected tissues. Glushkevich et al. (2022) found that high temperatures inhibit *PARG*, *NUDIX* and *ADPR-PPase* expression, which reinforces the hypothesis that PARylation influences the PVY-potato plant interaction. In *Malus x domestica* (apple), Balan et al. (2018) reported a role of the WYK TF in biotic stress responses, with WRKY40 involved in both fungal and bacterial (*E. amylovora*) pathogenic responses. WRKY40 interacts with WRKY6 and WRKY18 in fungal resistance, and with WRKY33 in response to *E. amylovora*. Pathogen-triggered MPK3/MPK6 signaling activates WRKY33, which increases camalexin production in *Arabidopsis* by altering gene expression (Mao et al. 2011). WRKY genes regulate signaling pathways in response to necrotrophic pathogens, controlling defense mechanisms that can either promote or suppress pathogen virulence. Infection with the bacterial pathogen *Pseudomonas syringae* pv. *tomato* DC3000 stimulated the upregulation of the *SlWRKY8* gene in tomato. Overexpressing *SlWRKY8* in tomato enhances pathogen resistance and increases the expression of the defense-related genes *SlPR1a1* and *SlPR7* (Gao et al. 2020).

The AP2/ERF TFs family significantly impacts plant's resistance to fungal infections. AP2/ERF TFs can activate defense-related genes in *Arabidopsis*, including pathogenesis-related proteins (PR), osmolyte genes, and enzymes like β-1,3-glucanase and chitinase, which are involved in plant defense responses. AP2/ERF TFs modulate the activity of PR response genes in Zhendao 88 (resistant) and Wuyujing 3 (susceptible) rice varieties after infection with rice stripe virus (RSV) (Liu et al. 2023). In *Arabidopsis*, ERF1 overexpression increases the resistance to *Botrytis cinerea*, *Plectosphaerella cucumerina* and *Fusarium oxysporum* (Huang et al. 2016). Overexpression of *AtERF14*, *AtERF5*, *AtERF6*, *AtERF15*, and *ORA59* enhances resistance to *Botrytis cinerea*, whereas mutants and silenced lines (such as *erf5 erf6*, *AtERF15*-RNAi, and *ORA59*-silenced plants) exhibit reduced resistance. In contrast, *ERF4* functions as a negative transcriptional regulator, and its overexpression in *Arabidopsis* increases susceptibility to *Fusarium oxysporum* infection (Zhang et al. 2016). Overexpressing ERF96 led to the upregulation of the JA/ET defense genes *PR-4*, *PDF1.2a* and *PR-3* (Catinot et al. 2015). ORA59 enhances the plant's immune response to *B. cinerea* by promoting the JA/ET pathway through direct binding to *PDF1.2* (*PATHOGEN-INDUCIBLE PLANT DEFENSIN*) gene promoter (Huang et al. 2024). AtERF9 is a transcriptional repressor that binds to the GCC-box of the PDF1.2 gene. Eliminating AtERF9 leads to the expression of *AtPDF1.2*, conferring enhanced resistance to *B. cinerea* (Maruyama et al. 2013). Zhang et al. (2016) compared AP2/ERF TF expression levels in the *Fusarium* wilt-resistant *Vernicia montana* to susceptible *Vernicia fordii*. *VmAP2/ERF036*, a core hub gene, exhibited repressed expression in *V. montana* during infection, while its ortholog, *VfAP2/ERF036*, had the opposite pattern in *V. fordii*. TIFY proteins are essential regulators of the jasmonic acid signaling cascade. The *VvTIFY9* gene encodes a protein with a conserved motif (TIF[F/Y]XG) and is primarily expressed in grape (*Vitis vinifera* L.) leaves. It is activated by SA, methyl jasmonate and *Erysiphe necator* infection. Overexpressing *VvTIFY9* in *Arabidopsis* increases *Golovinomyces cichoracearum* (powdery mildew fungus) resistance, upregulating *AtPR1* and *AtPDF1.2* (Yu et al. 2019).

4.3.2.2 Pest Resistance

Plant gene expression shifts during pest infestations, providing information about defense responses and the dynamics of plant-pest relationships. Analyzing these transcriptional variations can provide beneficial perspectives on plant resistance to develop potential strategies for crop protection. Coppola et al. (2019) reported that the *Trichoderma harzianum* strain T22 induces transcriptome and metabolome reprogramming in tomato plants, thereby priming and enhancing defense responses against aphids. By inducing the overexpression of several families of defense-related TFs (*WRKY, NAC, bZIP, AP2-ERF, MYB*), *Trichoderma* appears to aid in preparing the plants for attacks from pest insects. Insect attack induces both direct (chitinase, peroxidase, miraculin, kinases, polyphenol oxidase, glutathione-S-transferases (GST)) and indirect (sesquiterpene synthase, geranylgeranyl phosphate synthase) expression of defense genes. AtMYB44, via HrpNEato and PBD, regulates *EIN2* (a gene for ethylene biosynthesis) to enhance *Myzus persicae* aphid resistance in *Arabidopsis* (Lü et al. 2013). Expression of *Arabidopsis thaliana* AtMYB12 in tobacco (*Nicotiana tabacum*) enhanced flavonol accumulation by upregulating phenylpropanoid pathway genes. This increased rutin levels, conferring resistance to *Spodoptera litura* and *Helicoverpa armigera*. Inhibition of flavonoid synthesis using artificial microRNA reduced insect resistance in transgenic plants (Misra et al. 2010). Analysis of gene activity revealed that *Beauveria bassiana* triggers the SA pathway, a key component of plant defense against both pathogens and aphids. The activation of the SA pathway was evidenced by the elevated levels of SA markers, specifically *PR1*, *PR2*, *GRX480*, and *WRKY6*. *B. bassiana* additionally induced the expression of genes involved in JA and ET pathways, specifically *ERF1*, *PDF1.2*, *ORA59*, *ERF2*, and WRKY40, which are linked to wound responses and JA-mediated signaling (Raad et al. 2019).

The transcriptional response of cotton to *Bemisia tabaci* (whitefly) infestation includes genes associated with protein kinases, TFs, metabolite biosynthesis, and phytohormone signaling. *WRKY40* and copper transport proteins serve as key regulatory genes in cotton defense against whitefly. Silencing *GhMPK3* via virus-induced gene silencing disrupts various signaling pathways, including MPK, WRKY, JA, and ET, consequently making the plant more susceptible to whitefly infestation (Li et al. 2016). Pumpkin (*Cucurbita moschata* Duch.) exhibits a complex regulatory network for resistance to powdery mildew (PM), caused by *Podosphaera xanthii*, involving hormonal signaling pathways, transcription factors, and various defense mechanisms. Expression analysis in PM-affected varieties revealed differential expression of six key genes: *HSF*, *MLO3*, *bHLH87*, *ERF014*, *WRKY21*, and *SGT1*, which were substantially upregulated or downregulated in PM-resistant varieties (Guo et al. 2018). A study on the formation of systemic resistance in *Hemerocallis citrina* "Datong Huanghua" due to *Thrips palmi* feeding indicated the involvement of bHLH and WRKY gene families associated with biotic stress. Feeding by *T. palmi* led to a significant upregulation of AP2/ERF, bHLH, WRKY, and MYB TFs, suggesting that these four TF families are crucial for the induced defense response to *T. palmi* in daylilies (Sun et al. 2024).

4.4 Developmental Transcriptomics Reprogramming

4.4.1 Seed Germination and Dormancy

Seed germination is a complicated, multiphase developmental process. Germination of a seed starts when dry quiescent seeds begin water uptake and ends with the visible protrusion of embryonic tissue. Following the imbibition, dry quiescent seeds quickly resume fundamental metabolic processes, including respiration, the machinery involved in RNA and protein synthesis, and several enzyme activities. Several studies have shown the significance of hormonal signaling in seed germination and dormancy. Regulating dormancy and germination is mostly dependent on ABA and gibberellin (GA). Abscisic acid positively regulates and maintains seed dormancy, whereas gibberellic acid promotes seed germination. Thus, ABA and GA function antagonistically in the regulation of seed germination. GA promotes seed germination by stimulation of the embryo growth and also functions in weakening of the barrier tissues surrounding the embryo. GA levels are diminished by auxin (Liu et al. 2022). In the case of rice seed germination, GA signaling mediates the starch breakdown in the rice endosperm (Fu et al. 2025). As a result of this degradation, free sugars (like glucose, sucrose, and fructose) accumulate. The presence of free sugars suppresses ABA signaling during germination, while low levels of free sugars stimulate GA signaling. Research on photoreversible seed germination in *Arabidopsis* has also demonstrated antagonistically regulatory pathways for GA and ABA.

The activation of the GA pathway occurs as the GA receptor GID1, in the presence of GA, facilitates the degradation of DELLA repressors through the SLY1 complex. RGL2, a key DELLA protein, acts as a major inhibitor of seed germination (Zhao et al. 2022). RGL2 acts as a central integrator of gibberellin (GA) and ABA signaling pathways, with its regulation of seed germination mediated through the downstream action of ABI5 (ABA INSENSITIVE 5), a bZIP transcription factor. RGL2 activates ABI5 expression by both directly binding with the ABI5 promoter and working with NF-YC proteins to increase its activation (Liu et al. 2016). ABA stimulates *RGL2* expression and enhances its stability. ABA signaling works as antagonistic to GA, therefore delaying seed germination and achieving seed dormancy. FERONIA (FER), a receptor-like kinase, regulates early seed germination in the presence of ABA. FER interacts and phosphorylates the cytosolic ABA receptor kinase 1 (CARK1) protein, which is a receptor-like cytoplasmic kinase (RLCK) involved in the modulation of ABA signaling (Wang et al. 2024). When ABA binds to PYR1/PYL/RCAR receptors, it initiates a signaling cascade by forming a stable complex with type 2C protein phosphatases (PP2Cs), thereby modulating the expression of ABA-responsive genes. SnRK2s are released, and they phosphorylate downstream TFs AREB/ABF and ABI5 to regulate the expression of ABA-responsive genes. ABI5 is activated for transcription by SnRK2s through phosphorylation. TFs activate genes such as *EM1* (*EARLY METHIONINE*) and *EM6* by binding to ABRE sites in their promoters. This activation, especially by

ABI5, could be the final regulatory step in seed germination, mediating the balance between ABA and GA (Vishwakarma et al. 2017).

In light-regulated seed germination, the phytochrome pathway promotes GA accumulation by upregulating GA biosynthesis genes, *GA3ox1* and *GA3ox2* and downregulating the GA deactivation gene, *GA2ox2*. While promoting GA accumulation, the phytochrome pathway also controls ABA levels by suppressing the ABA synthesis gene *NCED6* and enhancing the expression of the ABA breakdown gene *CYP707A2* (Liu and Hou 2018). It has been reported that during seed germination, ABA reduces the activity of *GA3ox1*/ *GA3ox2*, and when GA levels are low, ABA production is increased by activating genes like *ABA1*, *NCED6* and *NCED9*, while simultaneously reducing ABA catabolism through the repression of *CYP707A2*. In *Arabidopsis*, *PIL5* (*PHYTOCHROME-INTERACTING FACTOR3-LIKE5*) regulates seed germination by both suppressing GA3ox1/2 and activating GA2ox2 GA genes, while also activating ABA1, NCED6/9 and repressing CYP707A2 (ABA genes). During light-dependent seed germination, the *GA3ox1* gene is directly downregulated by DOF AFFECTING GERMINATION1 (DAG1), a zinc finger protein that functions downstream of PHYTOCHROME INTERACTING FACTOR 5 (PIL5) (Liu and Hou 2018).

4.4.2 Flowering and Reproduction

Plants transition from vegetative growth to flowering, a process influenced by factors like light, temperature, age, and hormones. In Arabidopsis, flowering is controlled by a complex network involving gibberellin signaling, photoperiod, the autonomous pathway, and vernalization, all working together to regulate this critical developmental stage (Zhang et al. 2011). The integration of the promotion pathways occurs through the transcriptional regulation of *FT* (*FLOWERING LOCUS T*) and *SOC1* (*SUPPRESSOR OF OVEREXPRESSION OF CONSTANS 1*) which control flowering time, as well as *LFY* (*LEAFY*) and *AP1* (*APETALA1*) which determine floral meristem identity (Rehman et al. 2023). *FT* and *SOC1* are both modulated by *CO* (*CONSTANS*), which promotes long-day flowering, and *FLC* (*FLOWERING LOCUS C*), which inhibits flowering by directly binding to *SOC1* and *FT*, thereby linking the vernalization and autonomous pathways (Deng et al. 2011). The repression of *FLC* leads to the upregulation of floral meristem identity genes such as *AP1*. *LFY* has a key role in determining the identity of floral meristems. In *lfy* mutants, LFY significantly boosts the number of coflorescences due to the action of *AP1*. This master TF (LFY) drives the development of flowers by enhancing the expression of the floral commitment factor AP1. As a pioneer TF, LFY can bind to the AP1 target *locus* DNA even when it is wrapped into a nucleosome, in vitro (Jin et al. 2021).

Auxin is a critical regulator of plant morphogenesis, and it is involved in a variety of developmental processes. Flower formation occurs at the shoot tip in the peripheral zone (PZ) of the shoot apical meristem (SAM), where auxin accumulates and initiates organogenesis (Denay et al. 2017). The transcriptional response to auxin is primarily driven by the interaction between ARF and Aux/IAA regulators (Müllender et al. 2021). Larrieu et al. (2022) reported that two DOF TFs, *DOF3.4/OBP1* and *DOF5.8*, are activated during the initiation of primordia and likely play a role in connecting growth processes with the acquisition of floral fate. In addition to other regulatory factors, *LFY* contributes to the activation of both genes specifically within inflorescences and the *LFY* promoter is bound by the two DOFs, suggesting that *OBP1*, *DOF5.8*, and *LFY* are interconnected in a regulatory network controlled by auxin therefore suggesting their role in floral initiation. Drought severely impacts plant growth and fertility, especially during reproductive development, as seen in *Arabidopsis thaliana*, where limited water from flowering to seed maturation caused early floral arrest and sterility. Differential gene expression, including key flowering and stress-related genes, highlighted drought's effects on floral development, with specific roles for DEHYDRATION RESPONSE ELEMENT B1A and MYB DOMAIN PROTEIN 21 in regulating reproductive responses to drought (Su et al. 2013).

4.4.3 Leaf Senescence and Aging

Leaf senescence is the last stage of leaf growth. Within the leaf cells, major modifications occur in their structure, metabolism, and gene expression during this time. These changes lead to the breakdown of chlorophyll, the degradation of proteins, the distortion of cell structures, and even cell death. The primary characteristic of leaf senescence is the significant reprogramming of gene expression, followed by physiological alterations that convert leaves from nutrient-accumulating structures to nutrient-recycling organs (Fraga et al. 2021). The transcriptional level is the primary regulator of the expression of a variety of senescence-associated genes (SAGs) during leaf senescence. Regulating the initiation and progression of leaf senescence in plants has been attributed to the WRKY, NAC (NAM/ATAF/CUC2), bHLH, and MYB TF families (Kim et al. 2016).

Key TF families, containing ARF, NAC, bHLH, and WRKY, were found to regulate DEGs associated with senescence in *Zoysia japonica*. Seven TFs *ZjNAC1*, *ZjWRKY75*, *ZjNAC083*, *ZjNAP*, *ZjARF2*, *ZjPIL5* and *ZjARF1* were experimentally reported as regulators of senescence (Wang et al. 2023a). Members of the NAC TF family have been implicated in regulating transcriptional reprogramming linked to various developmental processes. Several genetic studies have revealed that NAC genes regulate leaf senescence. Subfamilies like *ANAC059/ORESARA1 SISTER1* (*ORS1*), *ANAC029/Arabidopsis* NAC-LIKE Activated by *AP3/PI* (*AtNAP*), *ANAC016* function as positive regulators, while *ANAC042/JUNGBRUNNEN1* (*JUB1*) and *ANAC083/VND-INTERACTING2* (*VNI2*) act as negative regulators of natural senescence (Kim et al. 2016).

A transcriptome analysis of Arabidopsis during senescence highlighted the role of WRKY genes. Functional studies indicate that various WRKY TFs regulate leaf senescence either positively or negatively. Leaf senescence is suppressed by the activity of WRKY54 *and* WRKY70 (Besseau et al. 2012). WRKY75 promotes leaf senescence through the gibberellin pathway by directly activating the expression of *SAG12* and *SAG29* as well as acts to activate *WRKY45* expression during leaf senescence (Zhang et al. 2021b). Overexpression of WRKY75 accelerates senescence, but *GAI* gain-of-function or *RGL1* overexpression (DELLA proteins) can partially reverse this effect. Niu et al. (2020) reported WRKY42 as a novel TF that positively regulates leaf senescence by directly upregulating expression of *ICS1*, *RbohF a*nd senescence-associated genes (*SAGs*). The WRKY42 TF upregulated leaf senescence in *Arabidopsis thaliana* by modulating the synthesis of SA and ROS.

In leaf senescence, WRKY6 acts upstream of SIRK. The WRKY6 function influences the transcriptional activation of *SIRK*. Zhang et al. (2021a) reported NPR1-WRKY46-WRKY6 signaling pathway mediates leaf senescence in *Arabidopsis thaliana* induced by probenazole/SA. *TaWRKY40-D* plays a role in promoting leaf senescence in wheat particularly by involving with the JA and ABA signaling pathways (Zhao et al. 2020). *SbWRKY50*, originating from *Sorghum bicolor*, is a direct target of ETHYLENE INSENSITIVE 3 within the ethylene signaling pathway and functions as a repressor of leaf senescence by preventing the breakdown of chlorophyll (Chen et al. 2023).

4.5 Integration of Hormonal Signaling in Transcriptomics Reprogramming

The integration of hormonal signaling in transcriptomic reprogramming is a complex and multifaceted process that plays a crucial role in regulating plant responses to environmental stimuli and developmental cues. Phytohormones, such as ET, SA, JA, and ABA, are key players in this process, interacting with each other in intricate networks to modulate gene expression and orchestrate plant growth and defense strategies. For example, ABA is primarily involved in stress responses, particularly drought and osmotic stress, while SA and JA are pivotal in defense against pathogens and pests (Khan 2025). Conversely, ethylene is involved in many different kinds of processes including senescence, fruit ripening, and responses to stress (Iqbal et al. 2017). The interaction between these hormones is critical, as they often act synergistically or antagonistically to plant responses. For example, JA and SA can coinduce certain genes while also suppressing others, illustrating the complex crosstalk between their signaling pathways. In order to adjust to evolving environmental conditions and maintain a balance between growth and development and defense strategies, plants require this crosstalk (Aerts et al. 2021). In stress responses, hormonal signaling pathways interact to modulate gene expression, leading to the activation of specific TFs and downstream effectors that enhance plant resilience. For example, ABA signaling involves the activation of SnRK2 protein kinases, which phosphorylate and activate TFs such AREB/ABF, resulting in the production of stress-responsive genes. Additionally, JA signaling involves the degradation of JAZ proteins, therefore releasing MYC TFs to initiate the expression of JA-responsive genes (Zhu 2016). The integration of these hormonal signals is mediated by key regulators such as the target of rapamycin (TOR), which acts as a central hub in integrating nutrient, energy, and environmental cues to modulate plant growth and metabolism. TOR signaling can interact with hormone pathways by regulating the activity of key enzymes involved in hormone biosynthesis or signaling, such as the regulation of ABA biosynthesis through the modulation of NCED3 expression (Shi et al. 2018). Transcriptomic analyses have revealed that hormonal signaling pathways lead to significant reprogramming of the plant transcriptome, involving the regulation of thousands of genes and the activation of specific modules of coexpressed genes. These modules are enriched in biological processes such as photosynthesis, protein transport, and stress responses, highlighting the role of hormones in orchestrating complex cellular processes (Bimpong et al. 2024). Moreover, the identification of hub genes and TFs within these modules will help to understand the molecular processes behind hormone crosstalk and how it affects plant physiology. The study of cis-acting elements and their interaction with TFs also sheds light on how hormonal signals are translated into gene expression changes, revealing motifs associated with key transcription factor families like WRKYs and bZIPs. For example, the W-box motif is recognized by WRKY TFs, which are involved in SA and JA signaling pathways, while the ABRE motif is recognized by bZIP TFs involved in ABA signaling (Li et al. 2024). Additionally, recent advances in high-throughput sequencing technologies have enabled the detailed mapping of chromatin modifications and histone marks associated with hormone-regulated gene expression, providing a deeper understanding of the epigenetic mechanisms involved in transcriptomic reprogramming. For example, histone acetylation and methylation play crucial roles in modulating chromatin accessibility and gene expression in response to hormonal cues. Histone acetyltransferases (HATs) and histone deacetylases regulate the acetylation status of histones, influencing the recruitment of TFs and RNA polymerase to specific genomic regions (Sena et al. 2024). Moreover, the role of noncoding RNAs, including miRNAs and lncRNAs, in mediating hormonal signaling and transcriptomic responses has become increasingly evident. By targeting certain mRNAs for degradation or translation inhibition, these noncoding RNAs may function as major regulators of gene expression, therefore changing the response of the plant to hormonal signals. For example, miR319 targets TCP TFs involved in JA signaling, while miR160 targets ARF TFs involved in auxin signaling (Ma et al. 2022). Overall, the integration of hormonal signaling in transcriptomic reprogramming is a dynamic and highly regulated process that allows plants to adapt to their environment while maintaining optimal growth and development. This complex interplay between different hormonal pathways underscores the importance of understanding how these signals are integrated at the molecular level to better appreciate plant biology and to develop strategies for improving plant resilience and productivity.

4.6 Transcriptomics Reprogramming in Plant Biotechnology

Transcriptomic reprogramming in plant biotechnology is a powerful tool for enhancing crop resilience and productivity by modulating gene expression in response to environmental stimuli. This approach involves comprehensive analysis of gene expression changes, allowing researchers to identify key genes and pathways crucial for stress tolerance and desirable traits (Abdulraheem et al. 2024). In crop improvement, transcriptomic reprogramming is instrumental in developing stress-tolerant crop varieties through genetic engineering and clustered regularly interspaced short palindromic repeats (CRISPR)-based gene editing. For example, CRISPR/Cas9 has been used to introduce pest and disease resistance in crops like wheat and tomato by modifying specific resistance loci. Transcriptomic studies help link phenotypic traits with genotypic variations, enabling the selection of superior genotypes for breeding purposes (Kumar et al. 2023). Synthetic biology approaches further expand the potential of transcriptomic reprogramming by engineering synthetic gene circuits that can modulate transcriptomic responses to achieve desired traits in plants. These circuits can enhance stress tolerance, improve nutritional content, or boost yield by controlling the expression of specific genes in response to environmental stress (Zhang et al. 2025). Furthermore, transcriptomic reprogramming has been used to understand the molecular mechanisms underlying pollination-dependent fruit set in tomato. The auxin-induced fruit set involves the activation of similar gene regulatory networks as the pollination-induced fruit set, with several transcription factor families showing significant transcript elevation during this process. Transcriptomics combined with other omics technologies like genomics, proteomics, and metabolomics provides a whole picture of plant responses to stress and developmental stimuli. This multiomics approach can guide the design of more effective genetic engineering strategies to improve crop performance under diverse conditions (Li et al. 2025).

Transcriptomics of identical plants under both optimal and stress circumstances typically helps in the recognition of abiotic stress candidate genes that support stress tolerance (Tahmasebi et al. 2019). Northern blots were part of conventional transcriptional studies. High-throughput technologies have allowed the elucidation of the complete transcriptome of both model and nonmodel plants. Cutting-edge methods for transcriptome explanation include microarray analysis and next-generation sequencing technologies. Microarrays provide a potential approach for the discovery and functional assessment of genes mediating multiple-stress tolerance by means of a genome-wide measurement of transcript levels (Pandey et al. 2020). Using a comparable microarray analytic technique, Zhu et al. (2013) investigated the transcriptome changes of cotton under five abiotic stressors. Their studies uncovered functional genes and stress-related pathways as well as suggested a crosstalk of responsive genes or pathways to more than one abiotic stimuli in cotton seedlings. A variety of bioinformatics tools, including Genevestigator, NASCArrays, ArrayExpress, the Gene Expression Omnibus, and Stanford Microarray Database, were utilized in the microarray-based research of the plant group (Pandey et al.

2020). RNA-seq is a useful tool for identifying related genes and their expression patterns in agricultural pests and plant species under many biotic and abiotic environments, according to recent studies. By modification of ion transporters, such as HAK, NHX1, SOS1-3, TPK1/KCO1, and HKT for Ca^{2+}, Na^{+}, and K^{+} transportation, the RNA-seq analysis of the salt-tolerant mutant (M4-73-30) line and wild-type (WT-Zarjou) cultivar of barley reveals particular genes supporting salinity tolerance. Among the salinity-responsive TFs found by RNA-seq data were HSF, NAC, AP2/ERF, MAD, CTR/DRE, and WRKY (Raza et al. 2023). Overall, transcriptomic reprogramming offers a powerful toolset for plant biotechnology, enabling the development of crops that are not only more resilient but also better suited to meet the challenges of a rapidly changing global environment. By harnessing the insights gained from transcriptomic analyses, researchers can engineer crops that thrive under stress, improve nutritional quality, and enhance yield, ultimately contributing to global food security. As the field continues to evolve, the integration of transcriptomics with synthetic biology and gene editing technologies holds immense promise for revolutionizing crop improvement and ensuring sustainable agricultural practices (Panahi et al. 2025).

4.7 Future Perspective and Challenges

As we delve into the future perspectives and challenges of transcriptomic reprogramming, particularly in plants, it becomes evident that this field is poised for significant advancements while also facing complex hurdles. One of the primary challenges in studying transcriptomic reprogramming is the intricate complexity of gene networks and the profound influence of environmental factors. Plants, being highly responsive to their surroundings, exhibit dynamic transcriptomic changes in response to stressors such as salt, drought, and temperature fluctuations, which complicates the analysis and prediction of gene expression patterns. Moreover, translating transcriptomic data into functional traits remains a significant challenge due to the gap between gene expression and phenotypic outcomes, necessitating more sophisticated bioinformatic tools and experimental designs to bridge this gap. Emerging areas of research include the role of lncRNAs and other newly discovered regulators in transcriptomic reprogramming. lncRNAs have been shown to play critical roles in guiding epigenetic decisions and modulating gene expression, but understanding their specific functions and interactions with effector proteins remains a challenge. Another promising area is the potential for machine learning (ML) and artificial intelligence (AI) in predicting transcriptomic responses based on environmental cues. These technologies can help analyze vast datasets from RNA-seq and other omics platforms to identify patterns and predict how plants might adapt to future environmental stresses, thereby enhancing our ability to engineer stress-resilient crops. However, integrating AI and ML into transcriptomics requires robust computational frameworks to handle the high-dimensional and noisy data generated from single-cell and bulk RNA sequencing experiments. Furthermore, the integration of transcriptomics with other omics data, such as proteomics, epigenomics, and genomics, will provide a comprehensive understanding of gene regulation and biological processes, enabling the identification of key regulatory mechanisms underlying stress adaptation in plants. Despite these advancements, challenges persist, including the need for more efficient sequencing technologies that can handle complex plant genomes and the development of standardized protocols for data analysis and interpretation across different species and environmental conditions. Additionally, ethical considerations and regulatory frameworks will become increasingly important as transcriptomic reprogramming technologies advance towards practical applications in agriculture and biotechnology. Overall, the future of transcriptomic reprogramming in plants holds great promise for improving crop resilience and productivity, but it will require continued innovation in both technological and analytical methodologies to fully realize its potential.

4.8 Conclusion

Transcriptomic reprogramming is essential for plants to respond to various environmental challenges by regulating gene expression via complex networks that include TFs, noncoding RNAs, epigenetic modifications, and hormone signaling. Advancements in high-throughput approaches like RNA-seq have enhanced our comprehension of these mechanisms, revealing pathways that regulate stress tolerance, growth, and development. Using this information provides interesting opportunities for improving crop resilience via genetic engineering and synthetic biology. Despite challenges like converting transcriptome insights into functional traits and integrating multiomics data, current research has significant possibilities for creating stress-resistant crops, enhancing sustainable agriculture, and resolving global food security during climate change.

Competing Interest All the authors declare that there is no conflict of interest.

Authors Contribution Statement
SS and YS: Literature collection and Writing. **VK:** Conceptualization, Editing and Supervision.

Funding The authors declare that no funding was received for this work.

References

Abdulraheem MI, Xiong Y, Moshood AY, Cadenas-Pliego G, Zhang H, Hu J (2024) Mechanisms of plant epigenetic regulation in response to plant stress: recent discoveries and implications. Plants 13(2):163. https://doi.org/10.3390/plants13020163

Acosta-Motos JR, Ortuño MF, Bernal-Vicente A, Diaz-Vivancos P, Sanchez-Blanco MJ, Hernandez JA (2017) Plant responses to salt stress: adaptive mechanisms. Agronomy 7(1):18. https://doi.org/10.3390/agronomy7010018

Aerts N, Pereira Mendes M, Van Wees SCM (2021) Multiple levels of crosstalk in hormone networks regulating plant defense. Plant J 105(2):489–504. https://doi.org/10.1111/tpj.15124

Agarwal M, Hao Y, Kapoor A, Dong C-H, Fujii H, Zheng X, Zhu J-K (2006) A R2R3 type MYB transcription factor is involved in the cold regulation of CBF genes and in acquired freezing tolerance. J Biol Chem 281(49):37636–37645. https://doi.org/10.1074/jbc.M605895200
An J-P, Wang X-F, Zhang X-W, Xu H-F, Bi S-Q, You C-X, Hao Y-J (2020) An apple MYB transcription factor regulates cold tolerance and anthocyanin accumulation and undergoes MIEL1-mediated degradation. Plant Biotechnol J 18(2):337–353. https://doi.org/10.1111/pbi.13201
Bai G, Xie H, Yao H, Li F, Chen X, Zhang Y, Xiao B, Yang J, Li Y, Yang D-H (2019) Genome-wide identification and characterization of ABA receptor PYL/RCAR gene family reveals evolution and roles in drought stress in Nicotiana tabacum. BMC Genomics 20(1):575. https://doi.org/10.1186/s12864-019-5839-2
Balan B, Marra FP, Caruso T, Martinelli F (2018) Transcriptomic responses to biotic stresses in Malus x domestica: a meta-analysis study. Sci Rep 8(1):1970. https://doi.org/10.1038/s41598-018-19348-4
Baldoni E, Genga A, Cominelli E (2015) Plant MYB transcription factors: their role in drought response mechanisms. Int J Mol Sci 16(7):15811–15851. https://doi.org/10.3390/ijms160715811
Bang SW, Choi S, Jin X, Jung SE, Choi JW, Seo JS, Kim J-K (2022) Transcriptional activation of rice CINNAMOYL-CoA REDUCTASE 10 by OsNAC5, contributes to drought tolerance by modulating lignin accumulation in roots. Plant Biotechnol J 20(4):736–747. https://doi.org/10.1111/pbi.13752
Barragán V, Leidi EO, Andrés Z, Rubio L, De Luca A, Fernández JA, Cubero B, Pardo JM (2012) Ion exchangers NHX1 and NHX2 mediate active potassium uptake into vacuoles to regulate cell turgor and stomatal function in Arabidopsis. Plant Cell 24(3):1127–1142. https://doi.org/10.1105/tpc.111.095273
Besseau S, Li J, Palva ET (2012) WRKY54 and WRKY70 co-operate as negative regulators of leaf senescence in Arabidopsis thaliana. J Exp Bot 63(7):2667–2679. https://doi.org/10.1093/jxb/err450
Bharath P, Gahir S, Raghavendra AS (2021) Abscisic acid-induced stomatal closure: an important component of plant defense against abiotic and biotic stress. Front Plant Sci 12:615114. https://doi.org/10.3389/fpls.2021.615114
Bhat BA, Mir RA, Mir WR, Hamdani SS, Mir MA (2024) Transcription factors-golden keys to modulate the plant metabolism to develop salinity tolerance. Plant Stress 11:100409. https://doi.org/10.1016/j.stress.2024.100409
Bimpong D, Zhao L, Ran M, Zhao X, Wu C, Li Z, Wang X, Cheng L, Fang Z, Hu Z, Fan C, Gyebi-Nimako B, Luo Y, Wang S, Zhang Y (2024) Transcriptomic analysis reveals the regulatory mechanisms of messenger RNA (mRNA) and long non-coding RNA (lncRNA) in response to waterlogging stress in rye (Secale cereale L.). BMC Plant Biol 24:534. https://doi.org/10.1186/s12870-024-05234-x
Birhanu MW, Kissoudis C, van der Linden CG (2020) WRKY gene silencing enhances tolerance to salt stress in transgenic tomato. J Biol Agric Healthc 10:14–25
Cai M, Yu H, Sun E, Zuo C (2024) Receptor-like proteins: decision-makers of plant immunity. Phytopathol Res 6(1):58. https://doi.org/10.1186/s42483-024-00279-0
Campos JF, Cara B, Pérez-Martín F, Pineda B, Egea I, Flores FB, Fernandez-Garcia N, Capel J, Moreno V, Angosto T, Lozano R, Bolarin MC (2016) The tomato mutant ars1 (altered response to salt stress 1) identifies an R1-type MYB transcription factor involved in stomatal closure under salt acclimation. Plant Biotechnol J 14(6):1345–1356. https://doi.org/10.1111/pbi.12498
Catinot J, Huang J-B, Huang P-Y, Tseng M-Y, Chen Y-L, Gu S-Y, Lo W-S, Wang L-C, Chen Y-R, Zimmerli L (2015) ETHYLENE RESPONSE FACTOR 96 positively regulates Arabidopsis resistance to necrotrophic pathogens by direct binding to GCC elements of jasmonate – and ethylene-responsive defence genes. Plant Cell Environ 38(12):2721–2734. https://doi.org/10.1111/pce.12583
Chen W, Zheng Y, Wang J, Wang Z, Yang Z, Chi X, Dai L, Lu G, Yang Y, Sun B (2023) Ethylene-responsive SbWRKY50 suppresses leaf senescence by inhibition of chlorophyll degradation in sorghum. New Phytologist 238(3):1129–1145. https://doi.org/10.1111/nph.18757

Coppola M, Diretto G, Digilio MC, Woo SL, Giuliano G, Molisso D, Pennacchio F, Lorito M, Rao R (2019) Transcriptome and metabolome reprogramming in tomato plants by Trichoderma harzianum strain T22 primes and enhances defense responses against aphids. Front Physiol 10: 745. https://doi.org/10.3389/fphys.2019.00745

Cui MH, Yoo KS, Hyoung S, Nguyen HTK, Kim YY, Kim HJ, Ok SH, Yoo SD, Shin JS (2013) An Arabidopsis R2R3-MYB transcription factor, AtMYB20, negatively regulates type 2C serine/threonine protein phosphatases to enhance salt tolerance. FEBS Lett 587(12):1773–1778. https://doi.org/10.1016/j.febslet.2013.04.028

Denay G, Chahtane H, Tichtinsky G, Parcy F (2017) A flower is born: an update on Arabidopsis floral meristem formation. Curr Opin Plant Biol 35:15–22. https://doi.org/10.1016/j.pbi.2016.09.003

Deng W, Ying H, Helliwell CA, Taylor JM, Peacock WJ, Dennis ES (2011) FLOWERING LOCUS C (FLC) regulates development pathways throughout the life cycle of Arabidopsis. Proc Natl Acad Sci USA 108(16):6680–6685. https://doi.org/10.1073/pnas.1103175108

Ding Y, Shi Y, Yang S (2020) Molecular regulation of plant responses to environmental temperatures. Mol Plant 13(4):544–564. https://doi.org/10.1016/j.molp.2020.02.004

Ding T, Li W, Li F, Ren M, Wang W (2024) microRNAs: key regulators in plant responses to abiotic and biotic stresses via endogenous and cross-kingdom mechanisms. Int J Mol Sci 25(2): 1154. https://doi.org/10.3390/ijms25021154

EL Sabagh A, Islam MS, Hossain A, Iqbal MA, Mubeen M, Waleed M, Reginato M, Battaglia M, Ahmed S, Rehman A, Arif M, Athar H-U-R, Ratnasekera D, Danish S, Raza MA, Rajendran K, Mushtaq M, Skalicky M, Brestic M, Soufan W, Fahad S, Pandey S, Kamran M, Datta R, Abdelhamid MT (2022) Phytohormones as growth regulators during abiotic stress tolerance in plants. Front Agron 4. https://doi.org/10.3389/fagro.2022.765068

Feng W, Kita D, Peaucelle A, Cartwright HN, Doan V, Duan Q, Liu M-C, Maman J, Steinhorst L, Schmitz-Thom I, Yvon R, Kudla J, Wu H-M, Cheung AY, Dinneny JR (2018) The FERONIA receptor kinase maintains Cell-Wall integrity during salt stress through Ca2+ signaling. Curr Biol CB 28(5):666–675.e5. https://doi.org/10.1016/j.cub.2018.01.023

Fidler J, Graska J, Gietler M, Nykiel M, Prabucka B, Rybarczyk-Płońska A, Muszyńska E, Morkunas I, Labudda M (2022) PYR/PYL/RCAR receptors play a vital role in the abscisic-acid-dependent responses of plants to external or internal stimuli. Cells 11(8):1352. https://doi.org/10.3390/cells11081352

Fraga OT, de Melo BP, Quadros IPS, Reis PAB, Fontes EPB (2021) Senescence-associated Glycine max (Gm)NAC genes: integration of natural and stress-induced leaf senescence. Int J Mol Sci 22(15):8287. https://doi.org/10.3390/ijms22158287

Fu D, Wu W, Mustafa G, Yang Y, Yang P (2025) Molecular mechanisms of rice seed germination. New Crops 2:100051. https://doi.org/10.1016/j.ncrops.2024.100051

Gan J, Qiu Y, Tao Y, Zhang L, Okita TW, Yan Y, Tian L (2024) RNA-seq analysis reveals transcriptome reprogramming and alternative splicing during early response to salt stress in tomato root. Front Plant Sci 15:1394223. https://doi.org/10.3389/fpls.2024.1394223

Gao Y-F, Liu J-K, Yang F-M, Zhang G-Y, Wang D, Zhang L, Ou Y-B, Yao Y-A (2020) The WRKY transcription factor WRKY8 promotes resistance to pathogen infection and mediates drought and salt stress tolerance in Solanum lycopersicum. Physiol Plant 168(1):98–117. https://doi.org/10.1111/ppl.12978

Glushkevich A, Spechenkova N, Fesenko I, Knyazev A, Samarskaya V, Kalinina NO, Taliansky M, Love AJ (2022) Transcriptomic reprogramming, alternative splicing and RNA methylation in potato (Solanum tuberosum L.) plants in response to potato virus Y infection. Plants 11(5):635. https://doi.org/10.3390/plants11050635

Guo W-L, Chen B-H, Chen X-J, Guo Y-Y, Yang H-L, Li X-Z, Wang G-Y (2018) Transcriptome profiling of pumpkin (Cucurbita moschata Duch.) leaves infected with powdery mildew. PLoS One 13(1):e0190175. https://doi.org/10.1371/journal.pone.0190175

Guo X, Zhang D, Wang Z, Xu S, Batistič O, Steinhorst L, Li H, Weng Y, Ren D, Kudla J, Xu Y, Chong K (2023) Cold-induced calreticulin OsCRT3 conformational changes promote OsCIPK7 binding and temperature sensing in rice. EMBO J 42(1):e110518. https://doi.org/10.15252/embj.2021110518

Hu Z, Song N, Zheng M, Liu X, Liu Z, Xing J, Ma J, Guo W, Yao Y, Peng H, Xin M, Zhou D-X, Ni Z, Sun Q (2015) Histone acetyltransferase GCN5 is essential for heat stress-responsive gene activation and thermotolerance in Arabidopsis. Plant J Cell Mol Biol 84(6):1178–1191. https://doi.org/10.1111/tpj.13076

Huang P-Y, Catinot J, Zimmerli L (2016) Ethylene response factors in Arabidopsis immunity. J Exp Bot 67(5):1231–1241. https://doi.org/10.1093/jxb/erv518

Huang Y, Guo Y, Liu Y, Zhang F, Wang Z, Wang H, Wang F, Li D, Mao D, Luan S, Liang M, Chen L (2018) 9-cis-Epoxycarotenoid dioxygenase 3 regulates plant growth and enhances multi-abiotic stress tolerance in rice. Front Plant Sci 9:162. https://doi.org/10.3389/fpls.2018.00162

Huang J, Liu F, Chao D, Xin B, Liu K, Cao S, Chen X, Peng L, Zhang B, Fu S, Xia J (2022) The WRKY transcription factor OsWRKY54 is involved in salt tolerance in Rice. Int J Mol Sci 23(19):11999. https://doi.org/10.3390/ijms231911999

Huang L-J, Zhang J, Lin Z, Yu P, Lu M, Li N (2024) The AP2/ERF transcription factor ORA59 regulates ethylene-induced phytoalexin synthesis through modulation of an acyltransferase gene expression. J Cell Physiol 239(10):e30935. https://doi.org/10.1002/jcp.30935

Iqbal N, Khan NA, Ferrante A, Trivellini A, Francini A, Khan MIR (2017) Ethylene role in plant growth, development and senescence: interaction with other phytohormones. Front Plant Sci 8. https://doi.org/10.3389/fpls.2017.00475

Islam F, Khan MSS, Ahmed S, Ikram AU, Hannan F, Jan M, Muhammad S, Chen J (2024) Transcriptomic reprogramming of rice cultivars in response to herbicide, salt and their combined stresses. Plant Stress 12:100504. https://doi.org/10.1016/j.stress.2024.100504

Jiang Y, Zhang H, Li Y, Chang C, Wang Y, Feng H, Li R (2022) A novel transcriptional regulator HbERF6 regulates the HbCIPK2-coordinated pathway conferring salt tolerance in halophytic Hordeum brevisubulatum. Front Plant Sci 13:927253. https://doi.org/10.3389/fpls.2022.927253

Jin R, Klasfeld S, Zhu Y, Fernandez Garcia M, Xiao J, Han S-K, Konkol A, Wagner D (2021) LEAFY is a pioneer transcription factor and licenses cell reprogramming to floral fate. Nat Commun 12:626. https://doi.org/10.1038/s41467-020-20883-w

Joshi R, Wani SH, Singh B, Bohra A, Dar ZA, Lone AA, Pareek A, Singla-Pareek SL (2016) Transcription factors and plants response to drought stress: current understanding and future directions. Front Plant Sci 7:1029. https://doi.org/10.3389/fpls.2016.01029

Khan N (2025) Decoding phytohormone signaling in plant stress physiology: insights, challenges, and future directions. Environ Exp Bot 231:106099. https://doi.org/10.1016/j.envexpbot.2025.106099

Kim HJ, Nam HG, Lim PO (2016) Regulatory network of NAC transcription factors in leaf senescence. Curr Opin Plant Biol 33:48–56. https://doi.org/10.1016/j.pbi.2016.06.002

Kumar M, Prusty MR, Pandey MK, Singh PK, Bohra A, Guo B, Varshney RK (2023) Application of CRISPR/Cas9-mediated gene editing for abiotic stress management in crop plants. Front Plant Sci 14:1157678. https://doi.org/10.3389/fpls.2023.1157678

Larrieu A, Brunoud G, Guérault A, Lainé S, Hennet L, Stigliani A, Gildea I, Just J, Soubigou-Taconnat L, Balzergue S, Davies B, Scarpella E, Helariutta Y, Parcy F, Vernoux T (2022) Transcriptional reprogramming during floral fate acquisition. iScience 25(7):104683. https://doi.org/10.1016/j.isci.2022.104683

Lee HG, Seo PJ (2015) The MYB96-HHP module integrates cold and abscisic acid signaling to activate the CBF-COR pathway in Arabidopsis. Plant J Cell Mol Biol 82(6):962–977. https://doi.org/10.1111/tpj.12866

Lee SB, Kim HU, Suh MC (2016) MYB94 and MYB96 additively activate Cuticular wax biosynthesis in Arabidopsis. Plant Cell Physiol 57(11):2300–2311. https://doi.org/10.1093/pcp/pcw147

Li X-M, Chao D-Y, Wu Y, Huang X, Chen K, Cui L-G, Su L, Ye W-W, Chen H, Chen H-C, Dong N-Q, Guo T, Shi M, Feng Q, Zhang P, Han B, Shan J-X, Gao J-P, Lin H-X (2015) Natural alleles of a proteasome α2 subunit gene contribute to thermotolerance and adaptation of African rice. Nat Genet 47(7):827–833. https://doi.org/10.1038/ng.3305

Li J, Zhu L, Hull JJ, Liang S, Daniell H, Jin S, Zhang X (2016) Transcriptome analysis reveals a comprehensive insect resistance response mechanism in cotton to infestation by the phloem feeding insect Bemisia tabaci (whitefly). Plant Biotechnol J 14(10):1956–1975. https://doi.org/10.1111/pbi.12554

Li S, Lin Y-CJ, Wang P, Zhang B, Li M, Chen S, Shi R, Tunlaya-Anukit S, Liu X, Wang Z, Dai X, Yu J, Zhou C, Liu B, Wang JP, Chiang VL, Li W (2019) The AREB1 transcription factor influences histone acetylation to regulate drought responses and tolerance in Populus trichocarpa[OPEN]. Plant Cell 31(3):663–686. https://doi.org/10.1105/tpc.18.00437

Li M, Chen R, Jiang Q, Sun X, Zhang H, Hu Z (2021) GmNAC06, a NAC domain transcription factor enhances salt stress tolerance in soybean. Plant Mol Biol 105(3):333–345. https://doi.org/10.1007/s11103-020-01091-y

Li C, Xu M, Cai X, Han Z, Si J, Chen D (2022) Jasmonate signaling pathway modulates plant defense, growth, and their trade-offs. Int J Mol Sci 23(7):3945. https://doi.org/10.3390/ijms23073945

Li S, Khoso MA, Wu J, Yu B, Wagan S, Liu L (2024) Exploring the mechanisms of WRKY transcription factors and regulated pathways in response to abiotic stress. Plant Stress 12:100429. https://doi.org/10.1016/j.stress.2024.100429

Li X, He B, Djari A, Frasse P, Maza E, Regad F, Pirrello J, Hu G, Bouzayen M (2025) Transcriptomic reprogramming and epigenetic regulation underlying pollination-dependent and auxin-induced fruit set in tomato. Front Plant Sci 16:1495494. https://doi.org/10.3389/fpls.2025.1495494

Liao C, Zheng Y, Guo Y (2017) MYB30 transcription factor regulates oxidative and heat stress responses through ANNEXIN-mediated cytosolic calcium signaling in Arabidopsis. New Phytol 216(1):163–177. https://doi.org/10.1111/nph.14679

Liu X, Hou X (2018) Antagonistic regulation of ABA and GA in metabolism and signaling pathways. Front Plant Sci 9:251. https://doi.org/10.3389/fpls.2018.00251

Liu X, Hu P, Huang M, Tang Y, Li Y, Li L, Hou X (2016) The NF-YC-RGL2 module integrates GA and ABA signalling to regulate seed germination in Arabidopsis. Nat Commun 7:12768. https://doi.org/10.1038/ncomms12768

Liu J, Shi Y, Yang S (2018) Insights into the regulation of C-repeat binding factors in plant cold signaling. J Integr Plant Biol 60(9):780–795. https://doi.org/10.1111/jipb.12657

Liu C, Schläppi MR, Mao B, Wang W, Wang A, Chu C (2019) The bZIP73 transcription factor controls rice cold tolerance at the reproductive stage. Plant Biotechnol J 17(9):1834–1849. https://doi.org/10.1111/pbi.13104

Liu Z, Wang J, Zhou Y, Zhang Y, Qin A, Yu X, Zhao Z, Wu R, Guo C, Bawa G, Rochaix J-D, Sun X (2022) Identification of novel regulators required for early development of vein pattern in the cotyledons by single-cell RNA-sequencing. Plant J Cell Mol Biol 110(1):7–22. https://doi.org/10.1111/tpj.15719

Liu Y, Liu W, Li L, Francis F, Wang X (2023) Transcriptome analysis reveals different response of resistant and susceptible rice varieties to rice stripe virus infection. J Integr Agric 22(6):1750–1762. https://doi.org/10.1016/j.jia.2022.10.010

Liu F, Xi M, Liu T, Wu X, Ju L, Wang D (2024a) The central role of transcription factors in bridging biotic and abiotic stress responses for plants' resilience. New Crops 1:100005. https://doi.org/10.1016/j.ncrops.2023.11.003

Liu L, Yahaya BS, Li J, Wu F (2024b) Enigmatic role of auxin response factors in plant growth and stress tolerance. Front Plant Sci 15:1398818. https://doi.org/10.3389/fpls.2024.1398818

Lü B-B, Li X-J, Sun W-W, Li L, Gao R, Zhu Q, Tian S-M, Fu M-Q, Yu H-L, Tang X-M, Zhang C-L, Dong H-S (2013) AtMYB44 regulates resistance to the green peach aphid and diamond-back moth by activating EIN2-affected defences in Arabidopsis. Plant Biol Stuttg Ger 15(5):841–850. https://doi.org/10.1111/j.1438-8677.2012.00675.x

Lv Z, Jiang S, Kong S, Zhang X, Yue J, Zhao W, Li L, Lin S (2024) Advances in single-cell transcriptome sequencing and spatial transcriptome sequencing in plants. Plants 13(12):1679. https://doi.org/10.3390/plants13121679

Ma X, Zhao F, Zhou B (2022) The characters of non-coding RNAs and their biological roles in plant development and abiotic stress response. Int J Mol Sci 23(8):4124. https://doi.org/10.3390/ijms23084124

Ma J, Li C, Sun L, Ma X, Qiao H, Zhao W, Yang R, Song S, Wang S, Huang H (2023) The SlWRKY57-SlVQ21/SlVQ16 module regulates salt stress in tomato. J Integr Plant Biol 65(11): 2437–2455. https://doi.org/10.1111/jipb.13562

Magwanga RO, Lu P, Kirungu JN, Lu H, Wang X, Cai X, Zhou Z, Zhang Z, Salih H, Wang K, Liu F (2018) Characterization of the late embryogenesis abundant (LEA) proteins family and their role in drought stress tolerance in upland cotton. BMC Genet 19(1):6. https://doi.org/10.1186/s12863-017-0596-1

Mao G, Meng X, Liu Y, Zheng Z, Chen Z, Zhang S (2011) Phosphorylation of a WRKY transcription factor by two pathogen-responsive MAPKs drives phytoalexin biosynthesis in Arabidopsis. Plant Cell 23(4):1639–1653. https://doi.org/10.1105/tpc.111.084996

Mareri L, Parrotta L, Cai G (2022) Environmental stress and plants. Int J Mol Sci 23(10):5416. https://doi.org/10.3390/ijms23105416

Maruyama Y, Yamoto N, Suzuki Y, Chiba Y, Yamazaki K, Sato T, Yamaguchi J (2013) The Arabidopsis transcriptional repressor ERF9 participates in resistance against necrotrophic fungi. Plant Sci Int J Exp Plant Biol 213:79–87. https://doi.org/10.1016/j.plantsci.2013.08.008

Mikołajczak K, Kuczyńska A, Krajewski P, Kempa M, Nuc M (2023) Transcriptome profiling disclosed the effect of single and combined drought and heat stress on reprogramming of genes expression in barley flag leaf. Front Plant Sci 13:1096685. https://doi.org/10.3389/fpls.2022.1096685

Misra P, Pandey A, Tiwari M, Chandrashekar K, Sidhu OP, Asif MH, Chakrabarty D, Singh PK, Trivedi PK, Nath P, Tuli R (2010) Modulation of transcriptome and metabolome of tobacco by Arabidopsis transcription factor, AtMYB12, leads to insect resistance. Plant Physiol 152(4): 2258–2268. https://doi.org/10.1104/pp.109.150979

Müllender M, Varrelmann M, Savenkov EI, Liebe S (2021) Manipulation of auxin signalling by plant viruses. Mol Plant Pathol 22(11):1449–1458. https://doi.org/10.1111/mpp.13122

Nakashima K, Yamaguchi-Shinozaki K, Shinozaki K (2014) The transcriptional regulatory network in the drought response and its crosstalk in abiotic stress responses including drought, cold, and heat. Front Plant Sci 5:170. https://doi.org/10.3389/fpls.2014.00170

Nguyen NH, Vu NT, Cheong J-J (2022) Transcriptional stress memory and transgenerational inheritance of drought tolerance in plants. Int J Mol Sci 23(21):12918. https://doi.org/10.3390/ijms232112918

Niehl A, Heinlein M (2019) Perception of double-stranded RNA in plant antiviral immunity. Mol Plant Pathol 20(9):1203–1210. https://doi.org/10.1111/mpp.12798

Niu F, Cui X, Zhao P, Sun M, Yang B, Deyholos MK, Li Y, Zhao X, Jiang Y-Q (2020) WRKY42 transcription factor positively regulates leaf senescence through modulating SA and ROS synthesis in Arabidopsis thaliana. Plant J Cell Mol Biol 104(1):171–184. https://doi.org/10.1111/tpj.14914

Nunez-Vazquez R, Desvoyes B, Gutierrez C (2022) Histone variants and modifications during abiotic stress response. Front Plant Sci 13:984702. https://doi.org/10.3389/fpls.2022.984702

Ohama N, Kusakabe K, Mizoi J, Zhao H, Kidokoro S, Koizumi S, Takahashi F, Ishida T, Yanagisawa S, Shinozaki K, Yamaguchi-Shinozaki K (2016) The transcriptional Cascade in the heat stress response of Arabidopsis is strictly regulated at the level of transcription factor expression. Plant Cell 28(1):181–201. https://doi.org/10.1105/tpc.15.00435

Padilla YG, Gisbert-Mullor R, Bueso E, Zhang L, Forment J, Lucini L, López-Galarza S, Calatayud Á (2023) New insights into short-term water stress tolerance through transcriptomic and metabolomic analyses on pepper roots. Plant Sci 333:111731. https://doi.org/10.1016/j.plantsci.2023.111731

Panahi B, Hamid R, Mohammad Zadeh Jalaly H (2025) Deciphering plant transcriptomes: leveraging machine learning for deeper insights. Curr Plant Biol 41:100432. https://doi.org/10.1016/j.cpb.2024.100432

Pandey P, Srivastava S, Pandey AK, Dubey RS (2020) Chapter 22 – Abiotic-stress tolerance in plants-system biology approach. In: Tripathi DK, Pratap Singh V, Chauhan DK, Sharma S, Prasad SM, Dubey NK, Ramawat N (eds) Plant life under changing environment. Academic, pp 577–609

Park S-I, Kim J-J, Shin S-Y, Kim Y-S, Yoon H-S (2019) ASR enhances environmental stress tolerance and improves grain yield by modulating stomatal closure in Rice. Front Plant Sci 10: 1752. https://doi.org/10.3389/fpls.2019.01752

Pessa JC, Joutsen J, Sistonen L (2024) Transcriptional reprogramming at the intersection of the heat shock response and proteostasis. Mol Cell 84(1):80–93. https://doi.org/10.1016/j.molcel.2023.11.024

Praveen A, Dubey S, Singh S, Sharma VK (2023) Abiotic stress tolerance in plants: a fascinating action of defense mechanisms. 3 Biotech 13(3):102. https://doi.org/10.1007/s13205-023-03519-w

Qian H, Xu Z, Cong K, Zhu X, Zhang L, Wang J, Wei J, Ji P (2021) Transcriptomic responses to drought stress in Polygonatum kingianum tuber. BMC Plant Biol 21(1):537. https://doi.org/10.1186/s12870-021-03297-8

Qiao S, Song W, Hu W, Wang F, Liao A, Tan W, Yang S (2025) The role of plant DNA methylation in development, stress response, and crop breeding. Agronomy 15(1):94. https://doi.org/10.3390/agronomy15010094

Raad M, Glare TR, Brochero HL, Müller C, Rostás M (2019) Transcriptional reprogramming of Arabidopsis thaliana defence pathways by the entomopathogen Beauveria bassiana correlates with resistance against a fungal pathogen but not against insects. Front Microbiol 10:615. https://doi.org/10.3389/fmicb.2019.00615

Raza A, Tabassum J, Fakhar AZ, Sharif R, Chen H, Zhang C, Ju L, Fotopoulos V, Siddique KHM, Singh RK, Zhuang W, Varshney RK (2023) Smart reprograming of plants against salinity stress using modern biotechnological tools. Crit Rev Biotechnol 43(7):1035–1062. https://doi.org/10.1080/07388551.2022.2093695

Rehman A, Azhar MT, Hinze L, Qayyum A, Li H, Peng Z, Qin G, Jia Y, Pan Z, He S, Du X (2021) Insight into abscisic acid perception and signaling to increase plant tolerance to abiotic stress. J Plant Interact 16(1):222–237. https://doi.org/10.1080/17429145.2021.1925759

Rehman S, Bahadur S, Xia W (2023) An overview of floral regulatory genes in annual and perennial plants. Gene 885:147699. https://doi.org/10.1016/j.gene.2023.147699

Ren C, Li Z, Song P, Wang Y, Liu W, Zhang L, Li X, Li W, Han D (2023) Overexpression of a grape MYB transcription factor gene VhMYB2 increases salinity and drought tolerance in Arabidopsis thaliana. Int J Mol Sci 24(13):10743. https://doi.org/10.3390/ijms241310743

Sato H, Takasaki H, Takahashi F, Suzuki T, Iuchi S, Mitsuda N, Ohme-Takagi M, Ikeda M, Seo M, Yamaguchi-Shinozaki K, Shinozaki K (2018) Arabidopsis thaliana NGATHA1 transcription factor induces ABA biosynthesis by activating NCED3 gene during dehydration stress. Proc Natl Acad Sci USA 115(47):E11178–E11187. https://doi.org/10.1073/pnas.1811491115

Sena S, Prakash A, Van Staden J, Kumar V (2024) Epigenetic control of plant regeneration: unraveling the role of histone methylation. Curr Plant Biol 40:100408. https://doi.org/10.1016/j.cpb.2024.100408

Shao H, Wang H, Tang X (2015) NAC transcription factors in plant multiple abiotic stress responses: progress and prospects. Front Plant Sci 6:902. https://doi.org/10.3389/fpls.2015.00902

Sharma PD, Singh N, Ahuja PS, Reddy TV (2011) Abscisic acid response element binding factor 1 is required for establishment of Arabidopsis seedlings during winter. Mol Biol Rep 38(8): 5147–5159. https://doi.org/10.1007/s11033-010-0664-3

Shen Q, Fu L, Su T, Ye L, Huang L, Kuang L, Wu L, Wu D, Chen Z-H, Zhang G (2020) Calmodulin HvCaM1 negatively regulates salt tolerance via modulation of HvHKT1s and HvCAMTA4. Plant Physiol 183(4):1650–1662. https://doi.org/10.1104/pp.20.00196

Shi L, Wu Y, Sheen J (2018) TOR signaling in plants: conservation and innovation. Development 145(13):dev160887. https://doi.org/10.1242/dev.160887

Shkolnik D, Finkler A, Pasmanik-Chor M, Fromm H (2019) CALMODULIN-BINDING TRANSCRIPTION ACTIVATOR 6: a key regulator of Na+ homeostasis during Germination1 [OPEN]. Plant Physiol 180(2):1101–1118. https://doi.org/10.1104/pp.19.00119

Son S, Park SR (2023) Plant translational reprogramming for stress resilience. Front Plant Sci 14: 1151587. https://doi.org/10.3389/fpls.2023.1151587

Song C, Lee J, Kim T, Hong JC, Lim CO (2018) VOZ1, a transcriptional repressor of DREB2C, mediates heat stress responses in Arabidopsis. Planta 247(6):1439–1448. https://doi.org/10.1007/s00425-018-2879-9

Su Z, Ma X, Guo H, Sukiran NL, Guo B, Assmann SM, Ma H (2013) Flower development under drought stress: morphological and transcriptomic analyses reveal acute responses and long-term acclimation in Arabidopsis[C][W]. Plant Cell 25(10):3785–3807. https://doi.org/10.1105/tpc.113.115428

Sun Z, Shen H, Chen Z, Ma N, Yang Y, Liu H, Li J (2024) Physiological responses and transcriptome analysis of Hemerocallis citrina Baroni exposed to Thrips palmi feeding stress. Front Plant Sci 15:1361276. https://doi.org/10.3389/fpls.2024.1361276

Tahmasebi A, Ashrafi-Dehkordi E, Shahriari AG, Mazloomi SM, Ebrahimie E (2019) Integrative meta-analysis of transcriptomic responses to abiotic stress in cotton. Prog Biophys Mol Biol 146:112–122. https://doi.org/10.1016/j.pbiomolbio.2019.02.005

Vannini C, Locatelli F, Bracale M, Magnani E, Marsoni M, Osnato M, Mattana M, Baldoni E, Coraggio I (2004) Overexpression of the rice Osmyb4 gene increases chilling and freezing tolerance of Arabidopsis thaliana plants. Plant J Cell Mol Biol 37(1):115–127. https://doi.org/10.1046/j.1365-313x.2003.01938.x

Verma V, Ravindran P, Kumar PP (2016) Plant hormone-mediated regulation of stress responses. BMC Plant Biol 16(1):86. https://doi.org/10.1186/s12870-016-0771-y

Vishwakarma K, Upadhyay N, Kumar N, Yadav G, Singh J, Mishra RK, Kumar V, Verma R, Upadhyay RG, Pandey M, Sharma S (2017) Abscisic acid signaling and abiotic stress tolerance in plants: a review on current knowledge and future prospects. Front Plant Sci 8:161. https://doi.org/10.3389/fpls.2017.00161

Wang Y, Chen Z-H, Zhang B, Hills A, Blatt MR (2013) PYR/PYL/RCAR Abscisic acid receptors regulate K+ and Cl− channels through reactive oxygen species-mediated activation of Ca2+ channels at the plasma membrane of intact Arabidopsis guard Cells1[W][OPEN]. Plant Physiol 163(2):566–577. https://doi.org/10.1104/pp.113.219758

Wang R-K, Cao Z-H, Hao Y-J (2014) Overexpression of a R2R3 MYB gene MdSIMYB1 increases tolerance to multiple stresses in transgenic tobacco and apples. Physiol Plant 150(1):76–87. https://doi.org/10.1111/ppl.12069

Wang J, Li Q, Mao X, Li A, Jing R (2016) Wheat transcription factor TaAREB3 participates in drought and freezing tolerances in Arabidopsis. Int J Biol Sci 12(2):257–269. https://doi.org/10.7150/ijbs.13538

Wang C-T, Ru J-N, Liu Y-W, Yang J-F, Li M, Xu Z-S, Fu J-D (2018) The maize WRKY transcription factor ZmWRKY40 confers drought resistance in transgenic Arabidopsis. Int J Mol Sci 19(9):2580. https://doi.org/10.3390/ijms19092580

Wang H, Zhu Y, Yuan P, Song S, Dong T, Chen P, Duan Z, Jiang L, Lu L, Duan H (2021a) Response of wheat DREB transcription factor to osmotic stress based on DNA methylation. Int J Mol Sci 22(14):7670. https://doi.org/10.3390/ijms22147670

Wang X, Li Q, Xie J, Huang M, Cai J, Zhou Q, Dai T, Jiang D (2021b) Abscisic acid and jasmonic acid are involved in drought priming-induced tolerance to drought in wheat. Crop J 9(1): 120–132. https://doi.org/10.1016/j.cj.2020.06.002

Wang X, Niu Y, Zheng Y (2021c) Multiple functions of MYB transcription factors in abiotic stress responses. Int J Mol Sci 22(11):6125. https://doi.org/10.3390/ijms22116125

Wang L, Doan PPT, Chuong NN, Lee H-Y, Kim JH, Kim J (2023a) Comprehensive transcriptomic analysis of age-, dark-, and salt-induced senescence reveals underlying mechanisms and key regulators of leaf senescence in Zoysia japonica. Front Plant Sci 14:1170808. https://doi.org/10.3389/fpls.2023.1170808

Wang X, Li N, Zan T, Xu K, Gao S, Yin Y, Yao M, Wang F (2023b) Genome-wide analysis of the TIFY family and function of CaTIFY7 and CaTIFY10b under cold stress in pepper (Capsicum annuum L.). Front. Plant Sci 14:1308721. https://doi.org/10.3389/fpls.2023.1308721

Wang Y, Zhang G, Zhou H, Yin S, Li Y, Ma C, Chen P, Sun L, Hao F (2023c) GhPYL9-5D and GhPYR1-3 A positively regulate Arabidopsis and cotton responses to ABA, drought, high salinity and osmotic stress. BMC Plant Biol 23(1):310. https://doi.org/10.1186/s12870-023-04330-8

Wang X, Liu J, Wang M, Liu L, Liu X, Zhao C (2024) FERONIA controls ABA-mediated seed germination via the regulation of CARK1 kinase activity. Cell Rep 43(11):114843. https://doi.org/10.1016/j.celrep.2024.114843

Wu Y, Zhang L, Nie L, Zheng Y, Zhu S, Hou J, Li R, Chen G, Tang X, Wang C, Yuan L (2022) Genome-wide analysis of the DREB family genes and functional identification of the involvement of BrDREB2B in abiotic stress in wucai (Brassica campestris L.). BMC Genomics 23(1): 598. https://doi.org/10.1186/s12864-022-08812-1

Xu W, Tang W, Wang C, Ge L, Sun J, Qi X, He Z, Zhou Y, Chen J, Xu Z, Ma Y-Z, Chen M (2020) SiMYB56 confers drought stress tolerance in transgenic rice by regulating lignin biosynthesis and ABA signaling pathway. Front Plant Sci 11:785. https://doi.org/10.3389/fpls.2020.00785

Xu T, Patitaki E, Zioutopoulou A, Kaiserli E (2025) Light and high temperatures control epigenomic and epitranscriptomic events in *Arabidopsis*. Curr Opin Plant Biol 83:102668. https://doi.org/10.1016/j.pbi.2024.102668

Yadav A, Mathan J, Dubey AK, Singh A (2024) The emerging role of non-coding RNAs (ncRNAs) in plant growth, development, and stress response signaling. Non-Coding RNA 10(1):13. https://doi.org/10.3390/ncrna10010013

Yang J, Guo X, Mei Q, Qiu L, Chen P, Li W, Mao K, Ma F (2023) MdbHLH4 negatively regulates apple cold tolerance by inhibiting MdCBF1/3 expression and promoting MdCAX3L-2 expression. Plant Physiol 191(1):789–806. https://doi.org/10.1093/plphys/kiac512

Yu Y, Wan Y, Jiao Z, Bian L, Yu K, Zhang G, Guo D (2019) Functional characterization of resistance to powdery mildew of VvTIFY9 from Vitis vinifera. Int J Mol Sci 20(17):4286. https://doi.org/10.3390/ijms20174286

Yuan X, Wang H, Cai J, Li D, Song F (2019) NAC transcription factors in plant immunity. Phytopathol Res 1(1):3. https://doi.org/10.1186/s42483-018-0008-0

Zhang J-Z, Ai X-Y, Sun L-M, Zhang D-L, Guo W-W, Deng X-X, Hu C-G (2011) Transcriptome profile analysis of flowering molecular processes of early flowering trifoliate orange mutant and the wild-type [Poncirus trifoliata (L.) Raf.] by massively parallel signature sequencing. BMC Genomics 12:63. https://doi.org/10.1186/1471-2164-12-63

Zhang Q, Gao M, Wu L, Wang Y, Chen Y (2016) Divergent expression patterns in two Vernicia species revealed the potential role of the hub gene VmAP2/ERF036 in resistance to Fusarium oxysporum in Vernicia montana. Genes 7(12):109. https://doi.org/10.3390/genes7120109

Zhang H, Liu D, Yang B, Liu W-Z, Mu B, Song H, Chen B, Li Y, Ren D, Deng H, Jiang Y-Q (2020) Arabidopsis CPK6 positively regulates ABA signaling and drought tolerance through phosphorylating ABA-responsive element-binding factors. J Exp Bot 71(1):188–203. https://doi.org/10.1093/jxb/erz432

Zhang D, Zhu Z, Gao J, Zhou X, Zhu S, Wang X, Wang X, Ren G, Kuai B (2021a) The NPR1-WRKY46-WRKY6 signaling cascade mediates probenazole/salicylic acid-elicited leaf senescence in Arabidopsis thaliana. J Integr Plant Biol 63(5):924–936. https://doi.org/10.1111/jipb.13044

Zhang H, Zhang L, Wu S, Chen Y, Yu D, Chen L (2021b) AtWRKY75 positively regulates age-triggered leaf senescence through gibberellin pathway. Plant Divers 43(4):331–340. https://doi.org/10.1016/j.pld.2020.10.002

Zhang D, Xu F, Wang F, Le L, Pu L (2025) Synthetic biology and artificial intelligence in crop improvement. Plant Commun 6(2):101220. https://doi.org/10.1016/j.xplc.2024.101220

Zhao L, Zhang W, Song Q, Xuan Y, Li K, Cheng L, Qiao H, Wang G, Zhou C (2020) A WRKY transcription factor, TaWRKY40-D, promotes leaf senescence associated with jasmonic acid and abscisic acid pathways in wheat. Plant Biol Stuttg Ger 22(6):1072–1085. https://doi.org/10.1111/plb.13155

Zhao H, Zhang Y, Zheng Y (2022) Integration of ABA, GA, and light signaling in seed germination through the regulation of ABI5. Front Plant Sci 13:1000803. https://doi.org/10.3389/fpls.2022.1000803

Zhu J-K (2016) Abiotic stress signaling and responses in plants. Cell 167(2):313–324. https://doi.org/10.1016/j.cell.2016.08.029

Zhu Y-N, Shi D-Q, Ruan M-B, Zhang L-L, Meng Z-H, Liu J, Yang W-C (2013) Transcriptome analysis reveals crosstalk of responsive genes to multiple abiotic stresses in cotton (Gossypium hirsutum L.). PLoS One 8(11):e80218. https://doi.org/10.1371/journal.pone.0080218

Chapter 5
Transcriptomic Signatures of Somaclonal Variation

Felipe López-Hernández and Andrés J. Cortés

Abstract Somaclonal variation, arising from genetic and epigenetic modifications during clonal propagation, was once regarded as a drawback of clonality but is now recognized as a valuable source of phenotypic innovation for plant breeding. This chapter explores the transcriptomic imprints of somaclonal variation and their implications for crop improvement. Such variation may result from single-nucleotide polymorphisms (SNPs), DNA methylation, transposable element activation, and chromosomal rearrangements, all of which can reshape gene expression and influence metabolic pathways, development, and stress responses. Advances in RNA sequencing (RNA-Seq) enable the identification of differentially expressed genes (DEGs) linked to pathogen resistance, oxidative stress, and hormonal regulation, thereby providing candidate traits for breeding throughout controlled propagation. Case studies in oil palm and cucumber highlight how floral abnormalities are associated with DNA methylation changes, and how auxin signaling and oxidative stress-related genes contribute to phenotypic diversity. In addition, RNA-based markers offer a powerful approach to tracing clonal diversity by capturing dynamic gene expression changes, as exemplified in mint species. These markers, combined with marker-assisted selection (MAS), can accelerate the identification of elite clones with improved resilience, yield, and secondary metabolite production. Looking forward, the integration of omics technologies with machine learning, big data analytics, single-cell transcriptomics, long-read sequencing, and CRISPR-based tools promises new opportunities to harness somaclonal variation for precision breeding. Ultimately, transcriptomic approaches will transform this once-undesired phenomenon into a driver of agricultural resilience, sustainability, food security, and agrobiodiversity conservation.

F. López-Hernández
Corporación Colombiana de Investigación Agropecuaria – AGROSAVIA, Rionegro, Colombia

A. J. Cortés (✉)
Facultad de Ciencias Agrarias – Departamento de Ciencias Forestales, Universidad Nacional de Colombia – Sede Medellín, Medellin, Colombia
e-mail: ancortesv@unal.edu.co

L. M. Vaschetto (ed.), *Plant Transcriptomics and Epitranscriptomics*, Concepts and Strategies in Plant Sciences, https://doi.org/10.1007/978-981-95-5183-5_5

Keywords RNA Sequencing (RNA-Seq) · Differential Gene Expression · Plant Breeding · Plant Tissue Culture · Marker-Assisted Selection (MAS) · Transcriptomic Profiling · RNA-based Marker Development

Abbreviations

DEGs	Differentially expressed genes
MAS	Marker-assisted selection
PGRs	Plant growth regulators
RNA-Seq	RNA sequencing
SNPs	Single nucleotide Polymorphisms

5.1 Introduction

Somaclonal variation is an inevitable consequence of clonal propagation, in which genetic and epigenetic changes happen while regenerating plants (Macaya-Sanz et al. 2016). Although this variation has generally been seen as a barrier to clonality, it may result in novel segregants for plant breeding (Ferreira et al. 2023). The study and utilization of somaclones has further been enhanced by modern high-throughput sequencing technologies, informing on the molecular mechanisms underlying phenotypic modifications and stress responses in clonally propagated plants.

Somaclonal variation can arise from multiple factors, including single-nucleotide polymorphisms (SNPs), DNA methylation changes, transposable element activity, and chromosomal rearrangements (Ferreira et al. 2023). Such alterations can lead to transcriptomic reprogramming, influencing plant development and adaptation. For instance, in *Musa* spp., a clonally propagated fruit crop, transcriptome analysis of the somaclone 'Guijiao 9' resistant to *Fusarium oxysporum* f. sp. *cubense* TR4 revealed extensive gene expression changes linked to pathogen defense, including signal transduction pathways and secondary metabolism genes (Sun et al. 2019). In turn, *Phalaenopsis* orchids have shown differential expression of genes involved in endoreduplication and floral morphology (Lee et al. 2017). Gene expression changes in somaclonal variants can also inform how plants adapt to clonal propagation and tissue culture conditions (Sánchez-Romero 2024a), concretely, oxidative stress and hormonal imbalances.

The integration of transcriptomic data with functional genomics and epigenetics, therefore, is at the frontier of precision breeding in clonal systems prone to somaclonal variation (Linacero and Ballesteros 2024a). Specifically, transcriptomic expression analysis in somaclones can provide a deeper understanding of plant adaptation mechanisms (Brautigam et al. 2013; Chinnusamy and Zhu 2009; Kristensen et al. 2018) and genetic plasticity (Fox et al. 2019; Nicotra et al. 2015), key questions to be addressed as part of the plant breeding triangle (*i.e.*, phenotype *vs.* genotype *vs.* environment), while sourcing innovative phenotypic variation for plant improvement (Varshney et al. 2005). Selecting somaclones with desirable gene expression profiles and trait segregation fast-tracks the development and deployment of crops with improved traits such as enhanced nutritional content, disease resistance, abiotic stress tolerance, and water/nutrient use efficiency (Varshney et al. 2021a).

As the field progresses, leveraging transcriptomic tools will be crucial for optimizing somaclonal selection and harnessing its potential for more resilient and sustainable agriculture. Single-cell transcriptomics, long-read sequencing, multiomics, and machine learning approaches can further dissect the complex regulatory networks prevailing in somaclonal variation (Libbrecht and Noble 2015), a tempting perspective we will detail in the last section of this chapter.

Given these opportunities, this chapter aims to explore the transcriptomic consequences of somaclonal variation, emphasizing key regulatory networks, stress-responsive pathways, and epigenetic modifications. We examine case studies where transcriptomic profiling has been key to identifying candidate biomarkers responsible for traits under selection. We also embrace the utilization of RNA-based markers to assess transcriptomic variability and leverage plant breeding strategies.

5.2 The Role of Somaclonal Variation in Plant Breeding

Somaclonal variation was initially regarded as an obstacle to maintaining the genetic fidelity of vegetative material. However, it is now viewed as a powerful source of novelty for crop improvement (Ferreira et al. 2023) together with sexual interbreeding and homologous recombination. Spontaneous genetic and epigenetic somaclonal variations can induce trait segregation, including disease resistance, enhanced yield, and improved adaptability (Kim 2005), all crucial selection targets for modern agriculture. In this section, we discuss how somaclonal variation leverages plant breeding programs, to later on describe, in the following sections, the pivotal role of transcriptomics.

Since the dawn of domestication thousands of years ago, humanity has taken advantage, almost inadvertently, of phenotypic and somaclonal variation to select varieties (Meyer and Purugganan 2013; Purugganan 2022), some of which can better withstand pathogens, tolerate environmental stresses, and produce higher yields (Alam and Purugganan 2024; Purugganan and Jackson 2021). Propagation via cuttings (Frolich and Platt 1971; Primack et al. 2015) and grafting (Bartusch and Melnyk 2020; Fernández-Paz et al. 2021; Goldschmidt 2014; Henao et al. 2024) were perhaps the first human attempts to propagate plants clonally, more recently escalated via micropropagation, tissue culture techniques, and plant growth regulators (PGRs), all of which inadvertently induce somaclonal variation. Among the latter, the most used regulators include benzylaminopurine (BAP), dichlorophenoxyacetic acid (2,4-D), and naphthalene acetic acid (NAA), compounds that increase the likelihood of genetic and epigenetic somaclonal changes (Martignago et al. 2019) when applied in specific concentrations. Repetitive subculture may also induce genetic instability and somatic mutations, boosting somaclonal diversity during *in vitro* cultivation (Sánchez-Romero 2024a).

The discovery of novel somaclonal variants has been particularly valuable in major clonal crops such as sugarcane, banana and potato (Miyao 2024). It has also been widely applied in ornamental plants, where variations in color, shape, and growth patterns are the targets of selection for commercial markets (Ferreira et al. 2023). Specifically, sugarcane has recently benefited from somaclonal variants, as they have served to improve sucrose content and disease resistance (*e.g.*, VSI434 sugarcane somaclone is commercialized due to its superior yield and resistance to red rot disease). In banana, somaclonal variation has sourced resistance to *Fusarium* wilt, a devastating fungal disease that threatens banana plantations worldwide (Sun et al. 2019). Identifying the resistant somaclones was feasible after banana tissues were subjected to extended subculturing with growth regulators. These candidate resistant somaclones are a promising avenue to mitigate the economic impact of *Fusarium* wilt in banana plantations worldwide.

Other striking consequences of somaclonal variation include plant morphology. In some cases, somaclones may exhibit dwarfism, a trait that can be beneficial in crops like wheat, pineapple, and fruit trees (Rai et al. 2024; Sánchez-Romero 2024b), where shorter plants are less prone to damage due to wind lodging and easier to harvest (Warschefsky et al. 2016). Other alterations observed during clonality include changes in leaf pigmentation, which is particularly valuable in ornamental plants where uniquely colored or multicolored foliage may lead to higher market prices. Somaclonal variation has also led to increased grain yields, as seen in rice, sorghum, and wheat. This mosaic of clonally induced changes has been exploited during the screening phases of germplasm propagation and pre-breeding, as well as during the more downstream steps of the recurrent selection schemes.

Beyond structural and morphological changes, molecular studies confirm the genomic consequences of somaclonal variation. Recent studies have favored SNP marker discovery, which offers greater precision to trace *de novo* and standing somaclonal variation. These high-throughput molecular screening tools focus on heritable DNA somaclonal mutations as compared to those that may be more temporary, like epigenetic modifications persisting only a few generations (Boskovic and Rando 2018; Heard and Martienssen Robert 2014; Hu et al. 2018; Lacal and Ventura 2018). Some examples of heritable somaclonal variation include polymorphisms in the *PMADS4* gene and expansins, respectively linked with floral modifications and cell expansion/dwarfism, a key trait to harvest fruit tree crops (Warschefsky et al. 2016). Similarly, variation in OP J-06 plays a role in disease resistance, particularly in banana plants affected by *Fusarium* wilt (Sun et al. 2019).

While somaclonal variation has proven to be a key source of novelty for selective breeding, challenges persist. One of the primary difficulties lies in distinguishing between permanent genetic mutations and transient epigenetic changes. Epigenetic modifications, such as DNA methylation, can be reversed in subsequent offspring, making it difficult to predict whether a desirable trait selected during breeding will be inherited and stabilized throughout generations (Cañas-Gutiérrez et al. 2022; Reyes-Herrera et al. 2020). Another major challenge is somaclonal mutation load, as part of which clonally induced genetic and epigenetic mutations are deleterious, requiring screening for purifying selection to ensure beneficial mutations are selected.

The impact of somaclonal variation extends beyond agricultural productivity. Crops must adapt to progressively unpredictable environmental conditions (Cortés and López-Hernández 2021; Cortés et al. 2020a), urging the development of stress-tolerant varieties (Cortés 2024). In this context, somaclonal variation could also offer novel variation useful for abiotic stress tolerance. For instance, it has already contributed to the breeding of rice (Miyao 2024) tolerant to high salinity, as well as other crop varieties with drought tolerance (Cortés and Blair 2018; Cortés et al. 2013). As global temperatures continue rising and extreme weather events become even more common (Cortés et al. 2022b; López-Hernández and Cortés 2019), somaclonal diversity becomes a milestone for future food security, as it has already sourced adaptative variation (Kristensen et al. 2018; Nicotra et al. 2015) and plasticity in natural ecosystems (Brautigam et al. 2013; Fox et al. 2019). Another promising application of somaclonal variation is in biofuel production (Sannigrahi et al. 2010). As previously discussed, sugarcane, a major biofuel crop, already benefited from somaclonal variants with higher sucrose content, which translates into higher ethanol production (Studer et al. 2011). With global demand for renewable energy sources uprising, somaclonal variation in biofuel crops may lift sustainable energy.

In short, somaclonal variation embodies a paradigm shift in plant breeding, transforming what was once seen as an unpredictable byproduct of clonal propagation into a strategic tool for agricultural varietal advancement (Cortés and Barnaby 2023). Somaclonal variation has sourced, and will continue innovating, more high-yielding, resilient, and sustainable clonal varieties (Ferreira et al. 2023), helping to meet the modern demands of food security, climate change adaptability, and a more sustainable agri-food sector. After all, standing somaclonal variation linked with the plant domestication syndrome and the natural adaptation of crop wild relatives is already being harnessed as part of modern crop breeding schemes (Brautigam et al. 2013; Nicotra et al. 2015). Still, as molecular techniques continue evolving (Cortés et al. 2020b), the full potential of somaclonal diversification is yet to be seen (Cortés and Barnaby 2023; Cortés et al. 2022a). This is why in the next sections we depict transcriptomics as the first-line high-throughput tool to discover, trace, and utilize somaclonal genetic and epigenetic variation more efficiently (Fig. 5.1).

5.3 Transcriptomic Profiling of Somaclonal Variants

As discussed in the previous section, somaclonal variation can happen due to transcriptomic/epigenetic modifications (Boskovic and Rando 2018; Heard and Martienssen Robert 2014), genetic mutations (Barrell and Conner 2011), or both. Oxidative stress and hormone imbalance during prolonged *in vitro* culture can in

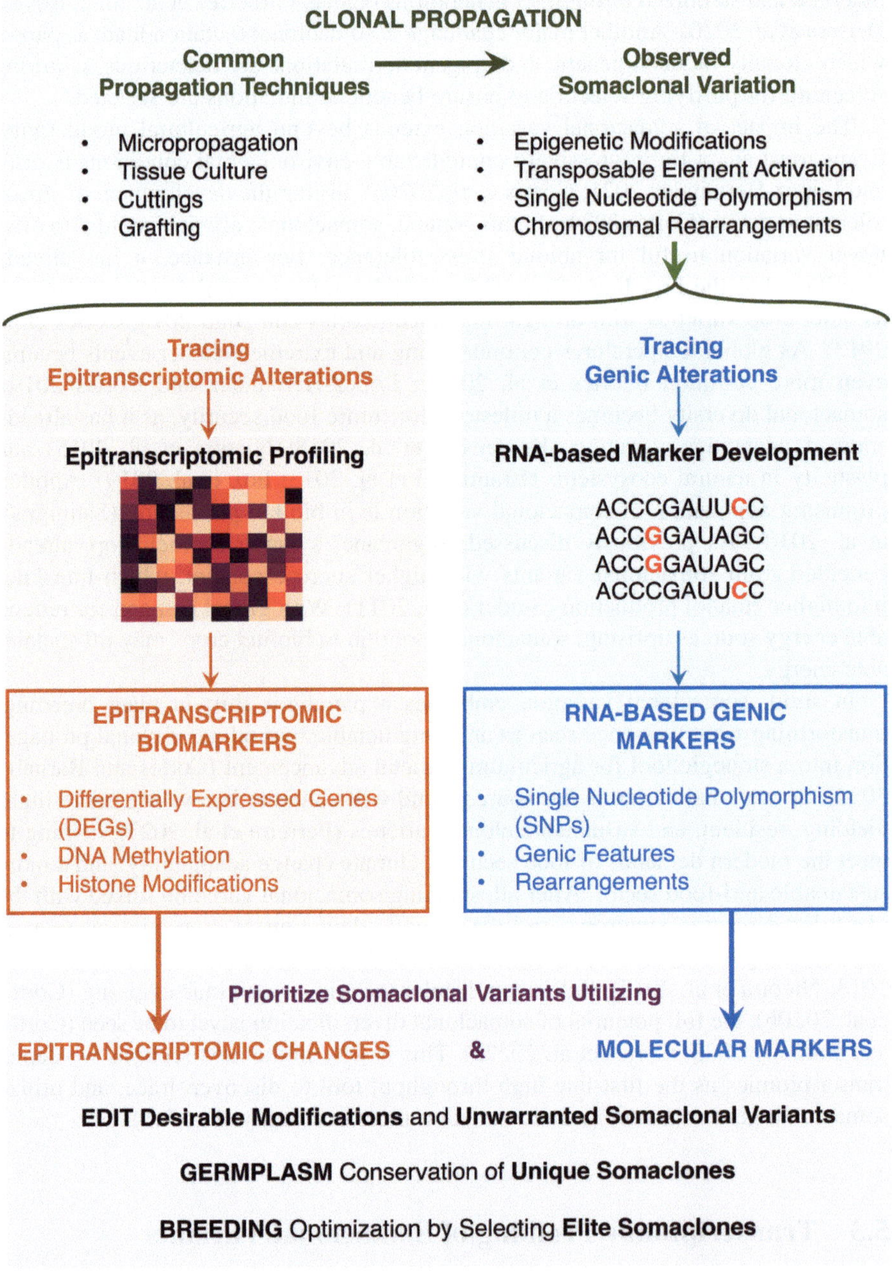

Fig. 5.1 Roadmap for a transcriptomic-based study of somaclonal variation. The diagram outlines the role of epitranscriptomic profiling and RNA-based marker development in identifying and prioritizing somaclonal variants for germplasm conservation and crop improvement. The green box highlights common propagation techniques and multiple ways somaclonal variation could arise. The roadmap emphasizes the discovery of epitranscriptomic biomarkers (orange box) and RNA-based genic markers (blue box), and their potential utilization and escalation for downstream applications (purple box) like germplasm traceability, indirect selective predictive breeding, and molecular editing

turn boost these changes (Sánchez-Romero 2024a). Tissue culture itself stresses cells (Chinnusamy and Zhu 2009), inducing DNA replication errors, chromosomal rearrangements, and epigenetic modifications including DNA methylation (Kong et al. 2022; Roudbar et al. 2020) and histone modifications. All these changes are the ultimate cause that may alter gene expression patterns and result in phenotypic variation.

Reconstructing the architecture of somaclonal variation is key because while some somaclonal variations may be responsible for detrimental traits such as physical abnormalities, decreased fertility, or altered flowering time (Beulé et al. 2006), others could enhance advantageous characteristics like higher biomass, better fruit quality, or more tolerance to infections (Ferreira et al. 2023). Transcriptomic profiling and RNA-based markers both allow identifying differentially expressed genes (DEGs) and regulatory networks that drive these variations (Linacero and Ballesteros 2024b).

As we exemplify in the next two sub-sections throughout a couple of case studies in oil palm and cucumber, transcriptomics are now disclosing the expression changes underlying clonal propagation. Specifically, sequencing and comparing RNA transcripts across diverse tissues and contrasting conditions enables pinpoint coding variation and inherited alterations during transcription in clonally propagated plants. We detail in each case pathways enriched for transcriptomic alterations after intense clonality.

5.3.1 *Case Study I: Profiling Undesired Somaclonal Variants in Oil Palm*

Oil palm is a major source of edible oil worldwide (Kwong et al. 2017) that is recently implementing micropropagation via somatic embryogenesis. However, a bottleneck for its scalation is detrimental somaclonal variants, notably the "mantled" phenotype (Beulé et al. 2006). It is characterized by floral abnormalities that result in feminized flowers and reduced fruit production, negatively impacting economic profit for the whole oil palm industry. Inferring the molecular basis of this undesired segregant (Linacero and Ballesteros 2024b) is key to ensuring the viability of clonal propagation.

Transcriptomic studies have unraveled the molecular mechanisms underlying this phenotype. Unlike conventional genetic mutations, mantled variants exhibit altered expression of genes involved in epigenetic regulation (Beulé et al. 2006). DNA methylation changes alter floral development genes, leading to homeotic transformations. Downregulation of DNA methyltransferases and DGE of chromatin-remodeling genes suggest epigenetic reprogramming is behind the mantled phenotype.

Genes involved in transcriptional regulation, hormone metabolism, and cell differentiation are differentially expressed, suggesting complex regulatory interactions driving the mantled phenotype. In particular, ethylene-related genes are affected, as ethylene signaling plays a role in stress responses (Cortés et al. 2012). Deviations in ethylene perception and signaling pathways may therefore be a proximal cause for the abnormal floral development observed in mantled oil palm.

Transcriptome profiling has also identified biomarkers to flag normal and variant clones (Loupit et al. 2022), which is crucial for screening and purging undesirable variants in oil (Cros et al. 2015) and date palm (Mirani et al. 2024) nurseries. Tracing DEGs in embryogenic callus forecasts whether a plant will be mantled, allowing for indirect selection and early removal of affected lines before transplanting (Beulé et al. 2006).

5.3.2 *Case Study II: Profiling Somaclonal Variation in Cucumber*

Just like oil palm, cucumber is a widely cultivated vegetable crop prone to somaclonal modifications during micropropagation (Davoudi et al. 2022). As part of automated hydroponic indoor vertical farming, cucumber tissue culture via somatic embryogenesis and organogenesis is becoming more common for clonal propagation. Nonetheless, such methods can introduce epi- and genetic changes altering cucumber plant development, as well as fruit yield and quality (Rai et al. 2024; Sarmast et al. 2024).

Transcriptomic analyses of somaclonal cucumber variants have once again revealed significant gene expression changes associated with cell cycle regulation, hormonal pathways, and stress responses (Pawełkowicz et al. 2021). Genes involve in auxin signaling, like AUX/IAA and ARF transcription factors, also show altered expression profiles, suggesting a disruption in hormone homeostasis. Auxins play a critical role in plant development, influencing cell division, elongation, differentiation (Busch et al. 2011), and overall resistance (Blair et al. 2016; Ghanashyam and Jain 2009). Alterations in auxin-related gene expression may therefore be responsible for stem elongation, abnormal leaf shape, or altered root architecture in somaclones (Lopez-Marin et al. 2009; Uga et al. 2013). Among DEGs involved in oxidative stress responses, the upregulation of superoxide dismutase and catalase genes are indicative of an adaptive response to oxidative damage – after all, accumulation of reactive oxygen species (ROS) in tissue culture can cause DNA damage and alter cellular homeostasis, risking regeneration (Barrell and Conner 2011).

Moreover, differential expression of DNA methyltransferases and histone-modifying enzymes are indicative of epigenetic alterations responsible for somaclonal diversity in cucumber (Chinnusamy and Zhu 2009). Epigenetic modifications can reshape chromatin structure and gene expression patterns, making heritable changes for a few generations without DNA sequence alterations (Boskovic and Rando 2018; Heard and Martienssen Robert 2014).

Overall, despite differences in generation time, mating systems, genetic backgrounds and propagation methods, oil palm and cucumber exhibit convergent transcriptomic patterns of somaclonal variation. For instance, in both examples, transcriptomic analyses unveil genetic and epigenetic candidates that can be targeted to reduce undesirable somaclonal variants, improving the efficiency of micro-cloning, *in vitro* propagation and hydroponic systems (Chinnusamy and Zhu 2009). These crops showed altered expression of hormone-related genes, particularly those associated with auxin and ethylene signaling pathways. Hormonal dysregulation was also linked to developmental abnormalities, while the upregulation of oxidative stress-related genes suggests a conserved response to tissue culture-induced stress. Strategies to control these undesired somaclonal alterations in microculture include reducing oxidative stress and optimizing hormone levels to maintain a stable hormonal balance. Interestingly, genes associated with DNA methylation and histone modification were also differentially expressed, reinforcing the role of epigenetic alterations in somaclonal variation. However, the extent to which optimized culture conditions can mitigate stress and epigenetic instability in regenerated plants remains to be validated and escalated (Sánchez-Romero 2024a).

5.4 RNA-Based Marker Development for Somaclonal Variation Tracing

RNA-based markers have been employed to identify and characterize clonal variation across germplasm and ramet generations in key traits like flower number and symmetry, self-fertility, and floral initiation (Ally et al. 2010), offering crucial insights into gene expression changes and mutational events at the transcriptomic level (Martín and González-Benito 2024). Development of such markers enables the detection of functional variations that influence plant phenotype, stress responses, and adaptation. In this section, we explore the role of RNA-Seq in marker development and its uses for germplasm conservation programs in clonal plants and pre-breeding.

As we discussed in the previous section, the causes of somaclonal variation span from transcript regulation, DNA methylation changes, transposable element activation (Martin et al. 2009; Slotkin and Martienssen 2007), but also point mutations and chromosomal rearrangements (Ferreira et al. 2023). Having exemplified in the two previous case studies the first type of regulatory modifications (*i.e.*, those that affect gene expression without altering the DNA sequence), now we focus on the latter, codifying and structural modifications (*i.e.*, genetic changes that are heritable sequence alterations). Among these, despite some may also lead to beneficial or detrimental traits, most will remain neutral (Nei 2010). Yet, only the first type of coding variations can be truly harnessed as part of modern plant selection schemes to develop new cultivars with enhanced traits, including but not limited to higher productivity, disease resistance, abiotic stress tolerance, and water and nutrient use efficiency.

Because RNA-based markers are diverse in their origin and consequences, major challenges still persist to fully utilize them, including the need to better disentangle stable (*i.e.*, heritable) and advantageous variants from those that are transitory, neutral (a.k.a., non-informative), or detrimental. Once this gap of confounding effects is bridged, RNA-based markers will enable us to: *(i)* understand how clonal variations manifest at the molecular level, and *(ii)* translate such variation during somaclonal divergence into applicable biomarkers for indirect selection (Cortés and Du 2023) and fast breeding (Varshney et al. 2021b). In the future, though RNA-based markers focus on coding mutations, advancements in epigenetic marker development may offer an expanded pool of markers (Kapazoglou et al. 2021).

5.4.1 RNA-Seq and Marker Discovery

RNA sequencing (RNA-Seq) has transformed the study of plant transcriptomes by offering unimaginable high-throughput data for gene expression analysis, inference of alternative splicing events, and genic single-nucleotide polymorphism (SNP) discovery (Badenes et al. 2016). Unlike DNA-based markers, RNA-based markers capture dynamic gene expression changes influenced by both epigenetic modifications and genetic (*i.e.*, coding). This makes them particularly useful for identifying and tracing functional somaclonal variation (López-Hernández and Cortés 2022).

Specifically, RNA-Seq offers a fast track to pinpoint SNPs in expressed genes, tracking variation in coding regions that may ultimately affect protein structure and function. Until some years back, simple sequence repeats (SSRs), particularly those expressed in transcribed regions (EST-SSRs), served as co-dominant markers for gene mapping and functional analysis (Alcaide et al. 2019). Nowadays, the role of these antique gene-based markers has been fully replenished by RNA-Seq technologies.

Regardless of the method of screening, one of the key advantages of RNA-based markers is their ability to identify functional coding variants in real-time, reflecting how plants respond to intensive clonal propagation without having to rely on massive linkage disequilibrium between the putative neutral non-genic marker and the candidate causal variant. Focusing on coding variation is particularly relevant for tracking variations that influence quantitative polygenic traits such as growth rate, stress responses, and secondary metabolite production in highly heterozygous allogamous plants with giant genomes (Henao-Rojas et al. 2022; López-Hernández and Cortés 2022). In crops where clonal uniformity is the main target, RNA-based markers also offer a rapid tool for target screening and selection of the most desirable clones, while minimizing undesirable trait segregation and over-saturation with non-informative markers (a likely scenario with conventional DNA marker discovery technologies, which enrich non-genic regions where the likelihood of stabilized substitutions at the population level is higher due to more frequency and lack of early purifying selection on detrimental genic mutations).

Meanwhile, the future of RNA-based marker development is soon going to expand beyond RNA-Seq to biomarkers capable of pinpointing key regulatory pathways affected during clonal propagation. For instance, DEGs between somaclonal variants and control plants may work as candidate biomarkers for early and more precise indirect selection. Overall, these advances in RNA-based biomarkers offer a major tool to monitor and harness somaclonal variation for more precise and sustainable crop improvement.

5.4.2 Leveraging RNA-Based Markers in Plant Breeding

Incorporating RNA-based markers into breeding programs will boost the early indirect selection of superior clones with desirable trait variation such as improved yield, disease resistance, and stress tolerance. Marker-assisted selection (MAS) using RNA markers is already enabling the rapid screening of plant populations for functional genetic diversity (Varshney et al. 2005). Transcriptome-wide association studies (eGWAS) that link RNA-based markers (eQTLs) to important agronomic traits (Sun 2012) are also making genome-wide prediction models and precision breeding more robust.

Moreover, the ability to monitor somaclonal diversity at the RNA level would benefit from extending marker development from coding variation to regulatory alterations. This would enable germplasm specialists and pre-breeders to detect early changes in plant development, leading to more efficient selection of elite clones. In the context of conservation, RNA-based markers can as well assist in preserving genetic diversity by identifying stable clones with unique expression profiles, which may contribute to transgenerational inherited adaptive traits in changing environmental conditions (Chinnusamy and Zhu 2009; Fox et al. 2019; Hu et al. 2018).

We should also acknowledge that the use of RNA-based markers in plant breeding extends beyond traditional crops and yield/adaptation-centered selection targets to medicinal and aromatic plants, where complex polygenic secondary metabolite production plays a significant commercial role, and is likely to be altered during intense clonal propagation (López-Hernández and Cortés 2022). Somaclonal variants with altered gene expression patterns can led to improved cultivars with higher content of bioactive compounds. For example, in crops such as mint, basil, and ginseng, RNA-Seq has revealed variations in genes associated with essential oil biosynthesis, paving the way for targeted RNA-based marker-assisted breeding strategies (Henao-Rojas et al. 2022). This is precisely the focus of the third case study, which we will detail in the next section. It complements the two industry- and horticulture-oriented oil palm and cucumber examples that were previously discussed. Back then, we focused on tracing detrimental transcriptional and epigenetic changes that occurred during clonal micropropagation. Now we discuss the utility of RNA-based markers in characterizing and conserving an aromatic plant crucial for the agro-industrial production of secondary metabolites.

5.4.3 Case Study III: RNA-Based Markers for Tracing Mint Somaclones

Recent studies on somaclonal variation in mint (*Mentha* spp.) using RNA-Seq serve us as a case study to illustrate the potential of RNA-based markers in clonally propagated aromatic crops (López-Hernández and Cortés 2022). In this work, whole transcriptome sequencing unveiled SNPs and differential gene expression patterns associated with secondary metabolite production. Discovering genic markers capable of tracing somaclonal variants with enhanced essential oil profiles will allow the industry to select improved cultivars with higher content of nutrients (Wu et al. 2022) and bioactive compounds (Henao-Rojas et al. 2022).

Besides its economic importance, mint is also an excellent model for studying somaclonal variation due to its widespread worldwide clonal propagation (Vining et al. 2020). Therefore, RNA-Seq has also provided new insights into how genic variation changes during clonal propagation in key metabolic pathways, such as those involved in menthol and carvone biosynthesis (López-Hernández and Cortés 2022). The RNA-based markers within these pathways, despite candidates to track variation affecting essential oil composition, are yet to be functionally validated before scaling them to marker-assisted selection (Cortés et al. 2020b) and conservation (Arenas et al. 2021) platforms.

Mint's adaptability to various environmental conditions also makes it a promising candidate for exploring how somaclonal expression contributes to stress tolerance (Vining et al. 2020) via transgenerational epigenetic inheritance (Hu et al. 2018). RNA-based biomarkers will then broaden the potential avenues for developing climate-resilient cultivars. Overall, these findings highlight broader implications of RNA-Seq in characterizing and utilizing somaclonal variation across diverse clonal crops.

5.5 Conclusions

Transcriptomic expression analysis of somaclonal variants deepens our comprehension of plant adaptation mechanisms and genetic plasticity (Brautigam et al. 2013; Fox et al. 2019; Kristensen et al. 2018). In turn, as plant transcriptomics advances, leveraging multi-omics will be vital to optimize the selection of somaclones and harness their potential for crop improvement (Cortés and López-Hernández 2021). So far, transcriptomic profiling has provided valuable insights into the molecular basis of somaclonal changes in oil palm (Beulé et al. 2006) and cucumber (Davoudi et al. 2022; Pawełkowicz et al. 2021). Utilizing RNA-Seq has further enabled the development of modern strategies to characterize somaclonal variation during propagation and trace it for plant improvement (Badenes et al. 2016). We envision that merging multi-omics layers will further enhance transcript-based applications in agriculture, contributing to the sustainable production of high-yield better-adapted crops (Cortés et al. 2023).

In other words, RNA-based markers hold great promise for characterizing somaclonal variation, offering mechanistic insights into gene expression dynamics and functional diversity. Pre-breeding programs can now utilize RNA-Seq technology to screen somaclonal variation, enhancing crop improvement, germplasm integrity, and conservation strategies (López-Hernández and Cortés 2022). As sequencing technologies evolve, RNA-based markers will continue playing a pivotal role in crop improvement, enabling precision breeding of conventional and novel selection targets. The integration of RNA-seq with emerging genomic tools will further expand our understanding of somaclonal variability and plant adaptation, ultimately contributing to global food security and agrobiodiversity conservation.

5.6 Perspectives

Through this chapter, we have briefed on frequent transcriptomic and epigenetic changes inherent to intense clonal propagation, some of which are responsible for somaclonal variation. We have also discussed the potential of RNA-based markers to identify, trace, and select somaclonal variants. Still, there are massive opportunities to further unveil and utilize the transcriptomic and regulatory signatures of somaclonal variation. After all, the integration of transcriptomic data with functional genomics and epigenetics (Boskovic and Rando 2018) is the perfect framework to implement last-generation approaches from the omics and big data disciplines (Libbrecht and Noble 2015), boosting fast and precision breeding (Crossa et al. 2025). For instance, we envision that pre-breeding phases could more rapidly and precisely select somaclones with desirable gene expression profiles, turbocharging the development of crops with improved traits such as enhanced nutritional content, disease resistance, and abiotic stress tolerance (Benitez-Alfonso et al. 2023). Screening traditionally clonally propagated crops for transcriptomic and epigenetic alterations will ultimately enable to pinpoint the exact regulatory mechanisms responsible for desirable trait variation, allowing for more targeted breeding strategies.

The identification of transcriptomic markers in oil palm (Beulé et al. 2006) and cucumber (Davoudi et al. 2022; Pawełkowicz et al. 2021) served us to illustrate, as case studies, how detecting somaclonal variation at early stages could minimize undesired consequences from clonal propagation. In turn, RNA-based marker development in mint exemplified a cost- and time-efficient method to trace somaclonal variants across diverse germplasm (López-Hernández and Cortés 2022). Yet, despite their advantages, RNA-based markers present several challenges, including dynamic, environmentally influenced gene expression, which requires functional validation across diverse conditions. Meanwhile, analyzing RNA-Seq data also demands more bioinformatic tools and computational resources capable of disentangling heritable from non-heritable variants (Boskovic and Rando 2018; Heard and Martienssen Robert 2014; Lacal and Ventura 2018).

Continue modernizing high-throughput sequencing technologies with machine learning approaches (Libbrecht and Noble 2015) will further enhance the forecasting, traceability, and utilization of somaclonal variants by integrating multiple omics layers (Cortés et al. 2023). Specifically, artificial intelligence (AI) algorithms can analyze RNA-Seq data and distinguish hidden patterns of somaclonal variation from those of tissue-specific DEGs and paralogy, improving the overall accuracy of RNA-based marker discovery and the downstream development of multi-omics predictive models of complex polygenic plant quantitative phenotypes. Combining transcriptomics with genomics, proteomics, and metabolomics throughout AI will ultimately offer a more comprehensive understanding and usability of somaclonal variants.

While the future of somaclonal variation in plant breeding is yet to be fully unleashed, especially when assisted with transcriptomics, big data, and AI, technological innovations on single-cell transcriptomics and long-read sequencing will continue refining the detection of transcriptomic variation, helping to dissect complex regulatory omnigenic networks (Boyle et al. 2017) underlying somaclonal diversity. Furthermore, applying systems biology techniques, including network analysis and pathway mathematical modeling, could help identify emergent properties of somaclonal transcriptomes, enabling the prediction of unforeseen phenotypic outcomes during clonal propagation (Carrera 2024).

Meanwhile, the rapid evolution of CRISPR-Cas and other gene-editing technologies (Doudna and Charpentier 2014) could also boost the utilization of somaclones by providing efficient strategies to stabilize beneficial somaclonal mutations and purify detrimental ones (Barrero et al. 2022). Specifically, selectively adjusting DNA methylation patterns (Kong et al. 2022; Roudbar et al. 2020) or histone modifications (Dunemann et al. 2022; Roca Paixão et al. 2019) could rapidly lead to first-generation stable clones with more predictable phenotypes. Similarly, optimizing hormone levels and mitigating oxidative stress during tissue culture could control the incidence of deleterious somaclonal variation, improving the reliability of micropropagation techniques. The application of CRISPR-based genome editing, and better refined clonal propagation techniques (Meyer et al. 2022), may both allow for targeted desirable modifications while controlling unwarranted variants, further expanding the utilization of transcriptomics and epigenetics in precision breeding (Barrera-Redondo et al. 2020).

Acknowledgments F.L.-H. and A.J.C. appreciate the editor's invitation to contribute to the book "Plant Transcriptomics and Epitranscriptomics – Decoding the RNA Landscape." They also appreciate his clarity and effort while providing detailed, much-valued feedback on the contents of this manuscript. A.J.C. thanks M.T. Vera, R.I.C.-Garnica, and M.J. Torres for support while drafting this chapter. This manuscript was stylistically refined with the assistance of the artificial intelligence (AI) tool Grammarly (Grammarly Inc., 2009). After its implementation, every section was carefully reviewed, revised, and approved by the authors to ensure the text's appropriateness concerning content and scientific correctness, while maintaining the integrity of the original content.

Author Contribution F.L.-H. and A.J.C. conceived the ideas drafted in this chapter. Both authors wrote the chapter and approved its final publication.

Conflict of Interest Disclosure The authors declare no conflict of interest.

Funding Information A.J.C. appreciates funding from Vetenskapsrådet (VR) through grants 2016-00418 and 2022-04411. Funding from the British Council (grant 527023146) and Kungliga Vetenskapsakademien (grant BS20170036) is also acknowledged.

References

Alam O, Purugganan MD (2024) Domestication and the evolution of crops: variable syndromes, complex genetic architectures, and ecological entanglements. Plant Cell 36(5):1227–1241. https://doi.org/10.1093/plcell/koae013

Alcaide F, Solla A, Mattioni C, Castellana S, Martín MA (2019) Adaptive diversity and drought tolerance in Castanea sativa assessed through genic markers EST-SSR. Forestry 92:287–296

Ally D, Ritland K, Otto SP (2010) Aging in a long-lived clonal tree. PLoS Biol 8:e1000454

Arenas S, Cortés AJ, Mastretta-Yanes A, Jaramillo-Correa JP (2021) Evaluating the accuracy of genomic prediction for the management and conservation of relictual natural tree populations. Tree Genet Genomes 17(1). https://doi.org/10.1007/s11295-020-01489-1

Badenes ML, Fernandez IMA, Rios G, Rubio-Cabetas MJ (2016) Application of genomic technologies to the breeding of trees. Front Genet 7:198

Barrell P, Conner A (2011) Facilitating the recovery of phenotypically normal transgenic lines in clonal crops: a new strategy illustrated in potato. Theor Appl Genet 122:1171–1177

Barrera-Redondo J, Pinero D, Eguiarte LE (2020) Genomic, transcriptomic and epigenomic tools to study the domestication of plants and animals: a field guide for beginners. Front Genet 11:742

Barrero LS, Willmann MR, Craft EJ, Akther KM, Harrington SE, Garzon-Martinez GA, Glahn RP, Piñeros MA, McCouch SR (2022) Identifying genes associated with abiotic stress tolerance suitable for CRISPR/Cas9 editing in upland rice cultivars adapted to acid soils. Plant Direct 6: e469

Bartusch K, Melnyk CW (2020) Insights into plant surgery: an overview of the multiple grafting techniques for Arabidopsis thaliana. Front Plant Sci 11:613442

Benitez-Alfonso Y, Soanes BK, Zimba S, Sinanaj B, German L, Sharma V, Bohra A, Kolesnikova A, Dunn JA, Martin AC, Khashi URM, Saati-Santamaria Z, Garcia-Fraile P, Ferreira EA, Frazao LA, Cowling WA, Siddique KHM, Pandey MK, Farooq M, Varshney RK, Chapman MA, Boesch C, Daszkowska-Golec A, Foyer CH (2023) Enhancing climate change resilience in agricultural crops. Curr Biol 33:R1246–R1261

Beulé T, Camps C, Dussert S, Jouannic S, Morcillo F, Tregear J (2006) A transcriptome approach to study epigenetic somaclonal variation in oil palm (Elaeis guineensis Jacq.) FESB congress Federation of European Societies of Plant Biology XV:17–21

Blair MW, Cortés AJ, This D (2016) Identification of an ERECTA gene and its drought adaptation associations with wild and cultivated common bean. Plant Sci 242:250–259

Boskovic A, Rando OJ (2018) Transgenerational epigenetic inheritance. Annu Rev Genet 52:21–41

Boyle EA, Li YI, Pritchard JK (2017) An expanded view of complex traits: from polygenic to omnigenic. Cell 169:1177–1186

Brautigam K, Vining KJ, Lafon-Placette C, Fossdal CG, Mirouze M, Marcos JG, Fluch S, Fraga MF, Guevara MA, Abarca D, Johnsen O, Maury S, Strauss SH, Campbell MM, Rohde A, Diaz-Sala C, Cervera MT (2013) Epigenetic regulation of adaptive responses of forest tree species to the environment. Ecol Evol 3:399–415

Busch W, Miotk A, Ariel FD, Zhao Z, Forner J, Daum G, Suzaki T, Schuster C, Schultheiss SJ, Leibfried A, Haubeiß S, Ha N, Chan RL, Lohmann JU (2011) transcriptional control of a plant stem cell Niche. Dev Cell 18:841–853

Cañas-Gutiérrez GP, Sepulveda-Ortega S, López-Hernández F, Navas-Arboleda AA, Cortés AJ (2022) Inheritance of yield components and morphological traits in Avocado cv. Hass from "Criollo" "Elite Trees" via Half-Sib seedling rootstocks. Front Plant Sci 13:843099

Carrera F (2024) Analysis of somaclonal variation by metabolite profiling. In: Somaclonal variation: basic and practical aspects, Springer Nature, Switzerland, pp 83–97

Chinnusamy V, Zhu JK (2009) Epigenetic regulation of stress responses in plants. Curr Opin Plant Biol 12:133–139

Cortés AJ (2024) Abiotic stress tolerance boosted by genetic diversity in plants. Int J Mol Sci 25(10):5367

Cortés AJ, Barnaby JY (2023) Editorial: harnessing genebanks: high-throughput phenotyping and genotyping of crop wild relatives and landraces. Front Plant Sci 14:1149469

Cortés AJ, Blair MW (2018) Genotyping by sequencing and genome—environment associations in wild common bean predict widespread divergent adaptation to drought. Front Plant Sci 9:128

Cortés AJ, Du H (2023) Molecular genetics enhances plant breeding. Int J Mol Sci 24:9977

Cortés AJ, López-Hernández F (2021) Harnessing crop wild diversity for climate change adaptation. Gene 12:783

Cortés AJ, This D, Chavarro C, Madriñán S, Blair MW (2012) Nucleotide diversity patterns at the drought-related DREB2 encoding genes in wild and cultivated common bean (Phaseolus vulgaris L.). Theor Appl Genet 125:1069–1085

Cortés AJ, Monserrate F, Ramírez-Villegas J, Madriñán S, Blair MW (2013) Drought tolerance in wild plant populations: the case of common beans (Phaseolus vulgaris L.). PLoS One 8:e62898

Cortés AJ, López-Hernández F, Osorio-Rodriguez D (2020a) Predicting thermal adaptation by looking into populations' genomic past. Front Genet 11:564515

Cortés AJ, Restrepo-Montoya M, Bedoya-Canas LE (2020b) Modern strategies to assess and breed forest tree adaptation to changing climate. Front Plant Sci 11:583323

Cortés AJ, Cornille A, Yockteng R (2022a) Evolutionary genetics of crop-wild complexes. Genes 13:1

Cortés AJ, López-Hernández F, Blair MW (2022b) Genome–environment associations, an innovative tool for studying heritable evolutionary adaptation in orphan crops and wild relatives. Front Genet 13:910386

Cortés AJ, Castillejo MÁ, Yockteng R (2023) 'Omics' approaches for crop improvement. Agronomy 13:1401

Cros D, Denis M, Sánchez L, Cochard B, Flori A, Durand-Gasselin T, Nouy B, Omoré A, Pomies V, Riou V, Suryana E, Bouvet JM (2015) Genomic selection prediction accuracy in a perennial crop: case study of oil palm (Elaeis guineensis Jacq.). Theor Appl Genet 128:397–410

Crossa J, Martini JWR, Vitale P, Perez-Rodriguez P, Costa-Neto G, Fritsche-Neto R, Runcie D, Cuevas J, Toledo F, Li H, De Vita P, Gerard G, Dreisigacker S, Crespo-Herrera L, Saint Pierre C, Bentley A, Lillemo M, Ortiz R, Montesinos-Lopez OA, Montesinos-Lopez A (2025) Expanding genomic prediction in plant breeding: harnessing big data, machine learning, and advanced software. Trends Plant Sci 30(7):756–774

Davoudi M, Song M, Zhang M, Chen J, Lou Q (2022) Long-distance control of pumpkin rootstock over cucumber scion under drought stress as revealed by transcriptome sequencing and mobile mRNAs identifications. Hortic Res 9:uhab033

Doudna JA, Charpentier E (2014) Genome editing. The new frontier of genome engineering with CRISPR-Cas9. Science 346:1258096

Dunemann F, Krüger A, Maier K, Struckmeyer S (2022) Insights from CRISPR/Cas9-mediated gene editing of centromeric histone H3 (CENH3) in carrot (Daucus carota subsp. sativus). bioRxiv 22:508489

Fernández-Paz J, Cortés AJ, Hernández-Varela CA, Mejía-de-Tafur MS, Rodriguez-Medina C, Baligar VC (2021) Rootstock-mediated genetic variance in Cadmium uptake by Juvenile Cacao (Theobroma cacao L.) genotypes, and its effect on growth and physiology. Front. Plant Sci 12: 777842

Ferreira MS, Rocha AJ, Nascimento FS, Oliveira WDS, Soares JMS, Rebouças TA, Morais Lino LS, Haddad F, Ferreira CF, Santos-Serejo JA, Fernández JS, Amorim EP (2023) The role of somaclonal variation in plant genetic improvement: a systematic review. Agronomy 13:730

Fox RJ, Donelson JM, Schunter C, Ravasi T, Gaitan-Espitia JD (2019) Beyond buying time: the role of plasticity in phenotypic adaptation to rapid environmental change. Philos Trans R Soc Lond Ser B Biol Sci 374:20180174

Frolich EF, Platt RG (1971) Use of the etiolation technique in rooting avocado cuttings. California Avocado Society 55:97–109

Ghanashyam C, Jain M (2009) Role of auxin-responsive genes in biotic stress responses. Plant Signal Behav 4:846–848

Goldschmidt EE (2014) Plant grafting: new mechanisms, evolutionary implications. Front Plant Sci 5:727

Heard E, Martienssen Robert A (2014) Transgenerational epigenetic inheritance: myths and mechanisms. Cell 157:95–109

Henao AM, Osorio T, Urrea AI (2024) Tissue culture of Cacao (Theobroma cacao L.): between somaclonal variation and mass propagation. In: Somaclonal variation: basic and practical aspects, Springer Nature, Switzerland, pp 143–168

Henao-Rojas JC, Osorio E, Isaza S, Madronero-Solarte IA, Sierra K, Zapata-Vahos IC, Betancur-Pérez JF, Arboleda-Valencia JW (2022) Towards bioprospection of commercial materials of Mentha spicata L. Using a combined strategy of metabolomics and biological activity analyses. Molecules 27(11). Springer Nature, Switzerland:3559

Hu L, Xiao P, Jiang Y, Dong M, Chen Z, Li H, Hu Z, Lei A, Wang J (2018) Transgenerational epigenetic inheritance under environmental stress by genome-wide DNA Methylation profiling in Cyanobacterium. Front Microbiol 9:1479

Kapazoglou A, Tani E, Avramidou EV, Abraham EM, Gerakari M, Megariti S, Doupis G, Doulis AG (2021) Epigenetic changes and transcriptional reprogramming upon woody plant grafting for crop sustainability in a changing environment. Front Plant Sci 11:613004

Kim Y (2005) Adaptation in Sexuals vs. Asexuals: clonal interference and the Fisher-Muller model. Genetics 171:1377–1386

Kong C, Su H, Deng S, Ji J, Wang Y, Zhang Y, Yang L, Fang Z, Lv H (2022) Global DNA Methylation and mRNA-miRNA variations activated by heat shock boost early Microspore Embryogenesis in Cabbage (Brassica oleracea). Int J Mol Sci 23:5147

Kristensen TN, Ketola T, Kronholm I (2018) Adaptation to environmental stress at different timescales. Ann N Y Acad Sci 1476(1):5–22

Kwong QB, Ong AL, Teh CK, Chew FT, Tammi M, Mayes S, Kulaveerasingam H, Yeoh SH, Harikrishna JA, Appleton DR (2017) Genomic selection in commercial perennial crops: applicability and improvement in oil palm (Elaeis guineensis Jacq.). Sci Rep 7. https://doi.org/10.1038/s41598-017-02602-6

Lacal I, Ventura R (2018) Epigenetic inheritance: concepts, mechanisms and perspectives. Front Mol Neurosci 11:292

Lee H-J, Yoon Y-J, Paek K-Y, Park S-Y (2017) Endoreduplication and gene expression in somaclonal variants of clonally propagated Phalaenopsis 'Wedding Promenade'. Hortic Environ Biotechnol 58:85–92

Libbrecht MW, Noble WS (2015) Machine learning applications in genetics and genomics. Nat Rev Genet 16:321–332

Linacero R, Ballesteros I (2024a) Epigenetic basis of somaclonal variation. In: Somaclonal variation: basic and practical aspects, Springer Nature, Switzerland, pp 21–35

Linacero R, Ballesteros I (2024b) Genetic basis of somaclonal variation. In: Somaclonal variation: basic and practical aspects, pp 1–20

López-Hernández F, Cortés AJ (2019) Last-generation genome–environment associations reveal the genetic basis of heat tolerance in common bean (Phaseolus vulgaris L.). Front Genet 10:22

López-Hernández F, Cortés AJ (2022) Whole transcriptome sequencing unveils the genomic determinants of putative somaclonal variation in Mint (Mentha L.). Int J Mol Sci 23:5291

Lopez-Marin HD, Rao IM, Blair MW (2009) Quantitative trait loci for root morphology traits under aluminum stress in common bean (Phaseolus vulgaris L.). Theor Appl Genet 119:449–458

Loupit G, Fonayet JV, Prigent S, Prodhomme D, Spilmont AS, Hilbert G, Franc C, De Revel G, Richard T, Ollat N, Cookson SJ (2022) Identifying early metabolite markers of successful graft union formation in grapevine. Hortic Res 9:uhab070

Macaya-Sanz D, Heuertz M, Lindtke D, Vendramin GG, Lexer C, Gonzalez-Martinez SC (2016) Causes and consequences of large clonal assemblies in a poplar hybrid zone. Mol Ecol 25:5330–5344

Martignago D, Rico-Medina A, Blasco-Escamez D, Fontanet-Manzaneque JB, Cano-Delgado AI (2019) Drought resistance by engineering plant tissue-specific responses. Front Plant Sci 10: 1676

Martín C, González-Benito ME (2024) Molecular markers for the detection and analysis of somaclonal variation. In: Somaclonal variation: basic and practical aspects, Springer Nature, Switzerland, pp 57–82

Martin A, Troadec C, Boualem A, Rajab M, Fernandez R, Morin H, Pitrat M, Dogimont C, Bendahmane A (2009) A transposon-induced epigenetic change leads to sex determination in melon. Nature 461:1135–1138

Meyer RS, Purugganan MD (2013) Evolution of crop species: genetics of domestication and diversification. Nat Rev Genet 14:840–852

Meyer CM, Goldman IL, Grzebelus E, Krysan PJ (2022) Efficient production of transgene-free, gene-edited carrot plants via protoplast transformation. Plant Cell Rep 41:947–960

Mirani AA, Harikrishna JA, Teo CH, Solangi N (2024) Factors influencing somaclonal variation in date palm, detection and selection for application in the plantation. In: Somaclonal variation: basic and practical aspects, Springer Nature, Switzerland, pp 169–197

Miyao A (2024) Somaclonal variation in rice: retrotransposon Tos17 and structural variation with NGS analysis. In: Somaclonal variation: basic and practical aspects, Springer Nature, Switzerland, pp 217–238

Nei M (2010) The neutral theory of molecular evolution in the genomic era. Annu Rev Genomics Hum Genet 11:265–289

Nicotra AB, Segal DL, Hoyle GL, Schrey AW, Verhoeven KJ, Richards CL (2015) Adaptive plasticity and epigenetic variation in response to warming in an Alpine plant. Ecol Evol 5:634–647

Pawełkowicz ME, Skarzyńska A, Mróz T, Bystrzycki E, Pląder W (2021) Molecular insight into somaclonal variation phenomena from transcriptome profiling of cucumber (Cucumis sativus L.) lines. Plant Cell Tissue Org Cult 145:239–259

Primack RB, Laube J, Gallinat AS, Menzel A (2015) From observations to experiments in phenology research: investigating climate change impacts on trees and shrubs using dormant twigs. Ann Bot 116:889–897

Purugganan MD (2022) What is domestication? Trends Ecol Evol 37:663–671

Purugganan MD, Jackson SA (2021) Advancing crop genomics from lab to field. Nat Genet 53(5): 595–601

Rai MK, Rathour R, Yadav S, Singh A, Kaushik S (2024) Somaclonal variation in fruit crop improvement. In: Somaclonal variation: basic and practical aspects, Springer Nature, Switzerland, pp 99–121

Reyes-Herrera PH, Muñoz-Baena L, Velásquez-Zapata V, Patiño L, Delgado-Paz OA, Díaz-Diez CA, Navas-Arboleda AA, Cortés AJ (2020) Inheritance of rootstock effects in avocado (Persea americana Mill.) cv. Hass Front Plant Sci 11:555071

Roca Paixão JF, Gillet FX, Ribeiro TP, Bournaud C, Lourenço-Tessutti IT, Noriega DD, Melo BP, de Almeida-Engler J, Grossi-de-Sa MF (2019) Improved drought stress tolerance in Arabidopsis by CRISPR/dCas9 fusion with a Histone AcetylTransferase. Sci Rep 9:8080

Roudbar MA, Momen M, Mousavi SF, Ardestani SS, Lopes FB, Gianola D, Khatib H (2020) Genome-wide methylation prediction of biological age using reproducing kernel Hilbert spaces and Bayesian ridge regressions. bioRxiv

Sánchez-Romero C (2024a) Factors affecting tissue culture-induced variations. In: Somaclonal variation: basic and practical aspects, Springer Nature, Switzerland, pp 37–56

Sánchez-Romero C (2024b) Somaclonal variation in olive. In: Somaclonal variation: basic and practical aspects, pp 199–216

Sannigrahi P, Ragauskas AJ, Tuskan GA (2010) Poplar as a feedstock for biofuels: a review of compositional characteristics. Biofuels Bioprod Biorefin 4:209–226

Sarmast MK, Ghaleh ZR, Alizadeh M (2024) Somaclonal variation in conifers. In: Somaclonal variation: basic and practical aspects, Springer Nature, Switzerland, pp 123–142

Slotkin RK, Martienssen R (2007) Transposable elements and the epigenetic regulation of the genome. Nat Rev Genet 8:272–285

Studer MH, DeMartini JD, Davis MF, Sykes RW, Davison B, Keller M, Tuskan GA, Wyman CE (2011) Lignin content in natural populus variants affects sugar release. Proc Natl Acad Sci 108(15):6300–6305

Sun W (2012) A statistical framework for eQTL mapping using RNA-seq Data. Biometrics 68:1–11

Sun J, Zhang J, Fang H, Peng L, Wei S, Li C, Zheng S, Lu J (2019) Comparative transcriptome analysis reveals resistance-related genes and pathways in Musa acuminata banana 'Guijiao 9' in response to Fusarium wilt. Plant Physiol Biochem 141:83–94

Uga Y, Sugimoto K, Ogawa S, Rane J, Ishitani M, Hara N, Kitomi Y, Inukai Y, Ono K, Kanno N, Inoue H, Takehisa H, Motoyama R, Nagamura Y, Wu J, Matsumoto T, Takai T, Okuno K, Yano M (2013) Control of root system architecture by DEEPER ROOTING 1 increases rice yield under drought conditions. Nat Genet 45:1097–1102

Varshney RK, Graner A, Sorrells ME (2005) Genomics-assisted breeding for crop improvement. Trends Plant Sci 10:621–630

Varshney RK, Barmukh R, Roorkiwal M, Qi Y, Kholova J, Tuberosa R, Reynolds MP, Tardieu F, Siddique KHM (2021a) Breeding custom-designed crops for improved drought adaptation. Adv Genet 2:e202100017

Varshney RK, Bohra A, Roorkiwal M, Barmukh R, Cowling WA, Chitikineni A, Lam H-M, Hickey LT, Croser JS, Bayer PE, Edwards D, Crossa J, Weckwerth W, Millar H, Kumar A, Bevan MW, Siddique KHM (2021b) Fast-forward breeding for a food-secure world. Trends Genet 37:1124–1136

Vining KJ, Hummer KE, Bassil NV, Lange BM, Khoury CK, Carver D (2020) Crop wild relatives as germplasm resource for cultivar improvement in Mint (Mentha L.). Front Plant Sci 11:1217

Warschefsky EJ, Klein LL, Frank MH, Chitwood DH, Londo JP, von Wettberg EJB, Miller AJ (2016) Rootstocks: diversity, domestication, and impacts on shoot phenotypes. Trends Plant Sci 21:418–437

Wu X, Cortés AJ, Blair MW (2022) Genetic differentiation of grain, fodder and pod vegetable type cowpeas (Vigna unguiculata L.) identified through single nucleotide polymorphisms from genotyping-by-sequencing. Mol Hortic 2(1):8

Chapter 6
Plant Development, Stress, and Transcriptomics

M. J. Nishanth

Abstract Plant developmental processes and stress-response pathways are interconnected. A holistic understanding of these molecular networks would help us appreciate the complexity of these interactions and also suggest novel molecular targets for crop improvement. In this regard, research so far has identified multiple molecular regulators of growth and stress responses, including hormones, transcription factors, microRNAs, and other gene regulatory components, and signaling pathways. Transcriptomic studies such as RNA-sequencing and microarray offer an in-depth understanding of these processes. The present chapter attempts to provide a comprehensive overview of the molecular regulation of major plant developmental stages (germination, shoot and root development, transition to flowering, and fruit and seed development). Subsequently, transcriptomic studies investigating stress-induced molecular changes during these developmental stages in various plant species are summarized.

Keywords Plant development · Transcriptomics · Abiotic stress · RNA-sequencing · Microarray

Abbreviations

ABA	Abscisic acid
ABA8OX	ABA 8′-hydroxylase
ABF	ABRE-binding factor
ABI	Abscisic acid insensitive
AHG	ABA-hypersensitive germination
AHP	*Arabidopsis* histidine phosphotransfer protein
AMS	Aborted microspores

M. J. Nishanth (✉)
Department of Biotechnology, School of Life Sciences, St Joseph's University, Bengaluru, Karnataka, India
e-mail: nishanth.mj@sju.edu.in

L. M. Vaschetto (ed.), *Plant Transcriptomics and Epitranscriptomics*, Concepts and Strategies in Plant Sciences, https://doi.org/10.1007/978-981-95-5183-5_6

AP1	Apetala1
ARF	Auxin response factor
BGAL	β-galactosidase
bHLH	Basic helix-loop-helix
BME3	Blue micropylar end 3
BOP	Blade-on-petiole
CAP	Chromatin-associated proteins
CDC	Cell division control protein
CDM	Callose-defective microspore
CKs	Cytokinins
CNR	Colorless nonripening
CRN	Coryne
CRY	Cryptochrome
CUC	Cup-shaped cotyledon
DRN	Dornröschen
DRNL	Dornröschen-like
DYT	Dysfunctional tapetum
ELF	Early flowering
EMS1	Excess microsporocytes
ERF	Ethylene response factor
FM	Floral meristem
FTIP	FT interacting protein
FUL	Fruitfull
GA3ox1	Gibberellin oxidases
GAI	Gibberellic acid insensitive
GAs	Gibberellins
GRAS	Gibberellin-acid insensitive (GAI), Repressor of GA1 (RGA), and Scarecrow (SCR)
GRFs	Growth Regulating Factors
HDA	Histone deacetylase
HY5	Elongated hypocotyl 5
IBH	Increased leaf inclination1 binding bHLH
KAN	Kanadi
KNOX	Knotted-like homeobox
LEA	Late embryogenesis abundant
LEC	Leafy cotyledon
LFS	Leafless
LFY	Leafy
LHY	Late elongated hypocotyl
LTP	Lipid transfer protein
MAPK	Mitogen-activated protein kinase
MEL	MAB4/ENP/NPY1-like
NCED	9-cis-epoxycarotenoid dioxygenase
NGA	Ngatha

NOR	Nonripening
PEG	Polyethylene glycol
PERK2	Proline-rich receptor-like protein kinase
PIF	Phytochrome interacting factor
PIN	Pinformed
PKL	Pickle
PLT	Plethora
RBE	Rabbit ears
RBPs	RNA-binding proteins
REV	Revoluta
RGA	Repressor of GA
RIN	Ripening inhibitor
RPL	Replumless
SAM	Shoot apical meristem
SAUR72	Small auxin upregulated
SCR	Scarecrow
scRNA-seq	single-cell RNA-sequencing
SHP	Shatter-proof
SLR	Solitary root
SMC	Structural maintenance of chromosomes
SnRK2	SNF1-related protein kinase
SPL	Squomosa promoter binding protein-like
SSP	Seed storage protein
STY	Stylish
SUP	Superman
TAG	Tomato agamous
TAGL	Tomato agamous-like
TCP	Teosinte branched1/Cycloidea/Proliferating cell factor
TDF	Defective in tapetum development and function
TFs	Transcription factors
TOLS2-RLK	Target of LBD sixteen 2-receptor-like kinase
TOR	Target of rapamycin
UPB	Upbeat
VAL	VP1/ABI3-like
WOX	Wuschel-related homeobox

6.1 Introduction

Plants, the primary energy source, are vital for the sustenance of animal life. Understanding the genetic and molecular basis of plant-environment interactions could benefit crop improvement and human welfare. Being sessile, plants have evolved biochemical mechanisms to overcome environmental conditions that adversely affect their growth (i.e., stress conditions). These growth-deterrents include extreme temperatures, drought or flooding, high ionic concentrations, pollutants, unfavorable light intensities and radiations (abiotic factors), as well as pests and herbivores (biotic factors). In response to stresses, plants recalibrate the metabolic flux to develop resilience (Sheshadri et al. 2016; Singh et al. 2020; Zhang et al. 2020). This process involves signaling cascades altering gene expression through transcription factors (TFs), RNA-binding proteins (RBPs), regulatory RNAs, etc., thereby leading to the production of stress-related hormones, sugars, amino acids, and specialized metabolites, which may confer resilience by modulating the cellular and physiological status (Meng and Zhang 2013; Verma et al. 2016; Singh et al. 2020; Chen et al. 2021; Mittler et al. 2022; Nishanth 2024). Resource allocation to stress-tolerance mechanisms is known to be concomitant with stunted growth and development. Thus, frequent exposure to stress conditions hinders plant development and crop yield. Understanding the molecular processes underlying plant responses to stress conditions could be valuable for devising novel tools to enhance crop productivity (Nguyen et al. 2016).

The biochemical adjustments associated with most physiological conditions, including plant stress responses, are orchestrated by widespread molecular processes (Nguyen et al. 2016; Singh et al. 2020). Examining these global molecular changes requires large-scale, high-throughput strategies such as transcriptomics, proteomics, and metabolomic approaches. Since altered gene expression levels are fundamental for further physiological modifications, quantifying these transcriptomic changes would help us delineate the molecular networks underlying stress responses. To this end, microarray and RNA-sequencing (RNA-Seq) are generally employed, which facilitate comparative analysis of gene expression levels. Attributing the increase/decrease in the gene expression levels to experimental (stress) conditions would suggest the mechanistic underpinnings of plant responses to the considered conditions (Depuydt and Vandepoele 2021). Moreover, since development and stress responses are closely intertwined, transcriptomic approaches would be valuable in studying these processes at the molecular level.

So far, a large number of transcriptomic studies have been conducted on various plant species under a broad spectrum of environmental conditions at different developmental stages. Focusing on angiosperms, the present chapter provides a comprehensive overview of (a) the molecular processes underlying plant developmental phases/transitions, (b) the effects of stresses on developmental processes, and (c) the transcriptomic analyses of stress responses in developing plant tissues.

6.2 Molecular Regulation of Plant Development

Seed germination, the critical process marking the beginning of a plant's life, is under tight control by various hormones, the most prominent being abscisic acid (ABA) and gibberellins (GAs) (Shu et al. 2016). After water uptake, the balance between dormancy and germination pivots on the environmental cues. The optimal levels of moisture, temperature, oxygen, and light, along with the absence of stress conditions, lead to the progression of germination and the development of seedlings. The regulatory interplay between ABA and GAs influences this process, wherein ABA induces dormancy and represses germination, while GAs promote germination (Sajeev et al. 2024). In seeds, important targets of ABA include TF genes such as *ABA INSENSITIVE 3* (*ABI3*), *ABI4*, and *ABI5*, which are among the major repressors of seed germination. Also, ABI3, ABI5, and DELLA TFs inhibit germination by modulating the expression of ABA biosynthesis and GA catabolism genes (Nakashima and Yamaguchi-Shinozaki 2013; Li et al. 2022). Phosphorylation of ABI5 by SNF1-RELATED PROTEIN KINASE 2 (SnRK2), a major signal integrator, modulates its protein-protein interaction properties, potentially influencing desiccation tolerance in seeds (Nakashima et al. 2009; Belda-Palazón et al. 2020; Lin et al. 2021). GAs counter the ABA-induced effects and alter cell wall properties, thereby facilitating cell elongation. DELLA proteins are degraded by GA-dependent or independent mechanisms, contributing to the inhibition of ABA-induced dormancy (Schwechheimer 2012) and the development of young root and shoot tissues. Thus, together with other growth regulators, ABA and GAs drive the events associated with seed development, that is, embryogenesis, seed maturation, as well as the control of dormancy and germination. Upon sensing optimal environmental conditions, rupture of the testa (seed coat) is followed by radicle protrusion, thus beginning the subsequent cycle of plant life (Sajeev et al. 2024).

Further growth and development necessitate energy production via photosynthesis. This process requires the optimal development and functioning of the photosynthetic tissues (mainly the leaves) as well as the root system. The shoot apical meristem (SAM) and leaf primordia give rise to leaves in a process regulated by environmental signals (Mohammed et al. 2018). The maintenance of undifferentiated cells in the meristems is regulated by WUSCHEL-RELATED HOMEOBOX (WOX) TFs, which also potentially promote leaf blade growth by modulating auxin/cytokinin ratios (Bäurle and Laux 2005). The SAM consists of a central zone, comprising pluripotent stem cells, and a peripheral zone, from where the lateral organs arise. Leaf morphogenesis involves four steps: (i) founder cells are recruited from the SAM, which initiates morphogenesis (these founder cells detach from the stem cell niche and subsequently form the leaf primordium). (ii) the leaf primordium further develops in the distal direction (away from the SAM). (iii) the leaf blade begins to develop, and is subsequently distinguished from the petiole. (iv) the leaf blade further expands toward distal and lateral directions (intercalary growth), thus completing the process of leaf morphogenesis (Nakata and Okada 2013). The environmental signals, including light and sugars, through the activity of TARGET OF RAPAMYCIN (TOR) kinase, lead to the differentiation of the peripheral zone stem cells into leaf cells. This process involves the activity of auxins, wherein auxin maxima are established in the differentiating cells by the auxin efflux carrier, PINFORMED1 (PIN1). The auxin maxima also induce the ETHYLENE RESPONSE FACTOR-TYPE (ERF) TFs and LEAFLESS (LFS), central to the process of leaf development (Du et al. 2018). Subsequent steps of morphogenesis are controlled by several TFs including KNOTTED-LIKE HOMEOBOX 2 (KNOX

2; antagonizes KNOX 1, driving the maintenance of SAM cell fate), CUP-SHAPED COTYLEDON 1 (CUC1), CUC2, and CUC3 (Ali et al. 2020; Yang et al. 2023; Byrne et al. 2024). Further, the asymmetric distribution of cell and tissue types along the developmental axes is essential for the final leaf form, a process influenced by TFs such as REVOLUTA (REV), HD-ZIPIII, AS2 (determine adaxial polarity), KANADI1 (KAN1), and AUXIN RESPONSE FACTORS (ARF2, ARF3, and ARF4) (promote abaxial cell fate). Subsequently, the class II TEOSINTE BRANCHED1/CYCLOIDEA/PROLIFERATING CELL FACTOR (TCP) as well as NGATHA (NGA) TFs inhibit cell proliferation and promote differentiation, contributing to intercalary growth (Du et al. 2018).

The root system develops from the root apical meristem initiated during embryogenesis or the adventitious root apical meristems. Longitudinally, roots consist of the meristematic, elongation, and differentiation zones. The meristematic zone consists of the stem cell niche from which all cell types are derived. In *Arabidopsis thaliana*, the stem cells are situated around a quiescent center consisting of four cells. The quiescent center is involved in the organization of the root meristem, largely dependent on the activity of WOX5, which suppresses cell proliferation (Forzani et al. 2014; Matosevich and Efroni 2021). Also, the Plethora (PLT) family TFs are known master regulators of root organogenesis; these TFs form a regulatory network and influence root organogenesis by activating proliferation and repressing differentiation. Similarly, SCARECROW (SCR), a GRAS family TF is also known to interact with a number of other TFs, and this TF network regulates the differentiation of cells in endodermis. Further, PLT and SCR interact with plant-specific TCP TFs and activate WOX5 (Forzani et al. 2014; Shimotohno et al. 2018). The PLT TFs are also involved in the formation of lateral roots, whose initiation is regulated by GATA23, TARGET OF LBD SIXTEEN 2-RECEPTOR-LIKE KINASE7 (TOLS2-RLK7) signaling pathway, as well as SOLITARY ROOT (SLR)/IAA14-ARF7-ARF19 pathway (Toyokura et al. 2019). Thus, the development of both primary and lateral roots is regulated by highly coordinated transcriptional and signaling mechanisms controlling stem cell proliferation and differentiation (Motte et al. 2019).

As a plant transitions to the reproductive phase, the genetic programs driving shoot development are partially repressed, and the expression of flower-specific genes is activated. The floral meristems (FMs) are formed on specific regions of the inflorescence meristem (Smyth et al. 1990). This process is determined by the auxin maxima established by the auxin efflux carrier PIN1. Also, a few TFs, including DORNRÖSCHEN (DRN), DORNRÖSCHEN-LIKE (DRNL), PUCHI, BLADE-ON-PETIOLE 1 (BOP1), and BOP2, are known to promote the expression of LEAFY (LFY), the master regulator of floral meristem identity specification. Subsequently, LFY controls auxin-dependent development of FMs, and suppresses GA levels in the FMs. LFY also activates DELLA and SQUAMOSA PROMOTER BINDING PROTEIN-LIKE (SPL) TFs, which in turn activate APETALA1 (AP1), a MADS-box TF and an important regulator of FM establishment (Thomson and Wellmer 2019). The subsequent patterning of floral organs is explained by the (A)B(C) model (Pinyopich et al. 2003; Theißen et al. 2016). This model posits that the combinatorial activities of different groups of homeotic genes control the development of floral whorls: (A)-function genes regulating the sepal development, (A)- and B-genes influencing the petal formation, all three groups of genes controlling stamen formation, and the (A)- and (C)-group genes regulating the pistil development. The boundaries demarcating the individual whorls are regulated by CUC genes (involved in boundary formation throughout plant development) as well as other flower-specific factors, including SUPERMAN (SUP) and RABBIT EARS (RBE). After the initiation of the formation of all floral organs, the floral stem cell activity is terminated, mainly through the suppression of the activity of WUS, in addition to other regulators. Hence, flower-specific and general stem cell regulators control the development of floral organs (Thomson and Wellmer 2019). Post-fertilization, auxin signals generated from the seeds further stimulate GA signaling, activating fruit development. Subsequently, the ovary and ovules mature into fruit and seeds in a tightly coordinated process (Seymour et al. 2013).

The development of the gynoecium in *A. thaliana* is regulated by auxin, associated with auxin biosynthesis at the apex, driven by TFs such as STYLISH1 (STY1), STY2, and NGA3 (Kuusk et al. 2002; Alvarez et al. 2009; Magnus Eklund et al. 2010). A number of ARFs also play important roles in fruit development and cellular patterning. In tomato, pericarp development is regulated by the TF TOMATO AGAMOUS (TAG)-LIKE1 (TAGL1) (ortholog of *A. thaliana* SHATTERPROOF, SHP) (Vrebalov et al. 2009). In addition, FRUITFULL (FUL), REPLUMLESS (RPL), ASYMMETRIC LEAVES 1, ASYMMETRIC LEAVES 2, and BREVIPEDICELLUS are also known to influence fruit patterning. Ripening of tomato is controlled by TFs such as RIPENING INHIBITOR (RIN, a MADS-box TF, a known master regulator of ripening), NONRIPENING (NOR, a NAC domain TF), and COLOURLESS NONRIPENING (CNR, an SBP-box TF). These TFs may regulate genes involved in ethylene biosynthesis, cell wall metabolism, and carotenoid and aroma biosynthesis. Thus, a network of TFs coordinates the various processes associated with fruit development and ripening, mainly by modulating hormone metabolism. Simultaneously, TF-mediated regulatory interplay between ABA and GA, as well as other hormones, directs the maturation of seeds (Seymour et al. 2013).

Seed development involves embryo and endosperm development, followed by maturation phases. In turn, embryo and endosperm morphogenesis include zygote formation and growth, wherein the single-celled zygote develops into a heart-shaped structure, subsequently giving rise to radicle, cotyledons, and hypocotyl (Raz et al. 2001). The completion of embryo growth is associated with an arrest of cell division in the embryo and the beginning of the seed maturation phase. The maturation process is associated with an onset of dormancy, which increases till the completion of seed development. Subsequent events leading to primary dormancy include the accumulation of energy reserves, establishment of desiccation tolerance, as well as decreased moisture and increased ABA levels (Raz et al. 2001). This process of maturation is known to be controlled by interacting TFs, FUS3, ABI3, LEAFY COTYLEDON 1 (LEC1), and LEC2, which enhance the expression of a number of dormancy-inducing genes. (Holdsworth et al. 2008). The early- and mid-maturation phases are associated with the expression of seed storage proteins (SSP), while the later phases are characterized by the expression of LATE EMBRYOGENESIS ABUNDANT (*LEA*) genes important for the establishment of desiccation tolerance (Wobus and Weber 1999; Hoekstra et al. 2001; Nguyen et al. 2015; Leprince et al. 2017). The progressive increase in seed dormancy reaches a maximum in harvest-ripe seeds and then starts to decrease during after-ripening (Holdsworth et al. 2008). Subsequent germination (emergence of the radical from an imbibed seed) requires optimal environmental conditions to ensure the healthy development of the emerging seedling, necessitating a regulatory switch from dormancy to germination (as described earlier, controlled by ABA and GAs) (Holdsworth et al. 2008). De novo GA production in embryos driven by the biosynthetic genes, *GIBBERELLIN OXIDASES* (*GA3ox1* and *GA3ox2*) (preferentially expressed in embryos upon imbibition), facilitates germination completion (Olszewski et al. 2002; Mitchum et al. 2006). Subsequently, GAs control the expression of extensin-like gene *AtEPR* and BLUE MICROPYLAR END 3/BME3. These genes influence cell wall remodeling and endosperm rupture, respectively (Dubreucq et al. 2000; Liu et al. 2005). Further, GA-dependent repression of embryonic traits is also influenced by PICKLE (PKL), a chromatin remodeling factor, which works along with the VP1/ABI3-LIKE (VAL) family TFs (Köhler et al. 2003; Henderson et al. 2004; Suzuki et al. 2007; Park et al. 2017). In addition, histone deacetylases (HDA6 and HDA19) are also known to influence the repression of embryonic traits in the germinating seed (Tanaka et al. 2008). Subsequent to a reduction in dormancy levels, imbibition under favorable environmental conditions leads to germination completion. Thus, TF networks modulating hormone-mediated growth pathways are central regulators of all developmental stages (Holdsworth et al. 2008). A simplified view of major regulatory components driving plant developmental transitions is given in Fig. 6.1.

6.3 Effects of Stress on Developmental Processes

The developmental processes during all stages of plant growth are known to be influenced by stress conditions. Plant maintenance under controlled/optimal laboratory conditions promotes maximal growth and production, while in nature, there is a trade-off between growth and stress-response pathways. Stress conditions disturb the physiological balance (e.g., extreme temperatures lead to decreased enzyme activities) and necessitate resource allocation to stress-response pathways, thus hindering plant growth. In addition, these signaling and metabolic processes underlying stress-resilience are also known to actively suppress growth-promoting pathways (Zhang et al. 2020). Further, stress responses are suppressed by developmental processes in the absence of stress. This reciprocal inhibition is mediated by hormones, TFs, and miRNAs, among other regulatory factors (Sewelam et al. 2016; Verma et al. 2016; Skalak et al. 2021). ABA, known to be upregulated by stresses, represses the growth-promoting hormones, including GA, brassinosteroids, auxins, and cytokinins (Waadt et al. 2022). Similarly, the TOR signaling pathway, important for the

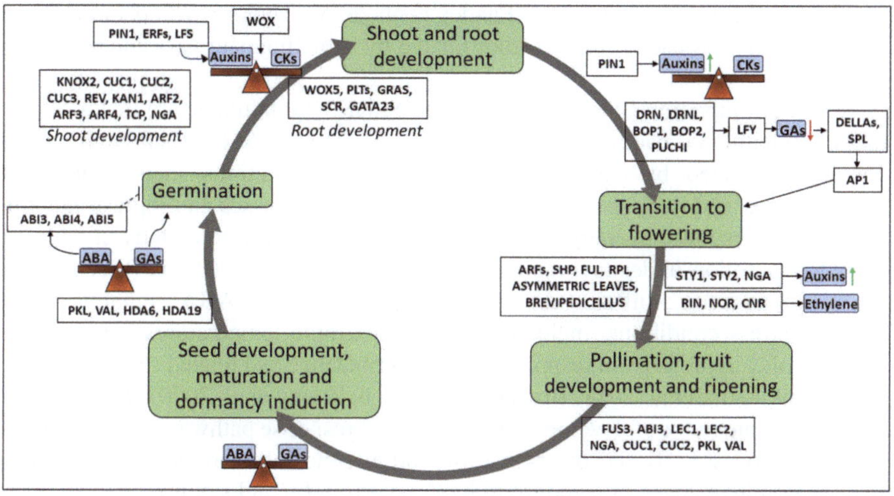

Fig. 6.1 A simplified view of the molecular regulation of plant developmental transitions
The major transcriptional regulatory factors known to drive developmental stage transitions are shown. Seed development, dormancy induction, and subsequent maturation and germination are controlled by abscisic acid (ABA) and gibberellins (GAs), which are in turn modulated by transcription factors. Regulatory proteins and hormones are shown in white and blue boxes, respectively. Green upward and red downward arrows represent an increase and a decrease in quantities. Black sharp arrows indicate positive influence, and the blunt arrow indicates inhibition. *PKL:* pickle, *VAL:* VP1/ABI3-like, *HDA:* histone deacetylase, *ABI*: abscisic acid insensitive, *KNOX*: knotted-like homeobox, *CUC:* cup-shaped cotyledon, *REV:* revoluta, kan: kanadi, *ARF*: auxin response factor, *TCP:* teosinte branched1/cycloidea/proliferating cell factor, *NGA:* ngatha, *PIN1:* pinformed1, *ERF:* ethylene response factor, *LFS:* leafless, *WOX:* wuschel-related homeobox, *CKs:* cytokinins, *PLT:* plethora, *GRAS:* gibberellin-acid insensitive (GAI), repressor of GA1 (RGA), and scarecrow (SCR), *SCR:* scarecrow, *DRN:* dornröschen, *DRNL:* dornröschen-like, *BOP*: blade-on-petiole, *LFY:* leafy, *SPL:* squomosa promoter binding protein-like; *AP1*: apetala1, *STY:* stylish, *SHP*: shatter-proof, *FUL*: fruitfull, *RPL:* replumless, *RIN*: ripening inhibitor, *NOR*: nonripening, *CNR:* colourless nonripening, *LEC:* leafy cotyledon

activation of cellular anabolic processes, is inhibited by the stress-responsive SnRKs. Further, the Growth Regulating Factors (GRFs; a family of TFs involved in growth promotion) are inhibited post-transcriptionally by the stress-induced microRNA miR396, demonstrating the reciprocal molecular repression of development and stress-response pathways (Zhang et al. 2020).

Plants under field conditions are simultaneously exposed to multiple stresses. Response to various stresses involves shared and unique adaptive strategies. For example, plants accumulate osmoprotectants in response to cold, drought, and salinity (Chinnusamy et al. 2004). Drought and salinity are also associated with decreased CO_2 diffusion into chloroplasts and reduced carbon metabolism (Munns 2002). Further, drought and salinity reduce photosynthetic rate and stomatal conductance (Ahmed et al. 2013). Morphologically, drought leads to a reduction in leaf number and area, and increased root growth, while heat stress is known to be associated with the opposite trend (Pandey et al. 2015). Thus, growth parameters are modulated in response to stresses, coordinated by signaling pathways and regulatory factors.

SnRKs, comprising SnRK1, SnRK2, and SnRK3 subfamilies, are major regulators of signaling cascades in response to various stresses. SnRK1 subfamily members are activated by energy deprivation due to conditions such as hypoxia, submergence, and darkness. They activate catabolic processes and repress anabolic pathways in response to energy depletion, while SnRK2s are potentially involved in growth promotion under favorable environmental conditions (Belda-Palazón et al. 2020). In addition, SnRK2s are induced in response to ABA signaling under multiple stress conditions, including drought, extreme temperatures, and salinity. SnRK3s play a role in plant responses to ionic imbalance and related stresses (Jossier et al. 2009; Kulik et al. 2011; Broeckx et al. 2016; Zhang et al. 2020). Thus, SnRKs are a major integrator of developmental and stress-response pathways and are pivotal for these processes. Based on these observations, the following sections focus on transcriptomic changes in response to hypoxia, drought, and salinity.

6.4 Transcriptomic Studies of Stress-Induced Effects on Plant Development

The forward and reverse genetic studies so far have successfully elucidated the major signaling cascades, regulatory factors, and hormones involved in developmental and stress-response pathways. The application of omics technologies in recent decades has not only corroborated these findings but has also provided a more comprehensive understanding of the large-scale, interconnected changes of gene expression patterns and metabolic cycles. The following sections provide a broad overview of the results from transcriptomic studies conducted on various plant species under waterlogging/hypoxia, drought, and salinity. The biological processes activated and repressed within germinating seeds, shoot and root meristems, and reproductive tissues are highlighted.

Cucumber seeds germinating under both drought and salinity stresses were found to have reduced transcript and protein levels of photosystem II components, cytochrome b6/f complex, and ATPase, suggesting a reduction in photosynthesis (Du et al. 2021). A comparative analysis of drought-resistant and susceptible varieties of *Vigna angularis* germinating under drought identified the potential molecular underpinnings of drought resistance (Zhu et al. 2020). In this study, the drought-resistant variety was associated with upregulated transcript levels of *AUXIN-INDUCED PROTEIN*, ERF family TFs, *bHLH041*, as well as carbohydrate, nitrogen, and lipid metabolism genes (Zhu et al. 2020). In *A. thaliana*, germination under drought stress led to an upregulation of genes encoding LEA and seed storage proteins. Further, genes involved in ABA biosynthesis and signaling, and carbohydrate catabolism were also induced (Maia et al. 2011). On the other hand, GA biosynthesis genes, GA-responsive genes, as well as energy metabolism genes were repressed (Maia et al. 2011), indicating growth inhibition. Transcriptomic analysis of *Brassica napus* seeds germinating under waterlogging stress showed an upregulation of the TF genes *ABI4, SMALL AUXIN UPREGULATED 72* (*SAUR72*), *ELONGATED HYPOCOTYL 5* (*HY5*), *PHYTOCHROME INTERACTING FACTOR 4* (*PIF4*), and *SPT*. The downregulated genes included those involved in auxin biosynthesis and transport, in addition to *EXPANSIN-A7*, *UPBEAT1* (*UPB1*), *CLV1*, *CORYNE* (*CRN*), Chromatin assembly factor 1 subunit *FAS2*, *GA20ox4*, *GA3ox1*, *SAUR32*, *CRYPTOCHROME-2* (*CRY2*), and basic helix-loop-helix (bHLH) TF gene *IBH1* (*INCREASED LEAF INCLINATION1 BINDING BHLH 1*) (Guo et al. 2021). These observations also point toward a complex modulation of transcriptomic patterns leading to the repression of growth pathways in response to waterlogging/hypoxia (Guo et al. 2021). Another study, also investigating the waterlogging-induced molecular changes in germinating *Brassica napus* reported an upregulation of genes encoding TFs and proteins involved in RNA metabolism, while a few genes encoding peroxidases, hydrolases, oxidoreductases, as well as those involved in ROS metabolism and cell wall metabolism were found to be downregulated (Li et al. 2021). Germinating Quinoa seeds subjected to salinity stress were associated with reduced expression of glucose and sucrose metabolism genes, with a concomitant decrease in the corresponding metabolite levels (including glucose, sucrose, trehalose, and fructose-6-phosphate) (Yan et al. 2022). Thus, stress conditions have a major impact on carbohydrate metabolism and hormone-mediated signaling processes in various species. Importantly, this inhibition of growth during the initial phases of germination could be deleterious for the seedling health and further growth.

Transcriptomic analysis of the shoot and root meristems would provide valuable insights into the effects of stress conditions on stem cell populations and pathways underlying morphogenesis. However, only a limited number of transcriptomic studies have been conducted on meristematic tissues. The coleoptile tissues from rice seedlings developing under hypoxic conditions displayed reduced expression of a number of photosynthesis-related genes, including *PHOTOSYSTEM II 10 kDA POLYPEPTIDE, CHLOROPLAST PRECURSOR* (Thapa et al. 2023). This observation corroborates a previous study on rice coleoptiles growing under hypoxia, which also reported downregulation of genes encoding CHLOROPHYLL A/B BINDING PROTEIN 2, photosystem I polypeptide subunits, PHOSPHOFRUCTOKINASE, and NADH-UBIQUINONE OXIDASE (Narsai et al. 2015). Downregulation of photosynthesis and carbohydrate metabolism within the highly active meristematic tissues highlights the mechanistic basis of stunted growth under stress.

In response to drought, unopened flower bud tissues of *A. thaliana* showed an upregulation of disaccharide-related genes, *β-GALACTOSIDASE 7* (*BGAL7*), *BGAL11*, and *BGAL13*, involved in pollen development and fertilization (Su et al. 2013). In addition, a number of genes important for ABA signaling and stress responses were also upregulated (including *ABI1*, *ABI2*, *ABI5*, *ABF1*, *ABF3*, *ABF4*, *DREB1A*, *ABF*, *NAC DOMAIN-CONTAINING PROTEIN19*, *RD20*, and *RD29A*). Additionally, genes encoding putative negative regulators of ABA signaling (*Protein phosphatase 2C*/*PP2CA*, *HIGHLY ABA-INDUCED 1*/*HAI1*, and *HAI2*) and *JASMONIC ACID CARBOXYL METHYLTRANSFERASE* involved in MeJA biosynthesis, flowering regulators (*LFY*, *SOC1*, and *FT*), as well as genes involved in pistil development (*AGL80*, *EOSTRE*, *RHF1a*, *RHF2a*, and *GPT1*) also showed differential expression (Su et al. 2013). Feng et al. (2022) reported a decrease in the expression of flowering regulator genes in the developing flower bud tissues of *Curcuma kwangsiensis* subjected to drought stress (Feng et al. 2022). In this study, *EARLY FLOWERING 3* (*ELF3*) and *LATE ELONGATED HYPOCOTYL* (*LHY*) were downregulated, while the floral integration factor *FT INTERACTING PROTEIN* (*FTIP*) was upregulated. In addition, putative *9-CIS-EPOXYCAROTENOID DIOXYGENASE 3* (*NCED*, involved in ABA biosynthesis), *PP2C51*, and *PP2C06* were also induced (Feng et al. 2022). The repression of glucan metabolism and photosynthesis was also observed in developing rice flowers under drought stress (Jin et al. 2013). Genes found to be downregulated in this study included the meiosis-related genes: orthologs of *STRUCTURAL MAINTENANCE OF CHROMOSOMES 3* (*SMC3*), *CHROMATIN-ASSOCIATED PROTEINS* (*CAP-E1* and *CAP-E2*), *ARABIDOPSIS HISTIDINE PHOSPHOTRANSFER PROTEIN2* (*AHP2*), *CELL DIVISION CONTROL PROTEIN 45* (*CDC45*) and *MAB4*/*ENP*/*NPY1-LIKE1* (*MEL1*), as well as *β-EXPANSIN* and *β-KETOACYL REDUCTASE* (involved in regulating pollen development) (Jin et al. 2013). Pollen development genes were also reported to be downregulated in developing tomato floral tissues, wherein *Solanum lycopersicum EXCESS MICROSPOROCYTES1* (*SlEMS1*), *ABORTED MICROSPORES* (*SlAMS*), *DEFECTIVE IN TAPETUM DEVELOPMENT AND FUNCTION1* (*SlTDF1*), *DYSFUNCTIONAL TAPETUM1* (*SlDYT1*), and *CALLOSE DEFECTIVE MICROSPORE1* (*SlCDM1*) were repressed, accompanied with defective development of tapetum and microspores (Lamin-Samu et al. 2021). In the study conducted by Jin and colleagues on rice flower development under drought, *LIPID TRANSFER PROTEIN* (*LTP*), MADS-box TFs, *OsMADS15*, and *OsMADS18* (known to play important roles in the transition to flowering, as well as fruit and seed maturation) were upregulated (Jin et al. 2013). Also, genes encoding ABA 8'-HYDROXYLASE (ABA8OX; involved in ABA catabolism) were downregulated, with a simultaneous increase in the expression levels of ABA-responsive genes such as *DEHYDRIN*, *DREB*, *LEA*, *MAPK*, and *ABRE-binding factor* (*ABF*) (Jin et al. 2013). Contrarily, Slcyp707a1, encoding ABA 8′--HYDROXYLASE in tomato, was found to be upregulated in another study involving drought-stressed floral tissues (Lamin-Samu et al. 2021). Further, the ABA signaling genes, including *SlPYL9*, *PP2C*, *ABA INSENSITIVE 2* (*SlABI2*) and *ABA-HYPERSENSITIVE GERMINATION 3* (*SlAHG3*), and *SnRK2* genes (*SlSRK2C and AY222455*) were also upregulated. Notably, this study revealed tissue-specific patterns of gene expression wherein opposite trends were observed in the expression patterns of several sugar metabolism and ABA metabolism genes in different reproductive tissues (Lamin-Samu et al. 2021), highlighting the potential tissue- and cell type-specific gene expression changes. In summary, stress conditions lead to altered metabolic flux and defective morphogenesis of floral organs, potentially affecting reproductive success and crop yield. These details are summarized in Table 6.1.

Table 6.1 Transcriptomic studies of plant stress responses at different developmental stages

S. No.	Plant species	Tissue	Stress treatments	Genes/metabolic pathways/cellular components/ biological processes modulated	Reference
1.	*Cucumis sativus* L. cv. Jinyou No.1 (Cucumber)	Germinating seeds	Polyethylene glycol (PEG)-induced drought and salinity	Porphyrin and chlorophyll metabolism, linoleic acid metabolism, degradation of limonene and pinene, glyoxylate and dicarboxylate metabolism, fatty acid degradation, carbon metabolism, carbon fixation and photosynthesis, biosynthesis of unsaturated fatty acids, biosynthesis of secondary metabolites	Du et al. (2021)
2.	*Vigna angularis* L. (Adzuki bean)	Germinating seeds	Mannitol-induced drought	Primary metabolism, plant hormone signal transduction, translation, ubiquitin proteasome system, and specialized metabolism	Zhu et al. (2020)
3.	*Arabidopsis thaliana*	Germinating seeds	PEG-induced drought	Induction of LEA, seed storage and dormancy related proteins, repression of photosynthesis, cell wall modification, and energy metabolism	Maia et al. (2011)
4.	Faba bean	Cotyledons with embryos	Mannitol-induced drought	Abscisic acid and mitogen-activated protein kinase (MAPK) pathways, sulfur metabolism, and sugar metabolism	Wen et al. (2024)
5.	*Setaria italica* (L.) P. Beauv. (Foxtail millet)	Germinating seeds	PEG-induced drought	Phenylpropanoid biosynthesis, plant hormone signal transduction, phenylalanine metabolism, PIP2–5 (aquaporin), cation/calcium exchanger 1, glycerol-3-phosphate transporter 4, bidirectional sugar transporter SWEET1b and mannan endo-1,4-beta-mannosidase 1, CIPK 16, SnRK regulatory subunit gamma-like PV42a, SnRK regulatory subunit gamma-1, two proline-rich receptor-like protein kinase 2 (PERK2), PERK9, phosphatase 2C, cation/calcium exchanger NIP2–2	Yu et al. (2020)
6.	*Brassica napus* (Rapeseed)	Germinating seeds	Hypoxia	Hormone signal transduction, phenylpropanoid biosynthesis, and starch and sucrose metabolism	Guo et al. (2020)
7.	*Oryza sativa* (Nipponbare) (Rice)	Germinating seeds	Hypoxia	Pentose phosphate pathway, glycolysis/gluconeogenesis, fructose and mannose metabolism, meiosis, DNA replication, and cell cycle	Chen et al. (2019)

(continued)

Table 6.1 (continued)

S. No.	Plant species	Tissue	Stress treatments	Genes/metabolic pathways/cellular components/ biological processes modulated	Reference
8.	*Brassica napus* L. (Rapeseed)	Germinating seeds	Hypoxia	Hormone-mediated signaling, response to organic substance, and microtubule-related processes	Li et al. (2021)
9.	*Oryza sativa* (Rice)	Germinating seeds	Hypoxia	Carbohydrate metabolic process, glucosyltransferase activity, protein metabolic process, lipid metabolic process, cellular nitrogen compound biosynthetic process, lipid biosynthetic process, and microtubule-based process	Thapa et al. (2023)
10.	*Zea mays* L. (Maize)	Embryos of germinating seeds	Hypoxia	Glycolysis, TCA cycle, and pentose phosphate pathway	Kim et al. (2024)
11.	*Populus euphratica* and *Populus pruinosa*	Germinating seeds	Salinity	Ribosome, proteasome, protein processing in endoplasmic reticulum, flavonoid biosynthesis, oxidative phosphorylation, spliceosome, carbon metabolism, glycolysis/gluconeogenesis, starch and sucrose metabolism, oxidative phosphorylation, photosynthesis, porphyrin and chlorophyll metabolism, and carotenoid biosynthesis	Zhang et al. (2019)
12.	*Chenopodium quinoa* (Quinoa)	Germinating seeds	Salinity	Starch and sucrose metabolism, amino sugar and nucleotide sugar metabolism, and ascorbate and aldarate metabolism	Yan et al. (2022)
13.	*Hordeum vulgare* L. (Barley)	Germinating seeds	Salinity	Protein processing in endoplasmic reticulum, arginine and proline metabolism, phenylpropanoid biosynthesis, carotenoid biosynthesis, carbon metabolism, plant hormone signal transduction, and biosynthesis of amino acids	Lai et al. (2020)
14.	*Panicum miliaceum* L. (Broomcorn millet)	Germinating seeds	Salinity	Phenylpropanoid biosynthesis, plant hormone signal transduction, starch and sucrose metabolism, and phenylalanine metabolism	An et al. (2022)

15.	*Oryza sativa* (Rice)	Germinating seeds	Salinity	Response to oxidative stress, cell wall organization, hydrogen peroxide catabolic process, auxin-activated signaling pathway, response to auxin, and metal ion transport	Li and Guo (2023)
16.	*Oryza sativa* f. *spontane* (weedy rice)	Coleoptile	Hypoxia	Carbohydrate metabolism, energy metabolism, and amino acid metabolism	Zhang et al. (2021)
17.	*Oryza sativa* (Rice)	Coleoptile	Hypoxia	Lipid transfer protein, carbohydrate kinase, receptor kinase-like protein, heat shock transcription factor, abiotic stress response, carbohydrate kinase, glutathione S-transferase, leucine-rich repeat, NB-ARC domain-containing protein, and peroxidase precursor	Thapa et al. (2023)
18.	*Oryza sativa* L. cv. Amaroo (Rice)	Coleoptile	Hypoxia	Hormone metabolism, nucleic acid metabolism, cell wall-related processes, abiotic stress responses, photosystem I	Narsai et al. (2015)
19.	*Hibiscus cannabinus* L (Kenaf)	Shoot tips	Salinity	Antioxidant activity, amino acid metabolism, and carbohydrate metabolism	Li et al. (2017)
20.	*Phaseolus vulgaris* (common bean)	Radicle	Salinity	Phenylpropanoid biosynthesis and phenylalanine metabolism, flavonoid biosynthesis	Zhang et al. (2023)
21.	*Arabidopsis thaliana*	Unopened flower bud	Drought	Glucosinolate biosynthesis, chlorophyll biosynthesis, RNA splicing, disaccharide catabolism, ABA signaling pathway, protein ubiquitination, lignin biosynthesis, fatty acid metabolism	Su et al. (2013)
22.	*Solanum lycopersicum* (tomato)	Anther	Drought	Carbohydrate metabolic process, lipid metabolic process, organonitrogen compound metabolic process, and amide biosynthetic process	Lamin-Samu et al. (2021)
23.	*Curcuma kwangsiensis*	Flower bud	Drought	Flowering regulation via photoperiod pathway, abscisic acid signaling	Feng et al. (2022)
24.	*Oryza sativa* subsp. *japonica* cv. Nipponbare (Rice)	Florets	Drought	Tapetum and microspore development, cell wall formation or expansion, starch synthesis, gibberellin signaling, abscisic acid catabolism	Jin et al. (2013)

6.5 Conclusions and Future Perspectives

Balancing stress responses and growth is an integral aspect of plant life, orchestrated by hormones and the underlying gene regulatory factors. Research studies have significantly advanced our understanding of how plants overcome environmental stresses. The recent transcriptomic analyses have provided a more comprehensive picture of the gene networks modulated in response to stress conditions. Under stress conditions, diverse plant species are known to upregulate ABA levels and the expression of SnRKs, while the central carbohydrate metabolism, photosynthesis, and developmental pathways are suppressed. This is associated with decreased plant growth. While these studies have provided valuable insights into the molecular changes in gene expression patterns, translating this knowledge to engineer stress-resilient plant varieties on a large scale is yet to be realized. To this end, modulating the major regulatory factors, including TFs and RBPs, which control multiple co-expressed genes in response to stress conditions, may be beneficial. In recent years, single-cell RNA-sequencing (scRNA-seq) studies have begun to explore the cellular heterogeneity associated with developing tissues. Computational approaches such as pseudotime trajectory analysis have facilitated a better understanding of the finer details of gene expression patterns during cellular development. Further studies employing techniques such as scRNA-seq and spatial transcriptomics could provide a more holistic understanding of plant growth and stress-response pathways, which could eventually benefit the development of stress-resilient crops.

Acknowledgments The infrastructural facilities provided by St Joseph's University, Bengaluru, India, are thankfully acknowledged.

Conflict of Interest Disclosure There are no conflicts of interest.

References

Ahmed IM, Cao F, Zhang M et al (2013) Difference in yield and physiological features in response to drought and salinity combined stress during Anthesis in Tibetan wild and cultivated barleys. PLoS One 8. https://doi.org/10.1371/journal.pone.0077869

Ali S, Khan N, Xie L (2020) Molecular and hormonal regulation of leaf morphogenesis in arabidopsis. Int J Mol Sci 21. https://doi.org/10.3390/ijms21145132

Alvarez JP, Goldshmidt A, Efroni I et al (2009) The NGATHA distal organ development genes are essential for style specification in arabidopsis. Plant Cell 21. https://doi.org/10.1105/tpc.109.065482

An P, Li X, Liu T et al (2022) The identification of broomcorn millet bZIP transcription factors, which regulate growth and development to enhance stress tolerance and seed germination. Int J Mol Sci 23. https://doi.org/10.3390/ijms23126448

Bäurle I, Laux T (2005) Regulation of WUSCHEL transcription in the stem cell niche of the Arabidopsis shoot meristem. Plant Cell 17. https://doi.org/10.1105/tpc.105.032623

Belda-Palazón B, Adamo M, Valerio C et al (2020) A dual function of SnRK2 kinases in the regulation of SnRK1 and plant growth. Nat Plants 6. https://doi.org/10.1038/s41477-020-00778-w

Broeckx T, Hulsmans S, Rolland F (2016) The plant energy sensor: evolutionary conservation and divergence of SnRK1 structure, regulation, and function. J Exp Bot 67. https://doi.org/10.1093/jxb/erw416

Byrne ME, Imlay E, Ridza NNB (2024) Shaping leaves through TALE homeodomain transcription factors. J Exp Bot 75:3220–3232. https://doi.org/10.1093/jxb/erae118

Chen MX, Zhu FY, Wang FZ et al (2019) Alternative splicing and translation play important roles in hypoxic germination in rice. J Exp Bot 70:885–895. https://doi.org/10.1093/jxb/ery393

Chen X, Ding Y, Yang Y et al (2021) Protein kinases in plant responses to drought, salt, and cold stress. J Integr Plant Biol 63. https://doi.org/10.1111/jipb.13061

Chinnusamy V, Schumaker K, Zhu JK (2004) Molecular genetic perspectives on cross-talk and specificity in abiotic stress signalling in plants. J Exp Bot 55:225–236. https://doi.org/10.1093/jxb/erh005

Depuydt T, Vandepoele K (2021) Multi-omics network-based functional annotation of unknown Arabidopsis genes. Plant J 108. https://doi.org/10.1111/tpj.15507

Du F, Guan C, Jiao Y (2018) Molecular mechanisms of leaf morphogenesis. Mol Plant 11:1117–1134. https://doi.org/10.1016/j.molp.2018.06.006

Du C, Li H, Liu C, Fan H (2021) Understanding of the postgerminative development response to salinity and drought stresses in cucumber seeds by integrated proteomics and transcriptomics analysis. J Proteome 232:104062. https://doi.org/10.1016/j.jprot.2020.104062

Dubreucq B, Berger N, Vincent E et al (2000) The Arabidopsis AtEPR1 extensin-like gene is specifically expressed in endosperm during seed germination. Plant J 23. https://doi.org/10.1046/j.1365-313X.2000.00829.x

Feng X, Zhou L, Sheng A et al (2022) Comparative transcriptome analysis on drought stress-induced floral formation of Curcuma kwangsiensis. Plant Signal Behav 17. https://doi.org/10.1080/15592324.2022.2114642

Forzani C, Aichinger E, Sornay E et al (2014) WOX5 suppresses CYCLIN D activity to establish quiescence at the Center of the root stem cell niche. Curr Biol 24. https://doi.org/10.1016/j.cub.2014.07.019

Guo Y, Chen J, Kuang L et al (2020) Effects of waterlogging stress on early seedling development and transcriptomic responses in Brassica napus. Mol Breed 40. https://doi.org/10.1007/s11032-020-01167-z

Guo J, Sun B, He H et al (2021) Current understanding of bhlh transcription factors in plant abiotic stress tolerance. Int J Mol Sci 22. https://doi.org/10.3390/ijms22094921

Henderson JT, Li HC, Rider SD et al (2004) Pickle acts throughout the plant to repress expression of embryonic traits and may play a role in gibberellin-dependent responses. Plant Physiol 134. https://doi.org/10.1104/pp.103.030148

Hoekstra FA, Golovina EA, Buitink J (2001) Mechanism of plant desiccation tolerance. Trends Plant Sci 6. https://doi.org/10.1016/s1360-1385(01)02052-0

Holdsworth MJ, Bentsink L, Soppe WJJ (2008) Molecular networks regulating Arabidopsis seed maturation, after-ripening, dormancy and germination. New Phytol 179. https://doi.org/10.1111/j.1469-8137.2008.02437.x

Jin Y, Yang H, Wei Z et al (2013) Rice male development under drought stress: phenotypic changes and stage-dependent transcriptomic reprogramming. Mol Plant 6:1630–1645. https://doi.org/10.1093/mp/sst067

Jossier M, Bouly JP, Meimoun P et al (2009) SnRK1 (SNF1-related kinase 1) has a central role in sugar and ABA signalling in Arabidopsis thaliana. Plant J 59. https://doi.org/10.1111/j.1365-313X.2009.03871.x

Kim JW, Hong S, Go J et al (2024) Transcriptomic changes reveal hypoxic stress response in submerged seeds of maize (Zea mays L.). Appl Biol Chem 67. https://doi.org/10.1186/s13765-024-00922-6

Köhler C, Hennig L, Spillane C et al (2003) The Polycomb-group protein MEDEA regulates seed development by controlling expression of the MADS-box gene PHERES1. Genes Dev 17. https://doi.org/10.1101/gad.257403
Kulik A, Wawer I, Krzywińska E et al (2011) SnRK2 protein kinases - key regulators of plant response to abiotic stresses. Omi A J Integr Biol 15. https://doi.org/10.1089/omi.2011.0091
Kuusk S, Sohlberg JJ, Long JA et al (2002) STY1 and STY2 promote the formation of apical tissues during Arabidopsis gynoecium development. Development 129. https://doi.org/10.1242/dev.129.20.4707
Lai Y, Zhang D, Wang J et al (2020) Integrative transcriptomic and proteomic analyses of molecular mechanism responding to salt stress during seed germination in hulless barley. Int J Mol Sci 21. https://doi.org/10.3390/ijms21010359
Lamin-Samu AT, Farghal M, Ali M, Lu G (2021) Morpho-physiological and transcriptome changes in tomato anthers of different developmental stages under drought stress. Cells 10. https://doi.org/10.3390/cells10071809
Leprince O, Pellizzaro A, Berriri S, Buitink J (2017) Late seed maturation: drying without dying. J Exp Bot 68. https://doi.org/10.1093/jxb/erw363
Li Y, Guo D (2023) Transcriptome and DNA Methylome analysis of two contrasting Rice genotypes under salt stress during germination. Int J Mol Sci 24. https://doi.org/10.3390/ijms24043978
Li H, Li D, Chen A et al (2017) RNA-seq for comparative transcript profiling of kenaf under salinity stress. J Plant Res 130:365–372. https://doi.org/10.1007/s10265-016-0898-9
Li J, Iqbal S, Zhang Y et al (2021) Transcriptome analysis reveals genes of flooding-tolerant and flooding-sensitive rapeseeds differentially respond to flooding at the germination stage. Plants 10. https://doi.org/10.3390/plants10040693
Li Z, Luo X, Wang L, Shu K (2022) ABSCISIC ACID INSENSITIVE 5 mediates light–ABA/gibberellin crosstalk networks during seed germination. J Exp Bot 73. https://doi.org/10.1093/jxb/erac200
Lin Z, Li Y, Wang Y et al (2021) Initiation and amplification of SnRK2 activation in abscisic acid signaling. Nat Commun 12. https://doi.org/10.1038/s41467-021-22812-x
Liu PP, Koizuka N, Martin RC, Nonogaki H (2005) The BME3 (blue Micropylar end 3) GATA zinc finger transcription factor is a positive regulator of Arabidopsis seed germination. Plant J 44:960–971. https://doi.org/10.1111/j.1365-313X.2005.02588.x
Magnus Eklund D, Ståldal V, Valsecchi I et al (2010) The Arabidopsis thaliana STYLISH1 protein acts as a transcriptional activator regulating auxin biosynthesis. Plant Cell 22. https://doi.org/10.1105/tpc.108.064816
Maia J, Dekkers BJW, Provart NJ et al (2011) The re-establishment of desiccation tolerance in germinated arabidopsis thaliana seeds and its associated transcriptome. PLoS One 6. https://doi.org/10.1371/journal.pone.0029123
Matosevich R, Efroni I (2021) The quiescent center and root regeneration, vol 72. J Exp, Bot, pp 6739–6745
Meng X, Zhang S (2013) MAPK cascades in plant disease resistance signaling. Annu Rev Phytopathol 51. https://doi.org/10.1146/annurev-phyto-082712-102314
Mitchum MG, Yamaguchi S, Hanada A et al (2006) Distinct and overlapping roles of two gibberellin 3-oxidases in Arabidopsis development. Plant J 45. https://doi.org/10.1111/j.1365-313X.2005.02642.x
Mittler R, Zandalinas SI, Fichman Y, Van Breusegem F (2022) Reactive oxygen species signalling in plant stress responses. Nat Rev Mol Cell Biol 23. https://doi.org/10.1038/s41580-022-00499-2
Mohammed B, Bilooei SF, Dóczi R et al (2018) Converging light, energy and hormonal signaling control meristem activity, leaf initiation, and growth. Plant Physiol 176. https://doi.org/10.1104/pp.17.01730
Motte H, Vanneste S, Beeckman T (2019) Molecular and environmental regulation of root development. Annu Rev Plant Biol 70. https://doi.org/10.1146/annurev-arplant-050718-100423

Munns R (2002) Comparative physiology of salt and water stress. Plant Cell Environ 25:239–250. https://doi.org/10.1046/j.0016-8025.2001.00808.x
Nakashima K, Yamaguchi-Shinozaki K (2013) ABA signaling in stress-response and seed development. Plant Cell Rep 32. https://doi.org/10.1007/s00299-013-1418-1
Nakashima K, Fujita Y, Kanamori N et al (2009) Three arabidopsis SnRK2 protein kinases, SRK2D/SnRK2.2, SRK2E/SnRK2.6/OST1 and SRK2I/SnRK2.3, involved in ABA signaling are essential for the control of seed development and dormancy. Plant Cell Physiol 50. https://doi.org/10.1093/pcp/pcp083
Nakata M, Okada K (2013) The leaf adaxial-abaxial boundary and lamina growth. Plants 2. https://doi.org/10.3390/plants2020174
Narsai R, Edwards JM, Roberts TH et al (2015) Mechanisms of growth and patterns of gene expression in oxygen-deprived rice coleoptiles. Plant J 82. https://doi.org/10.1111/tpj.12786
Nguyen TP, Cueff G, Hegedus DD et al (2015) A role for seed storage proteins in Arabidopsis seed longevity. J Exp Bot 66. https://doi.org/10.1093/jxb/erv348
Nguyen D, Rieu I, Mariani C, van Dam NM (2016) How plants handle multiple stresses: hormonal interactions underlying responses to abiotic stress and insect herbivory. Plant Mol Biol 91. https://doi.org/10.1007/s11103-016-0481-8
Nishanth MJ (2024) Transcriptome meta-analysis-based identification of hub transcription factors and RNA-binding proteins potentially orchestrating gene regulatory cascades and crosstalk in response to abiotic stresses in Arabidopsis thaliana. J Appl Genet 1–15. https://doi.org/10.1007/s13353-024-00837-4
Olszewski N, Sun TP, Gubler F (2002) Gibberellin signaling: biosynthesis, catabolism, and response pathways. Plant Cell 14. https://doi.org/10.1105/tpc.010476
Pandey P, Ramegowda V, Senthil-Kumar M (2015) Shared and unique responses of plants to multiple individual stresses and stress combinations: physiological and molecular mechanisms. Front Plant Sci 6. https://doi.org/10.3389/fpls.2015.00723
Park J, Oh DH, Dassanayake M et al (2017) Gibberellin signaling requires chromatin remodeler PICKLE to promote vegetative growth and phase transitions. Plant Physiol 173. https://doi.org/10.1104/pp.16.01471
Pinyopich A, Ditta GS, Savidge B et al (2003) Assessing the redundancy of MADS-box genes during carpel and ovule development. Nature 424. https://doi.org/10.1038/nature01741
Raz V, Bergervoet JHW, Koornneef M (2001) Sequential steps for developmental arrest in Arabidopsis seeds. Development 128. https://doi.org/10.1242/dev.128.2.243
Sajeev N, Koornneef M, Bentsink L (2024) A commitment for life: decades of unraveling the molecular mechanisms behind seed dormancy and germination. Plant Cell 36:1358–1376. https://doi.org/10.1093/plcell/koad328
Schwechheimer C (2012) Gibberellin signaling in plants—the extended version. Front Plant Sci 2. https://doi.org/10.3389/fpls.2011.00107
Sewelam N, Kazan K, Schenk PM (2016) Global plant stress signaling: reactive oxygen species at the cross-road. Front Plant Sci 7. https://doi.org/10.3389/fpls.2016.00187
Seymour GB, Ostergaard L, Chapman NH et al (2013) Fruit development and ripening. Annu Rev Plant Biol 64:219–241. https://doi.org/10.1146/annurev-arplant-050312-120057
Sheshadri SA, Nishanth MJ, Simon B (2016) Stress-mediated cis-element transcription factor interactions interconnecting primary and specialized metabolism in planta. Front Plant Sci 7. https://doi.org/10.3389/fpls.2016.01725
Shimotohno A, Heidstra R, Blilou I, Scheres B (2018) Root stem cell niche organizer specification by molecular convergence of PLETHORA and SCARECROW transcription factor modules. Genes Dev 32. https://doi.org/10.1101/gad.314096.118
Shu K, Liu XD, Xie Q, He ZH (2016) Two faces of one seed: hormonal regulation of dormancy and germination. Mol Plant 9. https://doi.org/10.1016/j.molp.2015.08.010
Singh AK, Dhanapal S, Yadav BS (2020) The dynamic responses of plant physiology and metabolism during environmental stress progression. Mol Biol Rep 47. https://doi.org/10.1007/s11033-019-05198-4

Skalak J, Nicolas KL, Vankova R, Hejatko J (2021) Signal integration in plant abiotic stress responses via multistep Phosphorelay Signaling. Front Plant Sci 12. https://doi.org/10.3389/fpls.2021.644823

Smyth DR, Bowman JL, Meyerowitz EM (1990) Early flower development in Arabidopsis. Plant Cell 2. https://doi.org/10.1105/tpc.2.8.755

Su Z, Ma X, Guo H et al (2013) Flower development under drought stress: morphological and transcriptomic analyses reveal acute responses and long-term acclimation in Arabidopsis. Plant Cell 25:3785–3807. https://doi.org/10.1105/tpc.113.115428

Suzuki M, Wang HHY, McCarty DR (2007) Repression of the leafy cotyledon 1/B3 regulatory network in plant embryo development by VP1/Abscisic acid insensitive 3-like B3 genes. Plant Physiol 143. https://doi.org/10.1104/pp.106.092320

Tanaka M, Kikuchi A, Kamada H (2008) The arabidopsis histone deacetylases HDA6 and HDA19 contribute to the repression of embryonic properties after germination. Plant Physiol 146. https://doi.org/10.1104/pp.107.111674

Thapa R, Tabien RE, Johnson CD, Septiningsih EM (2023) Comparative transcriptomic analysis of germinating rice seedlings to individual and combined anaerobic and cold stress. BMC Genomics 24:1–25. https://doi.org/10.1186/s12864-023-09262-z

Theißen G, Melzer R, Ruümpler F (2016) MADS-domain transcription factors and the floral quartet model of flower development: linking plant development and evolution. Development 143. https://doi.org/10.1242/dev.134080

Thomson B, Wellmer F (2019) Molecular regulation of flower development. Curr Top Dev Biol 131. https://doi.org/10.1016/bs.ctdb.2018.11.007

Toyokura K, Goh T, Shinohara H et al (2019) Lateral inhibition by a peptide hormone-receptor Cascade during Arabidopsis lateral root founder cell formation. Dev Cell 48. https://doi.org/10.1016/j.devcel.2018.11.031

Verma V, Ravindran P, Kumar PP (2016) Plant hormone-mediated regulation of stress responses. BMC Plant Biol 16. https://doi.org/10.1186/s12870-016-0771-y

Vrebalov J, Pan IL, Arroyo AJM et al (2009) Fleshy fruit expansion and ripening are regulated by the tomato SHATTERPROOF gene TAGL1. Plant Cell 21. https://doi.org/10.1105/tpc.109.066936

Waadt R, Seller CA, Hsu PK et al (2022) Plant hormone regulation of abiotic stress responses. Nat Rev Mol Cell Biol 23. https://doi.org/10.1038/s41580-022-00479-6

Wen X, Liu C, Yang F et al (2024) Accurate Long-read RNA sequencing analysis reveals the key pathways and candidate genes under drought stress in the seed germination stage in Faba bean. Int J Mol Sci 25. https://doi.org/10.3390/ijms25168875

Wobus U, Weber H (1999) Sugars as signal molecules in plant seed development. Biol Chem 380. https://doi.org/10.1515/bc.1999.116

Yan H, Nie Y, Cui K, Sun J (2022) Integrative transcriptome and metabolome profiles reveal common and unique pathways involved in seed initial imbibition under artificial and natural salt stresses during germination of halophyte quinoa. Front Plant Sci 13:1–18. https://doi.org/10.3389/fpls.2022.853326

Yang Q, Cong T, Yao Y et al (2023) KNOX genes were involved in regulating axillary bud formation of chrysanthemum × morifolium. Int J Mol Sci 24. https://doi.org/10.3390/ijms24087081

Yu A, Zhao J, Wang Z et al (2020) Transcriptome and metabolite analysis reveal the drought tolerance of foxtail millet significantly correlated with phenylpropanoids-related pathways during germination process under PEG stress. BMC Plant Biol 20:1–17. https://doi.org/10.1186/s12870-020-02483-4

Zhang C, Luo W, Li Y et al (2019) Transcriptomic analysis of seed germination under salt stress in two desert sister species (populus euphraticaand P. Pruinosa). Front Genet 10:1–16. https://doi.org/10.3389/fgene.2019.00231

Zhang H, Zhao Y, Zhu JK (2020) Thriving under stress: how plants balance growth and the stress response. Dev Cell 55:529–543. https://doi.org/10.1016/j.devcel.2020.10.012

Zhang G, Liu Y, Gui R et al (2021) Comparative multi-omics analysis of hypoxic germination tolerance in weedy rice embryos and coleoptiles. Genomics 113:3337–3348. https://doi.org/10.1016/j.ygeno.2021.07.021

Zhang Q, Wang S, Qin B et al (2023) Analysis of the transcriptome and metabolome reveals phenylpropanoid mechanism in common bean (Phaseolus vulgaris) responding to salt stress at sprout stage. Food Energy Secur 12:1–15. https://doi.org/10.1002/fes3.481

Zhu Z, Chen H, Xie K et al (2020) Characterization of drought-responsive transcriptome during seed germination in adzuki bean (Vigna angularis L.) by PacBio SMRT and Illumina sequencing. Front Genet 11:1–14. https://doi.org/10.3389/fgene.2020.00996

Chapter 7
Transcriptomics and Plant Stress Responses

Muhammad Qadir, Aarifa Nabi, and Long Wang

Abstract Under stress conditions, plants are highly susceptible to both biotic and abiotic stresses, which can impede growth and disrupt development. Upon encountering these stresses, plants activate cascade transcriptional responses, initiating key genes and regulatory networks to form core survival strategies for adapting to harsh environments. Transcriptomics, as a research methodology, effectively deciphers transcriptional response strategies in plants under diverse environmental challenges. This chapter begins by introducing several cutting-edge technologies for studying plant stress responses, including RNA-Seq, microarray-based transcriptomics, comparative transcriptomics, single-cell transcriptomics, Iso-Seq, and qPCR. It also elaborates on experimental design and analytical methodologies for transcriptomic studies. The focus lies on the role of transcriptomics in identifying stress-responsive genes and the regulatory mechanisms of alternative splicing (AS) during stress adaptation. By leveraging third-generation sequencing technologies, the chapter systematically reveals structural variations in stress-induced full-length transcripts, novel non-coding RNAs (lncRNAs), and functional roles of complex gene models in stress adaptation. Future applications of transcriptomics in deciphering plant stress responses are also prospected. This work provides a systematic understanding of transcriptomics in plant stress responses and offers novel insights to guide its future development in stress resilience research.

M. Qadir · A. Nabi
College of Biology, State Key Laboratory of Chemo/Biosensing and Chemometrics, and Hunan Province Key Laboratory of Plant Functional Genomics and Developmental Regulation, Hunan University, Changsha, People's Republic of China

L. Wang (✉)
College of Biology, State Key Laboratory of Chemo/Biosensing and Chemometrics, and Hunan Province Key Laboratory of Plant Functional Genomics and Developmental Regulation, Hunan University, Changsha, People's Republic of China

National Center of Technology Innovation for Saline-Alkali Tolerant Rice, Gerater by Area Institute For Innovation, Hunan University, Changsha, People's Republic of China

Interdisciplinary and Intelligent Seed Industry Equipment Research Department, Yue Lushan Laboratory, Changsha, People's Republic of China
e-mail: wanglong8591@hnu.edu.cn

L. M. Vaschetto (ed.), *Plant Transcriptomics and Epitranscriptomics*, Concepts and Strategies in Plant Sciences, https://doi.org/10.1007/978-981-95-5183-5_7

Keywords Environmental adaptation · Transcriptomic profiling · Phenotypic plasticity · Alternative splicing dynamics · Isoform diversity · Long-read sequencing technologies · RNA Regulation

Abbreviations

ABRE	Abscisic Acid Response Element
APA	Alternative polyadenylation
AQPs	Aquaporins
ARFs	auxin response factors
AS	Alternative Splicing
bHLH	Basic Helix-Loop-Helix
BR	Brassinosteroids
CDPK	Calcium-Dependent Protein Kinase
circRNAs	circular RNAs
CSD	Cu/Zn superoxide dismutase
DEGs	Differentially Expressed Genes
DRE	Dehydration-Responsive Element
DREB	Dehydration-Responsive Element-Binding
eIF4E	Eukaryotic Initiation Factor 4E
FER	FERONIA
FT	FLOWERING LOCUS T
GRNs	gene regulatory networks
GRPs	Glycine-Rich RNA-Binding Proteins
hnRNPs	heterogeneous nuclear ribonucleoproteins
HSP	Heat Shock Protein
HSTF	Heat Stress Transcription Factor
HTS	High-Throughput Sequencing
Iso-Seq	Isoform Quantification
JA	Jasmonic acid
LCR	LEAF CURLING RESPONSIVENESS
LEA	Late Embryogenesis Abundant
LncRNAs	non-coding RNAs
MAPK	Mitogen-Activated Protein Kinase
NGS	next-generation sequencing
PacBio	Pacific Biosciences sequencing technology
PEG	polyethylene glycol
PRRs	Pattern Recognition Receptors
PTR	post-transcriptional regulation
RAM	root apical meristem
RBP	RNA-Binding Protein
RIP	RNA Immunoprecipitation
RIP-seq	RNA Immunoprecipitation followed by sequencing
RLKs	receptor-like kinases
RNA-Seq	mRNA Sequencing

ROS	reactive oxygen species
RPA	RNA Polymerase II-associated Protein
SA	salicylic acid
scRNA-Seq	Single-cell RNA sequencing
siRNAs	small interfering RNAs
SPL	SQUAMOSA PROMOTER BINDING PROTEIN-LIKE
SR	Serine/Arginine-Rich
TFs	Transcription Factors
uORFs	Upstream Open Reading Frames
UTRs	5' untranslated regions
WGCNA	Weighted Gene Co-expression Network Analysis

7.1 Introduction

The phenotype of an organism arises from the dynamic interplay between its genotype and environmental factors, collectively shaping its observable traits. Neither the genetic makeup nor environmental influences alone can fully dictate phenotypic outcomes. Integrated multiomics approaches have emerged as indispensable tools for elucidating the molecular mechanisms of complex traits in biological systems. By synthesizing data across hierarchical levels—from environmental stimuli to genes, mRNAs, proteins, and metabolites—these approaches reconstruct gene regulatory networks (GRNs), elucidating causal relationships and molecular interactions that drive phenotypic plasticity (Boi 2011; Zarayeneh et al. 2017).

Transcriptomics, as a cornerstone methodology in multiomics frameworks, enables the systematic investigation of RNA transcripts at the cellular or tissue level. It enables the evaluation of gene transcription and provides a quantitative assessment of changes in gene expression at specific time points or under particular conditions. As sessile organisms, plants are inextricably tied to their environment and must endure adverse conditions that limit their growth, development, and productivity. These stresses can be broadly categorized as biotic or abiotic, both of which can have detrimental effects on plant health. Tolerance to such stresses involves the regulation of multiple genes and complex gene networks. At the transcriptional level, transcription factor families such as the NAC and WRKY families activate stress-responsive genes (e.g., *DREB2A* and *RD29A*) to establish an early warning system; posttranscriptional alternative splicing (AS) mediates rapid stress adaptation (Surówka et al. 2020).

7.2 Applications of Different Transcriptomic Methods

Since the first application of microarray technology in studies of *Arabidopsis thaliana* stress response in 1995, transcriptomics has become a "molecular microscope" for deciphering plant stress response mechanisms (Schena et al. 1995).

Several transcriptomic approaches have been developed, each with distinct advantages and applications:

7.2.1 Microarray-Based Transcriptomics

This pioneering technology utilizes the hybridization of fluorescently labeled cDNA to immobilize gene-specific probes. The signal intensity correlates with the transcript abundance, enabling cost-effective, high-throughput expression profiling (Conway and Schoolnik 2003). However, its dependence on predefined gene sets and limited sensitivity limit its application in novel transcript discovery.

7.2.2 mRNA Sequencing (RNA-Seq)

RNA-Seq is a next-generation sequencing (NGS) technology that sequences the entire transcriptome in a high-throughput manner, which produces an all-inclusive, computable profile of all genes expressed, i.e., coding and ncRNAs (non-coding RNAs). RNA-Seq is exceptionally valuable for de novo transcriptome assembly without a reference genome in species, analysis of gene expression under varying environmental conditions, detection of the AS, and the discovery of novel ncRNAs (Mutz et al. 2013).

7.2.3 Single-Cell RNA Sequencing (scRNA-Seq)

As an advanced version of RNA-Seq, single-cell RNA sequencing (scRNA-Seq) enables the analysis of gene expression at the single-cell level. This high-throughput procedure reveals heterogeneity in gene expression at the cellular level, which might have been obscured in the bulk RNA-Seq experiments. This technique is exceptionally useful for assessing tissue-specific responses, cellular differentiation and growth responses, and cell-type-specific gene expression patterns (Saliba et al. 2014).

7.2.4 Isoform Quantification (Iso-Seq)

Iso-Seq is a specified RNA-Seq procedure that emphasizes the reading of the full-length transcript along with its alternative isoforms. Such techniques are predominantly vital for studying transcriptional regulation and related AS events, enabling a deeper understanding of the intricacies of gene expression (Gao et al. 2019).

7.2.5 Comparative Transcriptomics

Comparative transcriptomics can be used to compare transcriptomes across different species and under different conditions, revealing the genetic basis of morphological diversity and morphogenesis in various organelles (Li et al. 2024). This approach is predominantly convenient for revealing the adaptive and evolutionary responses of different species to environmental stressors.

7.2.6 Quantitative PCR (qPCR)

This technique is not strictly considered a transcriptomic method but is broadly used to quantify the expression levels of a particular gene. qPCR is usually used as a follow-up technique to validate the outcomes from microarray or RNA-seq analysis. The qPCR technique is fine-tuned, precise, sensitive, and potentially cost-effective for assessing the expression of target genes; nevertheless, it lacks the breadth of a transcriptome-wide analysis (Buschmann et al. 2016).

7.3 Technical Characteristics of Transcriptomic Methods

7.3.1 Experimental Design

The experimental design is crucial and serves as the foundation and core of the transcriptomic research. A proper and well-structured experimental design enables the recording of dynamic plant responses to stress and the elucidation of gene functions based on temporal, dosage, and stress-type interactions. Some of the special experimental models and their applications are discussed below.

7.3.2 Time Series Model

A time series model is presented as a scientifically robust and widely adopted model to precisely dissect the dynamics of plant stress responses. By treating plants and collecting samples at multiple time points, sequential shifts in gene expression can be easily noted, revealing intricate regulatory mechanisms. For example, in a rice salt stress experiment, samples collected at 0, 6, 24, and 72 h posttreatment exhibited the prompt activation of early-response genes (e.g., *OsNHX1*) and deferred expression of mid-to-late adaptive genes (e.g., *SOS2*) (Almeida et al. 2017; Gong et al. 2001). Such an approach highlights the prime regulatory factors and adaptation mechanisms across stress phases, offering an approach to identify key regulatory factors and adaptation strategies across stress phases, offering an analytical understanding for the development and breeding of stress-tolerant crops. The time series model is exceptionally appropriate for studying GRNs in rapid-response and long-term adaptation processes.

7.3.3 Concentration Gradient Model

A concentration gradient model employs multiple concentrations of stressors (e.g., osmotic or salt) in a dose-dependent manner to identify stress thresholds and determine the dose-dependent effects of stressors in the expression of resistance genes (Yang et al. 2025). This model also estimates plant tolerance boundaries by incorporating physiological metrics with transcriptomic data, providing a complete view of adaptation mechanisms under sorted stress conditions.

7.3.4 Cross-Stress Model

The cross-stress model was designed to disentangle interactions between stress response pathways; it applies stressors in combinations and relates the consequences to single-stress responses. This methodology untangles cross-talk mechanisms in multiple stress circumstances, advancing our knowledge of plant flexibility in multiple environments and facilitating the development of crops tolerant to multiple stressors (Pandey et al. 2015).

7.3.5 Sampling Strategies

The legitimacy and reproducibility of the transcriptomic data of plants rely on scientifically sound and reproducible sampling strategies. Tissue-based responses to environmental stimuli require subsequent sampling protocols. In most cases, traditional bulk sampling (e.g., whole plants or leaves) frequently overlooks cellular heterogeneity, jeopardizing the exclusion of crucial genes and pathways responsible for the effects. Cutting-edge and modern techniques focus on tissue-specific or, in other cases, single-cell sampling to determine stress-specific regulatory networks (Cho et al. 2025).

7.3.6 Tissue-Specific Sampling

Unique stress response mechanisms are employed in different parts of the plants. For example, under saline conditions, roots exploit *SOS1*-mediated Na^+ localization and compartmentalization, whereas *SOS2*-activated H^+-ATPase transport to sustain ion homeostasis, and leaves control stomatal closure via *NCED3*-controlled ABA biosynthesis to reduce water loss. Specimens from roots, leaves, or other tissues independently reveal tissue-specific functional networks (Zhao et al. 2021a, b).

7.3.7 Single-Cell Sampling

Single-cell approaches are needed to scrutinize fine-grained regulatory mechanisms across cell subpopulations, and advances in scRNA-seq have increased the resolution of cellular heterogeneity. For example, scRNA-seq of maize roots revealed that *ZmWOX5b* and *ZmPIN1a* suppress F. verticillioides (Fv) infection in the root apical meristem (RAM) by regulating IAA biosynthesis, thereby revealing a global transcriptional landscape of cell fate determination during root development and delineating the immune regulatory network within major root tip cell types at single-cell resolution (Cao et al. 2023).

7.4 Data Analysis Workflow

7.4.1 Differentially Expressed Gene (DEG) Screening

DEG identification is pivotal for understanding the molecular mechanisms of stress responses. Tools such as DESeq2 (negative binomial model) and EdgeR (empirical Bayes) standardize the RNA-Seq count data and detect significant changes in expression. Biological replicates ($\geq$3 are recommended) mitigate batch effects and environmental noise, increasing the statistical rigor. Experimental designs should balance the replication frequency with cost efficiency to optimize precision and practicality.

7.4.2 Coexpression Network Construction

1. Weighted gene coexpression network analysis (WGCNA).
   WGCNA clusters genes into functional modules via weighted correlation networks. Topological overlap measure (TOM)-based hub genes often orchestrate core regulatory functions. For example, drought-responsive hub genes in plants may amplify stress signals through crosstalk among multiple pathways. Module correlations further link transcriptomic data to plant traits, guiding stress-resilient breeding (Yu et al. 2020).
2. Cytoscape visualization.
   Cytoscape maps regulatory networks using WGCNA-derived hub genes (kME $>$ 0.8). Node attributes and STRING-validated protein interactions

elucidate functional relationships (Yu et al. 2020). For instance, in the response of *Arabidopsis* to drought, the transcription factor *AREB1* forms a regulatory hub with the downstream gene *RD29A*. Integration with metabolic data results in the construction of gene-metabolite networks, revealing stress-induced metabolic complexity (Li et al. 2019).

7.4.3 Functional Annotation and Pathway Enrichment Analyses

GO analysis annotates DEGs across biological processes, cellular components, and molecular functions, whereas KEGG analysis focuses on metabolic and signaling pathways. PlantGSEA leverages plant-specific gene sets for functional mining. It identifies monocot-specific stress response pathways and novel genes or modules through integration with whole-genome data, advancing plant stress biology.

7.5 Abiotic Stress and Plant Transcriptome Responses

Through transcriptome analysis technology, we can delve into the dynamic changes in gene expression in organisms exposed to various stress conditions, and this method has become a key tool for understanding the mechanisms of biological adaptation. In the field of abiotic stress, the method has revealed how organisms respond to challenges posed by abiotic factors such as extreme temperatures, drought, salinity, and heavy metal pollution. Next, we introduce several application cases of transcriptomics in the context of abiotic stress in detail to demonstrate its analytical capabilities and practical application value (Wang et al. 2020a, b, c).

7.5.1 Drought Stress

Under conditions of drought stress, many drought-related genes are involved in the drought response mechanism to cope with stressful conditions. Transcriptome analysis has revealed that a subset of these proteins, whose expression levels are substantially altered, plays an active role in the adaptive mechanisms of host plants. These proteins, which often function as molecular osmoprotectants, or osmoregulators, and play an active role in the stress signaling pathways, exhibit differential expression patterns to mitigate the effects of water scarcity and cope with drought conditions. These transcriptomic findings provide valuable insights into the particular proteins and their regulatory mechanisms that are triggered or suppressed in response to drought stress, suggesting possible targets for the improvement of drought-tolerant crops and approaches for increasing plant elasticity in arid surroundings (Guo et al. 2025).

7.5.1.1 ABA-Responsive Genes

Abscisic acid (ABA) is considered a core hormone involved in plant stress signaling. Its responsive genes are involved in osmoregulation, stress resilience, and osmotic adjustments. For example, certain genes, such as *RD29A* and *RD22,* are stimulated by ABA, leading to the closure of the stomatal opening to mitigate water loss, increase the synthesis of osmoprotectants, and increase the expression of stress-related molecules such as LEA proteins. Xing et al. (2020) performed a relative transcriptomic investigation in sugar beet, revealing an ABA-responsive regulatory network connected to cell wall organization and oxidation-reduction processes. They identified 5,625 transcription elements contributing to ABA-responsive pathways, including key genes encoding AP2 and Dof domain-containing genes. Furthermore, transcriptomics revealed a novel abscisic acid response element in the rice aleurone layer, which was confirmed to affect ABA signaling (Watanabe et al. 2017). This breakthrough provides a molecular switch for the biotechnological manipulation of gene expression in crops to increase environmental stress resilience.

7.5.1.2 Dehydration-Responsive Element-Binding Proteins (DREB)

A class of transcription factors (TFs) called DREB proteins have the potential to act as dehydration-responsive elements (DREs) in the promoter regions of stress-responsive genes. These proteins play a crucial role in the adjustment of osmotic balance by inducing the expression of genes that actively promote the synthesis of osmoprotectants such as proline, low-molecular-weight sugars, and glycine betaine (Lata and Prasad 2011). Multiple studies have investigated the potential roles of *DREB* genes under stressful environmental conditions. Among the *DREB* genes, *DREB1/CBF*, which is particularly active under chilling conditions and *DREB2*, which is active under dehydration conditions, are the primary regulators of multiple stress-regulating pathways. In model plants such as rice, an investigation of *Arabidopsis DREB2* sequences revealed five *DREB2* homologs (Herath 2016). *OsDREB2A*, OsDREB2B, and *OsDREB2C* are expressed in response to abiotic stress. Among these genes, *OsDREB2B* is expressed in plants under cold, salt, and drought stress, whereas *OsDREB2A* and *OsDREB2C* are expressed only in plants under salt and drought conditions. These discoveries further support the role of *OsDREB2* genes in stress response and hormone regulation.

7.5.1.3 Aquaporins (AQPs)

AQPs are an important and vital class of membrane proteins that facilitate the transport of water and other neutral solute molecules across the plasma membrane. These membrane proteins play important roles in regulating tight water homeostatic conditions and serve as important baselines for the development of drought-resilient crops. Their expression is fine-tuned at numerous levels, such as the transcriptional and posttranslational levels. At the transcriptional level, their expression is fine-tuned by various hormones, e.g., brassinosteroids, ABA and GA, and environmental factors such as low temperature, drought, light, and saline conditions. At the posttranslational level, modifications such as phosphorylation and hetero-oligomerization further regulate AQP activity (Chaumont et al. 2005; Liu et al. 2013). Such coordination in the regulation of transcriptional and posttranslational levels confirm the fine-tuned expression and functionality of AQPs under varying physiological conditions.

When the abiotic conditions prevail, the expression of certain AQPs might undergo predominant changes to maintain a homeostatic water flux and osmotic balance. For example, certain AQPs are upregulated when plants face drought or saline environments to increase water uptake by roots, whereas others prevent dehydration by ceasing water loss from the aerial parts of the plants (Chaumont and Tyerman 2014). Additionally, the AQP subfamily members have different functions i.e., some primarily mediate water uptake by roots whereas the rest play critical roles in the leaf water homeostasis. This functional differentiation enables plants to achieve efficient water utilization in complex and dynamic environments.

7.5.1.4 Late Embryogenesis Abundant (LEA) Proteins

Under dehydration conditions, LEA proteins are particularly known to play a vital role in protecting cellular structures. They actively participate in osmotic adjustments by maintaining osmoregulation, stabilizing membrane integrity and preventing the denaturation of other vital proteins during osmotic stress (Priya et al. 2019). Several transcriptomic studies have revealed that *LEA* genes are overexpressed under conditions of osmotic imbalances (Banerjee and Roychoudhury 2016; Wang et al. 2014; Yu et al. 2016). Additionally, the transcriptomics has revealed the regulatory mechanisms by which the hormones such as ABA, which is critical for stress responses, regulate the expression of *LEA* genes. Pedrosa et al. (2015) reported that polyethylene glycol (PEG)-induced stress has upregulated the expression of CsLEA4 in underground parts and CsLEA63 in aerial parts but downregulated the expression of CsLEA11, CsLEA54, CsLEA58, CsLEA62, and CsLEA67 in leaves. Nevertheless, saline conditions downregulated the expression of CsLEA11 and CsLEA62 in roots. Under water-limited conditions, the mRNA expression of CsLEA55 and CsLEA60 in roots and that of CsLEA4 in leaves was distinctly upregulated. Researchers have proposed that these factors form multifaceted regulatory networks through interactions or coregulation. For instance, DREB may control ABA-responsive genes, although aquaporins and LEA proteins may cooperate to sustain cellular hydration.

7.5.2 Salinity Stress

When saline conditions prevail, plants accomplish adaptive strategies via the dynamic regulation of ion transporter gene expression. The primary transport system might include high-affinity potassium transporters, Na^+/H^+ antiporters, and potassium channels, whose expression levels vary substantially during stress conditions. Transcriptomic data have provided a systematic perspective to decipher the roles of these genes in maintaining homeostasis and increasing stress resilience. Additional studies have further validated that stress-responsive TFs specifically control ion transport and homeostasis networks by regulating transporter transcriptional activity via the perception of salt stress or plant hormones.

Integrated studies of DNA methylation and histone modifications, combined with transcriptomic analyses, have elucidated the molecular basis of salt stress memory. Yang and Guo (2018) revealed that salt stress induces the differential expression of histone acetyltransferases and DNA methyltransferases, enabling long-term activation of stress-responsive genes. For instance, dynamic DNA methylation regulates TF activity in soybean stress responses (Song et al. 2012). In maize roots, salt stress alters histone acetylation levels at promoter regions of cell cycle-related genes, thereby affecting cell division. Increased H3K9/H3K27 acetylation in both cultivated beet (*Beta vulgaris*) and wild beet (*Beta maritima*) activates peroxidase-encoding genes to increase their antioxidant capacity (Yolcu et al. 2016).

LncRNAs participate in salt stress responses through the targeted regulation of key genes. In *Arabidopsis*, miR394 modulates the cellular morphological adaptation by suppressing leaf curling responsiveness (LCR) gene expression (Song et al. 2013). Rice miRNAs (miR164e, miR408, and miR414) coordinate stress signaling by targeting DEAD-box helicase family genes (Macovei and Tuteja 2012). These findings reveal intricate RNA-level regulatory mechanisms within salt stress response networks.

7.6 AS and Isoform Diversity

AS at the transcriptional level in plants is a pivotal and crucial posttranscriptional mechanism that enables a single gene to synthesize different mRNA isoforms and hence increases the diversity at the translational level (Muhammad et al. 2023; Fig. 7.1). This diversity subsequently increases the functional capability of the host without increasing in the size of the host genome (Mastrangelo et al. 2012). During AS, based on the needs and circumstances, introns and exons are selectively included or excluded from the pre-mRNA, resulting in the production of isoforms whose structures and functions vary (Fig. 7.1; Rogalska et al. 2023). Such diversity at the translational level enables the host to effectively respond to and undergo morphological changes to withstand such environmental cues at different developmental stages and under different stress conditions, allowing plants to increase their ecological amplitude according to the changing environments (Mozgova et al. 2019). RNA-binding proteins and splicing factors are the guards that tightly check and regulate AS in plants through the recognition of specific sequence elements within the pre-mRNA (Rigo 2021). Any anomalies in AS could lead to vulnerability

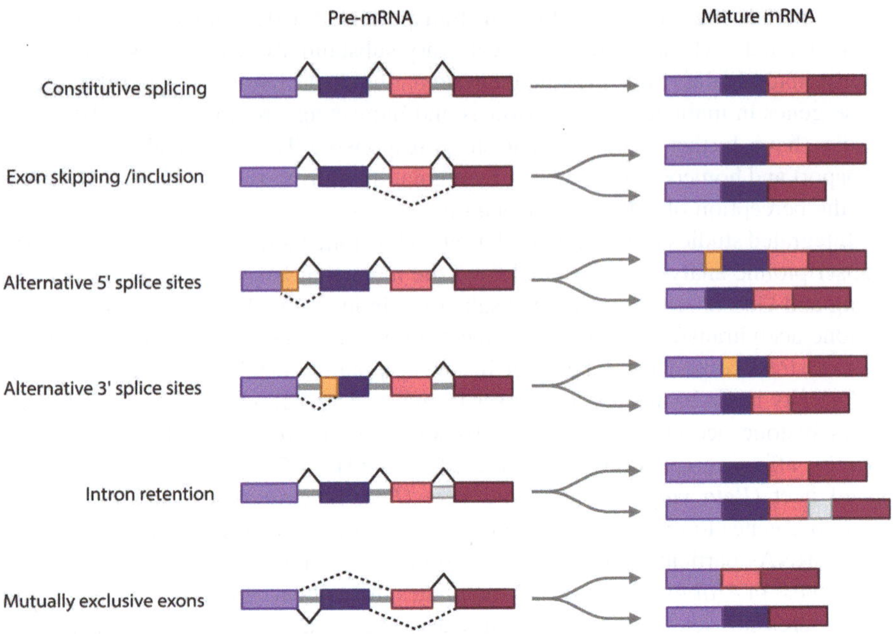

Fig. 7.1 Schematic representation of the RNA maturation from pre-mRNA, facing alternative splicing to synthesize a mature mRNA. The five forms of AS in plants
The diagram represents a critical step of mRNA splicing crucial for the maturation of pre-mRNA (precursor mRNA) into mature mRNA, following translation to produce proteins according to the needs of the cell. The left side represents pre-mRNA while the right side represents mature mRNA, showing the direction of mature from pre-mRNA to mature mRNA experiencing the splicing process. The pre-mRNA exons (the colored regions) represent the coding regions while the line in between represents the introns connecting the exons. The dotted line represents the possible alternative splicing events. On the right side, after the alternative splicing events, only the exon regions and introns have been removed. These only represent the coding regions, which will be translated into a protein according to the needs of the cell

to stress and subsequently reduce and impair growth, indicating that AS is more directly involved in plant resilience and stress tolerance (Muhammad et al. 2023).

In plants, stress-induced AS is considered pivotal because it predominantly contributes to proteome diversity and functional plasticity (Fig. 7.1), allowing plants to tailor the particular adaptive responses to a broad spectrum of environmental stress stimuli (Kjer-Hansen and Weatheritt 2023). In a condition when a plant senses certain stress stimuli such as drought, salinity, heavy metals, and pathogens, plants increase the activity of AS mechanisms to produce distinct isoforms with a particular role in precisely confronting a specific anomaly (Chaves and Oliveira 2004). For example, in the case of drought, *Arabidopsis thaliana* synthesizes more *DREB2B* isoforms with variable DNA-binding properties to effectively activate drought-responsive genes, which are pivotal for maintaining the osmoregulation and homeostatic conditions (Barbosa Amorim et al. 2016). In *Oryza sativa*, an isoform of the *OsbZIP73* gene is synthesized for acclimation to chilling stress, and its expression is regulated to stabilize the cellular function under chilling conditions (Usman et al. 2023). Similarly, pathogen-associated AS of *RPM1* leads to an enhanced and strong defense response (Zhao et al. 2021a, b), whereas *A. thaliana* produces heat-specific isoforms of *HSFA2* that induce the expression of heat shock proteins, protecting other crucial cellular proteins under conditions of thermal stress (Friedrich et al. 2021). Similarly, the interaction of the extracellular rapid alkalinization factor 1 (RALF1) peptide with its receptor FERONIA (FER) triggers a rapid and extensive RNA AS response in *A. thaliana*. This response involves the phosphorylation of glycine-rich RNA-binding protein 7 (GRP7) by FER, which increases the nuclear accumulation of GRP7. Phosphorylated GRP7 has an increased mRNA binding capacity and is associated with the spliceosome component U1-70K, facilitating precise splice site selection and modulating dynamic AS. Notably, the AS of GRP7 itself induces nonsense-mediated decay, creating a feedback loop within the RALF1-FER-GRP7 module to fine-tune stress responses and regulate cell growth. Furthermore, the genetic reversal of a RALF1-FER-dependent splicing target partially rescues GRP7-deficient mutants, underscoring the critical role of this regulatory module in maintaining cellular function under stress conditions (Wang et al. 2020a, b, c).

7.7 Complex Plant Genome and Long-Read Sequencing Technologies

7.7.1 The Application of Long-Read Sequencing Technology in Transcriptomics

With respect to plant adaptability to harsh environmental conditions, lncRNAs, including long lncRNAs and microRNAs, are being increasingly recognized for their potential and intricate regulatory roles. Nevertheless, the functional diversity characterization of ncRNAs is an immense challenge because of the limitations of the available short-read technologies. The advanced long-read sequencing technology, which is able to read entire RNA molecules without any obstruction, leads to the more accurate detection and characterization of ncRNAs together with all sequences responsible for the stress resilience of the host (Ma et al. 2022). Long-read sequencing technologies, including Oxford Nanopore and Pacific Biosciences (PacBio), have revolutionized genomics by enabling the reading of continuous DNA and RNA strands over tens of thousands of bases (Fig. 7.2). Unlike short-read platforms, which require sequence assembly of usually missing complex regions, an unprecedented access has been provided to whole genes, repetitive sequences, and structural variants by long-read technologies. The use of this technology in plant biology is fruitful, as large genomes and polyploidy, and repetitive sequences pose different challenges (Lang et al. 2020).

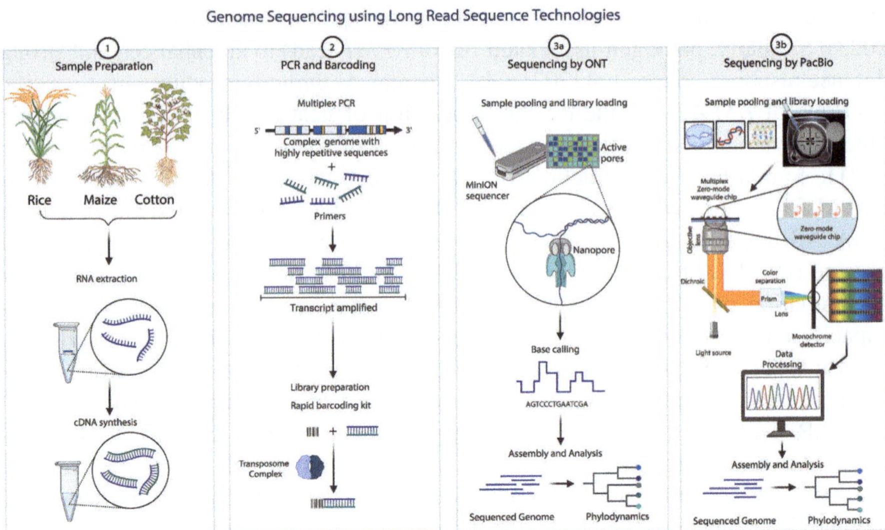

Fig. 7.2 The role of long read sequence technology of Oxford Nanopore technology and PacBio in sequencing the complex genome of polyploid plants
The above figure points out the intricate workflow of genome sequencing technologies, i.e., long read sequencing technologies, emphasizing two cutting-edge techniques: Oxford Nanopore Technology and Pacific Biosciences (PacBio). The whole process starts with the extraction of RNA from plants with complex and polyploid genomes with highly repetitive sequences, including rice, maize, and cotton. The RNA extraction is followed by the synthesis of cDNA. In the PCR and barcoding step, the highly repetitive and complex genome is amplified using multiplex PCR using specific primers. Following that, a library is developed through a rapid barcoding kit, generating a transposome complex. In the case of the ONT sequencing, the library is assembled and mounted onto an active pore of the MinION sequencer, where base calling is performed in real-time through the nanopore. The sequenced data is then assembled and analyzed to generate the genome and phylodynamic information. In the case of PacBio sequencing, the same techniques are followed as for ONT and but samples are loaded onto a sequencing chip with active pores. The analyzer of the PacBio sequencer consists of a set of a light source, objective lens, dichroic mirror, prism, and lens to image the fluorescence signals emitted during DNA synthesis. The data collected is further processed to produce a high-fidelity sequence, followed by its assembly and analysis to produce the genome and phylodynamic information

With respect to advanced and cutting-edge technologies, the Oxford Nanopore sequencing platform, which is based on nanopore technology, captures DNA or RNA molecules through biological nanopores embedded in a membrane (Branton et al. 2008). ONT produces distinct electrical signals when a nucleotide passes through its pore and subsequently determines the nucleotide sequence (Wanunu 2012). In plants, ONT has shown potential efficiency in resolving long, repetitive regions that are shared by agronomically important species, such as *Zea mays* (maize) and *Triticum aestivum* (wheat) (Fig. 7.2, Section 3a). ONT correspondingly facilitates whole-transcriptome studies in plants, which are primarily strategic for depicting AS and subsequent isoform diversity under stress conditions (Logsdon et al. 2020). For instance, in drought-stressed *O. sativa* (rice), direct sequencing of full-length mRNA isoforms was performed using ONT, providing support for the efficient recognition of stress-induced AS in genes responsible for osmoregulation. By understanding complete transcript isoforms, ONT sidesteps the assembly problems characteristic of short-read sequencing, guaranteeing that infrequent or tissue-specific isoforms are precisely depicted (Sun et al. 2020).

The former approach uses electrical signals to estimate and record the nucleotide sequences, whereas the latter uses single-molecule real-time sequencing, which is competent for constructing long reads with high precision due to the use of circular consensus sequencing. PacBio's high-fidelity sequence reads are characteristically approximately 15–25 kb long, which merge precision and read length, making it best suited for plant genomes that encompass highly repetitive sequences or polyploid complexity (Fig. 7.2 Section 3b) (Espinosa et al. 2024). For example, PacBio sequencing played a key role in assembling the complex genome of *Gossypium hirsutum* (cotton), which is polyploid and encompasses an enormous number of repetitive regions. The higher accuracy and read length permitted for a particular assembly of gene-enriched areas and resolution of homologous chromosomes, which further improved our understanding of the elasticity of the cotton genome to ecological pressure. PacBio also facilitates a systematic analysis of the plant transcriptome. Its ability to capture full-length cDNA sequences allows for very efficient classification of isoforms of stress-responsive genes (Wang et al. 2020a, b, c). In *A. thaliana*, PacBio has shown that stress-induced isoforms of *SR45*, a serine/arginine-rich (SR) protein involved in AS, play a role in fine-tuning gene expression in response to osmotic stress. By providing direct reads of these isoforms, PacBio reduces predispositions associated with assembly and transcriptional reconstruction, resulting in a more accurate picture of transcriptome complexity (Albaqami 2019).

Another notable application of long-read sequencing technologies is the potentially detailed classification and characterization of AS, a mechanism that expands protein isoform diversity and plays a central role in stress adaptation (van Rengs et al. 2022). In *Zea mays*, for instance, the activity of all the genes responsible for photosynthetic competence and drought resistance has been investigated using ONT. Such a technique of reading full-length transcripts allows scientists to classify the isoforms related to stress resilience, such as those that modify protein interactions or subcellular localization, helping plants to improve their ecological resilience, including under conditions of water deficiency, while modulating their photosynthetic pathways (Vikas et al. 2023). On the other hand, researchers have used PacBio technology to identify a previously uncharacterized isoform of the flowering locus T (FT) gene, providing insights into how AS of this gene influences flowering during different photoperiods. This discovery is crucial for developing maize varieties with optimized flowering times suited for specific growing conditions (Han et al. 2023).

Long-read sequencing technologies can be used to elucidate the mechanisms of AS that are critical for the development of numerous protein isoforms to help plants effectively respond to environmental cues. Nevertheless, Oxford Nanopore technology contributed effectively in identifying the stress-induced isoforms of genes involved in osmotic regulation in rice (Kamali and Singh 2023; Zhong et al.

2024). In some recent studies under drought stress conditions, several isoforms of the *OsDREB2* transcription factor have been found to be specifically expressed under drought conditions, regulating the expression of downstream targets and increasing root development. Root development and tight osmoregulation under such conditions demonstrate how AS increases host drought resilience (Kumar et al. 2023). In another study with *Gossypium hirsutum* (cotton), certain isoforms of *GhMYB* detected using long-read sequencing techniques were shown to play a potential role in fiber development and stress resilience. Specific isoforms of *GhMYB1* are upregulated in response to thermal stress. Their upregulation helps the plant to adjust physiological processes vital for sustaining fiber quality in the host and provides the plasticity for fiber development under varying environmental conditions, potentially explaining the importance of AS. The identification of full-length transcripts and stress-responsive isoforms is important for the potential adaptability and tolerance of plants (Su et al. 2024; Wang et al. 2024). Similarly, in poplar, an isoform of a potassium transporter gene has been discovered with the help of long-read sequencing technology; this isoform is differentially expressed under saline conditions and increases the ionic hemostatic and osmotic adjustments of the host, which are vital for flourishing under salt-stress conditions (Lv et al. 2024). In *Solanum lycopersicum*, lncRNAs were reported to alter gene expression in response to biotic stress, resulting in significantly altered expression upon infection with *Phytophthora infestans.* The modulation of gene expression by lncRNAs revealed the intricacy of the regulatory network after stimulus perception and revealed the roles of these lncRNAs in modulating the expression of defense-related genes (Su et al. 2023). In another study of *Helianthus annuus* (sunflower) using long-read sequencing technologies (PacBio in this case), variations in stress-responsive isoform expression linked to drought tolerance traits were mapped, which facilitated marker-assisted selection strategies aimed at developing drought-resistant sunflower varieties (Dwivedi et al. 2023).

7.7.2 Resolving Complex Gene Models

Plant genomes are best known for their intricacy, which is distinguished by the enormous number of repetitive sequences, polyploidy, and extensive AS. Conventional techniques often fail to read and capture this complexity, particularly for polyploid plants such as *Triticum aestivum* (wheat) and *Gossypium* spp. (cotton). Nevertheless, cutting-edge technologies produce full-length sequence reads that cover entire genes, enabling us to view continuous, high-resolution transcripts (Wang et al. 2023). For instance, PacBio-based sequencing of *Glycine max* (soybean) has provided precise annotations of intricate gene models, enabling host salinity stress resilience (Lackey 2024). The previously unannotated isoforms of transcription factor genes, i.e., *GmNAC,* were also included in these models, which potentially maintain the homeostatic conditions of ion transport and osmotic balance under salt stress. Using these cutting-edge technologies to read full-length transcripts, researchers have been able to elucidate the role of AS and isoform diversity in stress mitigation and host resilience, which are vital traits for crops in salinity-prone areas (Hoang et al. 2021). Similarly, in *A. thaliana*, the full-length structure of the genes responsible for the production of heat shock protein (*HSP*) family members revealed novel isoforms that are precisely overexpressed under thermal stress conditions. This information helps to elucidate the regulatory network that regulates protein stability and folding during thermal stress, a response essential for plant survival at high temperatures (Yu et al. 2024).

7.7.3 Identifying Novel ncRNAs and Functional Insights

The cutting-edge long-read sequencing technique has aided in the identification of novel ncRNAs that were formerly unable to be identified by conventional methods. For instance, in *Zea mays* (maize), a novel ncRNA was strongly linked to nutrient-deficient conditions like nitrogen and phosphorus deficiency. Certain lncRNAs discovered in this work have been linked to nitrogen uptake pathways, providing a possible link to the *Z. mays* strategies for nutrient acquisition under nutrient-limited conditions. Such functional ncRNAs can play a major role in breeding strategies designed to improve crop resilience in nutrient-poor soils (Ijaz et al. 2024). Furthermore, novel circular RNAs (circRNAs) that are responsive to both biotic and abiotic stresses, including pathogen attack and drought, have been discovered in *Solanum lycopersicum* (tomato). They tend to control homeostasis, likely functioning as molecular sponges for stress-related miRNAs, thereby stabilizing the expression of stress-defense genes. These insights deepen our understanding of the multifunctional roles that ncRNAs can play not only in directly regulating genes but also in safeguarding essential regulatory networks under stress (Li et al. 2023a, b; Wang et al. 2021).

7.8 Molecular Mechanisms Underlying Plant Stress Responses

Plants use intricate systems to perceive, signal, and respond accordingly to cope with biotic and abiotic stressors in the environment. When plants are exposed to certain environmental cues, i.e., drought, salinity, heavy metals, temperature, or phytopathogens, they trigger a cascade of molecular responses to maintain homeostatic conditions. These mechanisms include the perception of stress stimuli at specific sites on the cell membrane (Nykiel et al. 2022). Such perception involves molecules, particularly receptor-like kinases (RLKs) and ion channels that activate downstream signaling pathways, e.g., the mitogen-activated protein kinase (MAPK) and calcium-dependent protein kinase (CDPK) pathways (Nykiel et al. 2022). The perception of these stress gene signals is amplified by these cascades, which in turn activate the stress-related TFs, including DREB, WRKY, and NAC, which regulate the expression of stress-responsive genes (Chaffai et al. 2024). Similarly, plants also employ molecular mechanisms to detoxify reactive oxygen species (ROS), synthesize osmoprotectants, and regulate hormonal pathways (e.g., ABA, JA, and SA) to promote host defense mechanisms and modulate physiological and biochemical responses. In addition, gene expression during these responses is further fine-tuned by AS and epigenetic modifications to increase proteome diversity and adaptability. Collectively, these intricate mechanisms form a highly dynamic and integrated network that allows plants to modulate growth, survival, and resilience in response to diverse environmental challenges (Kesawat et al. 2023; Zhang et al. 2023).

7.8.1 Regulatory Networks Involved in Plant Stress Tolerance

Regulatory networks involved in plant stress tolerance involve intricate systems that include signaling pathways, TFs, and epigenetic modifications, all of which help plants sense, interpret, and respond to environmental challenges. When plants face stressors, such as salinity, drought, thermal stress, heavy metals, or pathogens, they activate a series of responses upon the perception of the stimuli (Zhang et al. 2023). Numerous investigations have been conducted on receptor-like kinases (RLKs) and pattern recognition receptors (PRRs) at the cellular surface, which sense changes in the environmental conditions and activate downstream cascades such as calcium (Ca^{2+}) signaling and mitogen-activated protein kinase (MAPK) pathways (Jose et al. 2020). These pathways are then driven by Ca^{2+} influxes, which accelerate signal transduction by altering the potential gradient across the membrane.

When exposed to environmental cues, plants release various phytohormones, including ABA, salicylic acid (SA), jasmonic acid (JA), and ethylene, which are central to triggering plant stress regulatory networks. Among the phytohormones, ABA is potentially vital for the response to abiotic stressors including drought and salinity, as it plays a role in the stomatal closure to maintain strict osmoregulation, minimizing water loss; for example, ABA-mediated activation of SnRK2 kinases, that triggers the phosphorylation of ion channels in guard cells, leads to the closure of stomata and hence decreases water loss (Liu et al. 2022a, b). In contrast, the phytohormones such as SA and JA are key mediators of the response to biotic stress, including pathogens and herbivores, and activate the defense mechanisms against pathogens and herbivores. For instance, JA actively triggers the defense mechanisms in response to wound perception, where it induces the expression of defense-related proteins to deter herbivores. Such interactions among these phytohormones, often referred to as “hormonal crosstalk” allow plants to balance growth and defense based on the prevailing environmental cues (Saha et al. 2023).

An additional layer of regulatory control includes the epigenetic modifications, which modulate plant gene expression profiles in response to stress without altering the genetic sequence to adapt best. For instance, DNA methylation has been shown to silence or activate genes involved in stress responses, particularly under recurring stress. For example, saline conditions can induce changes in DNA methylation in *Arabidopsis*, altering the expression of genes associated with cellular ionic homeostasis (Liu et al. 2022a, b). In addition, histone modifications, such as acetylation and methylation, modify the chromatin structure, thus modulating the availability of stress-related genes to the transcriptional machinery. lncRNAs, mostly microRNAs, are also essential; for example, miR398 modulates the expression of superoxide dismutase (SOD) genes under oxidative stress, ensuring that reactive oxygen species (ROS) are efficiently eliminated (Islam et al. 2022).

In addition to AS, posttranscriptional mechanisms such as RNA editing and translational control are also potentially important in plant stress responses. mRNA responses can be effectively modulated via RNA editing, resulting in the formation of proteins with modified amino acid compositions. For example, RNA editing in the *ndhB* gene of chloroplasts helps to stabilize the electron transport chain in photosynthesis during thermal stress conditions. In addition, translational reprogramming acts as an emergency response mechanism to allow plants to cope with stress factors through the production of stress-related proteins, such as ROS-scavenging enzymes, over other cellular proteins as a survival of the fittest mechanism during stress to increase resilience. An intricately organized system forms through the crosstalk among these regulatory layers that operates across time and space within the host plant. Spatially, numerous signaling molecules and TFs are every so often limited to a particular tissue or cell type, such as apical meristems and guard cells, in which immediate stress exposure occurs. Temporally, prompt responses such as ion fluxes and phosphorylation contribute to immediate protection, whereas longer-term adaptations like transcriptional reprogramming and epigenetic adjustments promote tolerance over prolonged stress periods (Fu et al. 2022).

7.8.2 TFs Regulate Stress-Responsive Genes

Stress resilience in plants is reinforced by a suite of TFs that smoothly manage a complex network of stress-related genes. In addition to the distinguished *NAC, AP2/ERF, bZIP, and MYB* families, numerous other transcription factors, such as *WRKY, bHLH, and HD-ZIP,* also play a role. The *WRKY* family, for example, is potentially important to certain biotic stress responses and mediates the response to certain abiotic stresses. In *Arabidopsis*, *WRKY33* acts as a defense against fungal phytopathogens, modulating the production of defensive plant hormones, e.g., jasmonic acid (JA) and salicylic acid (SA). *WRKY* TFs often act by binding to W-box elements in gene promoters, thereby activating or repressing defense genes (Chaffai et al. 2024; Jia et al. 2023). In rice, *OsWRKY45* increases resistance to blast disease, whereas *AtWRKY25* and *AtWRKY26* play a role in thermal stress tolerance in *Arabidopsis*, revealing their potential roles in the diverse ecological conditions (Saha et al. 2023).

7.8.3 Role of Cis-Regulatory Elements in Stress-Responsive Promoters and How Transcriptomic Data Aid in the Construction of GRNs

Plants are in a state of constant adaptation because of the continuous changes in the environment they face. Additional strategies used for adaptation by plants include the use of cis-regulatory elements (CREs) such as *ABREs* and W-boxes. They play a predominant role in stress adaptation by enabling precise transcriptional responses in plants exposed to various environmental cues. Specific TFs can bind to CREs located in gene promoters and synchronize the expression of stress-related genes, allowing plants to swiftly adapt to changing environmental conditions. The ABA signaling pathway is governed by ABA-responsive elements (ABREs) to perceive and respond to drought and salt stress. In *A. thaliana*, ABREs in the promoters of RD29A and RD29B are bound by AREB1 and AREB2 (ABA-responsive element binding proteins), increasing drought tolerance through osmoprotective gene expression (Nguyen et al. 2022). In the same manner, in rice, ABRE-containing promoters alter the expression of key genes that maintain osmoregulation under drought conditions. Modification of the ABREs in genes responsible for drought tolerance in rice has been used to improve drought resilience, with crops displaying increased water retention and reduced wilting under stress (Khan et al. 2021). Dehydration-responsive element (DRE) sequences in plant promoters are vital for the expression of genes to cope with chilling conditions, dehydration, and highly saline environments. The DRE is recognizable by dehydration-responsive element-binding (DREB) TFs, specifically *DREB1A* and *DREB2A,* in *A. thaliana*, which modify genes involved in osmoregulation, such as the genes encoding dehydrins and LEA proteins (Raza et al. 2020).

W-box elements, which are targeted by WRKYs, play a fundamental role in both biotic and abiotic stress responses. For instance, WRKY33 in *Arabidopsis* binds to W-box motifs in the promoters of defense-related genes, orchestrating a complex response to fungal pathogens by modulating the salicylic acid (SA) and jasmonic acid (JA) pathways. In soybean, *GmWRKY54* binds to W-boxes in the promoter regions of antioxidant enzyme-encoding genes, helping to mitigate oxidative damage under drought conditions. This dual role of W-boxes not only copes with pathogen attack but also relieves the drought stress in plants, exemplifying the versatility of *CRE*s in empowering plants to tolerate multiple stressors simultaneously. *GmWRKY21*, another *WRKY* TF in *Glycine max*, also targets W-boxes under saline conditions, where it activates a cascade that reduces damage at cellular levels, making W-boxes potential candidates for genetic alterations aimed at broad-spectrum stress tolerance (Chaffai et al. 2024; Jia et al. 2023; Saha et al. 2023).

A transformative role was determined using transcriptomic and high-throughput sequencing data to identify the CREs and their functions in different organisms. Stress-responsive gene expression profiles were constructed by scientists using cutting-edge RNA-seq analysis, which is particularly useful for revealing stress-responsive gene expression profiles, allowing scientists to construct GRNs that map TF-CRE interactions under different stress conditions. Transcriptomics studies with *A. thaliana* have shown that the combination of *ABRE* and *DRE* elements often appears in the promoters of drought-responsive genes and is involved in the regulation mediated by ABA and dehydration pathways. For instance, *RD29A* contains both *ABREs* and *DREs* elements in its promoter, leading to a synergistic activation by ABA-dependent and ABA-independent pathways. This combined regulation increases the ability of plants to respond to varying levels of water scarcity. Similarly, in maize, transcriptomic data have revealed that certain heat-responsive genes contain both heat shock element (HSE) and ABRE motifs, indicating crosstalk between heat shock proteins and ABA signaling, a finding that could be exploited in the development of heat-tolerant maize cultivars (Nguyen et al. 2022).

7.8.4 Mechanisms of RNA Metabolism in Response to Stress

With respect to plant stress response mechanisms, posttranscriptional regulation (PTR) plays a pivotal role in the precise control of RNA stability, processing, and translation, altering protein production swiftly without extensive transcriptional modulation. Such genome plasticity is pivotal for managing plant responses to environmental cues, including drought, heat, cold, and salinity, allowing the dynamic allocation of resources to address poor conditions. Such plasticity is achieved by the modulation of RNA stability, AS, RNA editing, translation, and polyadenylation, all of which contribute to the resilience of host plants to tolerance and adaptation via gene expression in a targeted manner.

7.8.4.1 RNA Stability

The modulation of RNA stability is the ultimate PTR, enabling the selective degradation and stabilization of transcripts at the cellular level based on their priorities and the stress plants face. For example, salinity conditions in *A. thaliana* activate the decapping enzymes such as CDP and suppress nonessential metabolic pathways to conserve energy to respond to stressful conditions more effectively (Panigrahi and Satapathy 2020). Similarly, thermal stress conditions trigger the accumulation of RNA-binding proteins (RBPs), such as RBP47, in stress granules that bind to mRNA, stabilizing them to produce heat shock proteins. Such selective mRNA stabilization permits the potential of the host to adapt and protect itself from the ever-changing, challenging environmental conditions through the production of essential chaperones and proteins that counteract cellular damage induced by thermal stress conditions (Londoño et al. 2022).

7.8.4.2 RNA Processing: AS and mRNA Editing

As a powerful RNA processing mechanism, AS generates diverse protein isoforms from only a single gene, increasing the functional plasticity of the host to cope with stress conditions. For instance, under thermal stress conditions, *Arabidopsis* heat shock transcription factor 2 (*HSFA2*) is modulated by AS to produce the potential isoforms that regulate heat shock proteins in different ways, resulting in versatile responses to various heat intensities (Friedrich et al. 2021). Similarly, in cold-stressed Populus, the *CBF2* gene produces AS isoforms with distinct DNA-binding capabilities, refining the host response to chilling conditions. RNA editing, primarily through cytidine-to-uridine (C-to-U) alterations, further enhances the adaptation of transcripts and their relative functions under stress without affecting the original DNA sequence. Cold stress in *Arabidopsis* induces C-to-U editing in RNA polymerase II subunits, resulting in the production of an isoform optimized for low-temperature gene expression, enabling a rapid adjustment to colder climates (Kumar et al. 2023).

7.8.4.3 Translation Control

Many mechanisms regulate the translation initiation, allowing plants to conserve energy for the production of stress-specific proteins to maintain their integrity and grow normally. In response to drought, plants usually reduce the expression of the general translation initiation factors, such as eIF4E, reducing the global protein production while sparing time and energy for the production of stress-adaptive transcripts. With open reading frames (uORFs) in their 5' untranslated regions (UTRs), specific mRNAs such as HSP101 in *A. thaliana* can circumvent this repression, thereby ensuring the translation of heat-responsive proteins even when protein synthesis is limited. Ribosome-associated RBPs also contribute to PTR under stress; during hypoxia, *Arabidopsis* poly(A)-binding protein (PABP) is concentrated in stress granules and binds selectively to anaerobic respiration-related transcripts, which helps optimize energy efficiency under low-oxygen conditions (Lohmann et al. 2024).

Another layer of translational control is performed by microRNAs, which alter gene expression by binding to specific mRNAs to either repress translation or promote degradation of the transcript. Under conditions of oxidative stress, for example, Cu/Zn superoxide dismutase (CSD) genes (*CSD1 and CSD2*) are targeted by miR398 in *A. thaliana*, maintaining ROS homeostasis with resource allocation. miRNAs perform dual functions of controlling ROS levels; they also allow plants to conserve energy and resources for vital metabolic functions as well (Islam et al. 2022). Nevertheless, alternative polyadenylation (APA) also contributes to mRNA stability and localization, endowing hosts with the ability to manage the isoforms of transcripts produced during stress exposure. For example, drought stress induces APA in *A. thaliana* NCED3, a gene involved in ABA biosynthesis. The resulting shorter 3' UTR enables the transcript to avoid the process of miRNA-mediated degradation, increasing ABA production, which is vital for drought tolerance. Such a strategy to use APA competently ensures the production of essential transcripts and their translated proteins and stability, supporting continuous defensive responses under adverse conditions. Together, these intricate PTR mechanisms potentially allow plants to instantly respond to perceived stimuli with tailored and efficient changes in gene expression. For instance, the intricacies of RNA stability modulation, AS, mRNA editing, translation control, and polyadenylation fine-tune responses to multiple stressors, increasing host resilience. Plants optimize resource use by dynamically managing gene expression, maintaining cellular integrity, and adapting to diverse stress environments (Truong et al. 2021).

7.9 Stress-Induced Changes in RBPs and Their Regulation of mRNA Splicing, Degradation, and Transport

In posttranscriptional gene regulation, stress-induced modulation of RBPs is vital, as these genes can modify the splicing, degradation, transport, and stability of mRNAs, potentially enhancing the ability of host plants to respond specifically to environmental cues. Under unfavorable conditions, RBPs undergo fine-tuned modifications, shift their localization within the cell, or form complexes with mRNAs, ensuring a responsive and selective layer of regulation of gene expression. These proteins, such as SR proteins and heterogeneous nuclear ribonucleoproteins (hnRNPs), act on numerous RNA targets, synchronizing the fate of stress-related transcripts and confirming that plants can rapidly adapt their protein production without a complete transcriptional overhaul (Lohmann et al. 2024; Londoño et al. 2022).

7.9.1 Regulation of mRNA Splicing by Stress-Responsive RBPs

When plants perceive abiotic stressors, such as drought or heat, RBPs play a crucial role in the modulation of mRNA splicing, which is a primary mechanism for generating diverse protein isoforms best suited for the perceived conditions (Fig. 7.3). For instance, in *A. thaliana,* the differential phosphorylation of SR proteins under heat stress alters their splicing activity to promote the production of stress-optimized protein isoforms. The SR45 protein, which modulates the AS of transcripts involved in stress responses, is phosphorylated during salt stress, altering its potential affinity for target mRNAs and changing its splicing patterns to ensure the production of specific proteins critical for ionic homeostasis and osmoregulation (Albaqami et al. 2019; Li et al. 2021). In rice, an RBP known as *OsGRP2* is upregulated under chilling conditions and interacts with pre-mRNAs, increasing the number of splicing events that trigger the production of cold-responsive isoforms and thereby increasing plants' resilience to chilling conditions (Cheng et al. 2023).

7.9.2 *Modulation of mRNA Stability and Degradation by RBPs*

RBPs regulate mRNA stability and degradation, a pivotal mechanism for the specific modulation of gene expression upon the perception of stress stimuli. The development of stress granules, which are membrane-less organelles containing RBPs and mRNAs, is often induced under stress conditions to control degradation and translation. For instance, the PABP (poly(A)-binding protein) binds specific mRNAs to stabilize transcripts that encode heat shock proteins during thermal stress conditions, while decapping enzymes are recruited to degrade all the accessory mRNAs (Fig. 7.3) (Panigrahi and Satapathy 2020). In *A. thaliana*, heat stress induces the accumulation of RBPs such as RBP47 in stress granules, which bind selectively to transcripts encoding protective proteins such as chaperone proteins, stabilizing that particular transcript on the one hand while downregulating other accessory transcripts. Similarly, during saline stress, Pumilio-type RBPs interact with certain mRNAs to effectively degrade them. They potentially reduce the levels of proteins involved in the growth and developmental processes and permit the resource allocation to develop better defense mechanisms to cope with saline stress conditions (Londoño et al. 2022).

7.9.3 *Control of mRNA Transport and Localization by Stress-Induced RBPs*

The intracellular localization of mRNAs is also influenced by RBPs, which deliver the target transcripts to specific cellular compartments for protein synthesis to meet instant stress demands for the stress response. For example, at relatively low oxygen concentrations, i.e., hypoxia, RBPs such as RBP45 in *A. thaliana* relocate transcripts for anaerobic respiration enzymes to stress granules in mitochondria, enabling rapid translation at the instant site of cellular respiration and optimizing the metabolic response of the plants to hypoxic conditions. Another example includes the RBP AtGRP7, which is activated under chilling conditions in *O. sativa* and interacts with mRNAs for encoding chilling-responsive proteins for transport to the locations of the cell where protein synthesis is essential, thus accelerating the process of enhanced adaptation. Similarly, RBPs such as RBP-P deliver specific mRNAs to chloroplasts in plants under light stress, ensuring the rapid translation of photosynthesis-related proteins to mitigate damage from excess light (Yang et al. 2024).

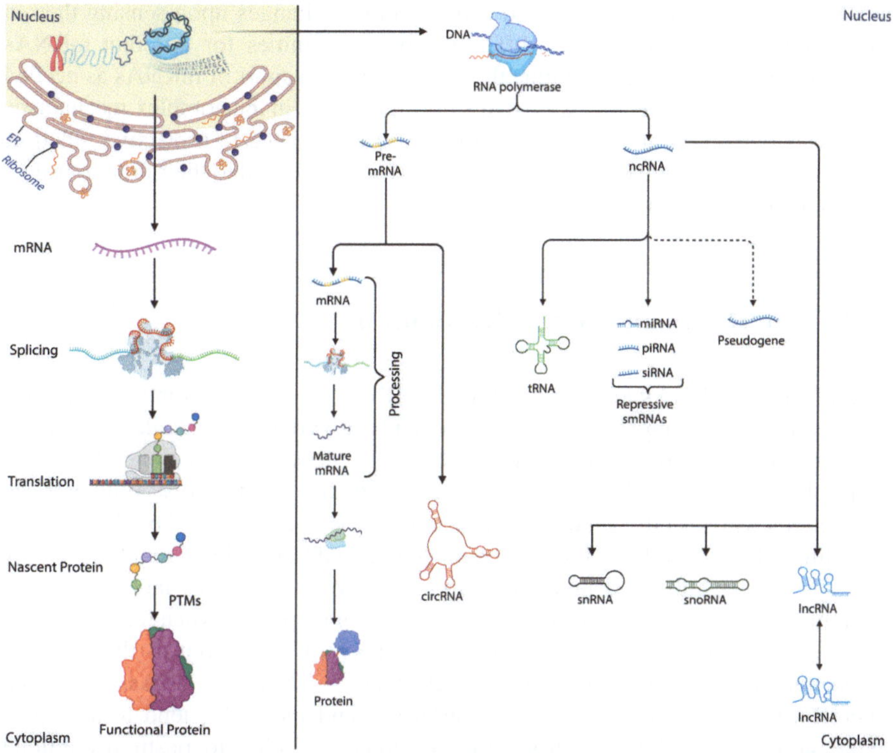

Fig. 7.3 The role of RNA-binding proteins in the fate of mRNA
The above diagram illustrates the detailed information about the transcription and subsequent degradation of mRNA, repression, or translation of a target gene to produce first mature mRNA and then the particular proteins according to the needs of the cell. In the nucleus, first, the DNA containing the information for the target gene to be expressed is transcribed into mRNA. The mRNA exits the nucleus through the nuclear pore, which follows either direct translation or several pathways, like undergoing splicing to remove introns and join exons, forming a mature mRNA molecule that is then translated into a protein, guided to be treated according to the needs of the cell and subsequently host. In the majority of cases, the protein undergoes posttranslational, such as glycosylation, to become a functional protein

7.9.4 Integrative Role of RBPs in Stress-Responsive Gene Regulation

RBPs play a role as hubs in maintaining the coordination in the mechanisms, such as integrating splicing, stability, degradation, and transport, when plants perceive stress stimuli. For instance, under conditions of increased ROS accumulation, the *Arabidopsis* RBP AtRZ-1a interacts with and binds to both mRNA and RNA polymerase II, linking the transcriptional and posttranscriptional regulation for abrupt changes in gene expression (Ma et al. 2021). In addition, RBPs such as hnRNP A1 in plants might undergo configurational changes upon sensing thermal stress and act as switches with different binding affinities for different mRNAs, permitting them to further fine-tune the process and transport of mRNAs as needed. Using these intricate mechanisms, RBPs play a key role in the complex mechanisms of mRNA regulation and translate all the transcripts related to protective functions on a priority basis while conserving resources that correspond to their primary role in the plant stress resilience (Bianchi 2023).

7.10 LncRNAs and Stress Regulation

Small RNAs (miRNAs and siRNAs) and lncRNAs are pivotal in the regulation of gene expression during stress responses in plants. A variety of versatile and intricate mechanisms are involved in the modulation of gene activity by noncoding RNA molecules, increasing plant adaptability and tolerance to stressors. With intricate interactions with specific genes, various cellular processes governed by these RNAs, including mRNA stability, chromatin remodeling, and translational regulation, allow plants to dynamically and precisely respond to stress (Fig. 7.4). lncRNAs play a predominant role in the regulation of the intricacy of responses when plants perceive stress signals, adding another sophisticated layer of gene control that operates independently of protein-coding mechanisms. These ncRNAs, including microRNAs, small interfering RNAs (siRNAs), and lncRNAs, tend to alter gene expression at several levels, from transcriptional silencing to posttranscriptional regulation, allowing plants to temper their physiological responses to ever-changing and multiple environmental cues, including salinity, drought, thermal stress, and pathogen invasion (Chao et al. 2022). For example, in *O. sativa* (rice), stress-responsive lncRNAs potentially modulate gene expression in response to drought and salinity conditions (Jain et al. 2021). Among several lncRNAs, the lncRNA OsDR2 has been shown to modulate the expression of drought-responsive genes by acting as a scaffold for RNA-binding proteins, thus modulating the transcriptional landscape under drought conditions. These findings reveal an additional layer of gene regulation, in which lncRNAs mediate specific stress response pathways, increasing the tolerance of rice to water scarcity (Behera et al. 2021; Jain et al. 2021).

Among the lncRNAs, the microRNAs (miRNAs 21–24 nucleotides in length) control gene expression at the posttranscriptional level by binding to the targets in the complementary mRNA, which leads to either the degradation of the mRNA or subsequent suppression at the translational level. Host plants activate specific miRNAs that alter key stress-responsive mechanisms when they are exposed to abiotic stressors such as drought, salinity, and cold. Under drought conditions, the overexpression of miR164 has been observed in *A. thaliana* and it targets the NAC1 transcription factor, which regulates root development under these conditions. Such conditions allow the host plant to reorganize its root growth architecture, thereby increasing its potential to conserve water (Gupta et al. 2020). CSD1 and CSD2 (Cu/Zn superoxide dismutase) are potential ROS scavengers, are targeted by

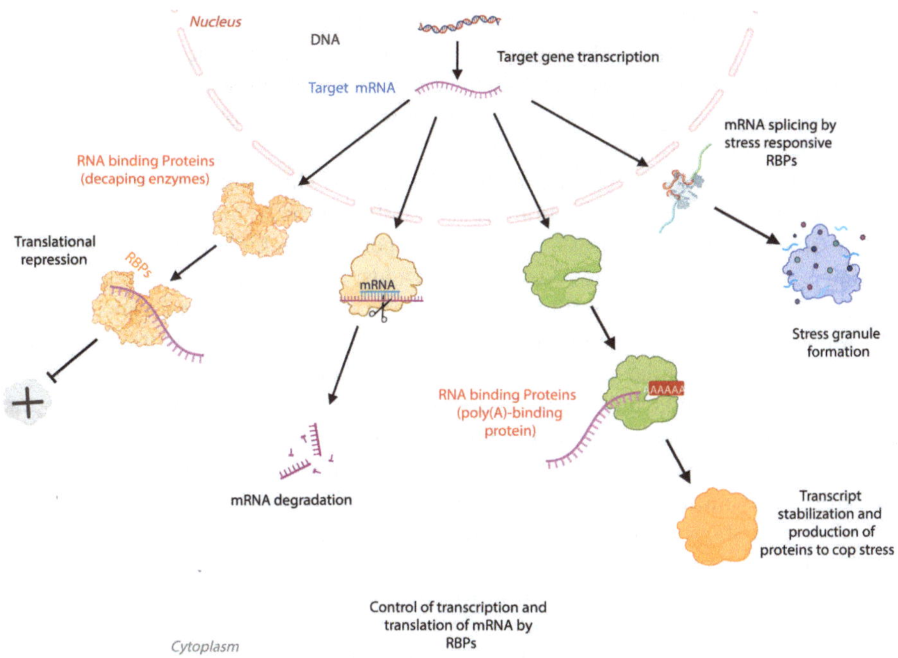

Fig. 7.4 RNA splicing and its roles in the control of mRNA expression and silencing
The figure represents the overview of the central dogma of molecular biology, emphasizing the role of various types of RNA in the expression and gene regulation. Inside the nucleus, transcription of the particular gene occurs to synthesize the particular pre-mRNA. The pre-mRNA synthesis is followed by its processing and modification, i.e., splicing to form a mature mRNA, which subsequently exits the nucleus and enters the cytoplasm. In the cytoplasm, the mature mRNA serves important information, serving as a template to undergo translation, forming a nascent protein by the ribosome, undergoes processing, including splicing, to mature into mRNA, which then exits the nucleus and enters the cytoplasm. Here, mRNA serves as a template for translation into a nascent protein by ribosomes. After the production of the nascent protein, it undergoes important changes of posttranslational modifications (PTMs) to refine the nascent protein further into a functional protein. In other cases, instead of protein formation, different non-coding RNAs (ncRNAs) from DNA transcription, including miRNA, piRNA, siRNA, and snoRNA, play a key role in gene regulation and mRNA processing

miR398 to cope with saline conditions. By modulating these enzymes, miR398 secondarily alters ROS levels, promoting the expression of other antioxidants that are more suitable for high-salinity surroundings (Islam et al. 2022).

In addition to miRNA, siRNAs likewise play a role in the regulation of responses via RNA-directed DNA methylation, thereby silencing genes at the transcriptional level by increasing DNA methylation at precise genomic regions. Upon phytopathogen attack, siRNAs modulate the defense-related silencing of genes such as PR1, protecting plants from pathogen attack and ensuring rapid and reversible stress responses and adaptations. Under the condition in which a pathogen attack occurs, these siRNAs increase the expression of defense-related genes that are vital for defense and resistance to pathogens (Mierziak and Wojtasik 2024). For instance,

TAS3-derived siRNAs in *Arabidopsis* downregulate the expression of auxin response factors (ARFs), which play a role in developmental processes and defense responses. When pathogen invasion is detected by plants, these siRNAs target ARF mRNAs, suppressing auxin signaling to lower the susceptibility to pathogens (López-Ruiz et al. 2020; Luo et al. 2022).

In addition, among lncRNAs, the long lncRNAs, which are typically 200 nucleotides long, have garnered attention because of their intricate roles in regulating gene expression in response to stress. lncRNAs can perform multiple functions, including acting as molecular sponges, decoys, scaffolds, or guides, and can manipulate chromatin structure, transcriptional activity, and posttranscriptional processing. In *O. sativa,* under long-day conditions, the long-day-specific male fertility-associated RNA is vital for regular pollen development. Disruption by thermal stress affects pollen viability, indicating how temperature-dependent lncRNA regulation can affect reproductive success under environmental stress (Shi et al. 2024). Additionally, while lncRNAs interact with certain TFs, they tend to fine-tune plant responses to stress. This intricate relationship between lncRNAs and TFs highlights the tight regulation of stress adaptation at the transcriptional and epigenetic levels (Li et al. 2023a, b).

The complicated interactions of miRNAs, siRNAs, and lncRNAs drive an intricate web of networks to regulate stress responses at multiple levels. Under conditions of excessive ROS accumulation, certain TFs are targeted by miRNAs to effectively regulate the downstream stress-responsive genes; moreover, lncRNAs alter the chromatin architecture, reinforcing changes in gene expression. siRNAs and miRNAs work synergistically to downregulate defense-suppressive genes during pathogen attack while activating pathogen resistance pathways. These unified techniques enable the host to recalibrate the mechanisms of gene expression more quickly according to the environmental conditions, ensuring resilience to multiple environmental cues (Jain et al. 2021; Ma et al. 2022).

7.11 Challenges in Plant Stress Transcriptomics

Remarkable progress has been made in the field of plant stress transcriptomics; however, key challenges need to be overcome to improve our knowledge of all the possible mechanisms underlying plant resilience to environmental cues. One of the main emerging issues is the functional annotation of genes and transcripts, predominantly in the non-model plant species. Although model plants such as *A. thaliana* and *O. sativa* have provided valuable insights due to their extensively studied genomes, the majority of wild species lack such extensive studies and detailed genetic resources, hindering the interpretation and biological significance of the transcriptomics data. Such limitations are predominant in the case of the stress-induced genes, the conservation of which is limited across species. Considering these aspects, an expansion of the horizon of such research on non-model plants is urgently needed through advanced experimental approaches such as mutant analysis, gene knockouts, and overexpression studies. Such experimental studies are vital for elucidating the functional roles of these transcripts while discovering their specific contributions to enhance the stress resistance of the host plant. Additionally, functional validation must extend beyond merely identifying the stress-responsive genes. It should also encompass the intricate regulatory mechanisms that orchestrate the stress regulatory pathway. By integrating transcriptomic data with functional studies, scientists can bridge the gap between gene expression and phenotypic manifestations.

Equally critical is the incorporation of transcriptomics with complementary omics approaches, including epigenomics, metabolomics, and proteomics, to obtain an inclusive understanding of plant stress adaptations. While the patterns of gene expression are revealed by transcriptomics, they still fail to decode the posttranscriptional and posttranslational modifications of transcripts and proteins, respectively, that play predominant roles in protein production, configuration, stability, and functionality. This intricate connection between the proteome and metabolome will contribute deeper insights into the biochemical and molecular processes underpinning stress responses. For instance, changes in transcript levels under stress do not necessarily correspond to the production of particular proteins in a particular configuration and their accumulation because of the involvement of a regulatory mechanism. These regulatory mechanisms include RNA-binding proteins, translation initiation factors, and posttranslational modifications. Correspondingly, metabolomic profiling highlights the accumulation of stress-specific metabolites, including enzymatic and nonenzymatic antioxidants, ROS quenchers, osmolytes, and compatible solutes, that are potentially important for cellular homeostatic conditions. By adding transcriptomic, proteomic and metabolomic analyses, scientists can discover novel molecular markers for stress resilience and identify potential targets for genetic improvements in crops.

Improvements in bioinformatics techniques include fast-tracking this multiomics intricacy, offering a computational method for the management of highly complex datasets. In terms of the labor of artificial intelligence and machine learning algorithms, intricate regulatory networks are now being identified, helping to govern the stress adaptation at the system level. Such innovative steps have allowed scientists to identify the key regulatory hubs, feedback mechanisms, loops, and crosstalk between different signaling pathways, paving the way for predictive modeling and the development of more tolerant crop varieties.

Laboratory-based studies have improved our understanding of precise plant stress responses, yet they often fail to elucidate the intricate dynamics of natural ecosystems. Transcriptomic data from field-based experiments are crucial for addressing these limitations, presenting a more realistic view of how plants respond to stress in their native environments. However, performing transcriptomic investigations is challenging because of the genetic variation in natural populations and the ever-changing nature of environmental stressors. In contrast to laboratory experiments, where aspects such as temperature, humidity, and light are specifically regulated, field surroundings are integrally vibrant and manipulated by a multitude of interacting variables. This intricacy is auxiliary intensified by the genetic diversity within plant populations, where distinct genotypes may employ exceptional stress tolerance mechanisms, leading to variability in the expression of genes even under similar stress conditions. Furthermore, ecological factors such as the soil composition, microclimatic variations, and biotic interactions, including those involving pathogens and herbivores, add supplementary layers of intricacy that are challenging to reproduce in controlled environments.

Despite these limitations, field-based transcriptomics has enormous potential for revealing genotype–environment interactions and recognizing molecular pathways that are vital for stress tolerance under field conditions. With the integration of cutting-edge RNA sequencing with precise environmental monitoring, scientists can create more precise and ecologically relevant models of stress responses. Such techniques not only increase our knowledge of how host plants adapt to the diverse and arbitrary stimuli in their surroundings but also provide greater insights for new breeds and engineered crops with an increased ability to cope with stress. Field-based trials have enormous potential for reshaping agricultural practices by permitting the documentation of molecular biomarkers in conjunction with stress resilience in crops. Such biomarkers can effectively guide the selection and development of different varieties of crops tailored to particular surroundings, paving the way for the breeding of tolerant plants that will have the potential to thrive under extreme environmental conditions. Nevertheless, several methodological problems must be addressed to understand this process. Key challenges include developing consistent approaches for RNA extraction from plants grown under field conditions that are subjected to multiple environmental conditions developing standardized data analysis protocols that account for both environmental fluctuations and genetic diversity. Furthermore, the incorporation of cutting-edge remote sensing technologies, such as unmanned aerial vehicles (UAVs) and satellite imaging, with RNA sequencing methods has the potential to transform field-based transcriptomics. These revolutions would empower scientists to screen plant stress responses across widespread and diverse landscapes, providing high-resolution knowledge of plant adaptation under real-world conditions. By focusing on these challenges, field-based transcriptomics can not only improve our knowledge of plant stress physiology but also provide an understanding to increase crop efficiency and flexibility in the face of global environmental change.

7.12 Perspectives

The importance of plant stress transcriptomics can be distinguished by addressing existing boundaries and enhancing current technologies. The prime urgency lies in proceeding with the functional annotation of genes, especially in non-model plant species. Numerous stress-related genes in plants with few genetic resources remain poorly understood. Transcriptomic data must be obtained with functional genomics using experimental methodologies such as CRISPR/Cas9 knockouts, overexpression of genes, and RNA interference to address this knowledge gap.

These cutting-edge approaches can confirm gene functionality and identify regulatory mechanisms responsible for stress resilience under unfavorable conditions. Furthermore, the state-of-the-art techniques of long-read sequencing technologies such as PacBio and Oxford Nanopore represent innovative strategies to capture full-length transcripts, providing insights into AS events and transcript isoforms that play pivotal roles in the response to stress conditions.

An additional focus is the incorporation of the multiomics data. Although transcriptomics has provided a substantial understanding of the modulation of gene expression during stress, a holistic knowledge of plant stress responses necessitates the combination of transcriptomics with proteomics, metabolomics, and epigenomics. Contemporary progress in high-throughput technologies now facilitates concurrent investigations across these omics layers, providing a deeper understanding of how plants react to stress at the systems level. The application of sophisticated bioinformatics tools and machine learning algorithms is also vital for managing multidimensional data and fabricating predictive models to elucidate the intricate regulatory mechanisms governing stress tolerance.

The shift in results to field-based transcriptomics is becoming substantially important for addressing the challenges of working in natural, variable environments. This approach comprises innovative RNA extraction methods developed for field settings and robust analytical approaches to acclimatize to inconsistent environmental factors and the genetic diversity of natural populations. Such progress in the field will present a more precise and sophisticated representation of the responses by plants and help to pinpoint the genotypes with the ability to tolerate climate-related stressors. In addition, incorporating field-based transcriptomics with remote sensing technologies, comprising UAVs and satellite imaging, will enable the large-scale monitoring of stress responses across distinct landscapes. This methodology will expedite the mapping of genotype–environment interactions and the documentation of stress-tolerant plants, providing valuable insights for refining crop flexibility and sustainability in a changing climate.

Acknowledgments This work was supported by Guangdong Basic and Applied Basic Research Foundation (2024A1515012807), supported by grants from the Science and Technology Innovation Program of Hunan Province (2024JJ5070).

Conflict of Interest Disclosure The authors have no conflicts of interest to declare that are relevant to the content of this chapter.

References

Albaqami M., Laluk et al. (2019) The Arabidopsis splicing regulator SR45 confers salt tolerance in a splice isoform-dependent manner. In: Plant Mol Biol 100:379–390

Almeida DM, Gregorio GB, Oliveira MM et al (2017) Five novel transcription factors as potential regulators of OsNHX1 gene expression in a salt tolerant rice genotype. Plant Mol Biol 93:61–77

Banerjee A, Roychoudhury A (2016) Group II late embryogenesis abundant (LEA) proteins: structural and functional aspects in plant abiotic stress. Plant Growth Regul 79:1–17

Barbosa Amorim LL, Bezerra-Neto JP, da Fonseca do Santos R et al (2016) Transcription factors involved in plant drought tolerance regulation. In: Drought stress tolerance in plants, Vol 2: molecular and genetic perspectives. Springer International Publishing, Cham, pp 315–358

Behera L, Sekhar S, Mohanty S et al (2021) Genomics and other omics approaches for rice improvement. In: Advances in Rice breeding: stress tolerance, climate resilience, quality and high yield, pp 369–426

Bianchi G. (2023) Compaction and condensation properties of intrinsically disordered model proteins

Boi L (2011) Plasticity and complexity in biology: topological organization, regulatory protein networks, and mechanisms of genetic expression. In: Information and living systems: philosophical and scientific perspectives, pp 205–250

Branton D, Deamer DW, Marziali A et al (2008) The potential and challenges of nanopore sequencing. Nat Biotechnol 26(10):1146–1153

Buschmann D., Haberberger A., Kirchner et al. (2016) Toward reliable biomarker signatures in the age of liquid biopsies-how to standardize the small RNA-Seq workflow. Nucleic Acids Res 44(13):5995–6018

Cao Y, Ma J, Han S, Hou M, Wei X, Zhang X, Zhang ZJ, Sun S, Ku L, Tang J, Dong Z, Zhu Z, Wang X, Zhou X, Zhang L, Li X, Long Y, Wan X, Duan C (2023) Single-cell RNA sequencing profiles reveal cell type-specific transcriptional regulation networks conditioning fungal invasion in maize roots. Plant Biotechnol J 21(9):1839–1859

Chaffai R., Ganesan M., Cherif et al. (2024) Transcriptional regulation of gene expression in plant abiotic stress response. In: Chaffai R., Ganesan M., Cherif A. (eds) Plant adaptation to abiotic stress: from Signaling pathways and microbiomes to molecular mechanisms. Springer Nature Singapore, Singapore, pp. 303–343

Chao H, Hu Y, Zhao L et al (2022) Biogenesis, functions, interactions, and resources of non-coding RNAs in plants. Int J Mol Sci 23 (7):3695

Chaumont F, Tyerman SD (2014) Aquaporins: highly regulated channels controlling plant water relations. Plant Physiol 164(4):1600–1618

Chaumont F, Moshelion M, Daniels MJ et al (2005) Regulation of plant aquaporin activity. Biol Cell 97(10):749–764

Chaves MM, Oliveira MM (2004) Mechanisms underlying plant resilience to water deficits: prospects for water-saving agriculture. J Exp Bot 55(407):2365–2384

Cheng K, Zhang C, Lu Y et al (2023) The glycine-rich RNA-binding protein is a vital post-transcriptional regulator in crops. Plants 12(19):3504

Cho Y, Kadam U et al (2025) Recent progress in single-cell transcriptomic studies in plants. Plant Biotechnology Reports:1–13

Conway T, Schoolnik KG (2003) Microarray expression profiling: capturing a genome-wide portrait of the transcriptome. Mol Microbiol 47(4):879–889

Dwivedi S.L., Quiroz L.F., Reddy et al. (2023) Alternative splicing variation: accessing and exploiting in crop improvement programs. Int J Mol Sci 24 (20)

Espinosa E., Bautista R., Larrosa et al. (2024) Advancements in long-read genome sequencing technologies and algorithms. Genomics 116 (3):110842

Friedrich T., Oberkofler V., Trindade et al. (2021) Heteromeric HSFA2/HSFA3 complexes drive transcriptional memory after heat stress in Arabidopsis. Nat Commun 12 (1):3426

Fu M, Lin X, Zhou Y et al (2022) OTP970 is required for RNA editing of chloroplast ndhB transcripts in *Arabidopsis thaliana*. Genes 13(1):139

Gao Y., Xi F., Zhang H., Liu et al. (2019) Single-molecule real-time (SMRT) isoform sequencing (Iso-Seq) in plants: the status of the bioinformatics tools to unravel the transcriptome complexity. Curr Bioinforma 14 (7):566–573

Gong Z, Koiwa H, Cushman MA et al (2001) Genes that are uniquely stress regulated in salt overly sensitive (sos) mutants. Plant Physiol 126(1):363–375

Guo X, Lv L et al (2025) Integrated transcriptome and metabolome analysis revealed differential drought stress response mechanisms of wheat seedlings with varying drought tolerance. BMC Plant Biol 25(1):571

Gupta A, Patil M, Qamar A et al (2020) Ath-miR164c influences plant responses to the combined stress of drought and bacterial infection by regulating proline metabolism. Environ Exp Bot 172: 103998

Han L, Zhong W, Qian J et al (2023) A multi-omics integrative network map of maize. Nat Genet 55(1):144–153

Herath V (2016) Small family, big impact: in silico analysis of DREB2 transcription factor family in rice. Comput Biol Chem 65:128–139

Hoang XL, Chuong NN, Hoa TTK et al (2021) The drought-mediated soybean GmNAC085 functions as a positive regulator of plant response to salinity. Int J Mol Sci 22(16):8986

Ijaz S., Iqbal J., Abbasi et al. (2024) Nc-RNA: a key player in stress response pathways, a molecular perspective. In: Prakash C.S., Waseem M., Fiaz S. (eds) ncRNAs: mediated regulation: the unarguable biological function to environmental constrains. Springer Nature Switzerland, Cham, pp. 139–163

Islam W., Tauqeer A., Waheedn et al. (2022) MicroRNA mediated plant responses to nutrient stress. Int J Mol Sci 23 (5), 2562

Jain P., Hussian S., Nishad et al. (2021) Identification and functional prediction of long non-coding RNAs of rice (*Oryza sativa* L.) at reproductive stage under salinity stress. Mol Biol Rep 48(3): 2261–2271

Jia C, Wang J, Guo B et al (2023) A group III WRKY transcription factor, SlWRKY52, positively regulates drought tolerance in tomato. Environ Exp Bot 215:105513

Jose J, Ghantasala S, Choudhury R et al (2020) Arabidopsis transmembrane receptor-like kinases (RLKs): a bridge between extracellular signal and intracellular regulatory machinery. Int J Mol Sci 21(11):4000

Kamali S, Singh A (2023) Genomic and transcriptomic approaches to developing abiotic stress-resilient crops. Agronomy 13(12):2903

Kesawat MS, Satheesh N, Kherawat BS et al (2023) Regulation of reactive oxygen species during salt stress in plants and their crosstalk with other Signaling molecules—current perspectives and future directions. Plants 12(4):864

Khan MIR, Palakolanu SR, Chopra P et al (2021) Improving drought tolerance in rice: ensuring food security through multi-dimensional approaches. Physiol Plant 172(2):645–668

Kjer-Hansen P, Weatheritt RJ (2023) The function of alternative splicing in the proteome: rewiring protein interactomes to put old functions into new contexts. Nat Struct Mol Biol 30(12): 1844–1856

Kumar P, Kumar P, Suniti et al (2023) Transcriptional regulatory network involved in drought and salt stress response in Rice. In: Kumar A, Dhansu P, Mann A (eds) Salinity and drought tolerance in plants: physiological perspectives. Springer Nature Singapore, Singapore, pp 237–274

Lackey SA (2024) Genome-wide analysis contributes to and promotes adaptation of the soybean, Glycine max, to Canadian agricultural landscapes. Dissertation, Carleton University

Lang D, Zhang S, Ren P et al (2020) Comparison of the two up-to-date sequencing technologies for genome assembly: HiFi reads of Pacific biosciences sequel II system and ultralong reads of Oxford Nanopore. Gigascience 9(12):123

Lata C, Prasad M (2011) Role of DREBs in regulation of abiotic stress responses in plants. J Exp Bot 62(14):4731–4748

Li S, Lin Y-CJ, Wang P et al (2019) The AREB1 transcription factor influences histone acetylation to regulate drought responses and tolerance in Populus trichocarpa. Plant Cell 31(3):663–686

Li Y., Guo Q., Liu P., Huang et al. (2021) Dual roles of the serine/arginine-rich splicing factor SR45a in promoting and interacting with nuclear cap-binding complex to modulate the salt-stress response in Arabidopsis. New Phytol 230(2):641–655

Li J, Zhou X, Wang Y et al (2023a) Inhibition of the maize salt overly sensitive pathway by ZmSK3 and ZmSK4. J Genet Genomics 50(12):960–970

Li Q, Shen H, Yuan S, Dai X et al (2023b) miRNAs and lncRNAs in tomato: roles in biotic and abiotic stress responses. Front Plant Sci 13:1094459

Li P, Liu Q, Wei Y et al (2024) Transcriptional landscape of cotton roots in response to salt stress at single-cell resolution. Plant Commun 5(2):100740

Liu C, Li C, Liang D et al (2013) Aquaporin expression in response to water-deficit stress in two Malus species: relationship with physiological status and drought tolerance. Plant Growth Regul 70:187–197

Liu H, Song S, Zhang H et al (2022a) Signaling transduction of ABA, ROS, and Ca^{2+} in plant stomatal closure in response to drought. Int J Mol Sci 23(23):14824

Liu Y, Wang J, Liu B et al (2022b) Dynamic regulation of DNA methylation and histone modifications in response to abiotic stresses in plants. J Integr Plant Biol 64(12):2252–2274

Logsdon GA, Vollger MR, Eichler EE et al (2020) Long-read human genome sequencing and its applications. Nat Rev Genet 21(10):597–614

Lohmann J, Herzog O, Rosenzweig K et al (2024) Thermal adaptation in plants: understanding the dynamics of translation factors and condensates. J Exp Bot 75(14):4258–4273

Londoño VV, Alquraish F, Tarbiyyah I et al (2022) Landscape of biomolecular condensates in heat stress responses. Front Plant Sci 13:1032045

López-Ruiz BA, Juárez-González VT, Gómez-Felipe A et al (2020) tasiR-ARFs production and target regulation during in vitro maize plant regeneration. Plants 9(7):849

Luo L, Yang X, Guo M et al (2022) TRANS-ACTING SIRNA3-derived short interfering RNAs confer cleavage of mRNAs in rice. Plant Physiol 188(1):347–362

Lv J, Zhou F, Wei Q, Long X et al (2024) An alternative 3′ splice site of PeuHKT1;3 improves the response to salt stress through enhancing affinity to K^+ in Populus. Plant Physiol Biochem 212: 108776

Ma L, Cheng K, Li J, Deng Z et al (2021) Roles of plant glycine-rich RNA-binding proteins in development and stress responses. Int J Mol Sci 22(11):5849

Ma X, Zhao F, Zhou B (2022) The characters of non-coding RNAs and their biological roles in plant development and abiotic stress response. Int J Mol Sci 23(8):4124

Macovei A, Tuteja N (2012) microRNAs targeting DEAD-box helicases are involved in salinity stress response in rice (*Oryza sativa* L.). BMC Plant Biol 12:183

Mastrangelo AM, Marone D, Laidò G et al (2012) Alternative splicing: enhancing ability to cope with stress via transcriptome plasticity. Plant Sci 185:40–49

Mierziak J, Wojtasik W (2024) Epigenetic weapons of plants against fungal pathogens. BMC Plant Biol 24(1):175

Mozgova I, Mikulski P, Pecinka A et al (2019) Epigenetic mechanisms of abiotic stress response and memory in plants. In: Epigenetics in plants of agronomic importance: fundamentals and applications: transcriptional regulation and chromatin remodeling in plants, pp 1–64

Muhammad S, Xu X, Zhou W (2023) Alternative splicing: an efficient regulatory approach towards plant developmental plasticity. Wiley Interdiscip Rev RNA 14(3):e1758

Mutz K-O, Heilkenbrinker A, Lönne M et al (2013) Transcriptome analysis using next-generation sequencing. Curr Opin Biotechnol 24(1):22–30

Nguyen NH, Vu NT, Cheong J-J (2022) Transcriptional stress memory and transgenerational inheritance of drought tolerance in plants. Int J Mol Sci 23(21):12918

Nykiel M, Gietler M, Jea F (2022) Signal transduction in cereal plants struggling with environmental stresses: from perception to response. Plants 11(8):1009

Pandey P, Ramegowda V et al (2015) Shared and unique responses of plants to multiple individual stresses and stress combinations: physiological and molecular mechanisms. Front Plant Sci 6: 723

Panigrahi GK, Satapathy KB (2020) Arabidopsis DCP5, a decapping complex protein interacts with ubiquitin-5 in the processing bodies. Plant Arch 20(1)

Pedrosa AM, Martins CPS, Goncalves LP et al (2015) Late embryogenesis abundant (LEA) constitutes a large and diverse family of proteins involved in development and abiotic stress responses in sweet orange (*Citrus sinensis* LOsb.). PLoS One 10(12):e0145785

Priya M, Dhanker OP, KHM S et al (2019) Drought and heat stress-related proteins: an update about their functional relevance in imparting stress tolerance in agricultural crops. Theor Appl Genet 132:1607–1638

Raza A, Charagh S, Sadaqat N et al (2020) *Arabidopsis thaliana*: model plant for the study of abiotic stress responses. In: The plant family Brassicaceae: biology and physiological responses to environmental stresses. Springer Singapore, Singapore, pp 129–180

Rigo R. (2021) Role of the NSR-ASCO ribonucleoprotein complex in the regulation of plant development. PhD diss., Université Paris-Saclay

Rogalska ME, Vivori C, Valcárcel J et al (2023) Regulation of pre-mRNA splicing: roles in physiology and disease, and therapeutic prospects. Nat Rev Genet 24(4):251–269

Saha B, Nayak J, Srivastava R et al (2023) Unraveling the involvement of WRKY TFs in regulating plant disease defense signaling. Planta 259(1):7

Saliba A-E, Westermann AJ, Gorski SA et al (2014) Single-cell RNA-seq: advances and future challenges. Nucleic Acids Res 42(14):8845–8860

Schena M, Shalon D et al (1995) Quantitative monitoring of gene expression patterns with a complementary DNA microarray. Science 270(5235):467–470

Shi D, Huang H, Zhang Y et al (2024) The roles of non-coding RNAs in male reproductive development and abiotic stress responses during this unique process in flowering plants. Plant Sci 341:111995

Song Y, Ji D, Li Q et al (2012) The dynamic changes of DNA methylation and histone modifications of salt responsive transcription factor genes in soybean. PLoS One 7(7):e41274

Song JB, Gao S, Sun D et al (2013) miR394 and LCR are involved in Arabidopsis salt and drought stress responses in an abscisic acid-dependent manner. BMC Plant Biol 13:1–16

Su C, Wang Z, Cui J et al (2023) Sl-lncRNA47980, a positive regulator affects tomato resistance to Phytophthora infestans. Int J Biol Macromol 248:125824

Su J, Zhan N, Cheng X et al (2024) Genome-wide analysis of cotton MYB transcription factors and the functional validation of GhMYB in response to drought stress. Plant Cell Physiol 65(1): 79–94

Sun M, Huang D, Zhang A et al (2020) Transcriptome analysis of heat stress and drought stress in pearl millet based on Pacbio full-length transcriptome sequencing. BMC Plant Biol 20(1):323

Surówka E, Rapacz M, Janowiak F et al (2020) Climate change influences the interactive effects of simultaneous impact of abiotic and biotic stresses on plants. In: Plant ecophysiology and adaptation under climate change: mechanisms and perspectives I: general consequences and plant responses. Springer Singapore, Singapore, pp 1–50

Truong HA, Lee S, Trịnh CS et al (2021) Overexpression of the HDA15 gene confers resistance to salt stress by the induction of NCED3, an ABA biosynthesis enzyme. Front Plant Sci 12:640443

Usman B, Derakhshani B, Jung KH et al (2023) Recent molecular aspects and integrated omics strategies for understanding the abiotic stress tolerance of rice. Plants 12(10):2019

van Rengs WMJ, Schmidt MHW, Effgen S et al (2022) A chromosome scale tomato genome built from complementary PacBio and Nanopore sequences alone reveals extensive linkage drag during breeding. Plant J 110(2):572–588

Vikas KS, Saksham P, Deepti C et al (2023) Enhancing maize (*Zea mays* L.) crop through advanced techniques: a comprehensive approach. In: Prashant K (ed) New prospects of maize. IntechOpen, Rijeka. p Ch. 7

Wang M, Li P, Cea L (2014) SiLEA14, a novel atypical LEA protein, confers abiotic stress resistance in foxtail millet. BMC Plant Biol 14:1–16

Wang X, Li N et al (2020a) Advances in transcriptomics in the response to stress in plants. Glob Med Genet 7(2):30–34

Wang D, Lu X, Chen X et al (2020b) Temporal salt stress-induced transcriptome alterations and regulatory mechanisms revealed by PacBio long-reads RNA sequencing in Gossypium hirsutum. BMC Genomics 21(1):838

Wang L, Yang T, Wang B et al (2020c) RALF1-FERONIA complex affects splicing dynamics to modulate stress responses and growth in plants. Sci Adv 6(21):1–13

Wang Z, Li N, Qea Y (2021) Genome-wide characterization of salt-responsive miRNAs, circRNAs and associated ceRNA networks in tomatoes. Int J Mol Sci 22(22):12238
Wang Y, Yu J, Jiang M, Lei W et al (2023) Sequencing and assembly of Polyploid genomes. In: Van de Peer Y (ed) Polyploidy: methods and protocols. Springer, New York, NY, pp 429–458
Wang Z, Peng Z, Sea K (2024) Unveiling the power of MYB transcription factors: master regulators of multi-stress responses and development in cotton. Int J Biol Macromol 276:133885
Wanunu M (2012) Nanopores: a journey towards DNA sequencing. Phys Life Rev 9(2):125–158
Watanabe KA, Homayouni A, Lea G (2017) Transcriptomic analysis of rice aleurone cells identified a novel abscisic acid response element. Plant Cell Environ 40(9):2004–2016
Xing W, Pi Z, Jea L (2020) Comparative transcriptome analysis reveals an ABA-responsive regulation network associated with cell wall organization and oxidation reduction in sugar beet. Plant Growth Regul 91:127–141
Yang Y, Guo Y (2018) Elucidating the molecular mechanisms mediating plant salt-stress responses. New Phytol 217(2):523–539
Yang C, Luo A, Lu H-P (2024) Diurnal regulation of alternative splicing associated with thermotolerance in rice by two glycine-rich RNA-binding proteins. Sci Bull 69(1):59–71
Yang X, He J, Lifeng X et al (2025) Salt gradient-driven adaptation in okra: uncovering mechanisms of tolerance and growth regulation. Front Plant Sci 16:1648092
Yolcu S, Ozdemir F, Güler A (2016) Histone acetylation influences the transcriptional activation of POX in *Beta vulgaris* L. and *Beta maritima* L. under salt stress. Plant Physiol Biochem 100:37–46
Yu J, Lai Y, Xea W (2016) Overexpression of OsEm1 encoding a group I LEA protein confers enhanced drought tolerance in rice. Biochem Biophys Res Commun 478(2):703–709
Yu B, Liu J et al (2020) Weighted gene coexpression network analysis-based identification of key modules and hub genes associated with drought sensitivity in rice. BMC Plant Biol 20(1):478
Yu G, Huang Z, Cea G (2024) Heat shock factor HSFA6b mediates mitochondrial unfolded protein response in *Arabidopsis thaliana*. Plants 13(22):3116
Zarayeneh N, Hejazi S, Lin S (2017) Integration of multi-omics data for integrative gene regulatory network inference. Int J Data Min Bioinform 18(3):223–239
Zhang H, Zhang X, Xiao J (2023) Epigenetic regulation of nitrogen Signaling and adaptation in plants. Plants 12(14):2725
Zhao G, Guo D, Lea W (2021a) Functions of RPM1-interacting protein 4 in plant immunity. Planta 253(1):11
Zhao S, Zhang Q et al (2021b) Regulation of plant responses to salt stress. Int J Mol Sci 22(9):4609
Zhong Y, Luo Y, Jea S (2024) Pan-transcriptomic analysis reveals alternative splicing control of cold tolerance in rice. Plant Cell 36(6):2117–2139

Chapter 8
The Dawn of Plant Spatial Transcriptomics Research

Chao Qin and Keke Xia

Abstract Plant spatial transcriptomics has emerged as a transformative technology in life sciences, enabling the simultaneous acquisition of gene expression data and spatial localization while preserving tissue architecture. This chapter reviews recent advancements in plant spatial transcriptomics, highlighting its evolution from tissue-level to single-cell and subcellular resolution. Key innovations include the integration of cell wall staining with sequencing-based approaches (e.g., Stereo-seq), enabling precise cell boundary delineation and subcellular transcriptomic profiling. Applications of this technology span diverse plant species and tissues, addressing critical questions in cell type discovery, functional annotation, micro-environment analysis, and developmental lineage reconstruction. Spatial transcriptomics also facilitates multi-omics integration, providing deeper insights into plant–microbe interactions and metabolic processes. Despite its transformative potential, challenges remain in optimizing spatial information utilization, reducing costs, and developing advanced analytical tools. Future directions include enhancing resolution, enabling simultaneous multi-omics data capture, and fostering global large-scale scientific cooperation to address complex biological questions. By bridging spatial and transcriptional information, spatial transcriptomics offers unprecedented opportunities to advance plant biology research and address global challenges in agriculture and ecology.

Keywords Plant · Spatial transcriptomics · Tools · Application · Challenge

Supplementary Information The online version contains supplementary material available at https://doi.org/10.1007/978-981-95-5183-5_8.

C. Qin · K. Xia (✉)
State Key Laboratory of Genome and Multi-omics Technologies, BGI Research, Shenzhen, China

BGI Research, Beijing, China
e-mail: xiakeke@genomics.cn

L. M. Vaschetto (ed.), *Plant Transcriptomics and Epitranscriptomics*, Concepts and Strategies in Plant Sciences, https://doi.org/10.1007/978-981-95-5183-5_8

Abbreviations

AI	Artificial intelligence
ATAC	Assay for Transposase-Accessible Chromatin
CNS	Central nervous system
DNB	DNA nanoball
FB28	Fluorescent Brightener 28
FISH	Fluorescence in situ hybridization
HDST	High-definition spatial transcriptomics
Hi-C	High-throughput chromosome conformation capture
IMN	Immature neurons
MERFISH	Multiplexed error-robust FISH
MZ	Marginal zone
nptxEX	*Nptx*$^{+}$ lateral pallium excitatory neurons
osmFISH	Sequential hybridization FISH
PCA	Principal component analysis
PRIMER	Primary immune responder
QC	Quality control
reaEGCs	Reactive ependymoglial cells
rIPC1	Regeneration intermediate progenitor cells 1
scStereo-seq	Single-cell spatial enhanced resolution omics sequencing
seqFISH	Sequencing-based FISH
SmT	Spatial metatranscriptomics
STOC Plant	Plant Spatiotemporal Omics Consortium
UMAP	Uniform manifold approximation and projection

8.1 Introduction

The rapid advancement in genomics has significantly enhanced our understanding of life. However, a considerable gap still exists between DNA sequences and biological functions. To bridge this gap, biomacromolecules such as RNA and proteins, which are encoded by DNA, need to be precisely compartmentalized in space and time at the cellular level. This enables a deeper understanding of the regulatory mechanisms underlying biological processes. As stated by the Cellular Dogma (Quake 2024), life must be analyzed at the single-cell level. Systematic integration across temporal and spatial dimensions is also crucial. The orderly progression of biological activities is likely governed by a Spatiotemporal Dogma, which plays a central role in orchestrating development, differentiation, and function execution. The emergence of spatial omics technologies offers unprecedented opportunities to decipher this enigmatic spatiotemporal code (Bressan et al. 2023; Liu et al. 2024).

Spatial transcriptomics, a cutting-edge field in life sciences, combines traditional transcriptomic analysis with spatial localization. This approach allows researchers to maintain the native spatial architecture of tissues or cells while conducting molecular-level profiling, capturing both the physical spatial organization and molecular signatures of biological samples. Its derivative technologies span genomics, transcriptomics, proteomics, and metabolomics, enabling a comprehensive understanding of biological systems from single-molecule to whole-organism levels. These tools provide robust support for deciphering molecular information in biological specimens.

Spatial transcriptomics generates high-resolution spatial molecular maps, empowering scientists to precisely address complex biological questions such as cell-type identification, function characterization, cell–cell interactions, and tissue micro-environment dynamics. This technology deepens the understanding of organizational principles and regulatory mechanisms in living systems. As spatial transcriptomics technology continues to evolve, it is revolutionizing plant biology research (Yin et al. 2023). However, its full application and potential are still faced with significant challenges (Bressan et al. 2023), including the optimization of spatial information utilization and the development of analytical tools for spatial data integration. Addressing these challenges will require global teamwork to realize the transformative potential of spatial transcriptomics in both basic research and agricultural applications.

8.2 Plant Spatial Transcriptomics Technologies

8.2.1 Spatial Transcriptomics Technology

Spatial transcriptomics is a cutting-edge technology that enables the simultaneous acquisition of gene expression data and spatial localization of cells while preserving tissue architecture. Its core principle integrates transcriptomic analysis with spatial mapping, capturing, and sequencing RNA directly within intact tissue sections or cells to achieve spatially resolved gene expression profiling. Early explorations of spatial transcriptomics date back to the late twentieth century, relying on traditional in situ hybridization techniques such as fluorescence in situ hybridization (FISH) to localize specific genes in tissue sections. While these methods provided spatial information, their low throughput limited simultaneous detection of multiple genes. Recent advancements in multiplexed hybridization-based imaging technologies—including osmFISH (sequential hybridization FISH), MERFISH (multiplexed error-robust FISH), and seqFISH (sequencing-based FISH)—have significantly enhanced throughput by enabling multi-gene detection through iterative hybridization and signal decoding. These imaging-based spatial omics technologies utilize fluorescence microscopy to directly visualize RNA localization and expression levels within cells (Liu et al. 2024).

Parallel breakthroughs in high-throughput sequencing and microarray technologies have revolutionized spatial transcriptomics, improving sensitivity, resolution, and scalability. Sequencing-based approaches, such as Slide-seq, HDST (high-definition spatial transcriptomics), Seq-Scope, and Stereo-seq, employ spatially barcoded oligonucleotide arrays. Tissue sections are placed on these arrays, allowing RNA molecules to hybridize with location-specific barcoded probes. After reverse transcription and sequencing, the spatial origin of each RNA molecule is decoded. For example, Stereo-seq leverages rolling circle amplification to generate DNA nanoball (DNB) arrays with 500 nm center-to-center spacing, achieving subcellular resolution (Chen et al. 2022).

Currently, spatial transcriptomics is advancing toward multi-omics integration and broader applications. By combining spatial transcriptomics with proteomics, metabolomics, and other omics layers, researchers can holistically dissect biological complexity. In plant research, spatial transcriptomics has evolved from basic cell-type identification to probing intricate processes such as development, environmental responses, and disease mechanisms. As a transformative tool in modern life sciences, continuous innovation in spatial transcriptomics is driving plant biology toward more precise and systemic exploration.

8.2.2 Plant-Adapted Spatial Transcriptomics Technologies

Plant spatial transcriptomics has undergone a resolution revolution, transitioning from tissue-level (200 μm) to single-cell and even subcellular precision. In 2017, Giacomello et al. pioneered plant spatial transcriptomics using microarray-based methods, which provided tissue-level gene expression profiles but lacked cellular resolution (Giacomello and Lundeberg 2018; Giacomello et al. 2017). By 2022, Xia et al. achieved a milestone by integrating cell wall staining with Stereo-seq, enabling precise delineation of cell boundaries and upgrading Stereo-seq to single-cell spatial transcriptomics (scStereo-seq) (Xia et al. 2022). This breakthrough allowed subcellular resolution (500 nm between two DNBs), revealing cell-type-specific gene expression patterns and spatial heterogeneity in plant tissues.

Another critical advancement is the expansion of detection targets from plant-specific genes to host–microbe dual profiling. Traditional studies focused solely on plant gene expression, overlooking microbial interactions crucial for plant health and ecosystem dynamics. In 2023, Saarenpää et al. developed spatial metatranscriptomics (SmT) (Saarenpää et al. 2024), a technology that concurrently captures plant host and microbial gene expression. SmT enables direct observation of microbial spatial distribution within plant tissues and uncovers spatial correlations between plant and microbial transcriptional activities.

In summary, the continuous refinement of plant spatial transcriptomics—in resolution, multi-omics integration, and cross-species detection—is unlocking new frontiers in plant biology. Future innovations will further empower this technology to address complex biological questions, driving plant science toward deeper mechanistic insights and systemic understanding.

8.3 Applications of Plant Spatial Transcriptomics

Spatial transcriptomics has emerged as a transformative tool in plant biology, successfully applied to study tissue and organ development in diverse species, including *Arabidopsis* leaves (Pour and Yanai 2022), tomato shoot regeneration (Song et al. 2023), maize ears (Wang et al. 2024), nodules (Fan et al. 2025; Liu et al. 2023b; Ye et al. 2024), orchid petals (Liu et al. 2022a), stems (Du et al. 2023; Li et al. 2023; Liu et al. 2022b), rhizome (Lian et al. 2024), root (Fan et al. 2025; Lian et al. 2024; Lyu et al. 2024), and seed development (Fu et al. 2023; Peirats-Llobet et al. 2023; Yao et al. 2024; Zhang et al. 2025). However, fully leveraging spatial information to address scientific questions remains a major challenge. Current applications highlight its unique value in four key areas: novel cell type discovery, cell functional annotation, cellular micro-environment analysis, and developmental lineage reconstruction.

8.3.1 Spatial Transcriptomics in Novel Plant Cell Type Discovery

Spatial transcriptomics demonstrates immense potential for discovering novel cell types in plants, particularly through scStereo-seq, which integrates anatomical imaging of plant tissues with transcriptomic data to identify cell types and resolve their molecular characteristics. By simultaneously capturing gene expression profiles and spatial coordinates, this technology provides a powerful tool for discovering novel cell types, characterizing subtypes, and investigating their functions. In *Arabidopsis* thaliana leaves, Xia et al. employed scStereo-seq to distinguish transcriptionally similar epidermal and mesophyll cell subtypes—upper and lower epidermal cells, as well as palisade and spongy mesophyll cells—which conventional single-cell technologies failed to resolve (Xia et al. 2022). Molecular profiling revealed that upper epidermal cells were enriched in biological processes such as photosynthesis, fatty acid biosynthesis, and photorespiration, while lower epidermal cells were associated with innate immune responses and mechanical stress adaptation. In another case of *Arabidopsis* thaliana leaves, Nobori et al. infected *Arabidopsis* leaves with bacterial pathogens and observed a rare cell population, primary immune responder (PRIMER) cells, through integrating single-nucleus multiomic and spatial transcriptomic data. PRIMER cells emerged at the nexus of immune-active hotspots and have non-canonical immune signatures (Nobori et al. 2025). In tomato callus, Song et al. utilized spatial omics to uncover epidermal cell heterogeneity, identifying two subtypes: Epidermis-like 1 (proximal to air, with marker genes linked to epidermal development, and lipid transport) and Epidermis-like 2 (proximal to culture medium, enriched in polysaccharide metabolism, glycolysis, and proton transmembrane transport, suggesting roles in nutrient uptake and metabolism) (Song et al. 2023). In maize ears, Wang et al. applied

Stereo-seq to identify four novel cell types, including three localized in inflorescence meristems and one at the center of floral meristems (Wang et al. 2024).

Through precise spatial cellular mapping and gene expression analysis, spatial transcriptomics has enabled researchers to uncover cell subtypes that were previously indistinguishable using traditional methods. These studies highlight the cellular diversity and functional specialization within plant tissues, showcasing the technology's unparalleled capability in resolving plant cell subtypes and providing novel perspectives and tools for plant biology research.

8.3.2 Spatial Transcriptomics in Cell Functional Annotation

Spatial transcriptomics enhances the functional annotation of plant cell types by simultaneously capturing gene expression profiles and spatial coordinates, providing more reliable insights into cellular roles. This technology enables researchers to precisely decipher the functional specialization and synergistic interactions among distinct cell types within plant tissues.

In *Arabidopsis* leaves, Xia et al. utilized scStereo-seq to resolve functional differences between palisade and spongy mesophyll cells (Xia et al. 2022). Palisade mesophyll cells exhibited significant enrichment of photosynthesis-related genes, whereas spongy mesophyll cells were enriched in stress-responsive genes, indicating a distinct division of labor in photosynthetic efficiency and environmental adaptation. Hao et al. applied Stereo-seq to analyze onion bulb development (Hao et al. 2023), constructing a spatiotemporal transcriptomic atlas of genes involved in bulb expansion. They identified epidermal cells as central hubs for flavonoid and cuticular wax synthesis, offering critical insights into water retention and pest resistance mechanisms. In animal studies, spatial transcriptomics has similarly advanced functional annotation. For example, in the mouse spleen, researchers mapped the spatial distribution and functional features of immune cell types (Zhang et al. 2023). By integrating single-cell RNA sequencing with spatial transcriptomics, they precisely localized and annotated immune cells, elucidating their roles in immune responses.

These studies exemplify the power of spatial transcriptomics in uncovering cell-type-specific functions and micro-environmental regulatory mechanisms across plant and animal systems. Through precise spatial and molecular profiling, the technology reveals cellular features and interactions that are inaccessible to conventional methods, providing transformative perspectives for understanding tissue development and functionality.

8.3.3 Spatial Transcriptomics in Cellular Micro-environment Analysis

Spatial transcriptomics provides a robust tool for dissecting plant cellular micro-environments by simultaneously capturing gene expression profiles and spatial coordinates. This technology enables researchers to precisely unravel cell–cell interactions, micro-environmental characteristics, and functional synergies within plant tissues.

In animal and human studies, this technology has been pivotal in resolving cellular micro-environments. For example, in a study of mouse spleen, researchers uncovered a well-organized hierarchical architecture of immune cells, including marginal zone (MZ) macrophages, MZ B cells, follicular B cells, and T cells, which are spatially stratified from outer to inner regions (Zhang et al. 2023). In human cancer research, spatial transcriptomics has been employed to dissect cellular interactions and spatial heterogeneity within tumor micro-environments. For instance, in breast cancer studies, the technology revealed dynamic crosstalk between tumor cells and immune cells at the tumor periphery and interior. By analyzing cellular composition and gene expression patterns in the tumor micro-environment, researchers identified cell subpopulations and molecular markers linked to tumor progression and prognosis (Wu et al. 2021).

Plant studies of cellular micro-environments remain relatively limited. In tomato callus, Song et al. utilized Stereo-seq to investigate the spatial micro-environment between chlorenchyma cells and shoot primordia cells (Song et al. 2023). They demonstrated that nascent shoot primordia are consistently surrounded by chlorenchyma cells expressing photosynthesis-related genes. These chlorenchyma cells produce sugars through photosynthesis and transfer them to shoot primordia, activating the glucose-TOR signaling pathway to promote shoot regeneration. This spatial metabolic coupling highlights the critical supportive role of chlorenchyma cells in organ regeneration. Nobori et al. designated cells surrounding PRIMER cells as bystander cells, which could activate genes for long-distance cell-to-cell immune signaling after *Arabidopsis* leaves were infected with bacterial pathogens (Nobori et al. 2025).

These studies underscore the power of spatial transcriptomics in resolving plant cellular micro-environments, offering novel perspectives on intercellular interactions and regulatory mechanisms. Through precise spatial mapping and transcriptional profiling, the technology unveils micro-environmental features and cooperative relationships that are otherwise undetectable by conventional methods, advancing our understanding of plant tissue organization and functionality.

8.3.4 Spatial Transcriptomics in Plant Developmental Lineage Reconstruction

Spatial transcriptomics plays a pivotal role in reconstructing plant cell developmental trajectories and lineages, enabling researchers to resolve differentiation processes and developmental pathways within a spatial framework.

In animal and human spatial omics studies, developmental lineages have been critical for addressing scientific questions. For example, in axolotl brain regeneration research, Wei et al. combined spatial transcriptomics with cellular spatial positioning to uncover a lineage conversion mechanism during telencephalon regeneration (Wei et al. 2022). They demonstrated that reactive ependymoglial cells (reaEGCs) proliferate to cover injury sites and subsequently transition into regeneration intermediate progenitor cells 1 (rIPC1) and immature neurons (IMN-nptxEX), forming a lineage cascade: reaEGC → rIPC1 → IMN → nptxEX, which rebuilds damaged neural circuits.

In plant research, however, the integration of spatial information into lineage reconstruction still requires deeper exploration. In *Arabidopsis* leaves, Xia et al. applied Stereo-seq to map vascular cell differentiation lineages, revealing that less-differentiated vascular cells predominantly cluster around midvein regions (Xia et al. 2022). In tomato callus, Song et al. employed spatial omics to delineate developmental relationships between shoot primordia cells and adjacent chlorenchyma cells, highlighting the essential role of photosynthesis in shoot regeneration (Song et al. 2023). For poplar stems (Du et al. 2023), Du et al. analyzed developmental trajectories in primary and secondary vascular tissues during stem growth. They identified that procambium-derived stem cells differentiate into phloem stem cells and vascular cambium stem cells, proposing a novel model where two stem cell niches—located in phloem and cambium—drive secondary vascular system development in woody plants.

These studies demonstrate how spatial transcriptomics, through precise spatial mapping and gene expression profiling, uncovers cell lineage relationships and developmental pathways that are inaccessible to conventional methods. By bridging spatial context with transcriptional dynamics, this technology provides transformative insights into plant developmental mechanisms and cellular differentiation.

By integrating spatial and transcriptional information through the workflow of "cell discovery - spatial localization - developmental trajectory reconstruction - gene network analysis - regulator mining", we can effectively clarify the biological functions of plant cells. This approach allows for the precise mapping of gene regulation within specific cellular regions and across time. It helps unravel the developmental paths of target cells, addressing where they originate and where they migrate. Moreover, it enables the identification of key regulatory mechanisms and factors that govern plant growth and development.

8.4 Plant Spatial Transcriptomics Data Analysis

8.4.1 Single-Cell Spatial Transcriptomics

Image processing is essential for analyzing spatial transcriptomics data, providing crucial support for subsequent gene expression analysis by processing and analyzing tissue section images. While animal studies typically rely on nuclear staining and spatial RNA capture distribution differences to delineate cell boundaries (Wei et al. 2022), plant spatial omics research can precisely and efficiently distinguish single-cell boundaries using cell wall staining imaging, enabling true spatial single-cell gene expression analysis (Xia et al. 2022). This section will detail the principles and processes of cell wall staining imaging, image registration, quality control, and cell segmentation, highlighting the advantages of single-cell segmentation technology based on cell wall imaging in plant spatial omics research.

8.4.1.1 Cell Wall Staining Imaging

Cell wall staining imaging is pivotal for identifying tissue microstructures in plant spatial omics research. Xia et al. employed ssDNA and Fluorescent Brightener 28 (FB28) staining on *Arabidopsis* thaliana leaves to delineate cell nuclei and walls, accurately identifying six major cell types (Xia et al. 2022). Adjacent tissue sections were stained consistently on glass slides, with ssDNA highlighting nuclei and FB staining cell walls. This dual-staining method enhances cell localization precision, facilitating subsequent image analysis.

8.4.1.2 Image Registration and Quality Control

Image registration and quality control are crucial in Stereo-seq technology, ensuring reliable staining imaging data and precise spatial alignment with sequencing data for robust downstream analysis. Prior to fluorescence imaging, clear staining channel trajectories, well-defined tissue boundaries, and non-overexposed images are required to capture high-quality raw data of cell nuclei and walls using high-precision microscopy. ImageStudio software (https://en.stomics.tech/products/stomics-software/stomics-offline-software/list.html) plays a vital role in image processing, automatically recognizing fluorescence signals and aligning them with sequencing data. It adjusts image position and rotation based on staining channel trajectories and tissue morphology to achieve optimal matching, enhancing data processing efficiency and accuracy. The "QC" model of ImageStudio performs automated quality control on captured images, detecting and scoring abnormalities such as blur or exposure issues. If necessary, researchers manually inspect and adjust image brightness and contrast to ensure high-quality data. For large tissue sections, ImageStudio's automatic stitching feature seamlessly integrates multiple images into a complete tissue image, facilitating comprehensive analysis of cell distribution and gene expression patterns.

8.4.1.3 Tissue Segmentation and Cell Segmentation

Following image data preprocessing, tissue and cell segmentation are conducted to precisely identify key tissues and cell populations. After image quality control using ImageStudio, the "TissueSeg" function segments tissues by identifying regions and separating them from background areas, defining the tissue contour. Subsequently, the "CellSeg" function performs cell segmentation based on cell wall staining images, determining cell boundaries and generating positional and morphological data. Manual adjustments can be made if automatic segmentation is suboptimal. These standardized tools enable rapid transformation of spatial omics raw data into accurate cellular expression data, outputting a spatial gene expression matrix essential for downstream analysis. In addition to ImageStudio, tools like Cellpose (Stringer and Pachitariu 2022), Stereopy (Fang et al. 2025), and Spateo (Qiu et al. 2024) employ deep learning algorithms for accurate cell boundary identification and segmentation. In *Arabidopsis* thaliana leaf studies, Xia et al. developed a comprehensive in situ single-cell spatial omics analysis method. Cluster and differential expression analyses, along with spatial expression pattern analysis, identified mesophyll and epidermal cells and revealed subtle transcriptomic differences between upper and lower epidermal cells. Trajectory analysis uncovered cell type-specific gene expression gradients from the midvein to the leaf margin and distinct spatial developmental trajectories of vascular and guard cells (Xia et al. 2022). These tools facilitate the integration of image and sequencing data.

8.4.1.4 Integration of Image and Gene Expression Analysis

In spatial transcriptomics, the integration of image data, spatial location, and gene expression is crucial for elucidating tissue structure and function. This multidimensional approach provides both morphological insights and spatial gene expression patterns, offering a holistic view of cell function and tissue development. The simplest integration method involves visualizing gene expression distribution on tissue sections, allowing researchers to analyze multiple genes simultaneously and annotate specific regions using anatomical knowledge. For instance, in tumor microenvironment studies, annotating the spatial distribution of tumor, stromal, and immune cells reveals intercellular interactions and signaling pathways, enhancing data accuracy and biological context.

Advances in technology have led to deep learning methods that integrate image and gene expression data, such as DeepST (Xu et al. 2022a) and NePSTA (Ritter et al. 2025). DeepST, a deep learning-based network integration algorithm, extracts image features from histological data (e.g., HE-stained images) and combines them with gene expression and spatial data to construct a spatially enhanced gene expression matrix. NePSTA integrates spatial transcriptomics with graph-based deep learning for comprehensive neuropathological diagnosis from a single tissue section. Trained on 130 central nervous system (CNS) malignancies and healthy donor samples, NePSTA demonstrated high accuracy in predicting histological features and methylation-based subclasses. These innovations improve spatial transcriptomics data analysis precision and provide new tools for pathological diagnosis and research, advancing biomedical applications.

8.4.2 Spatial Omics Data Analysis Tools

8.4.2.1 Data Preprocessing

The rapid advancement of spatial transcriptomics has revolutionized gene expression research. As the technology matures, innovative analysis methods and tools have emerged, enabling detailed dissection of gene expression patterns at the tissue level (Supplementary Table 8.1). In data preprocessing, SpaceRanger (https://www.10xgenomics.com/support/software/space-ranger/latest) has become essential for processing 10X Visium data. It employs sophisticated algorithms for quality control, noise removal, and low-quality data point elimination, ensuring data purity. SpaceRanger also corrects background signals and decodes spatial barcodes, providing precise spatial information for accurate gene expression localization on tissue sections, thus underpinning robust spatial analysis. Stereo-seq technology has further expanded the horizons of high-resolution spatial transcriptomics. For Stereo-seq data preprocessing, the SAW (Gong et al. 2024) tool has been developed. SAW focuses on the unique requirements of Stereo-seq data, enhancing data resolution and accuracy through precise spatial barcode correction. This ensures accurate spatial information and enables single-cell level gene expression analysis, offering deeper insights into cellular heterogeneity and interactions within tissues.

8.4.2.2 Clustering and Cell Annotation

In spatial transcriptomics, clustering and cell annotation are essential for identifying cell types and tissue structures. Traditional tools like Seurat (Hao et al. 2021; Hao et al. 2024) and Scanpy (Wolf et al. 2018) perform unsupervised clustering based on gene expression patterns, grouping similar cells into populations. However, these methods overlook spatial location information, which is crucial for understanding cell function and interactions. Recent advancements have led to tools such as Squidpy (Palla et al. 2022), Stereopy (Fang et al. 2025), and Giotto (Dries et al. 2021), which integrate spatial information with gene expression data to more accurately identify cell populations by considering both spatial distribution and gene expression similarity. Cell annotation methods have also evolved. Early approaches relied on differential expression of known marker genes (Luecken and Theis 2019) and GO enrichment analysis (Yu et al. 2012), which were limited in identifying unknown cell types. Modern tools like AUCell (Aibar et al. 2017) and decoupleR (Badia-i-Mompel et al. 2022) quantify the expression of cell type-specific gene sets, improving annotation accuracy. Automated annotation tools such as SingleR (Aran et al. 2019), scPred (Alquicira-Hernandez et al. 2019), and CellTypist (Xu et al. 2023) use reference datasets to annotate cell types efficiently and consistently across studies. For spatial transcriptomics, tools like Spatial-ID (Shen et al. 2022) integrate reference single-cell data with spatial information, achieving high-accuracy cell annotation through transfer learning and spatial embedding. This comprehensive approach enhances the understanding of cell types and their spatial distribution within tissues.

8.4.2.3 Cell Dynamics Analysis

After identifying plant cell types, researchers focus on the developmental trajectories and fate changes of key cell types in critical biological processes. For example, the induction and elimination of pluripotency in callus cells during regeneration, and the development and differentiation of vascular cells during plant growth are of particular interest. Tools used in these studies include developmental trajectory construction, RNA velocity analysis, and lineage tracing. In developmental trajectory construction, tools like Monocle2/3 (Cao et al. 2019; Trapnell et al. 2014), CytoTRACE (Gulati et al. 2020), Palantir (Setty et al. 2019), and stLearn (Pham et al. 2020) focus on core cell populations, performing pseudotime analysis based on gene expression similarity. This approach helps researchers understand state transitions and differentiation pathways by analyzing dynamic gene expression changes. For RNA velocity analysis, tools such as scVelo (Bergen et al. 2020), Velocyto (La Manno et al. 2018), and CellRank (Weiler et al. 2024) predict future cell states and directions based on unspliced and spliced mRNA abundance. These tools provide real-time insights into cellular dynamics by calculating RNA synthesis, splicing, and degradation rates, enhancing the understanding of cell differentiation. In trajectory construction across multiple time points or spatial tissue sections, tools like SLAT (Xia et al. 2023), spaTrack (Shen et al. 2025), and staVia (Stassen et al. 2024) build cell population trajectories based on gene expression profile similarity, continuity, and spatial distribution. This method reveals cell fate determination and developmental mechanisms under different temporal and spatial conditions, integrating multi-dimensional data to support spatial transcriptomics research.

8.4.2.4 Spatial Micro-Environment Analysis

In spatial transcriptomics, analyzing the cellular micro-environment is essential for understanding cell function and tissue structure. This micro-environment includes cell gene expression, cell-cell interactions, spatial positioning, and the types and states of neighboring cells. Through niche, neighborhood, and cell–cell communication analyses, researchers can elucidate cell function and dynamics within tissues. Niche analysis identifies and characterizes local cell micro-environments defined by cell type composition, state heterogeneity, and spatial organization. Tools like CellCharter (Varrone et al. 2024) and Squidpy (Palla et al. 2022) can identify and analyze spatial niches, revealing cell type distribution and interactions. Neighborhood analysis examines the local environment around cells, assessing how spatial proximity influences cell-cell interactions. By defining neighborhood ranges and analyzing cell type abundance and signaling activity, tools such as CellNeighborEX (Kim et al. 2023) can uncover interaction mechanisms. Cell–Cell communication analysis focuses on ligand–receptor interactions, using tools like CellPhoneDB (Efremova et al. 2020), CellChat (Jin et al. 2021), COMMOT (Cang et al. 2023), and stLearn (Pham et al. 2020) to analyze signaling pathways and interaction networks, elucidating how cells communicate via molecular signals. Additionally, the PlantPhoneDB database (Xu et al. 2022b), a specialized repository for plant ligand–receptor interactions, offers researchers a valuable resource for conducting ligand–receptor analyses in botanical studies.

8.4.2.5 Integration of Spatial Omics and Single-Cell Omics Data

Single-cell RNA sequencing and spatial transcriptomics are complementary technologies in molecular biology. Single-cell RNA sequencing provides detailed gene expression profiles at the single-cell level, revealing cell heterogeneity and functional states. In contrast, spatial transcriptomics retains cell spatial information, elucidating cell distribution and interactions within tissues. However, spatial transcriptomics cannot replace single-cell RNA sequencing entirely due to its high cost and challenges in balancing resolution and RNA capture efficiency. Integrating these technologies offers a comprehensive understanding of cell function and dynamics. Single-cell data can annotate cell types in spatial transcriptomics, identifying cell types and their states across different spatial regions. This integration also enhances spatial transcriptomic data through data augmentation, providing more detailed gene expression information and revealing cell interactions and functional changes. Tools such as Tangram (Biancalani et al. 2021; Li et al. 2022), Cell2location (Kleshchevnikov et al. 2022), and SPOTlight (Elosua-Bayes et al. 2021) facilitate this integration. Additionally, tools like scSpace (Qian et al. 2023) perform pseudo-spatial reconstruction of single-cell data to identify spatially specific cell subpopulations, offering new insights into spatial transcriptomics.

8.4.2.6 Integration of Spatial Omics with Multi-omics Data

The rapid advancement of spatial transcriptomics offers new insights into cell function and interactions within tissues. However, to fully understand the complexity of biological systems, integrating spatial transcriptomic data with other omics data (e.g., genomics, epigenomics, proteomics, metabolomics) is essential. This multi-omics integration reveals spatial associations between gene expression, epigenetic regulation, and protein expression, providing deeper biological insights. Integrating spatial transcriptomics with genomics will uncover spatial links between gene expression and genomic variations, elucidating how genomic changes affect gene expression patterns, cell function, and tissue structure. Combining spatial transcriptomics with ATAC-seq will reveal associations between gene expression and chromatin accessibility, highlighting epigenetic influences on spatial gene expression. Integrating spatial transcriptomics with proteomics will link gene expression to protein expression, shedding light on post-transcriptional regulation and its impact on cell function and tissue structure. To achieve multi-omics integration, tools like SpatialGlue (Long et al. 2024) and SIMO (Yang et al. 2025) have been developed to integrate spatial transcriptomic data with other omics data, revealing spatial associations between gene expression, epigenetics, and proteins. Additionally, technologies such as DBiT-seq (Liu et al. 2020) and Spatial-CITE-seq (Liu et al. 2023a) enable simultaneous detection of multi-omics data. Integrating spatial transcriptomics with genomics, epigenomics, and proteomics is a key direction in modern molecular biology. This approach provides a comprehensive understanding of spatial associations between gene expression, epigenetic regulation, and protein expression, offering new perspectives and methods for spatial transcriptomics research and aiding in the elucidation of cell functions and interactions within tissues.

8.4.3 General Analysis Workflow for Stereo-seq Applications in Plants

Stereo-seq (Chen et al. 2022) is a high-resolution spatial transcriptomics technology that could provide true single-cell gene expression spatial information, which has been widely used in plant research. Stereopy (Fang et al. 2025), a Python-based tool designed for spatial transcriptomics data (e.g., Stereo-seq), offers comprehensive functions from data preprocessing to visualization, clustering, and gene expression pattern identification (Fig. 8.1). Here, we introduce the common analysis pipeline for Stereo-seq data using Stereopy.

8.4.3.1 Data Reading and Preprocessing

Stereopy reads multiple spatial transcriptomics data formats, with Stereo-seq data typically stored in GEF files. The "read_gef" function imports GEF files into a StereoExpData object, the basic data structure for data storage and manipulation. Quality control (QC) is performed first, calculating metrics such as total counts, gene numbers, and mitochondrial gene proportions for each cell (or bin) using the "cal_qc" function. The "filter_cells" function was used to filter out low-quality cells and genes based on these metrics to enhance subsequent analysis accuracy.

8.4.3.2 Data Normalization and Dimensionality Reduction

Normalization is a key step in spatial transcriptomics data analysis. Stereopy provides methods like total normalization (normalize_total) and log transformation (log1p) to eliminate sequencing depth differences. It also supports identifying highly variable genes, which aids in dimensionality reduction and clustering.

Principal component analysis (PCA) serves as the initial dimensionality reduction step, and the max axes of greatest variation are identified with the "pca" function of Stereopy. Next, the "neighbors" function constructs the cell–cell neighborhood

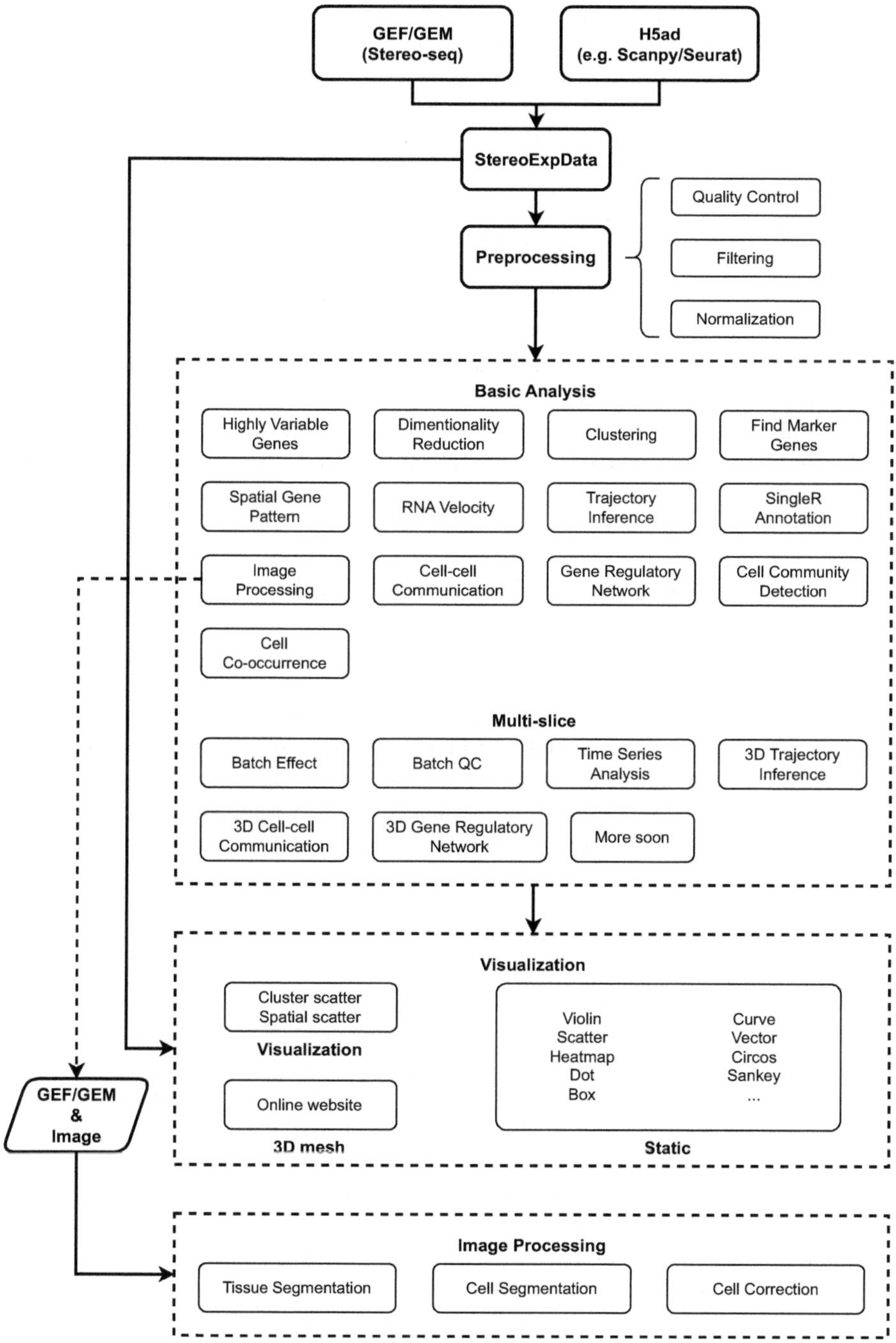

Fig. 8.1 The general analysis workflow of Stereo-seq data. (from Stereopy on Github)
Stereopy is a versatile analysis tool specifically designed for Stereo-seq data, and it is also compatible with other spatial transcriptomic data inputs. It encompasses a comprehensive suite of functions that cover the entire workflow, from preprocessing to basic analysis and visualization. This tool enables in-depth analysis of both single and multiple spatial slices, facilitating a thorough understanding of the spatial-temporal data of plants.

graph after PCA. When spatial coordinates are available, it is recommended to calculate the spatial neighbor graph for improved cell clustering and subsequent analysis using the "spatial_neighbor" function. Finally, Uniform Manifold Approximation and Projection (UMAP) embeds the graph in two or three dimensions, enabling intuitive exploring the data structure in low-dimensional space.

8.4.3.3 Clustering Analysis and Visualization

Based on the results of dimensionality reduction, Stereopy offers three clustering methods: Leiden, Louvain, and Phenograph to identify different cell types or states. During the clustering process, neighbors and spatial neighbors are utilized as inputs. The outcomes of the clustering are visualized through spatial scatter plots, facilitating an intuitive understanding of the spatial distribution of cell clusters. Additionally, gene expression patterns are further investigated using heatmaps and violin plots.

8.4.3.4 Format Conversion and Integration with Other Tools

To facilitate integration with other spatial transcriptomics tools (e.g., Seurat and Scanpy), Stereopy provides format conversion. The "stereo_to_anndata" function can convert StereoExpData objects to AnnData objects, enabling analysis with Scanpy. For working with Seurat, "h5ad2rds.R" from source package was needed to convert h5ad file into rds file after output h5ad file using the "stereo_to_anndata" function with flavor = "seurat". This enhances data portability and leverages the strengths of different tools for comprehensive analysis.

8.4.3.5 Advanced Analysis Functions

Regarding specific scientific issues, Stereopy also offers a range of advanced functions to address diverse research needs. At the gene level, this tool offers robust capabilities for spatial gene expression pattern analysis, including "spatial_hotspot" and "regulatory_network_inference" functions. These features enable the identification of informative genes that are differentially expressed in specific regions, thereby revealing tissue functional partitioning and cell activities.

At the cell level, Stereopy integrates some cell analysis tools, such as "gen_ccc_micro_envs", "cell_cell_communication", "co_occurrence", and "community_detection" functions. These tools facilitate the exploration of cell communication networks and micro-environment characteristics.

In addition to its aforementioned functions, Stereopy also provides a suite of other advanced tools, including 3D reconstruction and cell fate inference, among others. These additional tools can be explored and accessed through relevant literature (Fang et al. 2025) and the official documentation.

8.5 Future Directions and Challenges

8.5.1 *Innovation in Experimental Technology*

8.5.1.1 Cost Reduction

Spatial transcriptomics technology still holds great promise for plant research, yet its widespread use is hampered by high costs. Major cost drivers include specialized consumables and equipment (e.g., high-resolution microscopes, high throughput sequencers), data processing and analysis (large data volumes require significant computing resources and bioinformatics expertise), and sample preparation (labor-intensive procedures for collection, fixation, and sectioning). Cost reduction would bring immense benefits: democratizing access, enabling more labs to adopt the technology globally, especially in resource-limited settings; accelerating research by allowing more frequent use, thus speeding up cell-type identification, gene function studies, and cell interaction research; expanding applications into plant breeding, ecology, and plant–microbe interactions; and enabling larger-scale experiments for comprehensive data collection.

8.5.1.2 Further Resolution Enhancement

Advances in spatial transcriptomics resolution have significantly impacted plant research in cell-type identification, function studies, cell interactions, and micro-environment analysis. Higher resolution offers two main advantages. First, it enables more precise cell type distinction and functional analysis. For instance, in *Arabidopsis* leaves, it can differentiate palisade and spongy mesophyll cells and their subtypes, aiding in the discovery of new cell types and understanding of plant cell heterogeneity. It also allows more accurate localization of gene expression patterns, crucial for studying cell functions like photosynthesis and environmental adaptation. Second, it facilitates the revelation of cell signaling mechanisms and detailed micro-environment characterization. Enhanced resolution can pinpoint signal transmission gene expression locations in plant–microbe interactions, uncovering complex interaction networks. It also improves the mapping of microbial community spatial distributions in root zones, revealing interaction modes with root cells. In summary, further boosting spatial transcriptomics resolution will greatly benefit plant research. It will deepen our understanding of plant cell types and functions, overcome cell interaction and micro-environment research bottlenecks, advance plant development and environmental response mechanism studies, and offer innovative solutions for plant breeding and agricultural applications.

8.5.1.3 Simultaneous Capture of Multi-omics Data

Single-omics data (e.g., transcriptome) provides rich gene expression information but falls short in fully capturing cell function and biological processes, such as post-translational modifications of proteins, metabolite dynamics, and DNA structure. Cell function is governed by a complex interplay of multiple factors, including gene expression, protein synthesis and modification, metabolite turnover, and chromatin organization. Integrating multi-omics data is essential for a comprehensive understanding of cellular regulation. Simultaneous multi-omics data capture offers significant benefits: constructing a panoramic view of cell function by combining transcriptome, proteome, metabolome, ATAC, and Hi-C data, which is vital for studying plant photosynthesis; and gaining a more complete understanding of cell interactions and micro-environment regulation. For example, in plant–microbe interactions, capturing both parties' transcriptome, proteome, and metabolome data can reveal signal exchange, nutrient transactions, and immune responses. Additionally, it aids in uncovering key nodes in plant development and environmental responses, such as monitoring gene expression, protein synthesis, metabolite changes, and chromatin remodeling under drought stress to elucidate drought resistance mechanisms. It also provides valuable information for plant breeding and agricultural applications by analyzing omics differences among plant varieties to identify key genes and regulatory elements for molecular breeding and optimize agricultural practices.

In conclusion, enhancing spatial omics technologies to enable simultaneous multi-omics data capture is a crucial direction for future plant research. This approach overcomes the limitations of single-omics data, offers a comprehensive understanding of cellular function and regulation, and paves the way for innovative solutions in plant breeding and agricultural applications.

8.5.2 Innovation in Application Research

8.5.2.1 From Cellular Resolution to Ecological Evolution Scale

The formation of plant tissues and organs is crucial for their adaptation to specific environments. Spatial transcriptomics has shown its advantage in precisely identifying different cell types and their gene expression characteristics at the cellular level, and constructing cell differentiation and development trajectories to reveal the molecular mechanisms of key developmental events. Furthermore, research on cell and tissue functions from an ecological and evolutionary perspective is also necessary. For example, analyzing the cell types and gene expression patterns of specific tissues in plants at different evolutionary nodes or in diverse ecological environments can help understand how plants have adapted to environmental changes through cellular and molecular adjustments during long-term evolution. The integration of spatial transcriptomics in both macroscopic and microscopic research will better assist researchers in understanding how plants adapt and evolve under different environmental conditions and comprehend the biological functions of cells and tissues.

8.5.2.2 Addressing Scientific Questions on Plant–Environment Spatial Interactions

Plants encounter various environmental changes and stresses during growth and development. Spatial transcriptomics can analyze the specific responses of different cell types to stresses. For example, under drought and nutrient stress, it can precisely locate the expression changes of genes involved in drought and nutrient stress responses in different cell types, revealing the cellular heterogeneity mechanisms of plant stress tolerance. In studying plant responses to microbes and pests, spatial transcriptomics can uncover the spatial interaction patterns and molecular mechanisms between plants and microbes. For example, in root–microbe interactions, it can locate the spatial distribution of different microbial populations in the rhizosphere and reveal their interaction modes with plant root cells, and find novel cellular immunity state (Nobori et al. 2025). For various crop diseases, it can reveal the temporal and spatial heterogeneity of plant immune mechanisms across cell types, providing important theoretical support for understanding plant vascular immune responses. Spatial transcriptomics also helps understand the niche differentiation and competition mechanisms of plants in ecosystems. For example, analyzing the cell types and gene expression patterns of different plant species in the same ecological environment can reveal how they adapt to specific niches through molecular adjustments and how they compete or synergize with other species. In summary, spatial transcriptomics has broad application prospects in plant research. Through integrated research from subcellular precision to ecological evolution scales, it can address scientific issues in the evolution and development of plant organs; by analyzing plant-environment spatial interaction mechanisms, it can reveal plant response strategies to environmental changes; and by deciphering the spatial synthesis and transport mechanisms of important plant metabolites, it can provide a theoretical basis for metabolic engineering and crop improvement. The comprehensive application of spatial omics technologies will bring new breakthroughs to plant science research and offer innovative solutions for agricultural production and technological applications.

8.5.2.3 Deciphering the Spatial Synthesis and Transport Mechanisms of Plant Metabolites

Spatial transcriptomics can accurately locate the key cell types in plant cells responsible for metabolite synthesis. For example, in onion bulb research, it has revealed the central role of epidermal cells in flavonoid and cuticular wax synthesis. This cell-specific localization helps understand the complex inter-cellular coordinated regulation mechanisms of metabolite synthesis. Spatial transcriptomics can also map the transport pathways and regulatory mechanisms of metabolites within plant tissues. For example, during maize grain filling, it can reveal the loading and transport mechanisms of sucrose in phloem cells and its metabolic fate in different cell types. This mapping of metabolite transport pathways helps optimize crop nutrient use efficiency. Additionally, spatial transcriptomics can uncover the coordination mechanisms between metabolite synthesis and plant environmental adaptation. For example, in plant defense against pathogen infection, it can parse the synthesis and transport of defense-related metabolites in different cell types, revealing how plants regulate metabolites to counter pathogen attacks.

8.5.3 Innovation in Spatial Transcriptomics Data Analysis Tools

8.5.3.1 Standardization of Data Generation for Integration of Data from Different Platforms

Currently, data generation from different spatial transcriptomics platforms varies significantly in format, quality control, and analysis process, posing substantial challenges for cross-platform data integration. For example, different spatial transcriptomics platforms may use varying spatial barcode designs and sequencing strategies, limiting data comparability and reproducibility. The lack of unified data standards also complicates the integration and comparison of data from different sources. Standardizing data output will greatly facilitate the integration and analysis of data from different platforms. Unified data formats and quality control standards will allow researchers to more easily combine data from different technologies, gaining a more comprehensive and accurate understanding of biological systems. For example, in plant–microbe interaction studies, integrating spatial transcriptomics data from different platforms can provide a more complete picture of the complex interaction networks between plants and microbes, revealing their gene expression patterns and functional associations across different spatial locations.

8.5.3.2 Establishment of Standardized Open-Access Databases for Data Resource Sharing

Spatial transcriptomics generates large volumes of data that are complex and specialized. The absence of a unified open-access database restricts the sharing and utilization of these data resources. Researchers often need to invest significant time and effort in data acquisition and preprocessing, which reduces research efficiency and hampers collaboration within the field. Creating standardized open-access databases will significantly enhance the sharing and effective use of data resources. Through unified mechanisms for data storage, management, and sharing, researchers can more easily access and utilize high-quality data from various studies. This not only avoids redundant experiments and saves research costs but also promotes interdisciplinary and cross-sectoral research collaborations, accelerating the pace of scientific discovery.

8.5.3.3 Integration and Intelligence of Analysis Tools to Lower the Data Analysis Threshold and Enhance Widespread Data Utilization

Analyzing spatial transcriptomics data involves complex bioinformatics processes, including image processing, data preprocessing, dimensionality reduction, clustering, cell type annotation, and downstream analysis. Currently, these analysis tools are often fragmented and operationally complex, requiring specialized bioinformatics knowledge and programming skills. This limits the use of spatial transcriptomics data by non-specialist researchers. The integration and intelligence of analysis tools will significantly lower the data analysis threshold, enabling more researchers to easily perform spatial transcriptomics data analysis. Developing user-friendly graphical interfaces and automated analysis workflows allows researchers without extensive data analysis backgrounds to quickly learn and efficiently extract biological information from the data. Additionally, intelligent analysis tools can provide more accurate and in-depth analysis results, enhancing the value of data resource utilization.

Collectively, future innovations in spatial transcriptomics data analysis tools are essential in standardizing data generation, establishing open-access databases, and developing integrated and intelligent analysis tools. These advancements will promote the widespread application of spatial transcriptomics in plant and animal research, driving biological studies toward deeper and more systematic understanding.

8.6 Conclusions and Outlook

8.6.1 Big Data-Driven Plant Science Research

The rapid development of spatial omics, single-cell sequencing, and phenomics technologies has propelled plant science into a data-intensive era. The extensive application of spatial omics has generated massive high-dimensional data sets that comprehensively cover plant gene expression, protein levels, metabolite content, and cell spatial positioning, offering unprecedented opportunities for in-depth plant research.

Concurrently, the swift progress of artificial intelligence (AI) technology has provided robust tools for processing and analyzing these complex spatiotemporal omics big data. Machine learning and deep learning algorithms can automatically identify patterns, extract features, and make precise predictions and classifications from vast amounts of data. For instance, in cell-type identification, AI algorithms can automatically recognize new cell types and subtypes through deep learning of large-scale single-cell transcriptomic data, without relying on predefined marker genes, significantly enhancing the efficiency and accuracy of cell-type identification. In gene expression pattern analysis, AI technology can uncover the complex relationships between gene expression, cell location, and environmental factors, aiding researchers in discovering new biological mechanisms and gaining a deeper understanding of plant growth, development, and environmental adaptation strategies.

The deep integration of big data and AI is driving plant science research to transition from traditional "descriptive analysis" to "mechanistic prediction". This transformation not only deepens our understanding of plant life systems but also demonstrates transformative potential in addressing global challenges such as food security and ecosystem restoration. Looking ahead, ongoing algorithm innovation and interdisciplinary collaboration will usher in a new era of "AI-driven discovery" in plant science, providing fundamental technological support for human sustainable development.

8.6.2 Global Large-Scale Scientific Cooperation

Driven by the global large-scale scientific cooperation model, plant science research is entering a new era. Facing the challenges and opportunities of the big data age, this cooperation model is particularly important and feasible.

First, the global large-scale scientific cooperation model can effectively integrate dispersed research efforts and data resources, avoiding redundant work and competition. By establishing unified project designs, data sampling, and experimental protocols, it can minimize data bias caused by experimental condition differences, enhancing data comparability and reproducibility. This not only improves research efficiency but also ensures the efficiency and accuracy of data integration, providing robust data support for basic and applied research in plant science.

Second, international mega-science programs like STOC Plant (Plant Spatiotemporal Omics Consortium, https://www.cell.com/consortium/spatiotemporal-omics) and Plant Cell Atlas (Rhee et al. 2019) offer concrete implementation pathways for this cooperation model. STOC Plant focuses on initiating and implementing plant spatiotemporal omics megaprojects, dedicated to unifying data standards, sharing technical platforms, and coordinating resource allocation to solve critical problems in plant research. For example, the program integrates global laboratory resources to systematically construct high-resolution reference spatiotemporal maps for key crops such as wheat, corn, rice, potato, and tomato throughout their life cycles, as well as immune spatiotemporal maps and vascular system evolution and development spatiotemporal maps for multiple crops. These valuable reference data resources not only support basic research in plant science but also provide rich information for applied research.

Lastly, the global large-scale scientific cooperation model is not only a platform for technological collaboration but also an ecosystem for knowledge production and sharing. As more countries and institutions join, this model will drive plant science research from "isolated exploration" to "systematic dissection", offering a solid scientific foundation for addressing global challenges such as food crises and biodiversity loss. We call for global scientists to unite and join this new era of plant science research, working together to advance its innovative development and contribute to global sustainable development.

Acknowledgments This research was supported by the National Key R&D Program of China (2022YFC3400300), and the Science and Technology Major Special Project of Shenzhen (KJZD20230923114607016). We thank Dr. Shuangsang Fang (BGI Research, Beijing) for providing the workflow figure of Stereopy. We also thank the support of the Plant SpatioTemporal Omics Consortium (STOC Plant).

Competing Interests The authors have no conflicts of interest to declare that are relevant to the content of this chapter.

References

Aibar S et al (2017) SCENIC: single-cell regulatory network inference and clustering. Nat Methods 14:1083–1086. https://doi.org/10.1038/nmeth.4463

Alquicira-Hernandez J, Sathe A, Ji HP et al (2019) scPred: accurate supervised method for cell-type classification from single-cell RNA-seq data. Genome Biol 20:264. https://doi.org/10.1186/s13059-019-1862-5

Aran D et al (2019) Reference-based analysis of lung single-cell sequencing reveals a transitional profibrotic macrophage. Nat Immunol 20:163–172. https://doi.org/10.1038/s41590-018-0276-y

Badia-i-Mompel P et al (2022) decoupleR: ensemble of computational methods to infer biological activities from omics data bioinformatics. Advances 2:vbac016. https://doi.org/10.1093/bioadv/vbac016

Bergen V, Lange M, Peidli S et al (2020) Generalizing RNA velocity to transient cell states through dynamical modeling. Nat Biotechnol 38:1408–1414. https://doi.org/10.1038/s41587-020-0591-3

Biancalani T et al (2021) Deep learning and alignment of spatially resolved single-cell transcriptomes with tangram. Nat Methods 18:1352–1362. https://doi.org/10.1038/s41592-021-01264-7

Bressan D, Battistoni G, Hannon GJ (2023) The dawn of spatial omics. Science 381:eabq4964. https://doi.org/10.1126/science.abq4964

Cang Z et al (2023) Screening cell–cell communication in spatial transcriptomics via collective optimal transport. Nat Methods 20:218–228. https://doi.org/10.1038/s41592-022-01728-4

Cao J et al (2019) The single-cell transcriptional landscape of mammalian organogenesis. Nature 566:496–502. https://doi.org/10.1038/s41586-019-0969-x

Chen A et al (2022) Spatiotemporal transcriptomic atlas of mouse organogenesis using DNA nanoball-patterned arrays. Cell 185:1777–1792.e1721. https://doi.org/10.1016/j.cell.2022.04.003

Dries R et al (2021) Giotto: a toolbox for integrative analysis and visualization of spatial expression data. Genome Biol 22:78. https://doi.org/10.1186/s13059-021-02286-2

Du J, Wang Y, Chen W et al (2023) High-resolution anatomical and spatial transcriptome analyses reveal two types of meristematic cell pools within the secondary vascular tissue of poplar stem. Mol Plant 16:809–828. https://doi.org/10.1016/j.molp.2023.03.005

Efremova M, Vento-Tormo M, Teichmann SA et al (2020) CellPhoneDB: inferring cell–cell communication from combined expression of multi-subunit ligand–receptor complexes. Nat Protoc 15:1484–1506. https://doi.org/10.1038/s41596-020-0292-x

Elosua-Bayes M, Nieto P, Mereu E et al (2021) SPOTlight: seeded NMF regression to deconvolute spatial transcriptomics spots with single-cell transcriptomes. Nucleic Acids Res 49:e50. https://doi.org/10.1093/nar/gkab043

Fan J et al (2025) A large-scale integrated transcriptomic atlas for soybean organ development. Mol Plant. https://doi.org/10.1016/j.molp.2025.02.003

Fang S et al (2025) Stereopy: modeling comparative and spatiotemporal cellular heterogeneity via multi-sample spatial transcriptomics. Nat Commun 16. https://doi.org/10.1038/s41467-025-58079-9

Fu Y et al (2023) Spatial transcriptomics uncover sucrose post-phloem transport during maize kernel development. Nat Commun 14:7191. https://doi.org/10.1038/s41467-023-43006-7

Giacomello S, Lundeberg J (2018) Preparation of plant tissue to enable spatial Transcriptomics profiling using barcoded microarrays. Nat Protoc 13:2425–2446. https://doi.org/10.1038/s41596-018-0046-1

Giacomello S et al (2017) Spatially resolved transcriptome profiling in model plant species. Nat Plant 3:17061. https://doi.org/10.1038/nplants.2017.61

Gong C et al (2024) SAW: an efficient and accurate data analysis workflow for Stereo-seq spatial transcriptomics. Gigabyte 2024:1–12. https://doi.org/10.46471/gigabyte.111

Gulati GS et al (2020) Single-cell transcriptional diversity is a hallmark of developmental potential. Science. https://doi.org/10.1126/science.aax0249

Hao Y et al (2021) Integrated analysis of multimodal single-cell data. Cell 184:3573–3587.e3529. https://doi.org/10.1016/j.cell.2021.04.048

Hao F et al (2023) Chromosome-level genomes of three key Allium crops and their trait evolution. Nat Genet 55:1976–1986. https://doi.org/10.1038/s41588-023-01546-0

Hao Y et al (2024) Dictionary learning for integrative, multimodal and scalable single-cell analysis. Nat Biotechnol 42:293–304. https://doi.org/10.1038/s41587-023-01767-y

Jin S et al (2021) Inference and analysis of cell-cell communication using CellChat. Nat Commun 12:1–20. https://doi.org/10.1038/s41467-021-21246-9

Kim H et al (2023) CellNeighborEX: deciphering neighbor-dependent gene expression from spatial transcriptomics data. Mol Syst Biol 19:e11670. https://doi.org/10.15252/msb.202311670

Kleshchevnikov V et al (2022) Cell2location maps fine-grained cell types in spatial transcriptomics. Nat Biotechnol 40:661–671. https://doi.org/10.1038/s41587-021-01139-4

La Manno G et al (2018) RNA velocity of single cells. Nature 560:494–498. https://doi.org/10.1038/s41586-018-0414-6

Li B et al (2022) Benchmarking spatial and single-cell transcriptomics integration methods for transcript distribution prediction and cell type deconvolution. Nat Methods 19:662–670. https://doi.org/10.1038/s41592-022-01480-9

Li R, Wang Z, Wang J et al (2023) Combining single-cell RNA sequencing with spatial transcriptome analysis reveals dynamic molecular maps of cambium differentiation in the primary and secondary growth of trees. Plant Commun:100665. https://doi.org/10.1016/j.xplc.2023.100665

Lian X et al (2024) Spatiotemporal transcriptomic atlas of rhizome formation in Oryza longistaminata. Plant Biotechnol J 22:1652–1668. https://doi.org/10.1111/pbi.14294

Liu Y et al (2020) High-spatial-resolution multi-omics sequencing via deterministic barcoding in tissue. Cell 183:1665–1681.e1618. https://doi.org/10.1016/j.cell.2020.10.026

Liu C et al (2022a) A spatiotemporal atlas of organogenesis in the development of orchid flowers. Nucleic Acids Res 50:9724–9737. https://doi.org/10.1093/nar/gkac773

Liu Y et al (2022b) Spatial transcriptome analysis on peanut tissues shed light on cell heterogeneity of the peg. Plant Biotechnol J 20:1648–1650. https://doi.org/10.1111/pbi.13884

Liu Y et al (2023a) High-plex protein and whole transcriptome co-mapping at cellular resolution with spatial CITE-seq. Nat Biotechnol 41:1405–1409. https://doi.org/10.1038/s41587-023-01676-0

Liu Z et al (2023b) Integrated single-nucleus and spatial transcriptomics captures transitional states in soybean nodule maturation. Nat Plants 9:515–524. https://doi.org/10.1038/s41477-023-01387-z

Liu L, Chen A, Li Y et al (2024) Spatiotemporal omics for biology and medicine. Cell 187:4488–4519. https://doi.org/10.1016/j.cell.2024.07.040

Long Y et al (2024) Deciphering spatial domains from spatial multi-omics with SpatialGlue. Nat Methods 21:1658–1667. https://doi.org/10.1038/s41592-024-02316-4

Luecken MD, Theis FJ (2019) Current best practices in single-cell RNA-seq analysis: a tutorial. Mol Syst Biol 15:e8746. https://doi.org/10.15252/msb.20188746

Lyu X et al (2024) Tracing the evolutionary and genetic footprints of atmospheric tillandsioids transition from land to air. Nat Commun 15:9599. https://doi.org/10.1038/s41467-024-53756-7

Nobori T et al (2025) A rare PRIMER cell state in plant immunity. Nature 638:197–205. https://doi.org/10.1038/s41586-024-08383-z

Palla G et al (2022) Squidpy: a scalable framework for spatial omics analysis. Nat Methods 19:171–178. https://doi.org/10.1038/s41592-021-01358-2

Peirats-Llobet M, Yi C, Liew Lim C et al (2023) Spatially resolved transcriptomic analysis of the germinating barley grain. Nucleic Acids Res 51:7798–7819. https://doi.org/10.1093/nar/gkad521

Pham D et al (2020) stLearn: integrating spatial location, tissue morphology and gene expression to find cell types, cell-cell interactions and spatial trajectories within undissociated tissues. https://doi.org/10.1101/2020.05.31.125658

Pour M, Yanai I (2022) New adventures in spatial transcriptomics. Dev Cell 57:1209–1210. https://doi.org/10.1016/j.devcel.2022.04.021

Qian J et al (2023) Reconstruction of the cell pseudo-space from single-cell RNA sequencing data with scSpace. Nat Commun 14:2484. https://doi.org/10.1038/s41467-023-38121-4

Qiu X et al (2024) Spatiotemporal modeling of molecular holograms. Cell. https://doi.org/10.1016/j.cell.2024.10.011

Quake SR (2024) The cellular dogma. Cell 187:6421–6423. https://doi.org/10.1016/j.cell.2024.10.029

Rhee SY, Birnbaum KD, Ehrhardt DW (2019) Towards building a plant cell atlas. Trends Plant Sci 24:303–310. https://doi.org/10.1016/j.tplants.2019.01.006

Ritter M et al (2025) Spatially resolved transcriptomics and graph-based deep learning improve accuracy of routine CNS tumor diagnostics. Nat Cancer 6:292–306. https://doi.org/10.1038/s43018-024-00904-z

Saarenpää S, Shalev O, Ashkenazy H et al (2024) Spatial metatranscriptomics resolves host-bacteria-fungi interactomes. Nat biotechnol 42:1384–1393. https://doi.org/10.1038/s41587-023-01979-2

Setty M, Kiseliovas V, Levine J et al (2019) Characterization of cell fate probabilities in single-cell data with Palantir. Nat Biotechnol 37:451–460. https://doi.org/10.1038/s41587-019-0068-4

Shen R et al (2022) Spatial-ID: a cell typing method for spatially resolved transcriptomics via transfer learning and spatial embedding. Nat Commun 13:7640. https://doi.org/10.1038/s41467-022-35288-0

Shen X et al (2025) Inferring cell trajectories of spatial transcriptomics via optimal transport analysis. Cell Syst 16. https://doi.org/10.1016/j.cels.2025.101194

Song X et al (2023) Spatial transcriptomics reveals light- induced chlorenchyma cells involved in promoting shoot regeneration in tomato callus. Proc Natl Acad Sci 120:e2310163120. https://doi.org/10.1073/pnas.2310163120

Stassen SV, Kobashi M, Lam EY et al (2024) StaVia: spatially and temporally aware cartography with higher-order random walks for cell atlases. Genome Biol 25:224. https://doi.org/10.1186/s13059-024-03347-y

Stringer C, Pachitariu M (2022) Cellpose 2.0: how to train your own model. Nat Methods 19(12): 1634–1641. https://doi.org/10.1101/2022.04.01.486764

Trapnell C et al (2014) The dynamics and regulators of cell fate decisions are revealed by pseudotemporal ordering of single cells. Nat Biotechnol 32:381–386. https://doi.org/10.1038/nbt.2859

Varrone M, Tavernari D, Santamaria-Martínez A et al (2024) CellCharter reveals spatial cell niches associated with tissue remodeling and cell plasticity. Nat Genet 56:74–84. https://doi.org/10.1038/s41588-023-01588-4

Wang Y et al (2024) A spatial transcriptome map of the developing maize ear. Nat Plants 10:815–827. https://doi.org/10.1038/s41477-024-01683-2

Wei X et al (2022) Single-cell stereo-seq reveals induced progenitor cells involved in axolotl brain regeneration. Science 377:eabp9444. https://doi.org/10.1126/science.abp9444

Weiler P, Lange M, Klein M et al (2024) CellRank 2: unified fate mapping in multiview single-cell data. Nat Methods 21:1196–1205. https://doi.org/10.1038/s41592-024-02303-9

Wolf FA, Angerer P, Theis FJ (2018) SCANPY: large-scale single-cell gene expression data analysis. Genome Biol 19:15. https://doi.org/10.1186/s13059-017-1382-0

Wu SZ et al (2021) A single-cell and spatially resolved atlas of human breast cancers. Nat Genet 53: 1334–1347. https://doi.org/10.1038/s41588-021-00911-1

Xia K et al (2022) The single-cell stereo-seq reveals region-specific cell subtypes and transcriptome profiling in Arabidopsis leaves. Dev Cell 57:1299–1310. https://doi.org/10.1016/j.devcel.2022.04.011

Xia C-R, Cao Z-J, Tu X-M et al (2023) Spatial-linked alignment tool (SLAT) for aligning heterogenous slices. Nat Commun 14:7236. https://doi.org/10.1038/s41467-023-43105-5

Xu C et al (2022a) DeepST: identifying spatial domains in spatial transcriptomics by deep learning. Nucleic Acids Res 50:e131–e131. https://doi.org/10.1093/nar/gkac901

Xu C, Ma D, Ding Q et al (2022b) PlantPhoneDB: a manually curated pan-plant database of ligand-receptor pairs infers cell–cell communication plant. Biotechnol J 20:2123–2134. https://doi.org/10.1111/pbi.13893

Xu C et al (2023) Automatic cell-type harmonization and integration across human cell atlas datasets. Cell 186:5876–5891.e5820. https://doi.org/10.1016/j.cell.2023.11.026

Yang P et al (2025) Spatial integration of multi-omics single-cell data with SIMO. Nat Commun 16: 1–15. https://doi.org/10.1038/s41467-025-56523-4

Yao J et al (2024) Spatiotemporal transcriptomic landscape of rice embryonic cells during seed germination. Dev Cell 59:2320–2332.e2325. https://doi.org/10.1016/j.devcel.2024.05.016

Ye K et al (2024) Mapping the molecular landscape of Lotus japonicus nodule organogenesis through spatiotemporal transcriptomics. Nat Commun 15:6387. https://doi.org/10.1038/s41467-024-50737-8

Yin RL, Xia KK, Xu X (2023) Spatial transcriptomics drives a new era in plant research. Plant J 116:1571–1581. https://doi.org/10.1111/tpj.16437

Yu G, Wang L-G, Han Y et al (2012) clusterProfiler: an R package for comparing biological themes among gene clusters OMICS: a journal of. Integr Biol 16:284–287. https://doi.org/10.1089/omi.2011.0118

Zhang Y et al (2023) Microbiota-mediated shaping of mouse spleen structure and immune function characterized by scRNA-seq and stereo-seq. J Genet Genomics. = Yi chuan xue bao 50:688–701. https://doi.org/10.1016/j.jgg.2023.04.012

Zhang X et al (2025) A spatially resolved multi-omic single-cell atlas of soybean development. Cell 188:550–567.e519. https://doi.org/10.1016/j.cell.2024.10.050

Chapter 9
RNA Modifications: A Diverse Landscape

Peng Tian, Ruolin Gong, Guangze Li, and Jihong Hu

Abstract Epitranscriptomic chemical modifications of RNAs have emerged as potent regulatory mechanisms in plants. Owing to the technical advances in detecting RNA modifications and high-throughput sequencing, our understanding of the diversity and landscape of RNA modifications across organisms has been significantly improved, resulting in the birth of the epitranscriptome that focuses on deciphering the function of RNA modifications. Currently, more than 170 distinct chemical modifications have been identified in mRNAs, tRNAs, rRNAs, and non-coding RNAs (ncRNAs). RNA modifications include N6-methyladenosine (m^6A), 1-methyladenosine (m^1A), 5-methylcytosine (m^5C), 3-methylcytosine (m^3C), 7-methylguanosine (m^7G), pseudouridine (Ψ), and so on. Genetic and molecular studies have identified the genes responsible for the addition, interpretation, and removal of chemical modifications from RNA molecules, which are known as "writers", "readers", and "erasers", respectively. RNA modifications play important roles in gene expression regulation by regulating mRNA stability, protein translation efficiency, splicing of mRNA or pri-miRNAs, and mRNA nuclear exporting. RNA modifications are widely involved in plant growth, development, stress response, crop yield, and immune response. Thus, further investigation into the distribution patterns and functional mechanisms of RNA modifications will deepen our understanding of RNA modification biology from the perspective of epitranscriptomics and provide new strategies for potential applications in plants and future improvement of crops.

Keywords Epitranscriptomics · RNA modification · m^6A · m^5C · Writers · Readers · Erasers

P. Tian · R. Gong · G. Li · J. Hu (✉)
State Key Laboratory of Crop Stress Resistance and High-Efficiency Production, College of Agronomy, Northwest A&F University, Yangling, Shaanxi, China
e-mail: hujh05@nwafu.edu.cn

L. M. Vaschetto (ed.), *Plant Transcriptomics and Epitranscriptomics*, Concepts and Strategies in Plant Sciences, https://doi.org/10.1007/978-981-95-5183-5_9

Abbreviations

ACYR	N-ACETYLTRANSFERASEs FOR CYTIDINE IN RNA
AGO	Argonaute protein
ALKBH	AlkB homolog (demethylase family)
ALYREF	Aly/REF export factor
A-to-I	Adenosine-to-inosine editing
CBP	Cap-binding protein
CDS	Coding sequence
CPSF	Cleavage and polyadenylation specificity factor
DGCR	DiGeorge syndrome critical region gene
DNMT	DNA methyltransferase
dpCoA	Dephospho-CoA
DXO	Decapping exoribonuclease
ECT	Evolutionarily conserved C-terminal region
eIF	Eukaryotic Initiation Factor
EIF4E	Eukaryotic initiation factor 4E
FAD	Flavin adenine dinucleotide
FIONA1	Flowering locus D-interacting protein 1
FLC	FLOWERING LOCUS C
FLK	FLOWERING LOCUS K
FMRP	Fragile X mental retardation protein
5hmC	5-hydroxymethylcytosine
HNRNPC	Heterogeneous nuclear ribonucleoprotein
HOTAIR	HOX transcript antisense RNA
IGF2BP	Insulin-like growth factor 2 mRNA-binding proteins
lncRNA	Long Non-Coding RNA
m^1A	N1-methyladenosine
m^2C	N^2-Methylcytosine
m^3C	N^3-methylcytosine
m^5C	5-methylcytosine
m^6A	N^6-methyladenosine
m^6Am	N^6,2′-O-dimethyladenosine
m^7G	7-methylguanosine
MALAT1	Metastasis associated lung adenocarcinoma transcript 1
METTL	Methyltransferase-like
miRNA	MicroRNA
MORF	Multiple organelle RNA editing factor
mRNA	Messenger RNA
NAD^+	Nicotinamide adenine dinucleotide
NAT10	N-acetyltransferase 10
Nm	2′-O-methylation
NOP56	Nucleolar protein 56
NSUN	NOP2/Sun domain protein
ORF	Open reading frame
ORRM	Organelle RNA recognition motif protein

OZ1	Organelle zinc finger 1
PAB2/PAB4	Poly(A)-binding proteins 2 and 4
PABP1	Poly(A)-binding protein 1
PPO1	Protoporphyrinogen oxidase 1
PPR	Pentatricopeptide repeat protein
pre-mRNA	Precursor messenger RNA
pri-miRNA	Primary microRNA
PRRC2A	Proline-rich coiled-coil 2A
PUS	Pseudouridine synthase
qPCR	Quantitative polymerase chain reaction
RBPs	RNA-binding proteins
RFM	Rossmann fold methyltransferase
RIP	RNA immunoprecipitation
RNMT	RNA methyltransferase
RsmB	RNA methyltransferase B
SAM	S-adenosylmethionine
snoRNA	Small nucleolar RNA
snRNA	Small nuclear RNA
snRNP	Small nuclear ribonucleoprotein
SRSF	Serine/arginine-rich splicing factor
svRNA	Small vault RNA
TRDMT1	tRNA aspartic acid methyltransferase 1
TRM	tRNA methyltransferase
tRNA	Transfer RNA
TSS	Transcription start site
U6 snRNA	U6 small nuclear RNA
UDP-glucose	Uridine diphosphate glucose
VIR	VIRILIZER
vtRNA	Vault RNA
WDR4	WD repeat domain 4 protein
WTAP	Wilms tumor 1-associated protein
XIST	X-inactive specific transcript
YTH	YT521-B homology domain
YTHDC	YTH domain-containing protein
Ψ	Pseudouridine

9.1 Introduction

Ribonucleic acid (RNA) is a fundamental molecular component in all living organisms. Similar to post-translational modifications of protein and epigenetic changes of DNA, the four nucleotides adenine (A), uracil (U), guanine (G), and cytosine (C) of RNAs can be chemically modified, which may occur co- or post-transcriptionally and persist throughout the life cycle of most RNAs (Ge et al. 2025). With the development of recent technical advances in detecting RNA modification and high-throughput sequencing, our understanding of the diversity and abundance of RNA modifications within living organisms has been significantly enhanced, leading to the emergence of the epitranscriptome field dedicated to decipher the function of RNA modifications (Shen et al. 2019; Shen and Yu 2021; Sharma et al. 2023a, b).

Since the first discovery of chemical modification of RNA in calf liver RNA in the 1950s (Cohn 1951), RNA modifications have been documented to be ubiquitous in all organisms (Liang et al. 2020). RNA modifications refer to various forms of chemical modifications occurring on the four nucleotide bases (A, U, C, and G) of RNA. To date, more than 170 distinct RNA modifications have been identified in archaea, bacteria, viruses, and eukaryotes and distributed across various types of RNA, including messenger RNA (mRNA), transfer RNA (tRNA), ribosomal RNA (rRNA), small nuclear RNA (snRNA), small nucleolar RNA (snoRNA), microRNA (miRNA), and long non-coding RNA (lncRNA) and so on (Cappannini et al. 2024; Lang et al. 2024). In this chapter, we summarize the advances on the diverse landscape of RNA modifications, focusing on the various types, distribution characteristics, regulatory proteins, and functional roles in plant growth and development.

9.2 Diversity of RNA Modifications

9.2.1 Types of RNA Modifications

RNA modifications mainly occur at specific sites on the nitrogen and carbon atoms of the bases, through noncanonical base rearrangements, on the oxygen atom of the 2′OH group of the ribose (Fig. 9.1), and involve other more complex modifications (Motorin and Helm 2011; Xuan et al. 2018). In plants, RNA modifications play a vital role in growth, development, and responses to biotic and abiotic stresses (Shao et al. 2021; Cai et al. 2025; Ge et al. 2025). As research progresses, more RNA modification types and their functions are being uncovered, providing deeper insights into plant biology and potential applications in agriculture (Sharma et al. 2023a) (Fig. 9.1 and Table 9.1).

9.2.1.1 N6-Methyladenosine (m^6A)

N6-methyladenosine (m^6A) modification is the most common internal modification on eukaryotic mRNA and is also widely present in plants (Yue et al. 2019). This modification entails the addition of a methyl group to the N6 position of adenosine (A) by the RNA methyltransferase complex. The sequence surrounding m^6A modification sites on mRNA is highly conserved, primarily existing in the RRACH motif (R: purine; A: m^6A; H: non-guanine). However, m^6A modifications in rRNA, tRNA, and snRNA lack such conserved sequences. In plants, the m^6A methyltransferase complex mainly consists of subunits such as MTA (homolog of METTL3), MTB (homolog of METTL14), and FIP37 (homolog of WTAP) (Zhong et al. 2008; Yue et al. 2019). MTA and MTB form a heterodimer, which is the core enzyme

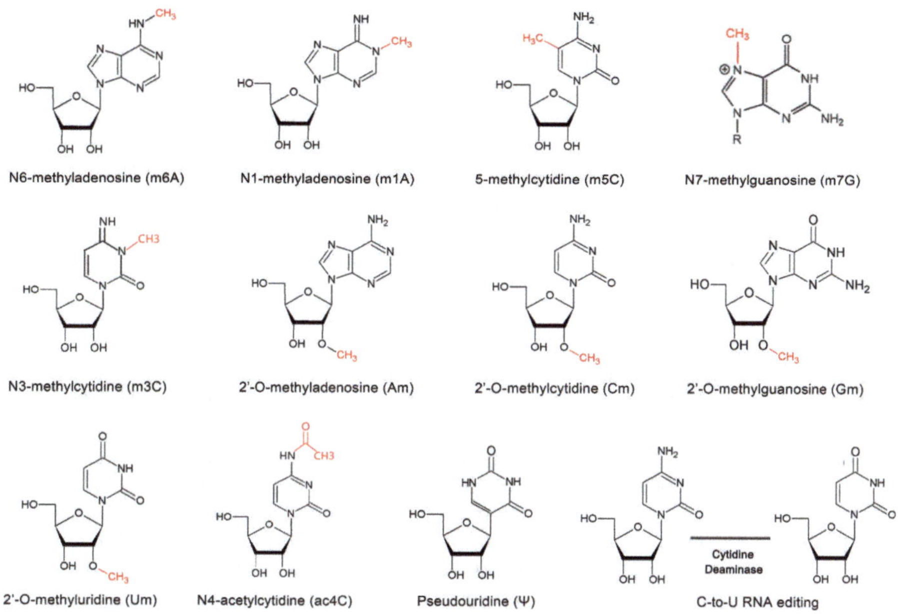

Fig. 9.1 The chemical structures of major types of RNA modifications in plants

catalyzing the m^{6}A modification, and FIP37 plays an auxiliary role, assisting the complex in localizing to specific RNA regions. In *Arabidopsis thaliana*, the *mta* mutant exhibits severe growth and developmental defects, including dwarf plants, smaller leaves, and delayed flowering (Zhong et al. 2008), indicating that m^{6}A modification is crucial for the normal growth and development of plants.

9.2.1.2 N1-Methyladenosine (m^{1}A)

The N1-methyladenosine (m^{1}A) modification, which involves the addition of a methyl group to the N1 position of adenosine (A), alters the base pairing properties of adenosine and significantly impacts the structure and function of RNA (Xiong et al. 2023). Initially discovered in non-coding RNAs such as rRNA and tRNA, m^{1}A is widely present in the mRNA of prokaryotes and various eukaryotes. Despite the relatively limited research on m^{1}A modification in plants, recent studies have shown that it exists in plant cells and performs important functions (Yang et al. 2020). Currently, some enzymes involved in plant m^{1}A modification have been identified, such as the TRMT6/61 complex, which is responsible for catalyzing m^{1}A modification at specific positions on tRNA and mRNA in *Arabidopsis thaliana* (Tang et al. 2020a).

9.2.1.3 5-Methylcytosine (m^{5}C)

5-methylcytosine (m^{5}C) modification adds a methyl group to the fifth carbon atom of cytosine (C) under the action of RNA methyltransferase. m^{5}C modification is

Table 9.1 RNA modification databases and prediction algorithms

Database	URL	Description	References
RNAMDB	http://rna-mdb.cas.albany.edu/RNAmods/	Provides detailed information on known RNA modifications, including their chemical structures, biosynthetic pathways, and other relevant aspects.	Cantara et al. (2010)
MODOMICS	https://iimcb.genesilico.pl/modomics/	A manually curated and centralized resource, storing and distributing comprehensive information about modified ribonucleosides	Boccaletto et al. (2022)
m6A-Atlas v2.0	http://rnamd.org/m6a	Provides high-confidence m6A sites and quantitative epitranscriptomic profiles.	Liang et al. (2024)
m7GHub v2.0	http://www.rnamd.org/m7GHub2	A database focused on m^7G modification, providing information on modification sites and functional annotations.	Wang et al. (2024)
m5C-Atlas	http://180.208.58.19/m5c-atlas/	A database dedicated to m5C modification, providing high-confidence m5C site information and functional annotations.	Ma et al. (2022)
MeT-DB v2.0	http://www.xjtlu.edu.cn/metdb2	Prediction based on ParCLIP-seq and MeRIP-seq data of m6A-related regulatory factors.	Liu et al. (2018)
RMBase v2.0	http://rna.sysu.edu.cn/rmbase/	A comprehensive database to integrate epitranscriptome sequencing data for exploring post-transcriptionally modifications of RNAs, as well as their relationships with microRNA binding events, disease-related SNPs and RNA-binding proteins (RBP).	Xuan et al. (2018)
REPIC	https://repicmod.uchicago.edu/repic	Integrate publicly available m6A-seq and MeRIP-seq data, along with histone ChIP-seq and DNA-seq data, for prediction.	Liu et al. (2020)
CVm6A	http://gb.whu.edu.cn:8080/CVm6A	A database focused on m6A modification in plants, providing information on modification sites and functional annotations.	Han et al. (2019)
RMVarn 2.0	http://rmvar.renlab.cn/	An updated database that houses functional variants involved in RNA modifications, aiming to provide insights into the relationship between genetic variants and RNA modifications.	Huang et al. (2025)
RMDisease	http://www.xjtlu.edu.cn/biologicalsciences/rmd	A database studying the relationship between RNA modifications and diseases, providing modification sites and disease association information.	Chen et al. (2021)
REDIdb	http://srv00.recas.ba.infn.it/redidb/	Provide information on RNA editing events, particularly A-to-I editing, covering editing sites across multiple species.	Picardi et al. (2017)
RADAR	http://rnaedit.com/	An RNA editing database providing information on A-to-I editing sites, covering species such as human and mice.	Ramaswami et al. (2014)

(continued)

Table 9.1 (continued)

Database	URL	Description	References
DARNED	http://darned.ucc.ie/	A database focused on RNA editing, providing information on known A-to-I and C-to-U editing sites.	Kiran et al. (2010)
REDIportal	http://srv00.recas.ba.infn.it/atlas/	Provide RNA editing site information for the human transcriptome based on high-throughput sequencing data.	Picardi et al. (2017)
m6A2Target	http://m6a2target.canceromics.org/	Integrate experimental validation and high-throughput sequencing data to predict m6A modification target genes.	Deng et al. (2021)
SRAMP	http://www.cuilab.cn/sramp/	Predict mammalian N6-methyladenosine (m6A) sites based on sequence features using a random forest algorithm.	Zhou et al. (2016)
PEA-m6A	https://github.com/cma2015/PEA-m6A	Simplify m6A-Seq data analysis to predict m6A-modified regions in plant genomes.	Song et al. (2024)
deepEA	http://deepea.omicstudio.cloud/	Interactive analysis for epitranscriptomic sequencing data.	Zhai et al. (2020)
PEA	https://hub.docker.com/r/malab/pea	PEA is a versatile epitranscriptome analysis pipeline covering CMR calling, prediction, and annotation.	Zhai et al. (2018)

common in plant RNA, found in tRNA, rRNA, mRNA, and other RNA types (Gao and Fang 2021). m^5C is catalyzed by the NSUN (NOP2/Sun domain protein 2) proteins in human (Yang et al. 2017). And its plant homolog, tRNA-specific methyltransferase 4B (TRM4B), belongs to the RNA (C5-cytosine) methyltransferase (RCMT) family has been identified as an m^5C mRNA methyltransferase in *Arabidopsis* (Cui et al. 2017; David et al. 2017), and OsNSUN2 is the major mRNA m^5C methyltransferase in rice (Tang et al. 2020b). In plants, the RCMT family can be classified into six subfamilies, all of which are present in species ranging from green algae to flowering plants, suggesting an ancient origin and widespread distribution of m^5C in green plants.

9.2.1.4 N3-Methylcytosine (m^3C)

Recently, a novel RNA methylation modification, m^3C, has been identified on mRNA. By integrating multiple techniques such as mass spectrometry, biochemistry, genetics, and cell biology, researchers have identified two tRNA m^3C methyltransferases, METTL2 and METTL6 (Ignatova et al. 2020), in mammals, as well as one mRNA m^3C methyltransferase, METTL8 (Xu et al. 2017). In plants, m^3C was initially identified in transcripts through computational methods, which were later validated on several candidate transcripts using RNA immunoprecipitation (RIP)–quantitative polymerase chain reaction (qPCR) using an m^3C-specific antibody (Vandivier et al. 2015). However, the regulatory mechanisms of m^3C in mRNA processing and metabolism, the identification of its methyltransferases, and their biological functions require further in-depth research in plants.

9.2.1.5 N7-Methylguanine (m^7G)

The m^7G modification in the 5′ cap structure of mRNA is the most common modification in eukaryotic mRNA (Furuichi 2015; Ramanathan et al. 2016). The methylation process of the mRNA cap structure in eukaryotic cells is accomplished by the coordinated action of RNA polymerase II, phosphatase, guanylyltransferase, and methyltransferase (Ramanathan et al. 2016). The m^7G cap structure performs multiple functions within the cell, as it can bind to the translation initiation factor EIF4F to promote mRNA translation and also prevent mRNA degradation by cellular exonucleases (Galloway and Cowling 2019). In addition to the 5′ end cap, m^7G is also found at internal positions in mRNA, tRNA, and rRNA (Haag et al. 2015; Tomikawa 2018; Malbec et al. 2019). In mammals, different methyltransferases mediate the m^7G modification of different RNAs: the m^7G modification of mRNA is believed to be catalyzed by the Rossmann fold methyltransferase (RFM) family member, methyltransferase RNMT, with methyltransferase BUD23 involved in the generation of m^7G in 18S rRNA (Zorbas et al. 2015); the homolog of methyltransferase METTL1 has been found to interact with the protein WDR4 in yeast, participating in the formation of m^7G in tRNA (Alexandrov et al. 2002).

9.2.1.6 N4-Acetylcytidine (ac^4C)

N4-acetylcytidine (ac^4C) modification is the only known acetylation modification on RNA. With the development of high-throughput sequencing and mass spectrometry technologies, ac^4C marks were confirmed to exist widely on the mRNAs of multiple plant species (including *Arabidopsis thaliana*, maize, rice, soybean, tomato, and tobacco) (Wang et al. 2023a). In mammals, RNA acetylation is mediated by the NAT10 protein (Ito et al. 2014), while the genomes of many plants contain two NAT10 homologous genes, named *N-ACETYLTRANSFERASEs FOR CYTIDINE IN RNA 1* (*ACYR1*) and *ACYR2* (Wang et al. 2023a). The complete deletion of *ACYR1* and *ACYR2* in *Arabidopsis thaliana* leads to embryo abortion, and partial knockout of *ACYRs* results in a significant decrease in mRNA acetylation abundance and affects the rosette leaf development of *Arabidopsis thaliana*. Integrated multi-omics analysis shows that the key gene for microRNA (miRNA) synthesis, *TOUGH* (*TGH*), is a target gene of *ACYR*, and there is a high abundance of ac^4C modification on its transcript. A decrease in cytosine acetylation level leads to rapid degradation and a decrease in the abundance of the *TGH* transcript, disrupting the miRNA homeostasis in plants and further affecting leaf development (Wang et al. 2023a).

9.2.1.7 2′-O-Methylation (Nm)

The 2′-O-methylation modification (also known as 2′-O-Me or Nm) involves the addition of a methyl group to the 2′-hydroxyl group of the ribose in RNA. Nm is the second most abundant RNA modification, occurring in various RNA species, including tRNA, rRNA, and some non-coding RNAs. In eukaryotic rRNA, Nm is coordinated by the action of the methyltransferase fibrillarin/NOP1, in conjunction with NOP56, NOP58, SNU13, and C/D snoRNA (Tollervey and Kiss 1997). C/D snoRNAs possess short-sequence motifs known as the C-box (5′-RUGAUGA-3′) and D-box (5′-CUGA-3′). These snoRNAs pair with their target RNA through 10–22 nucleotide-long complementary sequences. The nucleotide methylated is always recognized by the +5-guiding rule, where the nucleotide base pair to be methylated is located five nucleotides upstream of the D or D′ box (Rajan et al. 2019).

9.2.1.8 Pseudouridine (Ψ)

Pseudouridine (Ψ) modification represents a distinctive category of RNA modification. It is generated through the isomerization of uridine (U), wherein the glycosidic bond transitions from a C-N bond to a C-C bond. This structural transformation endows Ψ with unique chemical properties and functions. Ψ is relatively abundant in tRNA and rRNA and has also been detected in mRNA. It has the capacity to alter base-pairing properties, thereby influencing not only the structure of rRNA but also the coding capacity of mRNA. This suggests that Ψ has the potential to serve as a regulatory element in post-transcriptional regulatory processes. There are two primary mechanisms for the synthesis of Ψ across various organisms (Niu and Liu 2023). One involves the catalytic action of RNA-guided H/ACA ribonucleoproteins (H/ACA RNPs), which, under the guidance of RNA, isomerize specific uridines into Ψ in rRNA, U2 snRNA, and mRNA. The other mechanism involves stand-alone protein enzymes known as pseudouridine synthases (PUS), which operate independently of RNA (Hamma and Ferré-D'Amaré 2006).

9.2.1.9 RNA Editing

RNA editing is a post-transcriptional process that alters nucleotide sequences in RNA molecules, generating genetic diversity beyond DNA templates. In plants, RNA editing is particularly prevalent in mitochondria and chloroplasts, with emerging roles in nuclear transcripts. C-to-U editing is the most common form of RNA editing in plant organelles (mitochondria and chloroplasts). A number of factors are involved in plant RNA editing, including pentatricopeptide repeat (PPR) proteins, multiple organelle RNA editing factors (MORF, also known as RIP), organelle RNA recognition motif (ORRM) containing proteins, protoporphyrinogen IX oxidase

1 (PPO1), and organelle zinc finger 1 (OZ1) (Yan et al. 2018). Except for C-to-U, an additional U-to-C RNA editing reaction is observed in fern and moss organelles, suggesting some protist lineages seem to have acquired plant RNA editing factors and adapted them functionally for their own purposes (Knie et al. 2016; Knoop 2023).

Adenosine-to-inosine (A-to-I) editing is the most prevalent type of RNA editing in mammals, primarily occurring in double-stranded RNA regions. A-to-I editing widely occurs in pre-mRNA, non-coding RNA (such as miRNA, lncRNA), and tRNA (Fig. 9.2). This process is catalyzed by adenosine deaminases acting on RNA (ADARs), which deaminate adenosine residues in double-stranded RNA, converting them into inosine. Since inosine pairs with cytosine during base pairing, this modification results in an A-to-G transition, potentially leading to a range of functional changes. Editing in the coding region of pre-mRNA may alter the genetic code, resulting in amino acid substitutions. Editing within intronic regions may affect alternative splicing by generating or eliminating splice sites. Additionally, editing in the 3′ UTR may create or disrupt miRNA binding sites, influencing post-transcriptional regulation. A-to-I editing has been observed in fungi, specifically during sexual reproduction, such as wheat scab fungus *Fusarium graminearum* (Liu et al. 2017).

9.2.1.10 Noncanonical RNA Capping Modification

Noncanonical RNA capping in plants involves diverse 5′-end modifications beyond the canonical m^7G cap, including metabolites like NAD^+, FAD, and UDP-glucose (Hu et al. 2021; Doamekpor et al. 2022). These caps play roles in RNA stability, translation, and metabolic regulation. NAD^+ was first identified as a noncanonical RNA cap in plants, yeast, and mammals. NAD-RNAs can be full-length and polyadenylated, challenging earlier assumptions that noncanonical caps are limited to short or non-functional RNAs. Transcription start sites (TSS) for NAD-RNAs often differ from canonical m^7G-capped RNAs, suggesting distinct regulatory mechanisms. Besides, NAD-RNAs are preferentially translated in mitochondria, where they associate with mitochondrial ribosomes, indicating compartment-specific roles (Sharma et al. 2023b). NAD caps may link RNA metabolism to cellular redox states, as NAD^+ is a critical cofactor in redox reactions (Hu et al. 2021). FAD, UDP-glucose, and dephospho-CoA (dpCoA) also serve as RNA caps, though their prevalence and functions remain less studied. For example, UDP-N-acetylglucosamine-capped RNAs may connect glycosylation pathways to RNA regulation (Hu et al. 2021).

9.2.2 *Distributions of RNA Modifications*

Modifications on mRNA exhibit region-specific distribution patterns, with different modifications preferentially occurring in distinct transcript regions from the 5′ end to the 3′ untranslated region (3′ UTR), each playing unique functional roles (Fig. 9.2). The dynamic combination of these modifications regulates mRNA at the post-transcriptional level in multiple ways. 7-Methylguanosine (m^7G) cap at the 5′ end is essential for gene expression, transcript stability, and translation efficiency. N^6, 2′-O-dimethyladenosine (m^6Am) is located at the first nucleotide of the 5′ UTR, and is involved in mRNA decapping and miRNA-mediated degradation. N^2-Methylcytosine (m^2C) is enriched near the start codon, participating in the regulation of mRNA nuclear export.

9.2.2.1 m^6A Distribution

In terms of mRNA distribution, m^6A modifications show distinct positional preferences, primarily occurring in the coding sequence (CDS) and 3′ untranslated region (3′ UTR). Notably, m^6A is significantly enriched near the stop codon, a pattern that is highly conserved between animals and plants. Additionally, m^6A sites are enriched in exon regions near splicing sites, exhibiting significant spatial overlap with RNA-binding sites of certain splicing factors, suggesting a regulatory role in pre-mRNA splicing (Dominissini et al. 2016). The enrichment of m^6A modifications

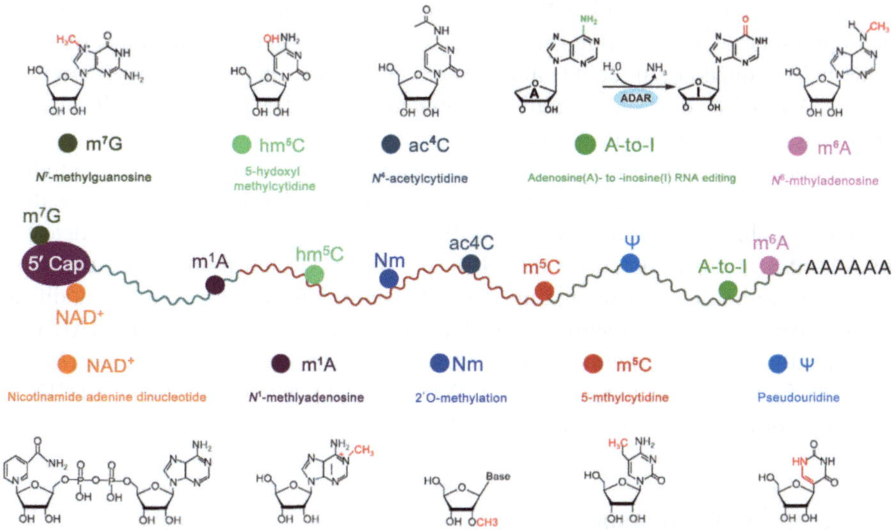

Fig. 9.2 The distribution of various known RNA modifications in different regions of mRNA in plants. Chemical structures of RNA modifications and their predominant locations in transcripts are divided into 5′ untranslated region (5′ UTR), coding sequence (CDS), and 3′UTR

in specific mRNA regions suggests its potential involvement in distinct RNA processing pathways. However, unlike exonic m^{6}A, m^{6}A modifications have not been detected in actively transcribed introns, suggesting unique regulatory mechanisms yet to be fully explored. Despite these findings, the dynamic regulation of m^{6}A modifications and its precise mechanistic roles require further in-depth investigation.

Although m^{6}A modifications are widespread, their levels vary across species, tissues, and developmental stages. In plants, studies on *Arabidopsis* have shown that m^{6}A levels in germinating tissues are higher than in roots and leaves, suggesting preferential enrichment in meristematic tissues (Ramakrishnan et al. 2022). These findings highlight that while m^{6}A modifications are ubiquitous across plant and animal tissues, they exhibit tissue-specific preferences—being particularly enriched in organs with active metabolic, developmental, or regulatory functions. The functional implications of this tissue-specific m^{6}A distribution remain to be fully elucidated.

Researchers have also explored the distribution characteristics and patterns of m^{6}A in plants like *Arabidopsis* and rice. In *Arabidopsis*, m^{6}A-modified genes are found to be associated with chloroplast-related plant-specific signaling pathways. m^{6}A is enriched around the start codon, stop codon, and 3′UTR regions, and the m^{6}A modification motif in the 3′UTR is consistent with that in mammals (RRACH). However, a new sequence pattern was identified in the m^{6}A-enriched region near the start codon. The unique characteristics of m^{6}A in *Arabidopsis* suggest that m^{6}A modifications may have plant-specific molecular mechanisms in post-transcriptional mRNA regulation. mRNAs rich in m^{6}A near the 3′UTR and stop codon are associated with mRNA degradation, while m^{6}A enrichment near the start codon is positively correlated with mRNA abundance, indicating that m^{6}A modifications play an important role in regulating gene expression in plants. In rice, deep transcriptome-wide m^{6}A sequencing in callus and leaf tissues revealed that each mRNA contains 2~3 m^{6}A modification sites, mainly distributed in the CDS region, 3′UTR, and the start codon region (Li et al. 2014).

Conserved m^{6}A modifications have been identified in various ncRNAs, including tRNA, rRNA, snRNA, and snoRNA, playing a crucial role in maintaining their structural stability and functional integrity. Additionally, researchers have detected m^{6}A modifications in precursor miRNAs (pre-miRNAs) and found that these modifications are involved in regulating miRNA maturation. Moreover, m^{6}A modifications have also been identified in lncRNAs such as XIST and MALAT1. Specifically, m^{6}A modifications in MALAT1 alter its stem-loop structure (Mauer et al. 2017), while m^{6}A modifications in XIST recruit the binding protein YTHDC1, thereby promoting X-chromosome gene transcriptional repression.

9.2.2.2 m^{1}A Distribution

Among the modified mRNA transcripts, an average of 1.4 m^{1}A peaks per transcript was detected and m^{1}A is primarily enriched in the first exon of the CDS. Unlike

m^6A, m^1A-modified sequences do not exhibit a distinct motif, but the regions surrounding m^1A modification sites are enriched in GC bases and have lower minimum free energy. Due to its charged nature, mRNA containing m^1A modifications is more likely to form secondary structures in the 5′UTR region, which is associated with enhanced translation efficiency. Studies have shown that m^1A responds dynamically to physiological conditions and can promote the translation of methylated mRNA.

9.2.2.3 m^5C Distribution

With the advancement of sequencing technologies, the identification of m^5C sites at the whole-transcriptome level has made significant progress, particularly in mRNA. Using bisulfite sequencing combined with transcriptome analysis, a total of 8495 sites were detected in mRNA. And of them, approximately 4000 were located in the CDS, around 1000 in the 5′ untranslated region (5′ UTR), and roughly 2500 in the 3′ UTR, indicating that m^5C sites are relatively enriched in the CDS and 3′UTR compared to introns and the 5′UTR. m^5C sites do not exhibit a clear sequence motif, but flanking sequences show a CG preference. Additionally, these sites are significantly enriched near AGO proteins, key components of the miRNA complex, suggesting a potential role of m^5C in miRNA-mediated mRNA stability regulation. This modification was found to be bound by the nuclear export protein ALYREF, suggesting a role in mRNA nuclear export. However, this enrichment pattern was not observed in *Arabidopsis* (Delatte et al. 2016). As a highly dynamic methylation modification in mRNA, m^5C research is still in its early stages compared to m^6A, and its distribution patterns and regulatory mechanisms require further exploration to provide insights into its functional roles.

m^5C modifications have been confirmed in the tRNA of many archaea and eukaryotes. These m^5C sites are primarily enriched in the variable arm and anticodon loop, where they help stabilize the secondary structure of tRNA, influence aminoacylation, and affect codon recognition. In rRNA, m^5C modifications are mainly located in regions where rRNA interacts with tRNA to exert translational activity, playing a role in ribosome synthesis and protein translation.

Besides tRNA and rRNA, m^5C modifications have also been detected in lncRNAs, such as vault RNA, scaRNA2, HOTAIR, and XIST. NSUN2 catalyzes m^5C modifications in vault RNA, regulating its processing into small RNAs that resemble miRNAs. In HOTAIR and XIST, m^5C modifications are mainly located in their functional domains, influencing their biological roles. For instance, m^5C sites in XIST inhibit PRC2 binding, thereby modulating X-chromosome inactivation.

9.2.2.4 Ψ Distribution

Pseudouridylation is one of the most abundant RNA modifications and is highly conserved across species. For a long time, due to the lack of effective detection methods, the presence and potential functional roles of pseudouridylation in mRNA have not been thoroughly investigated. Pseudouridine is widely present in tRNA, rRNA, and snRNA, playing a crucial role in the biogenesis and proper function of spliceosomal snRNA and rRNA. By examining pseudouridylation changes under different growth conditions, researchers have identified differentially modified pseudouridine sites in mRNA and ncRNA, revealing a pseudouridylation response to environmental signals.

9.2.2.5 Distribution of Other Types of RNA Modifications

Various other modifications also exist on mRNA, each with distinct distribution patterns. The m^7G capping process is a crucial step in the maturation of eukaryotic mRNA. In eukaryotic mRNA, the most common methylation modification is the methylation of the 5′ cap structure, m^7G (5′)ppp(5′)N, which plays an essential role in translation initiation and maintaining mRNA stability. Similar modifications have also been found in viral RNA. However, such methylation modifications do not exist in prokaryotic mRNA. 3-methylcytosine (m^3C) is another form of cytosine methylation present in mRNA. However, due to technological limitations, research on the identification and characterization of m^3C remains limited, necessitating further exploration. Additionally, mRNA contains 2′-*O*-methylation (2′-*O*Me or Nm) modifications, in which a methyl group is attached to the 2′-hydroxyl of the ribose. Studies have shown that Nm modifications are primarily found in mRNA and viral RNA, protecting the 3′ end of RNA from degradation and playing a role in the regulation of RNA decay.

9.3 Diversity of Regulatory Proteins in RNA Modifications

Many reports demonstrated that RNA modifications are highly dynamic and can sometimes perform contradictory functions even in the same cell, depending on the context and the specific modified RNA molecules (Sharma et al. 2023a, b). Cells contain various specific RNA methyltransferases, demethylases, methyl-binding proteins, and other types of RNA-modifying enzymes. Their collective action ensures the diversity of RNA modification sites and numbers, allowing different types of RNA to undergo dynamic modifications, thereby regulating various physiological processes. In order to understand the biological significance and context-specificity, identifying and characterizing RNA modification mechanisms is important, especially RNA-binding proteins (RBPs), which add, recognize, or remove chemical modifications, commonly referred to as "writers", "readers", and "erasers", respectively (Kramer et al. 2018). The m^6A marks are deposited by dedicated proteins, the "writers", is interpreted by "readers", and can be removed by enzymatic activities, the "erasers" (Meyer and Jaffrey, 2017). We describe the current knowledge about these proteins in the context of specific RNA modifications below.

9.3.1 Writers

Writers are proteins or protein complexes whose catalytic function is to chemically modify specific bases in various RNA molecules. In rice, OsTRM13 is a newly identified tRNA 2′-O-methyladenosine methyltransferase that adds the modification in the nucleus, whose expression increases under salt stress (Wang et al. 2017). As noted above, another common tRNA and rRNA modification is Ψ, which is primarily added by pseudouridine synthase (PUS) enzymes through both box-H/ACA-dependent and RNA-independent mechanisms. Furthermore, a comprehensive study of PUSs has identified six subfamilies: RluA, RsuA, TruA, TruB, PUS10, and TruD in *Arabidopsis* and maize (Xie et al. 2022). Similarly, the putative m^1A writer AtTRM61/AtTRM6 complex was identified to trigger *N*1-methylation at the A58 position on initiator methionyl-tRNA (tRNAiMet) (Wang et al. 2015).

The m^6A methyltransferase complex consists of METTL3, METTL14, WTAP, and KIAA29 proteins (Table 9.2). METTL3 was the first identified m^6A methyltransferase, containing an S-adenosylmethionine (SAM) binding site and a catalytic DPPW (Asp-Pro-Pro-Trp) functional domain. Through homologous sequence alignment with the METTL3 protein, METTL14 was identified to possess the same SAM binding site and DPPW functional domain required for m^6A catalysis. Meanwhile, MTA was subsequently identified as the main m^6A methyltransferase in plant systems (Zhong et al. 2008). Since then, the m^6A writer complex of proteins has expanded and included the core components of MTA, MTB, FIP37, and VIRILIZER (VIR) (Reichel et al. 2019; Růžička et al. 2017). Additionally, HAKAI protein in the m^6A methylation pathway as a writer protein was first discovered and demonstrated that loss of HAKAI led to a reduction of up to 35% of total m^6A levels, suggesting that the addition of this specific modification is a tightly regulated process (Růžička et al. 2017). Moreover, another m^6A methyltransferase, METTL16, has been uncovered as a member of the core complex to add m^6A to ncRNAs, lncRNAs, and mRNAs in eukaryotes (Warda et al. 2017). Lately, the ortholog of this protein in *Arabidopsis*, FIONA1, has been identified and confirmed to add m^6A onto the U6 snRNA and a small collection of mRNAs, in loss-of-function mutant plants (Wang et al. 2022). However, whether FIONA is truly independent of the larger m^6A writer core complex remains to be determined and further investigation is needed.

Although m^6A writer proteins have been identified widely, so far, there have been rarely reports on other RNA modifications such as m^1A and m^5C. Some potential candidates that could add these marks to tRNAs are two types of m^5C writers: tRNA aspartic acid methyltransferase 1 (TRDMT1) and tRNA-specific methyltransferase 4 (TRM4/NSUN2) (Burgess et al. 2015). Recently, a study demonstrated that *Arabidopsis* tRNA-specific methyltransferase, AtTRM4B, was likely to function as an mRNA m^5C writer in addition to its known function of adding m^5C methylations on tRNA at positions C48, C49, and C50. Identification and careful characterization of the writer proteins/complexes are needed to understand this specificity in modification deposition, which is important for our fundamental understanding of

Table 9.2 Characteristics of the putative m6A writers, erasers, and readers in 28 plant species

	m6A Writers				m6A Erasers		m6A Readers			
Arabidopsis thaliana	AtMTA	AtFIP37	AtVIR	AtHAKAI	AtALKBH9B	AtALKBH10B	AtECT2	AtECT3	AtECT4	AtCPSF30
	AtMTB									
	AtMTC									
Cicer arietinum	CaMTA	CaFIP37-1	CaVIR1	CaHAKAI	CaALKBH3B	CaALKBH16B	AtECT2	AtECT3	AtECT4	AtCPSF30
	CaMTB	CaFIP37-2	CaVIR2			CaALKBH17B	CaECT8	CaECT6		CaECT19
	CaMTC	CaFIP37-3	CaVIR3							
	CaFIP37-4									
Vitis vinifera	VvMTA	VvFIP37	VvVIR	VvHAKAI	VvALKBH6B	VvALKBH10B	VvECT9	VvECT1	VvECT11	VvECT3
	VvMTB									
	VvMTC									
Solanum lycopersicum	SlMTA	SlFIP37	SlVIR	SlHAKAI	SlALKBH3B	SlALKBH10B	SlECT1	SlECT2	–	SlECT4
	SlMTB				SlALKBH4B			SlECT9		SlECT6
	SlMTC				SlALKBH2					
	SlMTD									
Brassica rapa	BrMTA	BrFIP37-1	BrVIR	BrHAKAI	BrALKBH6B	BrALKBH12B	BrECT16	BrECT1	BrECT14	BrCPSF30-1
	BrMTB	BrFIP37-2						BrECT7	BrECT15	BrCPSF30-2
	BrMTC									
Gossypium hirsutum	GhMTA	GhFIP37-1	GhVIR1	GhHAKAI1	GhALKBH9B	GhALKBH11B	GhECT15	GhECT5	GhECT18	GhECT20
	GhMTB	GhFIP37-2	GhVIR2	GhHAKAI2	GhALKBH10B	GhALKBH23B	GhECT16	GhECT6		GhECT27
	GhMTC		GhVIR3	GhHAKAI3						
	GhMTD			GhHAKAI4						
	GhMTE			GhHAKAI5						
Chenopodium quinoa	CqMTA				CqALKBH13B	CqALKBH26B	CqECT14	CqECT15	CqECT9	CqCPSF30-1
	CqMTB	CqFIP37-1	CqVIR1	CqHAKAI1		CqALKBH27B				CqCPSF30-2

	CqMTC	CqFIP37-2	CqVIR2	CqHAKAI2						
	CqMTD									
	CqMTE									
	CqMTF									
Zea mays	ZmMTA	ZmFIP37-1	ZmVIR1	ZmHAKAI1	ZmALKBH1B	ZmALKBH5B	ZmECT7	ZmECT8	ZmECT11	ZmCPSF30-1
	ZmMTB	ZmFIP37-2	ZmVIR2	ZmHAKAI2		ZmALKBH10B	ZmECT23		ZmECT25	
	ZmMTC	ZmFIP37-3								
		ZmFIP37-4								
		ZmFIP37-5								
Triticum aestivum	TaMTA-D	TaFIP37-1D	TaVIR-D	TaHAKAI1-D	TaALKBH4B	TaALKBH6B	TaECT7	TaECT6	TaECT3	TaCPSF30-2
	TaMTA-A	TaFIP37-2D	TaVIR-A	TaHAKAI1-A		TaALKBH29B	TaECT21	TaECT22	TaCPSF30-4	TaCPSF30-5
	TaMTC-D	TaFIP37-2A	TaVIR-B	TaHAKAI1-B						
	TaMTB-A	TaFIP37-1B		TaHAKAI2-D						
	TaMTB-D									
	TaMTA-B									
	TaMTC-B									
	TaMTB-B									
	TaMTC-A									
Sorghum bicolor	SbMTA	SbFIP37	SbVIR1	SbHAKAI1	SbALKBH13B	SbALKBH4B	SbCPSF30-3	SbCPSF30-1	–	SbCPSF30-19
	SbMTB		SbVIR2	SbHAKAI2	SbALKBH14B	SbALKBH15B	SbCPSF30-4	SbCPSF30-2		
	SbMTC		SbCPSF30-8							
Oryza indica	OiMTA	OiFIP37	OiVIR	OiHAKAI						
	OiALKBH6B	OiALKBH5B	OiECT10	OiECT7	–	OiECT2				
	OiMTB					OiALKBH12B				
	OiMTC									
Hordeum vulgare	HvMTA	HvFIP37	HvVIR	HvHAKAI	HvALKBH1B	HvALKBH4B	HvECT2	HvECT4	HvECT1	
						HvALKBH8B	HvECT6	HvECT8		

(continued)

Table 9.2 (continued)

	m6A Writers				m6A Erasers		m6A Readers			
Selaginella moellendorffi	SmMTA	SmFIP37	–	–	SmALKBH4B	SmALKBH14B	–	SmECT2	–	SmECT1
	SmMTB					SmALKBH16B		SmECT4		SmECT3
Marchantia polymorpha	MapMTA	MapFIP37	MapVIR1	MpHAKAI	MapALKBH6B	MapALKBH10B	–	MapECT1	–	MapECT3
	MapMTB		MapVIR2					MapECT2		
	MapMTC									
Physcomitrella patens	PpMTA	PpFIP37-1	PpVIR	PpHAKAI1	PpALKBH5B	PpALKBH2B	–	PpECT1	–	PpECT4
	PpMTB	PpFIP37-2		PpHAKAI2	PpALKBH6B	PpALKBH4B		PpECT3		
	PpMTC			PpHAKAI3						
	PpMTD									
Cyanidioschyzon merolae	CmMTA	–	–	–	–	CmALKBH1B	–	–	–	–
	CmMTB									
Micromonas pusilla	MipMTA	MipFIP37	–	–	–	MipALKBH4B	MipECT1	–	–	MipCPSF30-1
										MipCPSF30-2
Emiliania huxleyi	EhMTA	EhFIP37-1	–	–	EhALKBH21B	EhALKBH18B	EhECT1	–	–	–
	EhMTB	EhFIP37-2				EhALKBH24B				
	EhMTC									
	EhMTD									
Volvox carteri	VcMTA	VcFIP37	–	–	VcALKBH1B	–	–	VcCPSF30-1	–	VcECT1
	VcMTB				VcALKBH4B					
	VcMTC									
Ectocarpus siliculosus	–	EsFIP37	–	–	EsALKBH2B	EsALKBH4B	–	EsECT1	–	EsECT2
						EsALKBH6B				EsCPSF30-2
Chlorella variabilis	CvMTA	CvFIP37	CvVIR	CvHAKAI	CvALKBH2B	–	–	CvCPSF30-1	–	CvECT1
	CvMTB									CvCPSF30-3
	CvMTC									

Brassica napus	BnMTA1	BnFIPa	BnVIR	BnHAKAI	BnALKBH9B1	BnALKBH10B	BnECT2a	BnECT3	BnETC8	BnCPSF30-L
	BnMTA2	BnFIPb	–	–	BnALKBH9B2	–	BnECT2b	BnECT4		
	BnMTB1	BnFIPc								
	BnMTB2	BnFIPd								
Brassica juncea					BjALKBH9B					
Citrullus lanatus	ClMTA									
	ClMTB									
	ClMTC									
	ClMTB									
	ClMT-A70	–	–	–	–	ClALKBH15				
Glycine max	GmMTAa	–	–	–	GmALKBH10B1	GmALKBH10B3				
	GmMTAb	–	–	–	GmALKBH10B2	GmALKBH10B4	–			
Oryza sativa	OsMTA	OsFIP	–	–	OsALKBH9	OsFTO	OsYTH07	–	OsEHD6	
	OsMT1-2	OsFIP37								
Phytophthora sojae	PsMTA1									
	PsMTA2									
Malus domestica	MdMTA	–	MdVIR1							
			MdVIR2							

these writer protein complexes and can also make them potential tools for future crop improvement.

Similar to m^6A, RNA m^5C modification is likely to be dynamically reversible. Methyltransferases use SAM as a donor to transfer a methyl group to cytosine (C), forming m^5C. RsmB was the first identified m^5C methyltransferase, primarily catalyzing methylation on bacterial rRNA. Subsequently, over 30 RNA m^5C methyltransferases have been discovered, which can be classified into four main groups: NOP2/NOL1, YebU/Trm4, RsmB/Yn1022c, and PH1991/NSUN. These enzymes exhibit high conservation in eukaryotes and the NSUN protein family has attracted significant attention in recent years. Among them, NSUN1 acts on 25S rRNA in yeast and *Arabidopsis*. NSUN2 has been reported as the primary mRNA m5C methyltransferase and can also modify rRNA and tRNA (Yang et al. 2017). NSUN3 catalyzes the formation of m^5C in mitochondrial tRNA, while NSUN4 primarily acts on the C911 site of 12S rRNA encoded by mitochondrial DNA. NSUN5 in nematodes, drosophila, yeast, and plants catalyzes the m^5C methylation of 28S and 25S rRNA. NSUN6 acts on $tRNA^{Cys}$ and $tRNA^{Thr}$, catalyzing methylation at the C72 position of their 3′ ends. Additionally, m^5C modification in enhancer RNA (eRNA) is likely catalyzed by NSUN7.

The formation of m^7G involves the catalytic actions of phosphatase, guanylyltransferase, and methyltransferase. In yeast, these enzymatic activities are carried out by three distinct proteins. Some viruses encode a single polypeptide that possesses all three enzymatic functions. In mammals, the methyltransferase RNMT is considered a member of the RFM family. In tomato, the *METTL1*, *METTL2*, *IF4E3* have been reported as the encoding enzymes to associate with the formation of m^7G (Zhao et al. 2025).

Under the catalytic action of pseudouridine synthase (PUS), uridine (U) undergoes sequence-specific isomerization to form pseudouridine (Ψ) (Carlile et al. 2014; Schwartz et al. 2014). In yeast, eight PUS enzymes have been identified, which directly catalyze pseudouridylation at specific sites on tRNA, U2 snRNA, and mitochondrial rRNA. Based on the genome-wide identification, 20 *AtPUSs* and 22 *ZmPUSs* from *Arabidopsis* and maize were found, which could be clustered into six subfamilies, RluA, RsuA, TruB, TruA, TruD, and PUS10 (Xie et al. 2022) (Table 9.3).

9.3.2 Readers

Currently, the progress on RNA modification interpreted proteins in plants mainly focuses on m^6A readers. Among the plant species, m^6A reader proteins containing the YTH domain have been identified (Table 9.2). Interestingly, one of the two YTHDC proteins in plants is a larger isoform of the highly conserved member of the eukaryotic polyadenylation complex CLEAVAGE AND POLYADENYLATION SPECIFICITY FACTOR 30 (CPSF30) (Hou et al. 2021). CPSF30 has been identified in the nucleus, while the members of the YTHDF clade ECT2, ECT3, and ECT4

Table 9.3 The known writers, erasers, and readers of RNA modifications

	m^6A	m^5C	I	Ψ	m^1A
Writers	VIR FLONA1 MTA MTB METTL3 METTL14 FIP37	NSUN2 TRDMT1 TRM4b	ADAR ADARB1 ADARB2	DKC1 PUS1 PUS3 PUS7 TRUB1	–
Erasers	ALKBH10B FTO ECT	TET1 TET2 TET3	–	–	ALKBH1 ALKBH3
Readers	YTH family HNRNP family eIF3 complex ECT YTP8	–	–	–	–

Note "–" represents the proteins that have not been reported

are found in the cytoplasm (Arribas-Hernández et al. 2019; Scutenaire et al. 2018; Song et al. 2021) (Table 9.3). In other plants, 12 YTH domain-containing protein-coding genes are identified in rice, 9 in tomato, and 39 in the hexaploid species common wheat (Yin et al. 2021).

9.3.2.1 YTH Family Proteins

The YTH domain contains approximately 16 amino acids and is capable of binding to single-stranded RNA. It is highly conserved in yeast, plants, and animals. And YTH domain family proteins can be categorized based on their subcellular localization: YTHDC2, YTHDF1, YTHDF2, and YTHDF3 are located in the cytoplasm, while YTHDC1 is located in the nucleus. The YTH at the C-terminal of YTH family proteins is primarily responsible for recognizing the m^6A sites (Table 9.3).

YTHDF1 can promote the translation of m^6A-modified mRNA. Ribosome profiling analysis shows that YTHDF1 enhances the binding of substrate mRNA to ribosomes, facilitating mRNA translation. The m^6A sites recognized by YTHDF1 are primarily located in the 3′ UTR. How m^6A promotes mRNA translation is thought to involve the interaction of YTHDF1 with eIF3, assisted by eIF4G, leading to the formation of a circular structure of the mRNA. YTHDF1 interacts with the translation initiation machinery, thereby promoting the translation of m^6A-containing mRNA, suggesting that m^6A is a dynamic mRNA marker to effectively regulate protein production (Wang et al. 2015).

YTHDF2 binds RNA types, mainly including mRNA and long non-coding RNA. YTHDF2 tends to bind near the stop codon and the 3′ UTR of mRNA and prefers to bind with a conserved motif containing GAC[U > A], which is highly similar to the m^6A binding motif. The mRNA half-life profile and ribosome profiling experiments show that the main function of YTHDF2 is to regulate the stability of m^6A-modified mRNA in the cytoplasm (Wang et al. 2014). Interestingly, m^6A modifications in the 3′ UTR of mRNA can lead to mRNA degradation, which can also occur in the open reading frame (ORF) of mRNA and lncRNAs. Thus, YTHDF2 first recognizes m^6A on mRNA and then brings the RNA substrate to the RNA degradation pathway, facilitating RNA degradation.

YTHDF3 was found to be a potential interacting protein with ribosomal proteins of the 40S small subunit and the 60S large subunit, suggesting that YTHDF3 may promote mRNA translation by recruiting ribosomal large and small subunit proteins (Li et al. 2017). YTHDF3 likely first binds to m^6A-modified mRNA, then promotes translation through YTHDF1, and finally recruits YTHDF2 to degrade the mRNA after translation is completed (Shi et al. 2017). YTHDF3 can also bind to m^6A-modified circular RNA, promoting its translation.

YTHDC1 is a nuclear m6A-binding protein with distinct biological functions compared to cytoplasmic binding proteins. YTHDC1 interacts with other splicing factors, suggesting its important regulatory role in RNA processing. Like YTHDF1–3, in vitro RNA oligonucleotide sedimentation assays have shown that YTHDC1 can specifically bind to m^6A-modified RNA. The distribution of YTHDC1 binding clusters on mRNA is similar to that of m^6A, tending to be located near the stop codon. Studies have revealed that YTHDC1 recognizes m^6A on exons and recruits the splicing factor SRSF3 while inhibiting the binding of SRSF10 to mRNA, thereby promoting exon retention. YTHDC1 and SRSF3 tend to promote exon retention, whereas SRSF10 promotes exon splicing (Xiao et al. 2016). YTHDC2 not only contains the YTH domain but also have multiple RNA helicase domains, influencing the translation efficiency and stability of mRNA containing m^6A modifications (Hsu et al. 2017).

9.3.2.2 HNRNP Family Proteins

HNRNPC is an RNA-binding protein located in the nucleus, and it tends to bind to consecutive uridine regions on single-stranded RNA. Previous studies have reported that m^6A modification weakens the hydrogen bond interaction between A-U, altering the local structure of mRNA and lncRNAs, exposing the single-stranded structure, and thereby enhancing the interaction between HNRNPC and RNA (Schwartz et al. 2014) (Table 9.3). Studies have found that the HNRNP protein family member HNRNPA2B1 is an m^6A-binding protein that can recognize m^6A modifications on pri-miRNA and regulate the splicing processing of pri-miRNA, revealing the important regulatory role of m^6A in miRNA biogenesis. Meanwhile, HNRNPA2B1 can interact with the pri-miRNA processing factor DGCR8 to regulate the splicing of pri-miRNA (Alarcon et al. 2015a; Alarcon et al. 2015b).

Studies have found that m^6A modifications can enhance the ability of surrounding RNA sequences to bind HNRNPG. HNRNPG binds to m^6A-modified RNA through its C-terminal low-complexity region, and the arginine-glycine-glycine repeat sequence within the low-complexity region is essential for binding to the

RNA motifs exposed by the m^6A modification. m^6A can recruit RNA-binding proteins such as HNRNPG by altering RNA structure, thereby regulating the metabolic process of mRNA processing (Liu et al. 2017).

9.3.2.3 EIF3

In the process of studying the mRNA 5′ UTR m^6A modification, researchers found that the m^6A modification on the 5′ UTR can promote cap-independent protein translation (Table 9.3). The specific mechanism is that eukaryotic initiation factor 3 (EIF3) can directly bind to the m^6A on the mRNA 5′ UTR of mRNA, and then recruit the 43S complex to initiate translation without EIF4E (Meyer et al. 2015).

9.3.3 Erasers

Compared with the advances made in understanding RNA modification writers, there is very limited knowledge of eraser proteins. The majority of research has focused on the removal of methyl marks with the ALKB family of nonheme Fe (II)/alpha-ketoglutarate-dependent dioxygenases (ALKBH family), catalyzing this demethylation process. In mammals, ALKBH1 and ALKBH3 act as erasers of m^1A in tRNAs (Seo and Kleiner 2020), while ALKBH5 is involved in demethylation of mRNA m6A. ALKBH9B and ALKBH10B in *Arabidopsis* and SIALKBH2 in tomato have been experimentally proven to remove mRNA methylation (Duan et al. 2017; Martínez-Pérez et al. 2017; Zhou et al. 2019) (Table 9.3). Intriguingly, the human demethylase enzyme FTO seems to function as a potent demethylase in plant species rice and potato, which mediates substantial m^6A demethylation of plant RNAs, causing a reduction in overall m^6A levels. Overexpression of FTO in rice and potato caused ~50% increases in yield and biomass in the field (Yu et al. 2021). In addition, the dynamic NAD+ cap modification on mRNA can also be removed by DXO1 acts as a de-NADing protein with regulating the stability of particular mRNAs in *Arabidopsis* (Jiao et al. 2017; Yu et al. 2021).

9.4 Functional Diversity of RNA Modifications in Plants

9.4.1 The Biological Functions of m^6A

N^6-methyladenosine (m^6A) is the most widespread mRNA modification, highly enriched near stop codons and splicing sites. It plays key roles in mRNA splicing, degradation, translation, and secondary structure modulation, as well as in regulating

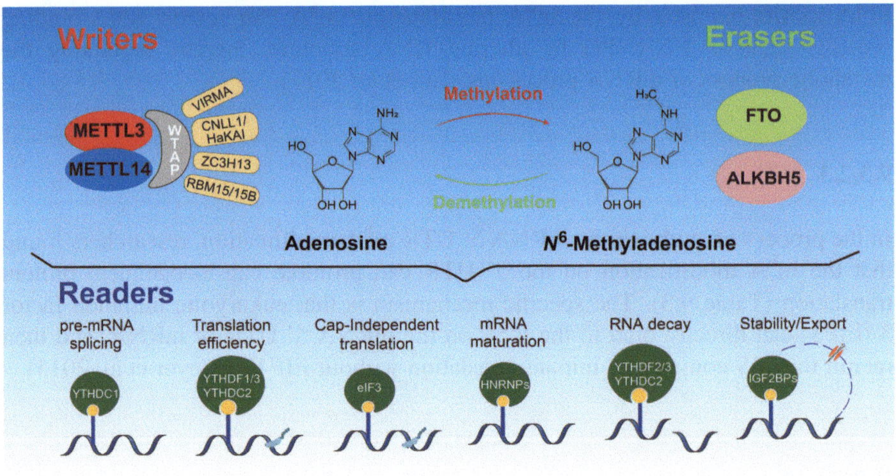

Fig. 9.3 The N6-methyladenosine (m^6A) methylation machinery and the biological functions of m^6A.The m^6A modification is installed onto mRNA by the methyltransferase complex (writers) comprising the methyltransferase-like 3 (METTL3)/METTL14 (methyltransferase-like 14) heterodimer core subunit and other cofactors, including WTAP, CNLL1, ZC3H13, and RBM15/RBM15b. The RNA m^6A modification can be reversibly removed by RNA demethylases (erasers), fat mass and obesity-associated protein (FTO), and AlkB family member 5 (ALKBH5). The biological functions of m^6A modification are achieved by specific recognition and binding by RNA-binding proteins (readers), which affect RNA fate by regulating RNA splicing, export, decay, stabilization, and translation

RNA-binding protein interactions (Fig. 9.3). Cells finely tune the dynamic regulation of mRNA modifications to orchestrate various physiological processes in an orderly manner (Huang et al. 2020).

The m^6A marks affect RNA fate in multiple aspects by recruiting different m6A-binding proteins (“readers”) (Fig. 9.3). The m^6A reader protein YTHDC1 with hnRNPC, hnRNPG, and hnRNPA2B1 may regulate mRNA splicing via recognizing m^6A on pre-mRNA in a direct or indirect manner using the manner of structural switch (Kramer et al. 2018). YTHDC1 could also mediate nuclear export of processed RNAs to the cytoplasm (Lauman and Garcia 2020). In the cytoplasm, transcripts containing m^6A are sorted to degrade in p-bodies when bound by YTHDF2, YTHDF3, or YTHDC2, while IGF2BP1/2/3, FMRP, and PRRC2A play an opposite role in stabilizing the m^6A-modified transcripts (Du et al. 2016; Liang et al. 2020; Liu et al. 2016; Liu et al. 2019). In addition to regulating RNA stability, binding of m^6A by reader proteins, including YTHDF1, YTHDF3, IGF2BP1/2/3, YTHDC2, and METTL3, also promotes mRNA translation (Du et al. 2016; Liang et al. 2020; Liu et al. 2013; Lu et al. 2020) (Fig. 9.3). Therefore, m^6A modification involves various aspects of gene regulation at the post-transcriptional level and RNA metabolism, depending on the functions of different readers.

9.4.1.1 mRNA Splicing and Nuclear Export

m^6A affects the interaction between RNA-binding proteins (such as the HNRNP family) and pre-mRNA, thereby regulating splicing site selection and leading to the production of different splicing variants. For example, METTL3-mediated m^6A modification has been shown to regulate alternative splicing of genes such as CD44 and SRSF. FIONA1 is an *Arabidopsis* U6 m^6A methyltransferase that adds m^6A at specific nucleotides in U6 snRNA, a core component of the spliceosome. While in mammals and fission yeast, the lack of this protein leads to changes in splicing, a study in *Arabidopsis* showed that this phenotype is not displayed by plants that lack the FIONA protein (Wang et al. 2022).

9.4.1.2 mRNA Translation

A single m^6A mark in the 5′ UTR of mRNA in mammals can directly bind with eIF3 to initiate translation (Wang et al. 2015). The m^6A reader protein YTHDF1 was shown to directly interact with eIF1 and promote translation (Meyer et al. 2015). However, the reader protein of m^6A for translation in plants is to date unknown and deserves more investigations. In *Arabidopsis*, m^6A-containing mRNAs generally had higher abundance and translation efficiency than non-m^6A-containing mRNAs under normal and low temperatures, indicating m6A has a positive role on translation efficiency (Wang et al. 2023b). In maize, global m^6A methylation was shown to be negatively correlated with translation, while transcripts with the m^6A mark on the 5′ end were positively correlated with translation, suggesting a position-specific role for m^6A in translation (Luo et al. 2020).

9.4.1.3 mRNA Stability

It has been documented that m^6A modification, enriched in the 3′ untranslated regions (UTRs) of eukaryotic mRNAs, could regulate mRNA degradation and stabilization (Liang et al. 2020). Recent studies have demonstrated that the m^6A-dependent stabilization and destabilization of mRNAs are influenced by specific m^6A reader proteins (Nguyen and Kang 2024). For instance, the *Arabidopsis* m^6A reader ECT2 enhances the stability of m^6A-modified target mRNAs by interacting with the poly(A)-binding proteins PAB2 and PAB4 (Wei et al. 2018; Song et al. 2023). By contrast, FLK, a novel m^6A reader in *Arabidopsis*, facilitates the decay of FLC mRNA by binding to the m^6A site in the 3′ UTRs of FLC transcripts (Amara et al. 2023). In plants, transcriptome-wide investigation of the effects of m^6A on RNA stability has been studied on loss mutation of the main m^6A proteins and their direct interacting cofactors, including MTA, ALKBH, FIP37, and VIR (Reichel et al. 2019; Seo and Kleiner 2020). The m^6A writer VIR-mediated m^6A methylation influences the stability of transcripts associated with stress-responsive genes or photoprotection to resist the salt stress or maintain photosynthetic efficiency, respectively (Hu et al. 2021; Zhang et al. 2022).

9.4.1.4 Plant Development

The m^6A of RNA modification varies from tissue to tissue and plays a critical role in plant growth and development (Ramakrishnan et al. 2022; Liang et al. 2020; Zhang et al. 2023) (Fig. 9.4). Previous studies have demonstrated the roles of m^6A in leaf morphology, trichome branching (Wei et al. 2018), root development (Arribas-Hernandez et al. 2019), seed development (Bodi et al. 2012), seed germination (Tang et al. 2021), floral transition and flower development (Duan et al. 2017), embryonic development (Ružička et al. 2017), fruit ripening (Zhou et al. 2019; Zhou et al. 2021), stem cell fate differentiation (Shen et al. 2016), nutrient utilization (Hou et al. 2021), crop yield (Yu et al. 2021), and stress development (Anderson et al. 2018). m^6A is also involved in maintaining circadian and seasonal rhythms in plants (Parker et al. 2019).

9.4.2 *The Biological Functions of m^1A*

Unlike m^6A, m^1A is enriched in the 5′ UTR region of mRNA transcripts and at splicing sites of exons containing the start codon, with a tendency to modify the translation initiation site. Studies have shown that m^1A is positively correlated with protein production, and mRNA with m1A modification exhibits higher translation efficiency. A known DNA/RNA demethylase, ALKBH3, can remove m1A from mRNA. The m^1A methylation can dynamically respond to stimuli (such as H_2O_2 and starvation), and under stress, hundreds of new m^1A sites can emerge. However, the roles of m^1A methyltransferases and binding proteins on mRNA are still unclear. In tRNA, the function of m^1A primarily regulates tRNA structure. For example, the stability of the three-dimensional structure of all eukaryotic $tRNA^{Met}$ requires the m^1A modification at position 58 (m^1A58). Studies have found that the absence of tRNA demethylase slows down the initiation and elongation of mRNA translation.

9.4.3 *The Biological Functions of m^5C*

9.4.3.1 mRNA m^5C Regulates mRNA Nuclear Export

m^5C sequencing results indicated that m^5C is enriched in the non-translated regions of mRNA (5′ UTR and 3′ UTR), suggesting that m^5C may be involved in protein translation and the stability of mRNA, but there is still no direct evidence. Recent studies have found that in vitro, the key protein involved in mRNA nuclear export, ALYREF, can specifically bind to oligonucleotides containing the m^5C modification. Further in vivo experiments also showed that ALYREF can bind to mRNA containing the m^5C modification and mediate its nuclear export (Yang et al. 2017). vtRNA is a non-coding RNA that exists in the nucleolar ribonucleoprotein complex

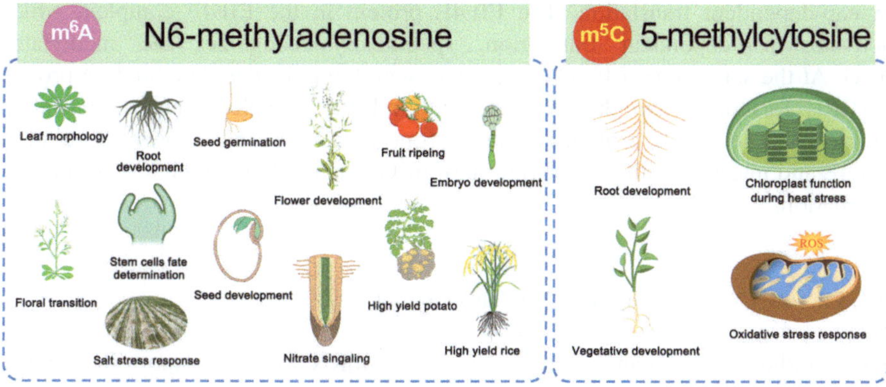

Fig. 9.4 Schematic diagram summarizing the physiological function of m^6A and m^5C RNA modifications in plants

and can be processed into small RNAs (svRNA), which regulate the expression of certain genes through the AGO-miRNA pathway. The presence of m^5C causes abnormalities in the processing of vtRNA into svRNA. m^5C modification is found in functional regions of long non-coding RNAs (lncRNAs), such as HOTAIR and XIST, that interact with chromatin. In vitro experiments have shown that m^5C can interfere with the interaction between RNA-binding proteins and RNA, suggesting that m^5C modification may increase the structural diversity of lncRNAs (Amort et al. 2013). Moreover, the m^5C of RNA modification plays an important role in plant growth and development, including root and vegetative development, chloroplast function, and oxidative stress response (Tang et al. 2020b; Ramakrishnan et al. 2022) (Fig. 9.4).

9.4.3.2 tRNA m^5C Regulates Protein Translation

In tRNA, m^5C is primarily found in the variable region and the anticodon loop, which stabilizes the secondary structure of tRNA and enhances its ability to recognize codons. Simultaneous knockout of the m^5C methyltransferase genes *DNMT2* and *NSUN2* leads to the complete loss of m^5C modification in tRNA, resulting in its instability and inhibition of protein translation. The m^5C modification in tRNA is believed to be associated with the fidelity of the translation process.

9.4.4 The Biological Functions of m^7G

9.4.4.1 mRNA Translation

In eukaryotes, protein translation can be divided into cap-dependent translation and cap-independent translation mediated by Internal Ribosome Entry Sites. Among these, cap-dependent translation is the predominant form. The m^7G cap is essential in cap-dependent translation. The EIF4E protein in the EIF4F complex, which activates cap-dependent translation, specifically recognizes the methylated m^7G. At the same time, EIF4G can interact with the polyadenylate-binding protein PABP1, forming a circular structure with mRNA, which enhances translation efficiency. The AGO2 protein contains a conserved motif that allows it to bind to EIF4E and interfere with its recognition of m^7G, leading to the inhibition of protein translation.

9.4.4.2 mRNA Splicing and Nuclear Export

Some studies have shown that m^7G can regulate mRNA nuclear export. In *Xenopus laevis* oocytes, the injection of dinucleotide containing the m^7G cap structure (m7GpppG) inhibits mRNA nuclear export, while mRNA without the cap is not affected. The regulatory effect of m^7G on pre-mRNA splicing is primarily mediated by the nuclear cap-binding complex (CBC), which consists of two cap-binding proteins, CBP80 and CBP20. However, the m^7G cap does not regulate RNA nuclear export in all species.

9.4.4.3 mRNA Stability

Another important function of the m^7G cap is to promote mRNA stability. Studies have shown that the m^7G methylated cap structure can prevent degradation of mRNA by the $5' \rightarrow 3'$ exonuclease, whereas uncapped RNA is rapidly degraded. The guanylation capping reaction is reversible, and guanylyltransferase, a decapping enzyme, can remove the unmodified guanosine cap. Uncapped mRNA is quickly degraded by the exonuclease, but when the guanosine cap is methylated to form the m^7G cap structure, it is not recognized by guanylyltransferase, thus enhancing the stability of the mRNA.

9.4.5 *Biological Functions of Pseudouridine (Ψ)*

It has been proven that, compared to uridine (U), pseudouridine (Ψ) pairing with any of A, G, U, or C enhances the stability of double-stranded RNA and more proteins are expressed, suggesting it may interfere with RNA-protein interactions. Studies have shown that artificially replacing U with Ψ in the nonsense codons of mRNA can cause the ribosome's decoding center to recognize it as a sense codon and undergo unconventional base pairing, allowing translation to continue. Based on these results, it can be predicted that Ψ may have two potential functions in mRNA: (1) Regulating mRNA processing and function by altering RNA structure; (2) In protein translation, besides converting nonsense codons into sense codons, Ψ may also diversify protein translation, leading to post-transcriptional gene encoding recombination. Compared to mRNA, research on Ψ in rRNA and snRNA is more extensive. In ribosomes, compared to normal ribosomes, ribosomes containing unmodified or low-modified rRNA show weakened interaction with tRNA, leading to reduced translation accuracy. The pseudouridylation of rRNA interferes with the function of ribosomes in regulating tRNA binding and translation accuracy, a function conserved in both yeast and humans (Li et al. 2015). In snRNA, Ψ plays a critical regulatory role in the biosynthesis and function of snRNPs.

9.5 Conclusions

Using high-throughput transcriptome-wide investigation, the diverse RNA modifications in plants and their roles in plant growth, development, and stress response have been gradually elucidated. Dynamic RNA modifications play important roles in RNA stability, translation, mRNA splicing, and nuclear export, ultimately regulating nearly all RNA-related processes for different RNA types. Identification of the chemical modification proteins of writers, readers, and erasers for m^6A and other RNA modifications could greatly enrich our understanding of their functions. However, this is a challenge to identify and characterize the key components responsible for the addition and removal of RNA modification marks. Future research on establishing the molecular links between RNA modifications and plant development and stress response still requires ongoing effort to unravel the mechanisms of RNA modifications in model plants and agricultural crops. Precise RNA modification or engineering RNA-modifying machinery could facilitate the breeding of smart crops with multiple desired traits. Integrating biotechnologies such as genomics, epigenomics, and epi-editing with artificial intelligence (AI) will provide new insights into crop improvement. With the advances in epitranscriptomics, programmable RNA methylation and demethylation tools are anticipated to be innovative options for plant stress resistance and breeding strategies.

Acknowledgments This work was supported by the National Natural Science Foundation of China (32370235) and the Graduate Teaching Reform Curriculum Project of Northwest A&F University (JXGG24041).

Declaration of Competing Interest The authors declare no competing financial interests and have not used any AI tools.

References

Alarcon CR, Goodarzi H, Lee H, Liu X, Tavazoie S, Tavazoie SF (2015a) HNRNPA2B1 is a mediator of m^6A-dependent nuclear RNA processing events. Cell 162:1299–1308

Alarcon CR, Lee H, Goodarzi H, Halberg N, Tavazoie SF (2015b) N^6-methyladenosine marks primary microRNAs for processing. Nature 519:482–485

Alexandrov A, Martzen MR, Phizicky EM (2002) Two proteins that form a complex are required for 7-methylguanosine modification of yeast tRNA. RNA 8:1253–1266

Amara U, Hu J, Cai J, Kang H (2023) FLK is an mRNA m^6A reader that regulates floral transition by modulating the stability and splicing of FLC in *Arabidopsis*. Mol Plant 16:919–929

Amort T, Soulière MF, Wille A, Jia XY, Fiegl H, Wörle H et al (2013) Long non-coding RNAs as targets for cytosine methylation. RNA Biol 10(6):1003–1008

Anderson SJ, Kramer MC, Gosai SJ, Yu X, Vandivier LE, Nelson ADL et al (2018) N^6-methyladenosine inhibits local ribonucleolytic cleavage to stabilize mRNAs in *Arabidopsis*. Cell Rep 25:1146–1157.e3

Arribas-Hernandez L, Brodersen P (2019) Occurrence and functions of m^6A and other covalent modifications in plant mRNA. Plant Physiol 182:79–96

Boccaletto P, Stefaniak F, Ray A, Cappannini A, Mukherjee S, Purta E et al (2022) MODOMICS: a databaseof RNA modification pathways. 2021 update. NuclC Acids Res 50: D231–D235

Bodi Z, Zhong S, Mehra S, Song J, Graham N, Li H et al (2012) Adenosine methylation in *Arabidopsis* mRNA is associated with the 3′ end and reduced levels cause developmental defects. Front Plant Sci 3:48

Burgess AL, David R, Searle IR (2015) Conservation of tRNA and rRNA 5-methylcytosine in the kingdom plantae. BMC Plant Biol 15:199

Cai J, Shen L, Kang H, Xu T (2025) RNA modifications in plant adaptation to abiotic stresses. Plant Commun 6:101229

Cantara WA, Crain PF, Rozenski J, McCloskey JA, Harris KA, Zhang X et al (2010) The RNA modificationdatabase, RNAMDB: 2011 update. NuclC Acids Res 39: D195–D201

Cappannini A, Ray A, Purta E, Mukherjee S, Boccaletto P, Moafinejad SN et al (2024) MODOMICS: a database of RNA modifications and related information. 2023 update. Nucleic Acids Res 52:D239–D244

Carlile TM, Rojas-Duran MF, Zinshteyn B, Shin H, Bartoli KM, Gilbert WV (2014) Pseudouridine profiling reveals regulated mRNA pseudouridylation in yeast and human cells. Nature 515(7525):143–146

Chen K, Song B, Tang Y, Wei Z, Xu Q, Su J et al (2021) RMDisease: a database of genetic variants thataffect RNA modifications, with implications for epitranscriptome pathogenesis. NuclC Acids Res 49: D1396–D1404

Cohn WE (1951) Some results of the applications of ion-exchange chromatography to nucleic acid chemistry. J Cell Physiol 38:21–40

Cui XA, Liang Z, Shen LS, Zhang Q, Bao SJ, Geng YK et al (2017) 5-Methylcytosine RNA methylation in *Arabidopsis thaliana*. Mol Plant 10:1387–1399

David R, Burgess A, Parker B, Li J, Pulsford K, Sibbritt T et al (2017) Transcriptome-wide mapping of RNA 5-methylcytosine in *Arabidopsis* mRNAs and noncoding RNAs. Plant Cell 29:445–460

Delatte B, Wang F, Ngoc LV, Collignon E, Bonvin E, Deplus R et al (2016) Transcriptome-wide distribution and function of RNA hydroxymethylcytosine. Science 351(6270):282–285

Deng S, Zhang H, Zhu K, Li X, Ye Y, Li R et al (2021) M6A2Target: a comprehensive database for targets ofm6A writers, erasers and readers. BriefS Bioinform 22

Doamekpor SK, Sharma S, Kiledjian M, Tong L (2022) Recent insights into noncanonical 5′ capping and decapping of RNA. J Biol Chem 298:102171

Dominissini D, Nachtergaele S, Moshitch-Moshkovitz S, Peer E, Kol N, Ben-Haim MS et al (2016) The dynamic N(1)-methyladenosine methylome in eukaryotic messenger RNA. Nature 530: 441–446

Du H, Zhao Y, He J, Zhang Y, Xi H, Liu M et al (2016) YTHDF2 destabilizes m^6A-containing RNA through direct recruitment of the CCR4-NOT deadenylase complex. Nat Commun 7: 12626

Duan HC, Wei LH, Zhang C, Wang Y, Chen L, Lu Z et al (2017) ALKBH10B is an RNA N^6-methyladenosine demethylase affecting *Arabidopsis* floral transition. Plant Cell 29:2995–3011

Furuichi Y (2015) Discovery of m^7G-cap in eukaryotic mRNAs. Proc Jpn Acad Ser B Phys Biol Sci 91:394–409

Galloway A, Cowling VH (2019) mRNA cap regulation in mammalian cell function and fate. Biochim Biophys Acta Gene Regul Mech 1862:270–279

Gao Y, Fang J (2021) RNA 5-methylcytosine modification and its emerging role as an epitranscriptomic mark. RNA Biol 18:117–127

Ge L, Pan F, Jia M, Pott DM, He H, Shan H et al (2025) RNA modifications in plant biotic interactions. Plant Commun 6:101232

Haag S, Kretschmer J, Bohnsack MT (2015) WBSCR22/Merml is required for late nuclear pre-ribosomal RNA processing and mediates N7-methylation of G1639 in human 18S rRNA. RNA 21:180–187

Hamma T, Ferré-D'Amaré AR (2006) Pseudouridine synthases. Chem Biol 13:1125–1135

Han Y, Feng J, Xia L, Dong X, Zhang X, Zhang S et al (2019) CVm6A: A Visualization and ExplorationDatabase for m6As in Cell Lines. Cells 8

Hou Y, Sun J, Wu B, Gao Y, Nie H, Nie Z et al (2021) CPSF30-L-mediated recognition of mRNA m^6A modification controls alternative polyadenylation of nitrate signaling-related gene transcripts in *Arabidopsis*. Mol Plant 14:688–699

Hsu PJ, Zhu Y, Ma H, Guo Y, Shi X, Liu Y et al (2017) YTHDC2 is an N^6-methyladenosine binding protein that regulates mammalian spermatogenesis. Cell Res 27:1115–1127

Hu H, Flynn N, Chen X (2021) Discovery, processing, and potential role of noncanonical caps in RNA. In: Jurga S, Barciszewski J (eds) Epitranscriptomics. Springer International Publishing, Cham, pp 435–469

Huang HL, Weng HY, Chen J (2020) The biogenesis and precise control of RNA m6Amethylation. Trends Genet 36:44–52

Huang Y, Zhang L, Mu W, Zheng M, Bao X, Li H et al (2025) RMVar 2.0: an updated database of functionalvariants in RNA modifications. Nucleic Acids Research 53: D275–D283

Ignatova VV, Kaiser S, Ho JSY, Bing X, Stolz P, Tan YX et al (2020) METTL6 is a tRNA m^3C methyltransferase that regulates pluripotency and tumor cell growth. Sci Adv 6:eaaz4551

Ito S, Horikawa S, Suzuki T, Kawauchi H, Tanaka Y, Suzuki T et al (2014) Human NAT10 is an ATP-dependent RNA acetyltransferase responsible for N4-acetylcytidine formation in 18S ribosomal RNA (rRNA). J Biol Chem 289:35724–35730

Jiao X, Doamekpor SK, Bird JG, Nickels BE, Tong L, Hart RP et al (2017) 5′ end nicotinamide adenine dinucleotide cap in human cells promotes RNA decay through DXO-mediated deNADding. Cell 168:1015–1027.e10

Kiran A, Baranov PV (2010) DARNED: a DAtabase of RNa EDiting in humans. Bioinform 26: 1772–1776

Knie N, Grewe F, Fischer S, Knoop V (2016) Reverse U-to-C editing exceeds C-to-U RNA editing in some ferns—a monilophyte-wide comparison of chloroplast and mitochondrial RNA editing suggests independent evolution of the two processes in both organelles. BMC Evol Biol 16:134

Knoop V (2023) C-to-U and U-to-C: RNA editing in plant organelles and beyond. J Exp Bot 74: 2273–2294

Kramer MC, Anderson SJ, Gregory BD (2018) The nucleotides they are a-changin': function of RNA binding proteins in post-transcriptional messenger RNA editing and modification in Arabidopsis. Curr Opin Plant Biol 45:88–95

Lang X, Yu C, Shen M, Gu L, Qian Q, Zhou D et al (2024) PRMD: an integrated database for plant RNA modifications. Nucleic Acids Res 52:D1597–d1613

Lauman R, Garcia BA (2020) Unraveling the RNA modification code with mass spectrometry. Mol Omics 16(4):305–315

Li Y, Wang X, Li C, Hu S, Yu J, Song S (2014) Transcriptome-wide N^6-methyladenosine profiling of rice callus and leaf reveals the presence of tissue-specific competitors involved in selective mRNA modification. RNA Biol 11(9):1180–1188

Li X, Zhu P, Ma S, Song J, Bai J, Sun F et al (2015) Chemical pulldown reveals dynamic pseudouridylation of the mammalian transcriptome. Nat Chem Biol 11:592–597

Li A, Chen YS, Ping XL, Yang X, Xiao W, Yang Y et al (2017) Cytoplasmic m^6A reader YTHDF3 promotes mRNA translation. Cell Res 27:444–447

Liang Z, Riaz A, Chachar S, Ding Y, Du H, Gu X (2020) Epigenetic modifications of mRNA and DNA in plants. Mol Plant 13:14–30

Liang Z, Ye H, Ma J, Wei Z, Wang Y, Zhang Y et al (2024) m6A-Atlas v2.0: updated resources for unravelingthe N6-methyladenosine (m6A) epitranscriptome among multiple species. NuclC Acids Res 52: D194–D202

Liu N, Parisien M, Dai Q, Zheng G, He C, Pan T (2013) Probing N^6-methyladenosine RNA modification status at single nucleotide resolution in mRNA and long noncoding RNA. RNA 19(12):1848–1856

Liu F, Clark W, Luo G, Wang X, Fu Y, Wei J et al (2016) ALKBH1-mediated tRNA demethylation regulates translation. Cell 167(3):816–828.e16

Liu N, Zhou KI, Parisien M, Dai Q, Diatchenko L, Pan T (2017) N^6-methyladenosine alters RNA structure to regulate binding of a low-complexity protein. Nucleic Acids Res 45:6051–6063

Liu H, Wang H, Wei Z, Zhang S, Hua G, Zhang S-W et al (2018) MeT-DB V2.0: elucidating context-specificfunctions of N6-methyl-adenosine methyltranscriptome. NuclC Acids Res 46: D281-D287

Liu H, Begik O, Lucas MC, Ramirez JM, Mason CE, Wiener D et al (2019) Accurate detection of m^6A RNA modifications in native RNA sequences. Nat Commun 10(1):4079

Liu S, Zhu A, He C, Chen M (2020) REPIC: a database for exploring the N6-methyladenosine methylome.Genome Biol 21

Lu L, Zhang Y, He Q, Qi ZX, Zhang G, Xu W et al (2020) MTA, an RNA m^6A methyltransferase, enhances drought tolerance by regulating the development of trichomes and roots in poplar. Int J Mol Sci 21(7):2462

Luo J, Wang Y, Wang M, Zhang L, Peng H, Zhou Y et al (2020) Natural variation in RNA m^6A methylation and its relationship with translational status. Plant Physiol 182(1):332–344

Ma J, Song B, Wei Z, Huang D, Zhang Y, Su J et al (2022) m5C-Atlas: a comprehensive database fordecoding and annotating the 5-methylcytosine (m5C) epitranscriptome. NuclC Acids Res 50: D196–D203

Malbec L, Zhang T, Chen YS, Zhang Y, Sun BF, Shi BY et al (2019) Dynamic methylome of internal mRNA N7-methylguanosine and its regulatory role in translation. Cell Res 29:927–941

Martínez-Pérez M, Aparicio F, López-Gresa MP, Bellés JM, Sánchez-Navarro JA, Pallás V (2017) *Arabidopsis* m^6A demethylase activity modulates viral infection of a plant virus and the m^6A abundance in its genomic RNAs. Proc Natl Acad Sci USA 114:10755–10760

Mauer J, Luo X, Blanjoie A, Jiao X, Grozhik AV, Patil DP et al (2017) Reversible methylation of m^6A_m in the 5′ cap controls mRNA stability. Nature 541:371–375

Meyer KD, Patil DP, Zhou J, Zinoviev A, Skabkin MA, Elemento O et al (2015) 5′ UTR m^6A promotes cap-independent translation. Cell 163:999–1010

Meyer K, Jaffrey S (2017) Rethinking m6A Readers, Writers, and Erasers. Annu Rev Cell Dev Biol 33: 319-342

Motorin Y, Helm M (2011) RNA nucleotide methylation. Wiley Interdiscip Rev RNA 2:611–631

Nguyen TKH, Kang H (2024) Reading m^6A marks in mRNA: a potent mechanism of gene regulation in plants. J Integr Plant Biol 66:2586–2599

Niu Y, Liu L (2023) RNA pseudouridine modification in plants. J Exp Bot 74:6431–6447

Parker MT, Knop K, Sherwood AV, Schurch NJ, Mackinnon K, Gould PD et al (2019) Nanopore direct RNA sequencing maps an Arabidopsis N^6-methyladenosine epitranscriptome. elife 9: e49658

Picardi E, D'Erchia AM, Lo Giudice C, Pesole G (2017) REDIportal: a comprehensive database of A-to-I RNAediting events in humans. NuclC Acids Res 45: D750–D757

Picardi E, Regina TMR, Verbitskiy D, Brennicke A, Quagliariello C (2011) REDIdb: An upgradedbioinformatics resource for organellar RNA editing sites. Mitochondrion 11: 360–365

Rajan KS, Chikne V, Decker K, Waldman Ben-Asher H, Michaeli S (2019) Unique aspects of rRNA biogenesis in Trypanosomatids. Trends Parasitol 35:778–794

Ramakrishnan M, Rajan K, Mullasseri S, Palakkal S, Kalpana K, Sharma A et al (2022) The plant epitranscriptome: revisiting pseudouridine and 2′-O-methyl RNA modifications. Plant Biotechnol J 20:1241–1256

Ramanathan A, Robb GB, Chan SH (2016) mRNA capping: biological functions and applications. Nucleic Acids Res 44:7511–7526

Ramaswami G, Li JB (2014) RADAR: a rigorously annotated database of A-to-I RNA editing. NuclC Acids Res 42: D109–D113

Reichel M, Köster T, Staiger D (2019) Marking RNA: m^6A writers, readers, and functions in Arabidopsis. J Mol Cell Biol 11(10):899–910

Ružička K, Zhang M, Campilho A, Bodi Z, Kashif M, Saleh M et al (2017) Identification of factors required for m^6A mRNA methylation in *Arabidopsis* reveals a role for the conserved E3 ubiquitin ligase HAKAI. New Phytol 215(1):157–172

Schwartz S, Bernstein DA, Mumbach MR, Jovanovic M, Herbst RH, Leon-Ricardo BX et al (2014) Transcriptome-wide mapping reveals widespread dynamic-regulated pseudouridylation of ncRNA and mRNA. Cell 159:148–162

Scutenaire J, Deragon JM, Jean V, Benhamed M, Raynaud C, Favory JJ et al (2018) The YTH domain protein ECT2 is an m^6A reader required for normal trichome branching in *Arabidopsis*. Plant Cell 30(5):986–1005

Seo KW, Kleiner RE (2020) YTHDF2 recognition of N1-methyladenosine (m1A)-modified RNA is associated with transcript destabilization. ACS Chem Biol 15:132–139

Shao YL, Wong CE, Shen LS, Yu H (2021) N6-methyladenosine modification underlies messenger RNA metabolism and plant development. Curr Opin Plant Biol 63:102047

Sharma B, Prall W, Bhatia G, Gregory BD (2023a) The diversity and functions of plant RNA modifications: what we know and where we go from here. Annu Rev Plant Biol 74:53–85

Sharma S, Yang J, Favate J, Shah P, Kiledjian M (2023b) NADcapPro and circNC: methods for accurate profiling of NAD and non-canonical RNA caps in eukaryotes. Commun Biol 6:406

Shen L, Yu H (2021) Epitranscriptome engineering in crop improvement. Mol Plant 14:1418–1420

Shen L, Liang Z, Gu X, Chen Y, Teo ZW, Hou X et al (2016) N^6-Methyladenosine RNA modification regulates shoot stem cell fate in *Arabidopsis*. Dev Cell 38:186–200

Shen LS, Liang Z, Wong CE, Yu H (2019) Messenger RNA modifications in plants. Trends Plant Sci 24:328–341

Shi H, Wang X, Lu Z, Zhao BS, Ma H, Hsu PJ et al (2017) YTHDF3 facilitates translation and decay of N^6-methyladenosine-modified RNA. Cell Res 27:315–328

Song P, Yang J, Wang C, Lu Q, Shi L, Tayier S et al (2021) *Arabidopsis* N^6-methyladenosine reader CPSF30-L recognizes FUE signals to control polyadenylation site choice in liquid-like nuclear bodies. Mol Plant 14(4):571–587

Song P, Wei L, Chen Z, Cai Z, Lu Q, Wang C et al (2023) m^6A readers ECT2/ECT3/ECT4 enhance mRNA stability through direct recruitment of the poly(A) binding proteins in *Arabidopsis*. Genome Biol 24:103

Song M, Zhao J, Zhang C, Jia C, Yang J, Zhao H et al (2024) PEA-m6A: an ensemble learning framework foraccurately predicting N6-methyladenosine modifications in plants. Plant Physiol 195: 1200–1213

Tang J, Jia P, Xin P, Chu J, Shi DQ, Yang WC (2020a) The *Arabidopsis* TRM61/TRM6 complex is a bona fide tRNA N1-methyladenosine methyltransferase. J Exp Bot 71:3024–3036

Tang Y, Gao CC, Gao Y, Yang Y, Shi B, Yu JL et al (2020b) OsNSUN2-mediated 5-methylcytosine mRNA modification enhances rice adaptation to high temperature. Dev Cell 53:272–286

Tang J, Yang J, Duan H, Jia G (2021) ALKBH10B, an mRNA m^6A demethylase, modulates ABA response during seed germination in *Arabidopsis*. Front Plant Sci 12:712713

Tollervey D, Kiss T (1997) Function and synthesis of small nucleolar RNAs. Curr Opin Cell Biol 9: 337–342

Tomikawa C (2018) 7-Methylguanosine modifications in transfer RNA (tRNA). Int J Mol Sci 19: 4080

Vandivier LE, Campos R, Kuksa PP, Silverman IM, Wang LS, Gregory BD (2015) Chemical modifications mark alternatively spliced and uncapped messenger RNAs in *Arabidopsis*. Plant Cell 27:3024–3037

Wang X, Lu Z, Gomez A, Hon GC, Yue Y, Han D et al (2014) N^6-methyladenosine-dependent regulation of messenger RNA stability. Nature 505:117–120

Wang X, Zhao BS, Roundtree IA, Lu Z, Han D, Ma H et al (2015) N^6-Methyladenosine modulates messenger RNA translation efficiency. Cell 161(6):1388–1399

Wang Y, Li D, Gao J, Li X, Zhang R, Jin X et al (2017) The 2-O-methyladenosine nucleoside modification gene OsTRM13 positively regulates salt stress tolerance in rice. J Exp Bot 68(7): 1479–1491

Wang C, Yang J, Song P, Zhang W, Lu Q, Yu Q et al (2022) FIONA1 is an RNA N^6-methyladenosine methyltransferase affecting *Arabidopsis* photomorphogenesis and flowering. Genome Biol 23(1):40

Wang W, Liu H, Wang F, Liu X, Sun Y, Zhao J et al (2023a) N4-acetylation of cytidine in mRNA plays essential roles in plants. Plant Cell 35:3739–3756

Wang S, Wang HY, Xu ZH, Jiang S, Shi Y, Xie H et al (2023b) m6A mRNA modification promotes chilling tolerance and modulates gene translation efficiency in Arabidopsis. Plant Physiol 192: 1466–1482

Wang X, Zhang Y, Chen K, Liang Z, Ma J, Xia R et al (2024) m7GHub V2.0: an updated database fordecoding the N7-methylguanosine (m7G) epitranscriptome. NuclC Acids Res 52: D203–D212

Warda AS, Kretschmer J, Hackert P, Lenz C, Urlaub H, Höbartneret C et al (2017) Human METTL16 is a N^6-methyladenosine (m^6A) methyltransferase that targets pre-mRNAs and various non-coding RNAs. EMBO Rep 18(11):2004–2014

Wei LH, Song P, Wang Y, Lu Z, Tang Q, Yu Q et al (2018) The m^6A reader ECT2 controls trichome morphology by affecting mRNA stability in *Arabidopsis*. Plant Cell 30:968–985

Xiao W, Adhikari S, Dahal U, Chen YS, Hao YJ, Sun BF et al (2016) Nuclear m^6A reader YTHDC1 regulates mRNA splicing. Mol Cell 61:507–519

Xie Y, Gu Y, Shi G, He J, Hu W, Zhang Z (2022) Genome-wide identification and expression analysis of pseudouridine synthase family in *Arabidopsis* and *maize*. Int J Mol Sci 23(5):2680

Xiong W, Zhao Y, Wei Z, Li C, Zhao R, Ge J et al (2023) N1-methyladenosine formation, gene regulation, biological functions, and clinical relevance. Mol Ther 31:308–330

Xu L, Liu X, Sheng N, Oo KS, Liang J, Chionh YH et al (2017) Three distinct 3-methylcytidine (m^3C) methyltransferases modify tRNA and mRNA in mice and humans. J Biol Chem 292: 14695–14703

Xuan JJ, Sun WJ, Lin PH, Zhou KR, Liu S, Zheng LL et al (2018) RMBase v2.0: deciphering the map of RNA modifications from epitranscriptome sequencing data. Nucleic Acids Res 46: D327–D334

Yan JJ, Zhang QX, Yin P (2018) RNA editing machinery in plant organelles. Sci China Life Sci 61: 162–169

Yang X, Yang Y, Sun BF, Chen YS, Xu JW, Lai WY et al (2017) 5-Methylcytosine promotes mRNA export—NSUN2 as the methyltransferase and ALYREF as an m^5C reader. Cell Res 27: 606–625

Yang W, Meng J, Liu J, Ding B, Tan T, Wei Q et al (2020) The N1-methyladenosine methylome of petunia mRNA. Plant Physiol 183:1710–1724

Yin S, Ao Q, Tan C, Yang Y (2021) Genome-wide identification and characterization of YTH domain-containing genes, encoding the m^6A readers, and their expression in tomato. Plant Cell Rep 40(7):1229–1245

Yu Q, Liu S, Yu L, Xiao Y, Zhang S, Wang XP et al (2021) RNA demethylation increases the yield and biomass of rice and potato plants in field trials. Nat Biotechnol 39:1581–1588

Yue H, Nie XJ, Yan ZG, Song WN (2019) N6-methyladenosine regulatory machinery in plants: composition, function and evolution. Plant Biotechnol J 17:1194–1208

Zhai J, Song J, Cheng Q, Tang Y, Ma C, Valencia A (2018) PEA: an integrated R toolkit for plantepitranscriptome analysis. Bioinform 34: 3747–3749

Zhai J, Song J, Zhang T, Xie S, Ma C (2020) deepEA: a containerized web server for interactive analysis ofepitranscriptome sequencing data. Plant Physiol 185: 29–33.2

Zhang M, Zeng Y, Peng R, Dong J, Lan Y, Duan S et al (2022) N6-methyladenosine RNA modification regulates photosynthesis during photodamage in plants. Nat Commun 13:7441

Zhang D, Guo W, Wang T, Wang Y, Le L, Xu F et al (2023) RNA 5-Methylcytosine modification regulates vegetative development associated with H3K27 trimethylation in *Arabidopsis*. Adv Sci 10:2204885

Zhao Z, Zeng S, Liu H, Chen M, Li D, Chen K (2025) Dynamic abundances of m^7G, m^5C and m^1A mRNA modifications are associated with tomato fruit quality following harvest. Postharvest Biol Tec 225:113522

Zhong S, Li H, Bodi Z, Button J, Vespa L, Herzog M et al (2008) MTA is an *Arabidopsis* messenger RNA adenosine methylase and interacts with a homolog of a sex-specific splicing factor. Plant Cell 20:1278–1288

Zhou Y, Zeng P, Li Y Zhang Z, Cui Q (2016) SRAMP: prediction of mammalian N6-methyladenosine (m6A)sites based on sequence-derived features. Nucleic Acids Research 44: e91–e91

Zhou L, Tian S, Qin G (2019) RNA methylomes reveal the m^6A-mediated regulation of DNA demethylase gene SlDML2 in tomato fruit ripening. Genome Biol 20:156

Zhou L, Tang R, Li X, Tian S, Li B, Qin G (2021) N^6-methyladenosine RNA modification regulates strawberry fruit ripening in an ABA-dependent manner. Genome Biol 22:168

Zorbas C, Nicolas E, Wacheul L, Huvelle E, Heurgué-Hamard V, Lafontaine DL (2015) The human 18S rRNA base methyltransferases DIMT1L and WBSCR22-TRMT112 but not rRNA modification are required for ribosome biogenesis. Mol Biol Cell 26:2080–2095

Chapter 10
Mapping and Interpreting RNA Modifications

Alexander Krohannon, Rebekah Eileen Lanning, Prasanth Kumar Thuthika, Swaraj Madan Thorat, and Sarath Chandra Janga

Abstract RNA modifications constitute a dynamic and multifaceted layer of post-transcriptional regulation with profound implications for cellular function, development, and disease. Once considered rare curiosities, over 170 distinct RNA modifications are now recognized as integral components of gene expression regulation, affecting processes from splicing and translation to RNA stability and localization. This chapter offers a comprehensive overview of the historical evolution, methodological advancements, and future directions in RNA modification detection. It traces the field's progress from classical chemical assays through the genomics era to modern direct sequencing and machine learning-driven approaches. Emphasis is placed on the convergence of high-throughput experimental platforms and computational innovations—including deep learning, signal processing, and multi-modal integration—that are transforming epitranscriptomic research. The chapter also explores the biological interpretation of RNA modifications, focusing on positional, structural, and comparative analyses across different biological contexts. Special attention is given to databases and bioinformatic resources essential for cataloging and analyzing modifications. Together, these developments are positioning RNA modification studies as a central component of plant genomics and metagenomics, with promising implications for both basic biology and therapeutic discovery.

A. Krohannon · R. E. Lanning · P. K. Thuthika · S. M. Thorat
Department of Biomedical Engineering and Informatics, Luddy School of Informatics, Computing, and Engineering, Indiana University Indianapolis, Indianapolis, IN, USA

S. C. Janga (✉)
Department of Biomedical Engineering and Informatics, Luddy School of Informatics, Computing, and Engineering, Indiana University Indianapolis, Indianapolis, IN, USA

Center for Computational Biology and Bioinformatics, Indiana University School of Medicine, Health Information and Translational Science (HITS), Indianapolis, IN, USA

Department of Medical and Molecular Genetics, Indiana University School of Medicine, Medical Research and Library Building, Indianapolis, IN, USA
e-mail: scjanga@iu.edu

L. M. Vaschetto (ed.), *Plant Transcriptomics and Epitranscriptomics*, Concepts and Strategies in Plant Sciences, https://doi.org/10.1007/978-981-95-5183-5_10

Keywords Epitranscriptomics · RNA modifications · m6A · Nanopore sequencing · Deep learning in genomics · RNA structure-function analysis · Post-transcriptional regulation · Transcriptome-wide mapping · Single-molecule sequencing · RNA modification databases

Abbreviations

2′-OMeA	2′-O-methyl-adenosine
5mC	5-methylcytosine
5moU	5-methoxyuridine
Am	2′-O-methyladenosine
ChIP-Seq	Chromatin ImmunoPrecipitation followed by sequencing
Cm	2′-O-methylcytidine
cryo-EM	cryo-electron microscopy
DART-seq	Deamination Adjacent to RNA modification Targets sequencing
DMS	DiMethyl Sulfate
DNA	DeoxyriboNucleic Acid
DRME	Detection of RNA Modifications from Expression
DRUMMER	Detection of Ribonucleic acid Modifications Manifested in Error Rates
f5C	5-formylcytosine
Gm	2′-O-methylgaunine
HAMR	High-throughput Analysis of Modified Ribonucleotides
HCC	hepatocellular carcinoma
hm5C	5-hydroxymethylcytidine
HPLC	High-Performance Liquid Chromatography
I	Inosine
m/z	mass-to-charge ratio
m1A	N1-methyladenosine
m1G	N1-methylguanosine
m22G	N2,N2-methylguanosine
m2G	N2-methylguanosine
m3C	N3-methylcytidine
m4Cm	N4, 2′-*o*-dimethylcytidine
m5C	N5-methylcytosine
m5U	N5-methyluradine
m62A	N6-dimethyladenosine
m6A	N6-methyladenosine
m6A-CLIP	m6A-CrossLinking ImmunoPrecipitation
m6AISH-PLA	m6A-specific In Situ Hybridization mediated Proximity Ligation Assay
m6A-LAIC-seq	m6A-Level And Isoform-Characterization sequencing
m6Am	N6–2′-O-dimethyladenosine

m7G	7-methylGuanosine
MACS	Model-based Analysis of ChIP-Seq
mcm5U	5-methoxycarbonylmethyl-2-uridine
Me-RIP-seq	Methylated RNA ImmunoPrecipitation sequencing
miCLIP	m6A individual-nucleotide-resolution CrossLinking and ImmunoPrecipitation
microRNA	micro RNA
mRNA	messenger RNA
NGS	Next-Generation Sequencing
Nm	2′-O-methylation
PABP	Poly(A)-Binding Protein
PA-m6A-seq	Photo-crosslinking-Assisted m6A sequencing
PIANO	Pseudouridine site Identification And fuNctional annOtation
picoMeRIP-seq	picogram-scale m6A RNA ImmunoPrecipitation and sequencing
RNA	RiboNucleic Acid
rRNA	ribosomal RNA
scDART-seq	single-cell Deamination Adjacent to RNA modification Targets sequencing
scm6A-seq	single-cell m6A sequencing
scNMT-seq	single-cell Nucleosome, Methylation, and Transcription sequencing
scRNA-seq	single-cell RNA sequencing
SHAPE	Selective 2′-Hydroxyl Acylation analyzed by Primer Extension
SMRT	Single-Molecule Real-Time
snRNA	small nuclear RNA
t6A	N6-threonylcarboyladenosine
TMG	2,27-trimethylguanosine
tRNA	transfer RNA
Um	2′-O-methyluracil
uORF	upstream open reading frame
UTR	UnTranslated Region
UV	UltraViolet
Ψ	pseudouridine

10.1 Introduction

The central dogma of molecular biology initially presented RNA as a simple intermediary between DNA and proteins. However, the understanding of RNA's role has evolved dramatically, revealing a complex and dynamic molecule that serves as a cornerstone of cellular function (Cech and Steitz 2014). RNAs participate in virtually every aspect of gene regulation and expression, from transcriptional control to protein synthesis, and from cellular defense to structural organization

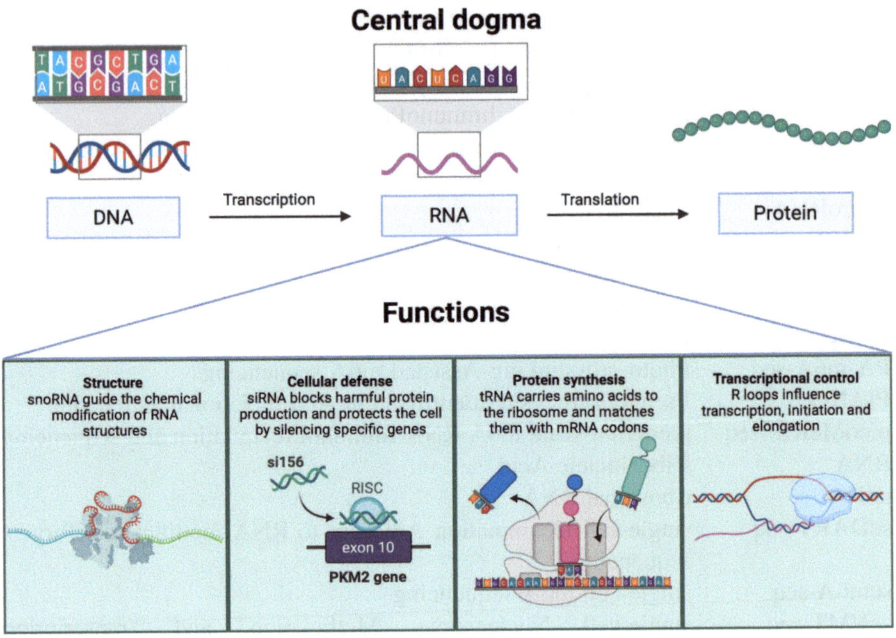

Fig. 10.1 The central dogma of molecular biology represents RNA merely as an intermediary signal between DNA and protein production. However, technological and scientific advances have shown RNA to play a role in nearly all cellular activities: serving as structural scaffolds for protein binding, deactivating and marking RNA for degradation, facilitating protein production, and modulating transcriptional activity. (Image created with BioRender)

(Roundtree et al. 2017). This functional diversity is achieved not only through different RNA species—such as messenger RNA (mRNA), transfer RNA (tRNA), ribosomal RNA (rRNA), and various regulatory RNAs—but also through an intricate system of chemical modifications that fine-tune their properties and functions (Gilbert et al. 2016) (Fig. 10.1).

RNA modifications represent a crucial layer of post-transcriptional regulation that has been conserved throughout evolution (Roundtree et al. 2017). Over 170 distinct chemical modifications have been identified across all domains of life, ranging from simple methylation to complex hypermodifications (Boccaletto et al. 2018). These modifications can be broadly categorized into several groups: base methylations (such as N6-methyladenosine, m6A), ribose methylations (2′-O-methylation, Nm), base isomerizations (pseudouridine, Ψ), and various other modifications, including acetylation, thiolation, and complex modifications involving multiple chemical changes. Each category of modification presents unique chemical properties that influence RNA structure, stability, and function in specific ways (Delaunay and Frye 2019; Shi et al. 2019).

The effects of RNA modifications on cellular processes are remarkably diverse and context-dependent (Frye et al. 2018). At the molecular level, modifications can alter RNA secondary structure, influence base-pairing properties, and modify interaction surfaces for protein binding (Liu et al. 2015). These molecular changes translate into broader functional impacts: modifications can affect RNA stability and degradation rates, regulate splicing decisions, influence translation efficiency, and even direct RNA localization within cells.

Understanding the precise locations of RNA modifications is crucial for several reasons. First, the position of a modification within an RNA molecule often determines its functional impact—the same modification type can have different effects depending on its location within the transcript (Li et al. 2017). Second, modification patterns are often dynamic and respond to cellular conditions, suggesting they play important regulatory roles (Nachtergaele and He 2017). Third, dysregulation of RNA modifications has been implicated in various diseases, from cancer to neurological disorders, making them potential therapeutic targets (Barbieri and Kouzarides 2020). The ability to accurately identify modification sites is therefore essential for understanding their biological roles and developing targeted interventions. Moreover, the emergence of epitranscriptomics—the study of RNA modifications and their biological functions—has revealed that these chemical changes form a complex regulatory network analogous to epigenetic modifications in DNA (Kan et al. 2022). This "RNA epigenetics" appears to be just as important for cellular function as its DNA counterpart, regulating gene expression patterns, cellular differentiation, and responses to environmental signals. The growing appreciation of RNA modifications' importance has driven the development of increasingly sophisticated methods for their detection and quantification, from traditional biochemical approaches to cutting-edge sequencing technologies (Zaccara et al. 2019; Delaunay et al. 2024).

This chapter will explore the current methodologies available for identifying RNA modification sites, discuss the challenges and considerations in interpreting modification data, and examine how this information can be integrated into the understanding of cellular regulation. This chapter will cover both established techniques and emerging technologies, providing a comprehensive overview of the tools available to investigators in this rapidly evolving field. Understanding these methods is crucial not only for basic research but also for clinical applications, as RNA modifications increasingly emerge as important biomarkers and therapeutic targets.

10.2 History of RNA Modification Detection

Human understanding of RNA modifications has dramatically evolved alongside detection technologies, revealing a complex regulatory landscape that fundamentally reshapes scientists' view of gene expression (Jonkhout et al. 2017; Kim et al. 2025). What began as an unexpected discovery of chemical modifications beyond canonical nucleosides has transformed into recognizing these marks as dynamic controllers of RNA function across all domains of life (Pan 2013; Starling 2017).

The journey of RNA modification detection progressed from basic analytical techniques to sophisticated high-throughput methods, uncovering a remarkable

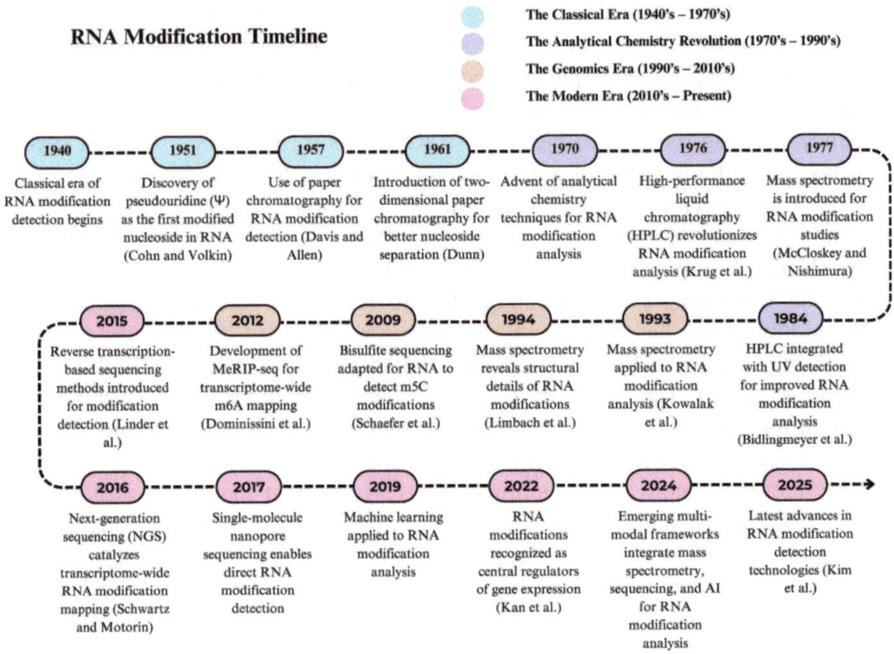

Fig. 10.2 Timeline highlighting significant events in the evolution of studies focusing on RNA modifications. Timeline is organized into four distinct eras: the Classical Era (1940s–1970s), the Analytical Chemistry Revolution (1970s–1990s), the Genomics Era (1990s–2010s), and the Modern Era (2010s–Present)

breadth of complexity (Fig. 10.2). From early discoveries using paper chromatography in the 1950s to modern single-molecule sequencing, scientists have identified over 170 different modification types, demonstrating that these chemical marks are not rare anomalies but pervasive regulatory mechanisms with profound biological significance.

10.2.1 The Classical Era (1940s–1970s)

The classical era of RNA modification detection began with a fundamental question: could RNA molecules contain chemical modifications beyond the four canonical nucleosides? Answering this question required innovative approaches that established the foundation for all future studies of RNA modifications. This period, spanning from the 1940s through the 1970s, saw the development of essential techniques that revealed the first glimpses of RNA's chemical complexity.

The watershed moment came in 1951 when James Cohn and Elliott Volkin at Oak Ridge National Laboratory discovered Ψ, the first modified nucleoside identified in RNA (Cohn and Volkin 1951). Their discovery emerged from careful analysis of

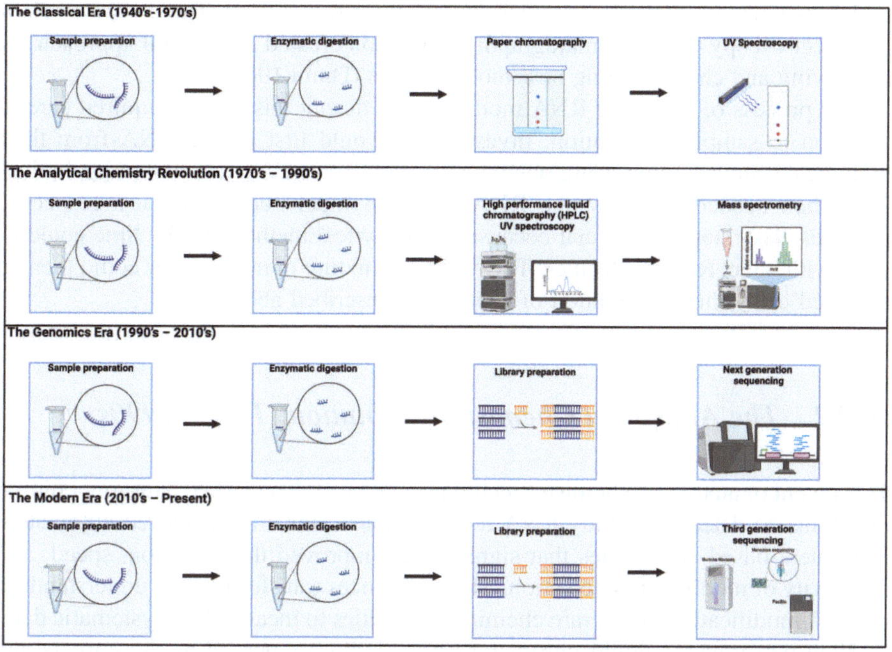

Fig. 10.3 The methodologies used for detecting RNA modifications have changed significantly since the first identification of Ψ. Each technological improvement has facilitated further advancements in the study of RNA modifications, and led to significant alterations in their experimental workflows. (Image created with BioRender)

RNA using paper chromatography, a technique that would become central to modification research for the next two decades (Davis and Allen 1957). This finding revealed the possibility that RNA molecules may harbor numerous chemical modifications, fundamentally transforming how scientists perceived RNA structure and function.

The detection methods of this period centered around two core technologies: paper chromatography and ultraviolet spectroscopy (Wyatt 1951) (Fig. 10.3). Each technique contributed unique capabilities to the growing field of RNA modification research. Paper chromatography allowed researchers to separate complex mixtures of nucleosides based on their chemical properties. By applying RNA samples to specialized paper sheets and running different solvents across them, scientists could separate modified nucleosides from their unmodified counterparts. Two-dimensional paper chromatography, which used different solvents in perpendicular directions, provided even better separation and became a standard approach for modification analysis (Davis and Allen 1957; Dunn 1961).

Ultraviolet spectroscopy completed the analytical toolkit of this period. Modified nucleosides often showed distinct UV absorption patterns compared to their unmodified counterparts, providing crucial structural information (Wyatt 1951). This technique helped researchers confirm the presence of modifications and gather initial insights into their chemical structures (Voet et al. 1963). The combination of UV spectroscopy with chromatographic separation created a powerful approach for identifying and characterizing new modifications (Dunn 1961).

The process of detecting RNA modifications during this period required careful attention to sample preparation. Investigators would first isolate RNA from their biological samples, often using phenol extraction followed by alcohol precipitation. They then needed to digest the RNA into individual nucleosides using specific enzymes. This step was crucial because the analytical methods of the time couldn't work with intact RNA molecules. The resulting mixture of nucleosides could then be analyzed using the combination of techniques described above.

10.2.2 The Analytical Chemistry Revolution (1970s–1990s)

The advent of analytical chemistry techniques in the 1970s marked a turning point in the ability to detect and study RNA modifications. This era introduced automated, instrument-based approaches that significantly improved the precision, speed, and sensitivity of modification detection. These advances transformed the understanding of RNA modifications from rare chemical curiosities to measurable, systematic data.

High-performance liquid chromatography (HPLC) emerged as the cornerstone technology of this revolution. Unlike the paper chromatography of the classical era, HPLC used high-pressure pumps to force samples through specialized columns packed with tiny particles (Bidlingmeyer et al. 1984). This seemingly simple change brought extraordinary improvements in separation power. What once took days could now be completed in hours, and researchers could distinguish between modifications that had previously appeared identical (Krug et al. 1976; McCloskey and Nishimura 1977).

HPLC's key advantage was its ability to seamlessly integrate with other analytical techniques. When combined with UV detectors, it allowed scientists to track modified nucleosides in real-time as they eluted from the column, each displaying a distinct absorption pattern. The integration of radioactive detection systems pushed sensitivity even further, allowing researchers to track modifications present at just a few molecules per cell. These hybrid approaches revealed the existence of rare modifications that had escaped detection using classical methods.

The introduction of mass spectrometry to RNA modification analysis represented another quantum leap forward. Early mass spectrometers could only analyze pure compounds, but the development of interfaces linking HPLC to mass spectrometers changed everything. By measuring the exact mass-to-charge ratio (m/z) of modified nucleosides, mass spectrometry enabled precise identification of chemical alterations, including methylations, acetylations, and thiolations. Scientists could now separate complex mixtures of nucleosides and immediately determine their molecular masses and structures (Kowalak et al. 1993). This combination proved particularly powerful for discovering new modifications, as mass spectrometry could reveal structural details that other detection methods missed (Limbach et al. 1994).

The methodology for studying RNA modifications during this period followed a careful sequence. Researchers would first isolate RNA from their biological samples and enzymatically digest it into individual nucleosides. These nucleoside mixtures would then be separated using HPLC, with modified nucleosides identified through their characteristic UV absorption patterns, mass spectra, or both. The addition of radioactive labeling or antibody-based detection provided additional ways to track specific modifications of interest.

10.2.3 The Genomics Era (1990s–2010s)

The rise of genomic technologies in the 1990s ushered in a new chapter in RNA modification research. While previous methodologies allowed scientists to identify and measure modifications, the genomics era introduced methods to map these chemical marks across entire transcriptomes. By enabling high-throughput profiling of modified RNA, these advances revealed the dynamic and context-specific nature of RNA modifications, linking them to cellular processes, development, and disease. This capability transformed the understanding of RNA modifications from isolated chemical features to components of a complex regulatory system.

The development of next-generation sequencing (NGS) technologies catalyzed this large-scale genomic transformation (Schwartz and Motorin 2016). These platforms could sequence millions of RNA molecules in parallel, generating unprecedented views of cellular RNA populations (Dominissini et al. 2012). However, standard RNA sequencing couldn't directly detect modifications. Scientists needed to develop clever approaches that could convert chemical modifications into sequence changes that NGS could detect (Zaringhalam and Papavasiliou 2016).

A breakthrough came with the adaptation of bisulfite sequencing for RNA (Schaefer et al. 2009). This technique, initially developed for DNA methylation analysis, was later adapted to identify m5C modifications in RNA. The method works by converting unmodified cytosines to uracils while leaving modified cytosines unchanged. When the RNA is sequenced, unmodified cytosines appear as C-to-T transitions, allowing researchers to map modified positions (Squires et al. 2012). This approach provided the first transcriptome-wide maps of RNA methylation and revealed that this modification was far more widespread than previously thought (Amort et al. 2017).

The computational analysis of bisulfite sequencing data required specialized algorithms that could account for the unique challenges of RNA modifications (Legrand et al. 2017). Scientists developed tools like meRanTK and HAMR (High-throughput Analysis of Modified Ribonucleotides) that could process the complex patterns of sequence changes (Rieder et al. 2016; Vandivier et al. 2019). These tools implemented sophisticated statistical models to distinguish genuine modifications from sequencing artifacts and account for the inherent noise in high-throughput data (Ries et al. 2019).

Another major advance came through antibody-based approaches. Scientists developed techniques like MeRIP-seq (Methylated RNA Immunoprecipitation Sequencing) that combined modification-specific antibodies with RNA sequencing (Meyer et al. 2012). In these methods, antibodies are used to enrich for RNA fragments containing specific modifications. Sequencing these enriched fragments reveals where modifications occur across the transcriptome (Dominissini et al. 2012). This approach proved particularly valuable for studying m6A, leading to the discovery that this modification plays crucial roles in controlling gene expression (Ke et al. 2015).

These methods generated their own computational challenges, particularly in normalizing signal intensity across different genomic regions and identifying significant peaks of modification enrichment. Tools like MACS (Model-based Analysis of ChIP-Seq), originally developed for DNA-protein interactions, were adapted for RNA modification analysis (Zhang et al. 2008). New algorithms like MeTPeak and DRME (Detection of RNA Modifications from Expression) were specifically designed to handle the unique characteristics of modification enrichment data (Liu et al. 2016; Xiaodong et al. 2016).

The introduction of reverse transcription-based detection methods broadened the methodological toolkit. Scientists discovered that certain modifications could cause reverse transcriptase to stop or make errors at modified positions (Hauenschild et al. 2015). By carefully analyzing these patterns in sequencing data, researchers could identify modification sites (Helm and Motorin 2017). Techniques like Pseudo-seq for Ψ detection and AlkAniline-seq for Nm emerged from this principle, each revealing new insights into modification distributions (Schwartz et al. 2014; Dai et al. 2017).

The large quantity of data generated by these new methods prompted another wave of computational innovation. Researchers developed machine learning algorithms that could recognize the subtle patterns of reverse transcriptase stops and errors that indicated modification sites. These algorithms needed to account for RNA secondary structure, which could also generate reverse transcriptase artifacts (Spitale et al. 2015; Sun et al. 2019). Tools like PIANO (Pseudouridine site Identification And fuNctional annOtation) and SRAMP incorporated structure prediction algorithms to improve modification site identification (Zhou et al. 2016; Song et al. 2020b).

The emergence of the genomics era brought computational analysis to the forefront of modification research. The massive datasets generated by sequencing approaches required sophisticated algorithms capable of distinguishing genuine modification sites from technical noise. These computational tools also enabled researchers to discover patterns in modification distribution, revealing connections to gene regulation that hadn't been apparent when studying individual RNA molecules.

10.2.4 The Modern Era (2010s–Present)

The modern era of RNA modification detection has been characterized by technological convergence, where multiple approaches combine to provide unprecedented insight into the epitranscriptome. This period, building on previous advances, has introduced techniques that enable single-molecule resolution analysis, simultaneous detection of multiple modification types, and reveal modification dynamics in living cells.

Third-generation sequencing technologies have revolutionized direct modification detection. Nanopore sequencing, which threads individual RNA molecules through tiny pores while measuring electrical current changes, can detect modifications without chemical conversion or enrichment steps (Garalde et al. 2018). Each modified base creates a distinctive disruption in the current signal, allowing real-time identification of multiple modification types in a single experiment (Workman et al. 2019). This approach has proven particularly valuable for studying modification patterns in individual RNA molecules, revealing heterogeneity that bulk methods had missed (Leger et al. 2021).

Modern deep learning models, particularly convolutional neural networks and transformers, have become essential for translating these electrical signatures into modification calls (Liu et al. 2019). Tools like Tombo and EpiNano implement advanced algorithms that can distinguish subtle differences in current disruptions caused by different modifications, while accounting for sequence context and molecular dynamics (Liu et al. 2021).

Similarly, single-molecule real-time (SMRT) sequencing has enabled direct detection of RNA modifications through analysis of polymerase kinetics (Flusberg et al. 2010). When the polymerase encounters a modified base, its speed changes in characteristic ways (Eid et al. 2009). By measuring these kinetic signatures, researchers can identify modification sites and types (Ardui et al. 2018). This method has been especially powerful for studying bacterial RNA modifications and has revealed new patterns in modification distribution (Beaulaurier et al. 2019).

The computational challenge of interpreting these kinetic signatures led to the development of sophisticated statistical frameworks. Tools like DRUMMER (Detection of Ribonucleic acid Modifications Manifested in Error Rates) employ probabilistic models to identify modification-induced changes in polymerase dynamics, while controlling for natural variation in enzyme behavior (Abebe et al. 2022). Machine learning approaches have proven particularly valuable in this context, as they can learn to recognize complex patterns in kinetic data that might be missed by traditional statistical methods (Xu and Seki 2020; Hassan et al. 2022).

During this era, mass spectrometry has experienced its own transformative evolution. Advancements in modern instrumentation now enable analysis of intact RNA molecules, allowing researchers to study how multiple modifications interact on individual RNA species (Taucher and Breuker 2012; Giessing and Kirpekar 2012; Wein et al. 2020; Amalric et al. 2022). The development of top-down proteomics approaches adapted for RNA analysis has enabled the mapping of multiple modifications on intact RNA molecules, preserving molecular connectivity information (Galloway et al. 2020). These advances have been particularly important for understanding how different modifications interact and influence RNA function (Amalric et al. 2024).

Antibody-based approaches for detecting RNA modifications have advanced significantly in both sensitivity and resolution. The development of highly specific antibodies combined with improved sequencing methods have enabled more precise mapping of modification sites (Hussain et al. 2013). Techniques like miCLIP can identify modification sites with single-nucleotide precision, while requiring less starting material than previous methods (Linder et al. 2015).

The rise of single-cell sequencing technologies has created new computational challenges in modification detection (Luecken and Theis 2019). Tools like scNMT-seq (single-cell Nucleosome, Methylation and Transcription sequencing) require novel computational approaches to handle sparse data and technical noise while maintaining sensitivity to modification signals (Clark et al. 2018). Machine learning methods, particularly deep learning architectures designed for sparse data, have proven instrumental for extracting meaningful patterns from single-cell modification data (Eraslan et al. 2019).

Artificial intelligence has become increasingly central to modification analysis in the modern era (Wu et al. 2024). Deep learning models can now predict modification sites with high accuracy, assess their functional impact, and even suggest potential regulatory roles (Tsukiyama et al. 2022). These models, exemplified by tools like DeepTorrent and m6Anet, can recognize subtle patterns in sequence and structure that indicate modification potential, while accounting for the complex interplay between different types of modifications (Liu et al. 2021, 2022).

The modern era has transformed RNA modification detection from a specialized field into a central component of RNA biology. The combination of single-molecule resolution, direct detection methods, and multi-modal analysis has revealed the dynamic and complex nature of RNA modifications. As these technologies continue to evolve, they promise to provide even deeper insights into how RNA modifications regulate cellular function, potentially revolutionizing the understanding of gene regulation and disease mechanisms.

10.3 Future of RNA Modification Detection and Interpretation

The identification and characterization of RNA modifications represent one of the most dynamic and rapidly evolving areas in molecular biology. As the understanding of RNA's role in cellular regulation deepens, the methods being used to study these crucial molecular marks are undergoing a remarkable transformation. The convergence of advanced computational approaches, particularly machine learning and sophisticated signal processing, with innovative experimental techniques is ushering in a new era of RNA modification analysis. This transformation is particularly evident in three key areas: the development of more powerful computational models, the integration of complementary analytical methods, and the evolution of experimental approaches that provide unprecedented insight into RNA modifications. These advancements enhance scientists' ability to detect and characterize known RNA modifications while simultaneously uncovering previously unrecognized patterns within the intricate landscape of RNA regulation (Table 10.1). Looking ahead, these developments may fundamentally change the understanding of how RNA modifications influence cellular processes and their roles in both normal physiology and disease.

10.3.1 Developments in Computational Models

The landscape of computational methods for RNA modification identification is undergoing a remarkable transformation, driven by advances in machine learning and improvements in signal-processing capabilities (El Allali et al. 2021; Furlan et al. 2021). These developments are not merely incremental improvements but represent fundamental shifts in how the scientific community approaches the challenge of detecting and characterizing RNA modifications.

The integration of deep learning into RNA modification detection marks a significant departure from traditional sequence-based prediction methods. While earlier approaches relied heavily on position weight matrices and simplistic statistical models, recent contemporary neural networks leverage their ability to capture intricate, nonlinear patterns within RNA sequences that often remain undetected by traditional analytical approaches (Zhang et al. 2019b). These advanced machine learning systems are particularly adept at recognizing the subtle sequence contexts and structural features that indicate the presence of modifications.

Modern deep learning architectures, especially those incorporating attention mechanisms, have demonstrated remarkable success in identifying modification sites (Wang et al. 2024). These models work by simultaneously analyzing multiple aspects of the RNA molecule. The attention mechanism allows the model to weigh different features dynamically, much like a skilled researcher might focus on different aspects of the data depending on the context.

The success of advanced machine learning methods stems from their ability to handle the inherent complexity and noise in biological data. Unlike traditional methods that might struggle with technical variations in sequencing data, neural networks can learn to distinguish genuine signals from experimental artifacts. This resilience to noise is particularly valuable when analyzing data from high-throughput sequencing experiments, where technical variations are common.

Another driver for the advancement of computational models is the advent of direct RNA sequencing technologies, such as those developed by Oxford Nanopore (Workman et al. 2019). These devices produce intricate electrical signals embedded

Table 10.1 List of recent (published since 2020) RNA modification detection methods, the specific modifications they detect, the model system that was used for their development, the type of method, a summary of the method, and the corresponding reference

Method	Modification(s)	Model system	Type	Summary	Reference
AcIM SHAPE + Direct RNA sequencing	m4Cm, m5C, Nm, SHAPE Adducts, etc.	*Escherichia coli* and *Saccharomyces cerevisiae*	Experimental	Mixed experimental/computational protocol for detecting RNA modifications using Nanopore signal shifts, dwell times, and direct Nanopore sequencing data	Stephenson et al. (2022)
BID-seq	Ψ	HeLa, HEK293T, A549	Experimental	Experimental technique for profiling Ψ modifications in mammalian mRNA	Dai et al. (2023)
CHEUI	m6A, m5C	*Mus musculus*, HEK293, and HeLa	Computational	Neural network model on raw Nanopore signals to detect m6A and m5C	Acera Mateos et al. (2024)
DAMM-seq	m1A, m3C, m1G, m2,2G	Human mitochondrial RNA	Experimental	Experimental technique that can detect multiple RNA methylations at a time in steady-state and nascent RNA	Zhang et al. (2021)
DENA	m6A	*Arabidopsis thaliana*	Computational	Neural network model on Nanopore sequencing data to detect m6A modifications	Qin et al. (2022)
DeepEdit	A-to-I RNA editing	*Homo sapiens* and *Schizosaccharomyces pombe*	Computational	Neural network model on raw Nanopore reads to detect A-to-I editing events	Chen et al. (2023a)
DeepPromise	m1A, m6A	*Arabidopsis thaliana*, *Escherichia coli*, *Homo sapiens*, *Mus musculus*, and *Saccharomyces cerevisiae*	Computational	Deep learning based method that predicts m1A and m6A	Chen et al. (2020b)
Differr	m6A	*Arabidopsis thaliana*	Computational	Statistical framework for identification of m6A modifications between samples	Parker et al. (2020)

DRUMMER	m6A	*Mus musculus*	Computational	Package used to detect RNA modifications by analyzing base-calling error patterns in Oxford Nanopore DRS data	Abebe et al. (2022)
ELIGOS2	5moU, m6A, m1A, hm5C, f5C, m7G, Ino, psU	*Homo sapiens*, *Mus musculus*, *Saccharomyces cerevisiae*, and H460	Computational	Tool to detect RNA modifications using sequencing error patterns on Oxford Nanopore DRS data	Jenjaroenpun et al. (2021)
EpiNano	m6A	*Saccharomyces cerevisiae*	Computational	Algorithm for detecting m6A modifications from dsRNA-seq data	Liu et al. (2021)
eTAM-seq	m6A	*Mus musculus*, and HeLa	Experimental	Enzyme-based m6A mapping via transcriptome-wide adenosine deamination	Xiao et al. (2023)
GLORI-seq	m6A	HCC (hepatocellular carcinoma)	Experimental	Sequencing technique for profiling m6A in RNA at single-base resolution	Liu et al. (2023)
iM6A	m6A	*Mus musculus*	Computational	Deep learning based model that predicts m6A using flanking nucleotide sequences	Luo et al. (2022)
JACUSA2	m6A, Ψ	HEK293	Computational	Computational framework on Illumina or Nanopore sequencing to detect RNA modifications	Piechotta et al. (2022)
m5C-meRIP-seq	m5C	RB Y79, WERI-Rb1, ARPE-19	Experimental	Transcriptome-wide method for mapping m5C RNA modifications	Zuo et al. (2023)
m6Anet	m6A	*Arabidopsis thaliana*, HCT116, and HEK293T	Computational	Method utilizing multiple instance learning for detecting m6A from Nanopore RNA sequencing	Hendra et al. (2022)
m6A-SAC-seq	m6A	HEK293, HeLa, and HepG2,	Experimental	Transcriptome-wide quantitative experimental method for profiling m6A at single-base resolution	Hu et al. (2022)

(continued)

Table 10.1 (continued)

Method	Modification(s)	Model system	Type	Summary	Reference
m7G-quant-seq	m7G	HEK293T, and HeLa	Experimental	Technique for detecting m7G RNA modifications using a combination of chemical and enzymatic steps	Zhang et al. (2022)
MultiRM	m6A, m1A, m5C, m5U, m6Am, m7G, Ψ, I, Am, Cm, Gm, and Um	*Homo sapiens*	Computational	Deep learning framework for predicting 12 different RNA modifications	Song et al. (2021)
MspA-PBA Nanopore sensor	m6A, m5C, m7G, m1A, I, Ψ, etc.	Synthetic miRNA/tRNA	Experimental	Engineered pore plus machine learning classifies 11 modified nucleosides by current	Wang et al. (2022)
Nanocompore	m6A, I, m5C, Ψ, m62A, m1G, 2′-OMeA	*Saccharomyces cerevisiae*, *MOLM13*, and Synthetic oligonucleotides	Computational	Comparative GMM method on raw Nanopore electrical signals to detect seven RNA modifications	Leger et al. (2021)
nanoDoc2	many	*Escherichia coli*, *Saccharomyces cerevisiae*, and SARS COV-2	Computational	Deep learning tool for detecting various RNA modifications in raw Nanopore signals	Ueda et al. (2023)
Nano ICE-Seq	I	*Mus musculus*	Experimental	Targeted workflow for detecting inosine modification sites from Nanopore sequencing	Ramasamy et al. (2022)
Nanom6A	m6A	*Populus trichocarpa*	Computational	XGBoost-based tool to detect m6A modifications at single base resolution from Nanopore direct RNA sequencing data	Gao et al. (2021)
NanoRMA	Ψ	*Saccharomyces cerevisiae*	Computational	Algorithm that can predict Ψ modifications from Nanopore data	Begik et al. (2021)
Nano-tRNAseq	m1G, Ψ, m2G, m22G, mcm5U, t6A, I, m5U, etc.	*Saccharomyces cerevisiae*	Computational	Nanopore-based method for tRNA sequencing, detects tRNA modifications and abundances	(Lucas et al. (2024)

Nm-Mut-seq	Nm	HeLa	Experimental	Experimental mutation-based method that quantifies Nm RNA modifications	Chen et al. (2023b)
Nm-Nano	Nm	HEK293, HeLa	Computational	XGBoost and random forest-based machine learning method for detecting Nm modifications in Nanopore sequencing data	Hassan et al. (2024)
Penguin	Ψ	HEK293, HeLa	Computational	Machine learning-based tool that detects Ψ modifications from Nanopore direct RNA sequencing signals	Hassan et al. (2022)
picoMeRIP-seq	m6A	*Mus musculus*	Experimental	Experimental method for mapping m6A modifications in single cells using minimal RNA input	Li et al. (2024a)
scDART-seq	m6A	HEK293T	Experimental	Single-cell method for mapping m6A modifications transcriptome-wide	Tegowski et al. (2022)
scm6A-seq	m6a	*Mus musculus*	Experimental	Single-cell sequencing method for detecting m6A in embryos	Yao et al. (2023)
TandemMod	m6A, m5C, m7G, Ψ, and I	*Oryza sativa*	Computational	Deep learning method for profiling multiple different RNA modifications in a single DRS sample	Wu et al. (2024)
TDm6A	m6A	A549, CD8T, and HEK293	Computational	Deep learning model to predict m6A	Wang and Wang (2020)
UBS-seq	5mC, m5C	A549, HEK293T, HeLa, and mES	Experimental	A faster and low-input bisulfite sequencing method that maps 5mC in DNA and m5C in mRNA	Dai et al. (2024)
xPore	m6A	HEK293T	Computational	Gaussian Mixed Model on raw currents to quantify m6A	Pratanwanich et al. (2021)

with valuable information on RNA modifications; however, extracting and interpreting this data requires advanced signal processing techniques.

Modern signal processing algorithms are being developed to handle the unique characteristics of Nanopore data. These algorithms must process multiple levels of information simultaneously: the raw electrical signal, the temporal patterns in the signal, and the relationship between signals from different molecules. The complexity of this task has spurred the development of novel computational approaches that combine traditional signal processing techniques with machine learning.

A particularly promising development is the emergence of real-time analysis capabilities (Athanasopoulou et al. 2021). New algorithms can process Nanopore signals as they are generated, potentially allowing for immediate detection of modifications during the sequencing process. This real-time analysis capability could revolutionize the ability to study RNA modifications in various contexts, from basic research to clinical applications.

The signal processing challenge extends beyond simple detection to the quantification of modification levels. Modern algorithms are being developed to estimate the stoichiometry of modifications at specific sites, a crucial capability for understanding the biological significance of these modifications. These algorithms must account for various sources of uncertainty, including sequence context effects and technical variations in the sequencing process.

The future of signal processing in this field lies in the development of significantly more nuanced algorithms that can handle increasingly complex data types. For instance, new Nanopore configurations capable of generating multidimensional signals will require corresponding advances in signal processing capabilities. These developments will likely incorporate elements of both traditional signal processing and modern machine learning approaches, creating hybrid systems that can extract maximum information from sequencing data.

10.3.2 Integration of Complementary Computational Models

The future of RNA modification detection lies not just in improving individual computational methods, but in creating sophisticated systems that combine multiple approaches into unified analytical frameworks. This integration represents a crucial step forward, as individual methods often excel at detecting specific types of modifications while struggling with others. By combining complementary approaches, more robust and comprehensive detection systems can be created.

The development of unified detection platforms represents one of the most promising directions in RNA modification analysis. These platforms move beyond the traditional approach of using separate tools for different modifications, instead creating integrated systems that can simultaneously identify and characterize multiple types of RNA modifications. This integration is particularly important because different modifications often interact with and influence each other, creating complex patterns that can only be fully understood when examined together.

The ability to detect multiple modifications simultaneously remains a significant challenge that future experimental methods must address. While current methods often focus on one modification at a time, biological systems typically involve multiple modifications working in concert. This limitation is particularly evident in Nanopore-based sequencing approaches, which have traditionally been restricted to analyzing one or a few modification types simultaneously. This constraint becomes increasingly problematic as research reveals that RNA molecules frequently carry multiple types of co-occurring modifications within the same transcript, especially in tRNAs (Lorenz et al. 2017).

For example, analyzing both m6A and Ψ modifications in the same RNA molecule presents a significant challenge. These modifications have distinct chemical properties and typically require different detection methods. Traditional approaches would analyze these modifications separately, potentially missing important relationships between them. Modern unified platforms have revolutionized this process by enabling simultaneous detection of both modifications, providing crucial insights into their biological interactions (Huang et al. 2024).

These integrated platforms typically employ multiple machine learning models working in concert, each specialized for different types of modifications or different aspects of the analysis. The key innovation lies in how these models communicate and collaborate. When one model detects a potential modification site, it can share this information with other models, providing context that might improve their predictions. This cross-communication between models could lead to more accurate results than each model operating in isolation.

The challenge of resolving conflicting predictions presents another area where unified platforms show their value. When different detection methods disagree about the presence or nature of a modification, these platforms could employ sophisticated arbitration algorithms. These algorithms would consider the confidence levels of each prediction, the known strengths and weaknesses of each method, and the broader context of the RNA molecule to resolve conflicts and provide a consolidated prediction.

Beyond mere detection, modern computational frameworks are evolving to provide a more comprehensive analysis of the RNA modification landscape (Wu et al. 2024). These frameworks integrate modification detection with broader biological context, creating a more complete picture of RNA regulation and function. This integration operates at multiple levels, from individual molecules to entire cellular pathways.

At the molecular level, these frameworks analyze modification crosstalk—the ways in which different modifications influence each other's presence or function. This analysis requires sophisticated statistical methods that can detect patterns of correlation and causation among multiple modifications. For instance, the presence of one modification might enhance or inhibit the likelihood of another modification occurring nearby, creating complex patterns that these frameworks must unravel.

Integrating gene expression data is a key feature of these frameworks, enabling investigators to connect modification patterns with changes in RNA function or stability. Modern systems can correlate modification patterns with expression levels, splicing patterns, and other aspects of RNA regulation. This integration helps reveal how modifications influence gene expression and regulation, providing insights into their biological roles. For example, a framework might detect that certain modification patterns consistently correspond with increased expression of specific genes, suggesting a regulatory relationship.

Pathway and network analysis represents a higher level of integration, where modification data is examined in the context of known biological pathways. These frameworks can reveal how modifications cluster around genes involved in particular cellular processes, suggesting functional relationships. This level of analysis often employs graph theory and network analysis techniques to identify patterns that might not be apparent when examining individual molecules.

The development of these comprehensive frameworks requires careful attention to data integration and standardization. Different types of data often come in different formats and with different levels of uncertainty. Modern frameworks must handle these variations gracefully, ensuring that all data sources contribute appropriately to the final analysis. This often involves complex statistical methods for data normalization and uncertainty quantification.

10.3.3 Development of Experimental Methods

The experimental landscape of RNA modification detection is undergoing a remarkable transformation, driven by technological innovations and the improved understanding of RNA biology. While computational methods continue to advance, experimental approaches remain crucial for validating predictions and discovering novel modifications. The future of experimental methods lies in both enhancing existing technologies and developing entirely new approaches for studying RNA modifications.

One of the most promising directions in the field is the development of sequencing technologies specifically designed to detect RNA modifications. While conventional RNA sequencing provides valuable insights, it cannot reliably detect modifications without additional process steps. However, emerging technologies are beginning to overcome these limitations in innovative ways.

Nanopore sequencing stands at the forefront of this revolution. Unlike traditional sequencing methods that read DNA or RNA bases indirectly through synthesis reactions, Nanopore technology directly measures the electrical properties of nucleotides as they pass through a molecular pore. This direct measurement approach is particularly valuable for modification detection because modified nucleotides create distinctive electrical signatures. The future of Nanopore technology lies in developing pores specifically engineered to maximize sensitivity to RNA modifications (van Dijk et al. 2023). These next-generation pores might incorporate specific chemical groups or structural features that enhance their ability to distinguish between different modifications.

The challenge of resolution—the ability to distinguish between similar modifications—is driving the development of more sophisticated detection systems. Future Nanopore devices might employ multiple measurements of the same molecule, perhaps using different types of pores or varying electrical conditions, to generate more detailed modification profiles. This multi-measurement approach could help resolve ambiguities that currently plague modification detection, such as distinguishing between similar modifications like m6A and m1A.

While Nanopore technology represents a promising future direction, the field has already made significant strides in developing methods for precise modification detection. Achieving single-base resolution for cellular RNA modifications has become increasingly crucial as evidence mounts for their role in determining cellular states and developmental trajectories. The evolution of these technologies traces back to the pioneering development of m6A-REF-seq, which revolutionized m6A mapping capabilities (Zhang et al. 2019b). Since then, the field has witnessed a remarkable expansion in sequencing methodologies for comprehensive epitranscriptomic profiling.

These methodologies can be broadly categorized into two approaches: antibody-based and antibody-free techniques. The antibody-based techniques include sophisticated methods such as PA-m6A-seq, miCLIP, m6A-CLIP, and m6A-LAIC-seq (Chen et al. 2015; Linder et al. 2015; Ke et al. 2015; Molinie et al. 2016). Each of these methods has contributed unique capabilities to the field. Complementing these are the antibody-free approaches, including DART-seq, MAZTER-seq, m6A-REF-seq, and m6A-SEAL (Pandey and Pillai 2019; Meyer 2019; Wang et al. 2020; Gao et al. 2021). While most of these methods can achieve single-nucleotide resolution detection, quantifying modification levels accurately remains a challenge that only a limited number of techniques have successfully addressed.

Recent innovations have pushed the boundaries of what's possible in modification detection. The GLORI-seq technique, which cleverly employs glyoxal and nitrite-mediated deamination of unmethylated adenosines, has demonstrated remarkable capability in achieving single-nucleotide resolution (Chan et al. 2024). More advanced techniques like RBS-seq have taken this further by enabling the detection of multiple modifications simultaneously at single-nucleotide resolution (Khoddami et al. 2019). However, these advances also highlight a critical challenge in the field: the need for reproducible, transcriptome-wide profiling of multiple modifications simultaneously at single-nucleotide resolution.

Sample requirements present another area where experimental methods are likely to see significant advancement. Current methods often require substantial amounts of RNA, limiting their application in situations where biological material is scarce. Future technologies will likely achieve greater sensitivity through various innovations, such as improved molecular amplification methods or more sensitive detection systems. These advances would be particularly valuable for studying rare cell types or clinical samples where material is limited.

The development of such sensitive detection methods has already begun to transform scientists' understanding of RNA modifications at the single-cell level. While bulk sequencing technologies have been instrumental in identifying over

170 distinct RNA modifications, these approaches cannot capture the cellular heterogeneity and dynamic regulation of modifications present in complex biological systems. The integration of epitranscriptomic analysis with single-cell RNA sequencing (scRNA-seq) represents a crucial advancement in addressing this limitation (Lu et al. 2023). Through innovative protocols that maximize sensitivity and efficiency, researchers can now map modifications like m6A at single-cell resolution, revealing heterogeneous distribution patterns across different cell populations. For instance, techniques such as scm6A-seq and scDART-seq have demonstrated that many RNA modification sites are confined to specific cell subpopulations, a finding that would have been masked in bulk analysis approaches (Tegowski et al. 2022; Yao et al. 2023).

Despite this, significant technical challenges remain in the field of single-cell epitranscriptomics. Currently, few methods have been developed for single-cell profiling—m6AISH-PLA, scDART-seq, scm6A-seq, and picoMeRIP-seq—and these are limited to analyzing relatively few types of RNA modifications, namely m6A (Ren et al. 2021; Tegowski et al. 2022; Yao et al. 2023; Li et al. 2024a). The extremely low RNA content in individual cells continues to pose a substantial technical barrier, highlighting the ongoing need for even more sensitive detection methods. As these technologies evolve, parallel development of sophisticated computational algorithms and analytical tools will be essential for accurate interpretation of RNA modifications at the single-cell level.

The development of new biochemical techniques for studying RNA modifications represents another crucial frontier. These approaches complement sequencing-based methods and often provide unique insights into modification dynamics and function. The future of biochemical approaches lies in creating more specific and sensitive methods for detecting and studying modifications.

Engineered proteins represent a particularly promising direction. By understanding and modifying the natural proteins that recognize RNA modifications, researchers are developing new tools for studying these modifications in their cellular context (Rauch et al. 2019). These engineered proteins might be designed to recognize specific modifications with high affinity and specificity, perhaps even distinguishing between similar modifications that current methods struggle to differentiate (Sinnott et al. 2024). Future developments might include proteins that can be easily tagged for visualization or proteins that change their properties upon binding to modified RNA, enabling real-time monitoring of modification states.

Chemical probes for modification detection are also evolving rapidly (Khoddami et al. 2019; Zhang et al. 2019a; Janin et al. 2019). Traditional approaches often rely on relatively harsh chemical treatments that can damage RNA or create artifacts. The next generation of chemical probes is poised to leverage advances in biorthogonal chemistry, offering unprecedented selectivity and compatibility with native RNA structures (Liu et al. 2022; Li et al. 2024b). These improved probes might enable more accurate quantification of modification levels and better preservation of RNA structure and function during analysis.

The real-time study of RNA modification dynamics in living cells represents a cutting-edge frontier in experimental research. Current approaches often provide static snapshots of modification states, but biological processes are inherently dynamic. Future methods will likely enable real-time tracking of modification changes in living cells (Ma et al. 2019). This might involve fluorescent reporters that respond to modification states or methods for rapidly sampling and analyzing RNA modifications without disrupting cellular function (Chen et al. 2022, 2025).

The ability to detect multiple modifications simultaneously remains a significant challenge that future experimental methods must address. While current methods often focus on one modification at a time, biological systems typically involve multiple modifications working in concert. This limitation is particularly evident in Nanopore-based sequencing approaches, which have traditionally been restricted to analyzing one or a few modification types simultaneously (Hassan et al. 2022, 2024). This constraint becomes increasingly problematic as research reveals that RNA molecules frequently carry multiple types of co-occurring modifications within the same transcript. Future approaches will need to overcome these limitations through multiplexed detection systems, perhaps using different fluorescent labels or orthogonal chemical reactions to track multiple modifications in the same experiment.

The integration of experimental methods with spatial information represents another important direction. Current methods typically analyze modifications in bulk RNA extracted from cells or tissues, losing information about the spatial distribution of modifications. Advanced methodologies are emerging that combine spatial resolution with modification detection, integrating techniques from in situ analysis with spatial transcriptomics (Lu et al. 2022). This would reveal how modification patterns vary across different cellular compartments or tissue regions, providing new insights into their biological roles.

10.4 RNA Modification Databases

RNA modifications represent a complex molecular frontier that defies traditional genomic research methods, with single chemical changes capable of dramatically reshaping an RNA molecule's functional profile. These subtle yet transformative molecular annotations require specialized detection techniques that go beyond conventional sequencing approaches.

To address these challenges, dedicated RNA modification databases have emerged as critical computational solutions. These refined platforms systematically collect, curate, and contextualize scientific insights, enabling comparative analyses across diverse biological contexts and supporting advanced prediction models that illuminate the intricate language of cellular regulation. These databases can be broadly broken down into two categories: general and specialized databases.

10.4.1 General Databases

Among the myriad of RNA modification databases, 2 have emerged as the primary resources for RNA modification information, Modomics and RMBase (Dunin-Horkawicz et al. 2006; Sun et al. 2016). These general RNA modification databases collectively provide a robust foundation for investigating the diverse world of RNA modifications. When researchers encounter questions about specific modifications, their chemical structures, enzymatic machinery, genomic locations, or evolutionary significance, these two databases offer a natural starting point for exploration. Each database brings distinct strengths to the field while maintaining a shared commitment to advancing the understanding of RNA biology.

10.4.1.1 Modomics

Modomics, developed at the University of Warsaw, stands as a pioneering computational resource in the field of RNA biology. This database addresses a fundamental need in molecular biology by providing a centralized repository of RNA modification information with unprecedented chemical and structural detail.

At its core, Modomics distinguishes itself through its molecular-level precision, offering complete chemical structures for over 170 different RNA modifications. This goes far beyond simple cataloging to provide researchers with insights into the chemical mechanisms that drive RNA transformations. Each modification is meticulously annotated with comprehensive details, including precise chemical structures, responsible enzymatic machinery, taxonomic distribution across organisms, potential functional implications, and structural context within different RNA types.

The database's architectural approach combines breadth and depth, mapping modifications across tRNA, mRNA, and rRNA. This enables comparative analyses that reveal patterns of modification across different RNA categories and biological contexts. Particularly noteworthy is Modomics' commitment to manual curation in an era of automated data collection, with each entry undergoing rigorous expert review to ensure scientific accuracy.

By cataloging the full spectrum of RNA modifications—from methylations to complex alterations like pseudouridylation and acetylation—Modomics illuminates the sophisticated chemical language of RNA that extends beyond traditional genetic sequencing. This comprehensive approach has transformed our understanding of epitranscriptomic regulation, enabling researchers to explore more sophisticated questions about gene expression, cellular regulation, and the fundamental processes of life.

10.4.1.2 RMBase

RMBase emerged as a transformative resource in the field of RNA modification research, addressing a critical challenge in epitranscriptomics: the large-scale, comprehensive detection and annotation of RNA modifications across entire genomes. Developed as a response to the limitations of traditional sequencing approaches, the database represents a sophisticated computational platform that integrates multiple high-throughput sequencing technologies to map RNA modifications with unprecedented breadth and depth.

The fundamental innovation of RMBase lies in its approach to modification detection. By leveraging advanced technologies like MeRIP-seq (methylated RNA immunoprecipitation sequencing) and m6A-seq, RMBase captures modifications across entire transcriptomes. This genome-wide approach allows researchers to develop a more comprehensive understanding of RNA modification landscapes rather than relying on isolated observations.

RMBase offers a dynamic perspective on RNA modifications by comprehensively mapping their variations across different experimental conditions. The database employs sophisticated machine learning algorithms to predict and validate modification sites, enabling researchers to explore how RNA modifications change in response to cellular states, developmental stages, and environmental stimuli. This extends understanding beyond static genetic annotations to reveal the dynamic nature of the epitranscriptome.

The database's multidimensional coverage integrates data from various sequencing platforms across different RNA types (mRNA, tRNA, rRNA), organisms, and experimental conditions. Particularly noteworthy is RMBase's computational approach, which provides probability scores and confidence levels for each predicted modification. This nuanced methodology allows researchers to understand the certainty associated with each modification detection—a critical consideration in computational biology.

By integrating large-scale sequencing data with advanced computational methods, RMBase transforms how scientists conceptualize and investigate the complex chemical landscape of RNA molecules, establishing itself as a foundational resource in the rapidly evolving field of computational epitranscriptomics.

10.4.2 Specialized Databases

The landscape of RNA modification research has evolved beyond the foundational databases of Modomics and RMBase into a rich ecosystem of specialized resources (Table 10.2). These specialized databases reflect researchers' understanding of epitranscriptomics and its contextual complexity across different organisms and cellular environments.

In the realm of plant biology, PRMD, REDIdb, and Plant snoRNA database have emerged as essential platforms for understanding plant-specific epitranscriptomics (Brown et al. 2003; Picardi et al. 2007; Lang et al. 2024). These databases address crucial needs in the field by capturing modification patterns unique to photosynthetic organisms and plant cellular systems. Where generalist databases might overlook

Table 10.2 Table summarizing a list of RNA modification databases along with a succinct summary of the database, number of types of modifications, number of species represented in the database, and the number of identified sites

Database	Summary	Modifications	Species	Mapped sites
Modomics	Comprehensive catalogue of RNA modifications across various RNA types provides detailed chemical structure and molecular configurations	>190	118	NA
RMBase	Comprehensive platform with eight modules focused on exploring RNA modifications and transcripts	73	62	1,074,100
RNAmods (Liu and Gregory 2019)	Database containing chemical, structural, and phylogenetic information of RNA modifications	5	21	~500
M6A2Target (Deng et al. 2021)	Public repository of m6A WER target genes identified from both low-throughput and high-throughput studies	1	2	1034 WER-target pairs validated experimentally 124,772 WER-target pairs interacting with WERs 330,102 WER-target pairs with differential expression, methylation, translation efficiencies, or alternative splicing levels.
MiREDiBase (MeR-DB) (Marceca et al. 2021)	Curated information resource of editing events in miRNAs	2	4	3059
POSTAR (Hu et al. 2017)	Comprehensive resource of post-transcriptional regulation influenced by RNA-binding proteins (RBPs)	73	7	117,049 in humans 91,930 in mice
M6A-Atlas	Comprehensive database for investigating the m6A epitranscriptome	1	14	797,091
PRMD	Integrated database that annotates and visualizes plant RNA modifications	7	20	N/A
Plant snoRNA database	Database containing information about plant snoRNA, including modification sites	2	19	~120

(continued)

Table 10.2 (continued)

Database	Summary	Modifications	Species	Mapped sites
tMODBase (Lei et al. 2023)	Analytical database focused on tRNA modifications and their associated biological pathways	27	10	N/A
RM2Target (Bao et al. 2023)	A public repository focused on RNA modification (RM)-related WER-target associations, including gene information, RNA modification, RNA interactions, RNA protein interactions, and other useful RNA modification information	9	2	1, 619, 653 WER-target associations
RMDisease	Integrated database focused on genetic variants that can influence RNA modifications	16	20	873, 819
RMVar	Comprehensive database that hosts all RNA modification (RM)-associated variants	9	2	1,254,925
M6A-TSHub (Song et al. 2023)	Comprehensive platform focused on human cancer-specific and tissue-specific m6A modifications	1	1	683, 923
REPIC	Collection of publicly available m6A-seq and MeRIP-seq data, which allows users to explore m6A modifications	1	11	~10 million
CVm6A (Han et al. 2019b)	Visualization database of global m6A patterns across cell lines	1	4	520,151
M5C-Atlas	Comprehensive database focused on RNA 5-methylcytosine	1	13	166,540
PIANO	Web server focused on Ψ identification and functional annotation	1	1	4303
M7gHub	Integrated database that hosts mRNA m7G sites identified in the human transcriptome	1	23	430,898

(continued)

Table 10.2 (continued)

Database	Summary	Modifications	Species	Mapped sites
REDIdb	Interactive database that aims to catalog RNA editing events (substitutions, insertions, and deletions) occurring in various organisms	2	281	~26,000
REDIportal	Comprehensive database of RNA editing A-to-I events in human samples, tissues, and body sites	1	1	4668,50

plant-specific nuances, these databases provide detailed insights into how RNA modifications function within the distinct molecular environment of plant cells.

Other databases like M6A-Atlas, M5C-Atlas, or M7gHub complement these specialized resources by providing comprehensive coverage of a single type of mutation across multiple organisms (Song et al. 2020a; Tang et al. 2021; Ma et al. 2022). These databases serve as extensive repositories that bridge the gap between specialized and general knowledge, offering researchers both breadth and accessibility.

Still other databases like REPIC, REDIportal, RMVar, and RMDisease have focused on the functional and medical implications of RNA modifications (Picardi et al. 2017; Liu et al. 2020; Chen et al. 2020a; Luo et al. 2021). These resources track how modification patterns shift in various disease states, creating vital connections between basic molecular biology and clinical research. This development reflects the growing recognition of epitranscriptomic changes as potential therapeutic targets and diagnostic markers.

The proliferation of specialized RNA modification databases represents more than just data collection—it reflects a fundamental shift in epitranscriptomics. These databases reveal RNA modifications as a sophisticated molecular language that requires careful interpretation within specific biological contexts. Each specialized database offers researchers a unique analytical lens, providing the depth necessary for understanding specific molecular environments, organism-specific modifications, or unique chemical changes that might be overlooked in broader databases.

This evolution of specialized databases mirrors the field's progression from basic cataloging to sophisticated analysis and interpretation. Collectively, these resources create a comprehensive framework for investigating the complex roles of RNA modifications in biological systems, spanning fundamental molecular mechanisms to their implications in health and disease.

10.5 Interpreting RNA Modifications

RNA modifications represent a complex layer of post-transcriptional regulation that can profoundly influence RNA metabolism, function, and fate. While modern sequencing technologies have revolutionized the ability to detect these modifications, interpreting their biological significance remains a complex task requiring a systematic analytical approach combined with careful experimental validation.

10.5.1 Positional Context Analysis

The location of RNA modifications within transcripts provides crucial insights into their regulatory potential and biological functions. A total of six different transcript regions have been identified where RNA modifications appear to have distinct effects: 5′ cap, 5′ untranslated region (5′ UTR), coding regions, splice site regions, 3′ untranslated region (3′ UTR), and 3′ Poly(A) tails (Fig. 10.4).

10.5.1.1 5′ Cap Modifications

The 7-methylguanosine (m7G) cap attached at the 5′ end of eukaryotic mRNAs represents one of nature's most sophisticated molecular innovations in RNA regulation (Furuichi and Shatkin 2000). The cap structure can undergo extensive modifications, generating structurally diverse variants that can profoundly influence RNA fate and function. The first and second nucleotides following the cap can be

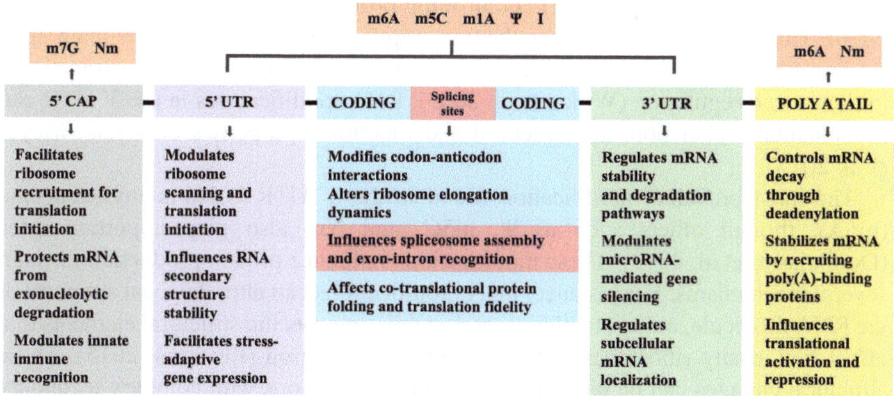

Fig. 10.4 Mature eukaryotic messenger RNA (mRNA) is made up of several distinct regions: 5′ cap, 5′ untranslated region (5′ UTR), coding regions, splice site regions, 3′ untranslated region (3′ UTR), and 3′ Poly(A) tails. Different types of RNA modifications are more common in different regions, and have been shown to have varying biological effects

methylated at the 2′-O position, creating Cap-1 and Cap-2 structures, respectively (Shuman 2015). Additionally, the guanosine cap itself can receive additional methyl groups at various positions, including the N2 position, generating unique cap variants such as 2,2,7-trimethylguanosine (TMG-cap) (Menezes et al. 2018). Each of these chemical modifications creates distinct recognition surfaces for RNA-binding proteins and influences how the cellular machinery interacts with the transcript (Topisirovic et al. 2011).

Cap modifications play a pivotal role in translation initiation by influencing the recruitment and activity of cap-binding proteins (Calero et al. 2002). The canonical cap-binding protein eIF4E, for instance, shows varying affinities for differently modified caps. When the first nucleotide contains a 2′-O-methylation (Cap-1), the binding stability of eIF4E typically increases, promoting more efficient translation initiation (Muthukrishnan et al. 1978). This enhancement results from the added methyl group, which optimizes the binding pocket and enhances cap stability by shielding it from decapping enzymes (Daffis et al. 2010).

Beyond translation regulation, cap modifications serve as crucial determinants of RNA stability and processing. The presence of Nm on the first transcribed nucleotide acts as a molecular signature of "self" RNA, helping cells distinguish their own transcripts from foreign RNA molecules (Daffis et al. 2010). This modification prevents the activation of innate immune sensors such as RIG-I and MDA5, which would otherwise trigger antiviral responses (Schuberth-Wagner et al. 2015). Furthermore, cap modifications can influence the recruitment of decapping enzymes, thereby affecting mRNA half-life and turnover rates (Łabno et al. 2016).

10.5.1.2 5′ UTR Modifications

The 5′ UTR of mRNAs represents a sophisticated regulatory hub where chemical modifications play crucial roles in controlling protein production (Hinnebusch et al. 2016). This region, spanning from the transcription start site (TSS) to the main open reading frame's (ORF's) start codon, serves as a critical checkpoint for translation initiation and regulation (Wilkie et al. 2003). RNA modifications in the 5′ UTR can fundamentally alter both the structural and functional properties of RNAs (Meyer et al. 2015).

The most prevalent modification found in the 5′ UTR is N6-methyladenosine (m6A), though others such as Ψ, m5C, and Nm also play important roles (Dominissini et al. 2012). These modifications influence protein production through several mechanisms. For instance, m6A modifications can alter the local structure of the RNA molecule, either stabilizing or destabilizing specific structural elements that affect how easily ribosomes can scan through the region (Liu et al. 2015). These structural changes can be particularly important in regions with complex secondary structures, where modifications might help unwind inhibitory structures that would otherwise impede translation (Spitale et al. 2015).

5′ UTR modifications often function as part of larger regulatory networks that help cells respond to environmental changes and stress conditions (Zhou et al. 2015).

These modifications can be dynamically added or removed, allowing rapid adjustments to protein synthesis patterns without the need for new transcription (Meyer and Jaffrey 2017). This dynamic nature is particularly important during cellular stress responses, where quick changes in protein production can be crucial for survival (Anders et al. 2018). The modification pattern in the 5′ UTR can help determine which mRNAs maintain efficient translation during stress conditions and which are temporarily suppressed (Wek 2018).

10.5.1.3 uORF Modifications

A subtype of 5′ UTR modification of particular interest to plant biologists is that found in Upstream Open Reading Frames (uORFs) (Johnstone et al. 2016). uORFs are alternative start codons found in the 5′ UTR of many eukaryotic mRNAs. These uORFs are another form of RNA regulation that often acts to reduce the translation efficiency of the transcript, either by stalling the ribosome, or through downstream interactions of the translated peptides (Johnstone et al. 2016). Modifications in uORF regions can either enhance or suppress their usage, creating a dynamic system for controlling gene expression (Zhou et al. 2022).

10.5.1.4 Coding Regions

The coding region of messenger RNA, which directly specifies the amino acid sequence of proteins, represents a particularly sensitive area for chemical modifications. Unlike modifications in untranslated regions, changes in the coding sequence must maintain the delicate balance between regulatory function and preservation of the genetic code (Shi et al. 2018). These modifications can influence multiple aspects of gene expression, from translation efficiency to protein folding, making them crucial players in gene regulation (Choi et al. 2016).

The impact of RNA modifications in coding regions depends heavily on their position within individual codons. In a typical codon, modifications in the first and second positions can have more dramatic effects as they're crucial for correct amino acid selection (Borchardt et al. 2020). Third-position modifications are sometimes considered less critical due to codon redundancy, but can significantly influence translation speed.

Modifications in the coding region can create translational "speed bumps" or "express lanes" for ribosomes (Peer et al. 2017). These alterations in translation speed can have profound effects on protein folding and function. When ribosomes slow down at specific locations, it can provide crucial pauses that allow proper protein folding (Tsai et al. 2008). Conversely, regions of faster translation might help prevent inappropriate protein interactions during synthesis.

The presence of modifications can also influence frame maintenance during translation. Some modifications help ensure that ribosomes maintain the correct reading frame, particularly in regions where the sequence might otherwise promote frameshifting (Hori 2014). This becomes especially important in genes where precise frame maintenance is crucial for proper protein production.

10.5.1.5 Splice Site Regions

The precise removal of introns and joining of exons represents one of the most complex and crucial steps in gene expression (Matera and Wang 2014). This process, known as splicing, occurs at specific sequences called splice sites (Wahl et al. 2009). When chemical modifications are added to these regions, it creates an additional layer of regulation that helps cells fine-tune splicing decisions (Pendleton et al. 2017).

RNA modifications show distinct patterns of distribution around splice sites, with different modifications exhibiting preferences for specific positions relative to splice junctions. m6A modifications often cluster near splice sites, particularly in the last exon of genes (Kim et al. 2025). This positioning is not random but rather represents an evolutionarily conserved pattern that helps orchestrate proper splicing decisions (Xiao et al. 2016).

The density of modifications typically varies between the 5′ splice site (donor site) and the 3′ splice site (acceptor site) (Dominissini et al. 2012). Modifications are often concentrated within 50–100 nucleotides of these splice junctions, suggesting they play important roles in splice site recognition and selection.

Modifications near splice sites can profoundly affect how the splicing machinery recognizes and interacts with these crucial sequences. For example, m6A modifications can alter local RNA structure, making splice sites either more or less accessible to the splicing machinery (Xiao et al. 2016). This structural modulation can determine whether a particular splice site is used frequently or skipped in favor of alternative sites.

The presence of modifications can also influence the binding of key splicing factors. Some modifications serve as direct recognition signals for specific RNA-binding proteins that help recruit or repel components of the splicing machinery (Huang et al. 2018). This creates a sophisticated system where the modification status of splice sites helps determine splicing outcomes. A prominent example of this is the m6A modification of U6 snRNA, significantly altering 5′ splice site selections in *Arabidopsis* (Parker et al. 2022).

10.5.1.6 3′ UTR Regions

The 3′ untranslated region represents a crucial regulatory domain in messenger RNA, serving as a binding platform for numerous regulatory factors that control mRNA fate. The 3′ UTR hosts a diverse array of RNA modifications, m6A being particularly prevalent, but also Ψ, m5C, and various 2-O-methylations (Chan et al. 2023).

One of the most significant impacts of 3′ UTR modifications involves their influence on mRNA stability (Mayr 2017). These modifications can either protect transcripts from degradation or mark them for rapid turnover, depending on the specific modification and its context (Wang et al. 2014). For example, m6A modifications can recruit specific reader proteins that influence the accessibility of decay-promoting factors to the transcript. This mechanism allows cells to modulate the lifetime of specific mRNAs in response to changing conditions, creating a dynamic system for controlling gene expression (Zhou et al. 2015).

The presence of modifications in the 3′ UTR can significantly affect microRNA-mediated regulation (Ali Syeda et al. 2020). Modifications can either enhance or inhibit microRNA binding by altering local RNA structure or directly interfering with base–pairing interactions. This regulatory mechanism becomes particularly important during development and stress responses, where precise control of gene expression is essential. For instance, modifications near microRNA binding sites can help determine whether a particular transcript will be susceptible or resistant to microRNA-mediated repression (Garcia et al. 2011).

Modifications in the 3′ UTR can create or eliminate binding sites for RNA-binding proteins, establishing a sophisticated system for post-transcriptional regulation. These modifications often work by altering local RNA structure or providing direct recognition sites for specific proteins. The resulting RNA–protein interactions can influence various aspects of mRNA metabolism, from localization within the cell to translation efficiency and stability (Shi et al. 2018).

The modification status of the 3′ UTR plays a crucial role in determining where mRNAs localize within cells. This becomes particularly important in highly polarized cells like neurons, where certain mRNAs must be transported to specific subcellular compartments for local translation (Holt and Schuman 2013). Modifications can influence this process by affecting the binding of transport factors or by modulating the RNA's ability to form transport-competent particles.

10.5.1.7 Poly(A) Tails

The poly(A) tail, a stretch of predominantly adenosine residues added to the 3′ end of most messenger RNAs, serves as more than just a simple polymer terminus (Passmore and Coller 2022). This region can be extensively modified, creating an additional layer of regulation that influences mRNA stability, translation efficiency, and ultimate fate (Slobodin et al. 2017). Understanding how chemical modifications affect poly(A) tail function has revealed new mechanisms of post-transcriptional gene regulation (Subtelny et al. 2014).

Several types of modifications can occur within poly(A) tails, with methylation being particularly prominent. m6A modifications can appear throughout the tail, while Nm often occurs at specific positions (Zhang et al. 2024a, b). The distribution of these modifications is not random but rather follows patterns that suggest functional significance in regulating mRNA metabolism.

Modifications in the poly(A) tail region significantly influence mRNA stability and turnover rates. These modifications can affect how efficiently deadenylases recognize and process the tail, thereby modulating mRNA decay kinetics (Yan 2014). For example, certain modifications can protect the tail from degradation by making it less accessible to deadenylases, while others might enhance deadenylation and promote rapid mRNA turnover.

The modification status of poly(A) tails plays a crucial role in determining how efficiently mRNAs are translated. Modified nucleotides can influence the binding of poly(A)-binding protein (PABP), a key factor in translation initiation (Khong and Parker 2018). This regulation becomes particularly important during cellular stress or developmental transitions, where rapid changes in translation patterns are necessary.

Modifications in the poly(A) tail region can affect how the tail length is regulated. Some modifications influence the activity of cytoplasmic deadenylases and poly (A) polymerases, creating a dynamic system for controlling tail length. This regulation is crucial because tail length often correlates with mRNA stability and translational efficiency (Jalkanen et al. 2014).

Different cell types often show distinct patterns of poly(A) tail modifications, suggesting these chemical marks help establish tissue-specific gene expression programs. These patterns can influence which mRNAs are most stable or efficiently translated in different cell types, contributing to tissue-specific functions and development.

10.5.2 Structural Context Analysis

The three-dimensional structure of RNA molecules plays a crucial role in determining both where modifications occur and how they affect RNA function. RNA modifications can either stabilize existing structures or induce structural changes, creating a complex interplay between chemical modification and molecular architecture that shapes RNA behavior (Dethoff et al. 2012).

Structure-sensitive probing techniques have revolutionized the understanding of this relationship. Methods like SHAPE (Selective 2′-Hydroxyl Acylation analyzed by Primer Extension) and DMS (Dimethyl Sulfate) probing can reveal how modifications alter local and global RNA structure (Mortimer et al. 2012; Rouskin et al. 2014). When combined with modification mapping techniques, these approaches provide a comprehensive view of how structure and modification interact. Studying the structural environment before and after m6A modification has revealed that this mark often appears in regions that can switch between single-stranded and double-stranded conformations, suggesting a role in regulatory structural transitions (Liu et al. 2013).

The structural context of modifications often determines their functional impact. Modifications in highly structured regions might disrupt base–pairing interactions, leading to local unfolding, while modifications in flexible regions might promote the formation of new structural elements. This context-dependent behavior explains why identical modifications can have different effects depending on their location (Vögele et al. 2023). Ψ modifications have been shown to stabilize RNA structure through improved base stacking or facilitate structural rearrangements by providing alternative hydrogen bonding opportunities (Carlile et al. 2014).

Advances in secondary structure prediction incorporate RNA modifications, enabling more accurate modeling of RNA folding dynamics (Wu et al. 2016). These computational tools evaluated how modifications influence base-pairing energetics and structural flexibility (Bose et al. 2024). Integration of machine learning with experimental structure probing data has enhanced researchers' ability to predict modification-induced structural changes in RNA, providing insights into their functional consequences (Homan et al. 2014; Tomezsko et al. 2020).

Recent advances in cryo-electron microscopy (cryo-EM) have provided atomic-resolution views of modified RNA structures, revealing how modifications can create unique structural features that serve as recognition elements for RNA-binding proteins (Stojković et al. 2020). These structural insights help explain how modifications can regulate RNA-protein interactions and influence RNA fate. For instance, structural analysis has shown how m6A modifications can create distinctive binding pockets for YTH-domain proteins, leading to downstream regulatory effects (Zhang et al. 2019b).

Time-resolved structural probing techniques have begun to reveal the dynamics of modification-induced structural changes (Rouskin et al. 2014). These approaches show that some modifications cause immediate structural alterations, while others might trigger slower, progressive changes in RNA architecture (Spitale et al. 2015). Understanding these temporal dynamics is crucial for deciphering how modifications regulate RNA function in response to cellular signals.

The study of RNA structure in cellular environments has highlighted the importance of cellular context in modification function. Factors like ionic conditions, molecular crowding, and protein binding can all influence how modifications affect RNA structure (Wan et al. 2011). This cellular context helps explain why some modifications might have different effects in vitro versus in vivo, and why their impact might vary across cellular compartments or conditions (Dethoff et al. 2012).

Integrative approaches combining structural analysis with functional studies have proven particularly powerful. By correlating structural changes with functional outcomes like altered stability, splicing, or translation, researchers can better understand how structural effects of modifications lead to biological consequences. This multilayered analysis helps establish causal relationships between modification-induced structural changes and their functional impacts.

10.5.3 Comparative Analysis

Comparative analysis serves as a powerful lens through which the biological significance and functional roles of RNA modifications can be understood. By examining how modifications vary across species, tissues, developmental stages, and disease states, the fundamental principles about their roles and regulation can be uncovered.

Evolutionary comparative analysis reveals the deep conservation of many RNA modification pathways, suggesting their fundamental importance in cellular function. Some modifications, such as Ψ in ribosomal RNA, show remarkable conservation from bacteria to humans, indicating their essential roles in basic cellular processes (Decatur and Fournier 2002). Other modifications appear to have evolved more recently or show lineage-specific patterns, highlighting their roles in specialized functions (Jeltsch et al. 2017; Bohnsack et al. 2019; Kuznetsova et al. 2019).

Cross-species comparison of modification sites often reveals conservation hotspots—regions where modifications persist across evolutionary time (Batista et al. 2014). These conserved sites frequently occur in functionally critical regions of RNAs, such as the peptidyl transferase center of ribosomes or splice sites in mRNAs (Decatur and Fournier 2002). The degree of conservation can indicate the functional importance of specific modifications, with highly conserved sites often playing crucial roles in RNA function or regulation. However, the absence of conservation doesn't necessarily indicate a lack of function—some modifications might have evolved independently in different lineages to serve similar purposes through different mechanisms (Taoka et al. 2018).

Tissue-specific comparative analysis provides another valuable perspective on modification function (Adetula et al. 2021). Examining modifications across diverse cell types reveals their contribution to both fundamental and tissue-specific gene regulation. While some modifications are conserved across many cell types, others exhibit distinct tissue-specific patterns that align with specialized functions (Xiong et al. 2021). For example, m6A modifications have been shown to be significantly altered among shoot cells in *Arabidopsis* (Shen et al. 2016). This tissue-specific variation often relates to the expression patterns of modification enzymes and their regulators, revealing the regulatory networks that control modification deployment (Tegowski et al. 2024).

Developmental comparative analysis examines how modification patterns change during organism development and cellular differentiation (Zhao et al. 2017). This temporal perspective reveals how modifications contribute to developmental regulation and cell fate decisions. For instance, changes in m6A modification patterns during embryonic development suggest roles in controlling gene expression programs that guide cellular differentiation (Batista et al. 2014). Understanding these developmental dynamics helps explain how modifications contribute to normal development and how their dysregulation might lead to developmental disorders (Weng et al. 2018).

Disease-state comparative analysis provides crucial insights into the medical relevance of RNA modifications (Han et al. 2019a). By comparing modification patterns between healthy and diseased tissues, modifications that might contribute to disease progression can be identified and could serve as diagnostic markers. This approach has revealed modifications in various cancers, neurological disorders, and genetic diseases, suggesting potential therapeutic strategies targeting modification pathways (Chen et al. 2021).

Comparative analysis across different cellular conditions reveals how modifications respond to environmental changes and cellular stress. By examining how modification patterns shift during stress responses, nutrient limitation, or other cellular perturbations, their roles in cellular adaptation can be understood (Zhou et al. 2015; Wilkinson et al. 2021). This dynamic perspective shows how cells use modifications to rapidly adjust gene expression programs in response to changing conditions (Meyer et al. 2015).

The integration of comparative analyses across these different dimensions—evolutionary, tissue-specific, developmental, disease-related, and condition-dependent—provides a comprehensive understanding of modification function. Finding that a modification site is both evolutionarily conserved and dynamically regulated during development suggests it plays important roles in both basic cellular function and developmental regulation. This integrative approach helps prioritize modifications for further study and suggests potential therapeutic targets.

10.6 Conclusion

The future of RNA modification identification stands at an exciting intersection of computational innovation and experimental advancement. The emergence of sophisticated machine learning approaches, particularly those incorporating deep learning and attention mechanisms, combined with improved signal processing capabilities, has dramatically enhanced the ability to detect and characterize RNA modifications. These computational advances are being complemented by revolutionary experimental methods, especially in nanopore sequencing and engineered protein systems, which provide increasingly detailed views of RNA modifications in their cellular context. Perhaps most significantly, the field is moving toward integrated approaches that combine multiple analytical methods into unified frameworks, enabling comprehensive analysis of the RNA modification landscape. As these technologies continue to evolve, even more sensitive and precise methods for studying RNA modifications can be expected. These methods may ultimately lead to deeper insights into their biological roles and potential therapeutic applications. The convergence of these various approaches suggests that the complete characterization of the epitranscriptome—including the dynamic interplay between different modifications and their spatial distribution within cells—may become a realistic possibility.

Acknowledgments The authors wish to thank the members of the Janga lab for providing feedback on previous versions of this chapter.

Conflict of Interest Disclosure The authors declare no conflicts of interest.

Text was refined with the assistance of artificial intelligence (AI) tools such as Anthropic, Claude, 3.7 Sonnet, 2025, and OpenAI, ChatGPT 4.0-o, 2025. Every section was carefully reviewed, revised, and approved by authors to ensure that the text is appropriate concerning content and scientific correctness and maintains the integrity of the original content.

References

Abebe JS, Price AM, Hayer KE et al (2022) DRUMMER-rapid detection of RNA modifications through comparative nanopore sequencing. Bioinformatics 38:3113–3115. https://doi.org/10.1093/bioinformatics/btac274

Acera Mateos P, Sethi AJ, Ravindran A et al (2024) Prediction of m6A and m5C at single-molecule resolution reveals a transcriptome-wide co-occurrence of RNA modifications. Nat Commun 15: 3899. https://doi.org/10.1038/s41467-024-47953-7

Adetula AA, Fan X, Zhang Y et al (2021) Landscape of tissue-specific RNA Editome provides insight into co-regulated and altered gene expression in pigs (sus-scrofa). RNA Biol 18:439–450. https://doi.org/10.1080/15476286.2021.1954380

Ali Syeda Z, Langden SSS, Munkhzul C et al (2020) Regulatory mechanism of MicroRNA expression in cancer. Int J Mol Sci 21:1723. https://doi.org/10.3390/ijms21051723

Amalric A, Bastide A, Attina A et al (2022) Quantifying RNA modifications by mass spectrometry: a novel source of biomarkers in oncology. Crit Rev Clin Lab Sci 59:1–18. https://doi.org/10.1080/10408363.2021.1958743

Amalric A, Attina A, Bastide A et al (2024) Mass spectrometry-based pipeline for identifying RNA modifications involved in a functional process: application to cancer cell adaptation. Anal Chem 96:1825–1833. https://doi.org/10.1021/acs.analchem.3c02635

Amort T, Rieder D, Wille A et al (2017) Distinct 5-methylcytosine profiles in poly(a) RNA from mouse embryonic stem cells and brain. Genome Biol 18:1. https://doi.org/10.1186/s13059-016-1139-1

Anders M, Chelysheva I, Goebel I et al (2018) Dynamic m6A methylation facilitates mRNA triaging to stress granules. Life Sci Alliance 1:e201800113. https://doi.org/10.26508/lsa.201800113

Ardui S, Ameur A, Vermeesch JR, Hestand MS (2018) Single molecule real-time (SMRT) sequencing comes of age: applications and utilities for medical diagnostics. Nucleic Acids Res 46:2159–2168. https://doi.org/10.1093/nar/gky066

Athanasopoulou K, Boti MA, Adamopoulos PG et al (2021) Third-generation sequencing: the spearhead towards the radical transformation of modern genomics. Life (Basel) 12:30. https://doi.org/10.3390/life12010030

Bao X, Zhang Y, Li H et al (2023) RM2Target: a comprehensive database for targets of writers, erasers and readers of RNA modifications. Nucleic Acids Res 51:D269–D279. https://doi.org/10.1093/nar/gkac945

Barbieri I, Kouzarides T (2020) Role of RNA modifications in cancer. Nat Rev Cancer 20:303–322. https://doi.org/10.1038/s41568-020-0253-2

Batista PJ, Molinie B, Wang J et al (2014) m6A RNA modification controls cell fate transition in mammalian embryonic stem cells. Cell Stem Cell 15:707–719. https://doi.org/10.1016/j.stem.2014.09.019

Beaulaurier J, Schadt EE, Fang G (2019) Deciphering bacterial epigenomes using modern sequencing technologies. Nat Rev Genet 20:157–172. https://doi.org/10.1038/s41576-018-0081-3

Begik O, Lucas MC, Pryszcz LP et al (2021) Quantitative profiling of pseudouridylation dynamics in native RNAs with nanopore sequencing. Nat Biotechnol 39:1278–1291. https://doi.org/10.1038/s41587-021-00915-6

Bidlingmeyer BA, Cohen SA, Tarvin TL (1984) Rapid analysis of amino acids using pre-column derivatization. J Chromatogr B Biomed Sci Appl 336:93–104. https://doi.org/10.1016/S0378-4347(00)85133-6

Boccaletto P, Machnicka MA, Purta E et al (2018) MODOMICS: a database of RNA modification pathways. 2017 update. Nucleic Acids Res 46:D303–D307. https://doi.org/10.1093/nar/gkx1030

Bohnsack KE, Höbartner C, Bohnsack MT (2019) Eukaryotic 5-methylcytosine (m^5C) RNA Methyltransferases: mechanisms, cellular functions, and links to disease. Genes (Basel) 10: 102. https://doi.org/10.3390/genes10020102

Borchardt EK, Martinez NM, Gilbert WV (2020) Regulation and function of RNA Pseudouridylation in human cells. Annu Rev Genet 54:309–336. https://doi.org/10.1146/annurev-genet-112618-043830

Bose E, Xiong S, Jones AN (2024) Probing RNA structure and dynamics using nanopore and next generation sequencing. J Biol Chem 300:107317. https://doi.org/10.1016/j.jbc.2024.107317

Brown JWS, Echeverria M, Qu L-H et al (2003) Plant snoRNA database. Nucleic Acids Res 31: 432–435. https://doi.org/10.1093/nar/gkg009

Calero G, Wilson KF, Ly T et al (2002) Structural basis of m7GpppG binding to the nuclear cap-binding protein complex. Nat Struct Biol 9:912–917. https://doi.org/10.1038/nsb874

Carlile TM, Rojas-Duran MF, Zinshteyn B et al (2014) Pseudouridine profiling reveals regulated mRNA pseudouridylation in yeast and human cells. Nature 515:143–146. https://doi.org/10.1038/nature13802

Cech TR, Steitz JA (2014) The noncoding RNA revolution—trashing old rules to forge new ones. Cell 157:77–94. https://doi.org/10.1016/j.cell.2014.03.008

Chan JJ, Tabatabaeian H, Tay Y (2023) 3'UTR heterogeneity and cancer progression. Trends Cell Biol 33:568–582. https://doi.org/10.1016/j.tcb.2022.10.001

Chan F-F, Kwan KK-L, Seoung D-H et al (2024) N6-Methyladenosine modification activates the serine synthesis pathway to mediate therapeutic resistance in liver cancer. Mol Ther 32:4435–4447. https://doi.org/10.1016/j.ymthe.2024.10.025

Chen K, Lu Z, Wang X et al (2015) High-resolution N(6) -methyladenosine (m(6) A) map using photo-crosslinking-assisted m(6) A sequencing. Angew Chem Int Ed Engl 54:1587–1590. https://doi.org/10.1002/anie.201410647

Chen K, Song B, Tang Y et al (2020a) RMDisease: a database of genetic variants that affect RNA modifications, with implications for epitranscriptome pathogenesis. Nucleic Acids Res 49: D1396–D1404. https://doi.org/10.1093/nar/gkaa790

Chen Z, Zhao P, Li F et al (2020b) Comprehensive review and assessment of computational methods for predicting RNA post-transcriptional modification sites from RNA sequences. Brief Bioinform 21:1676–1696. https://doi.org/10.1093/bib/bbz112

Chen X, Wang J, Tahir M et al (2021) Current insights into the implications of m6A RNA methylation and autophagy interaction in human diseases. Cell Biosci 11:147. https://doi.org/10.1186/s13578-021-00661-x

Chen M, Sui T, Yang L et al (2022) Live imaging of RNA and RNA splicing in mammalian cells via the dcas13a-SunTag-BiFC system. Biosens Bioelectron 204:114074. https://doi.org/10.1016/j.bios.2022.114074

Chen L, Ou L, Jing X et al (2023a) DeepEdit: single-molecule detection and phasing of A-to-I RNA editing events using nanopore direct RNA sequencing. Genome Biol 24:75. https://doi.org/10.1186/s13059-023-02921-0

Chen L, Zhang L-S, Ye C et al (2023b) Nm-Mut-seq: a base-resolution quantitative method for mapping transcriptome-wide 2'-O-methylation. Cell Res 33:727–730. https://doi.org/10.1038/s41422-023-00836-w

Chen F, Li X, Bai M, Zhao Y (2025) Visualizing epigenetic modifications and their spatial proximities in single cells using three DNA-encoded amplifying FISH imaging strategies: BEA-FISH, PPDA-FISH and cell-TALKING. Nat Protoc 20:220–247. https://doi.org/10.1038/s41596-024-01036-5

Choi J, Ieong K-W, Demirci H et al (2016) N(6)-methyladenosine in mRNA disrupts tRNA selection and translation-elongation dynamics. Nat Struct Mol Biol 23:110–115. https://doi.org/10.1038/nsmb.3148

Clark SJ, Argelaguet R, Kapourani C-A et al (2018) scNMT-seq enables joint profiling of chromatin accessibility DNA methylation and transcription in single cells. Nat Commun 9: 781. https://doi.org/10.1038/s41467-018-03149-4

Cohn WE, Volkin E (1951) Nucleoside-5'-phosphates from ribonucleic acid. Nature 167:483–484. https://doi.org/10.1038/167483a0

Daffis S, Szretter KJ, Schriewer J et al (2010) 2'-O methylation of the viral mRNA cap evades host restriction by IFIT family members. Nature 468:452–456. https://doi.org/10.1038/nature09489

Dai Q, Moshitch-Moshkovitz S, Han D et al (2017) Nm-seq maps 2'-O-methylation sites in human mRNA with base precision. Nat Methods 14:695–698. https://doi.org/10.1038/nmeth.4294

Dai Q, Zhang L-S, Sun H-L et al (2023) Quantitative sequencing using BID-seq uncovers abundant pseudouridines in mammalian mRNA at base resolution. Nat Biotechnol 41:344–354. https://doi.org/10.1038/s41587-022-01505-w

Dai Q, Ye C, Irkliyenko I et al (2024) Ultrafast bisulfite sequencing detection of 5-methylcytosine in DNA and RNA. Nat Biotechnol 42:1559–1570. https://doi.org/10.1038/s41587-023-02034-w

Davis FF, Allen FW (1957) Ribonucleic acids from yeast which contain a fifth nucleotide. J Biol Chem 227:907–915. https://doi.org/10.1016/S0021-9258(18)70770-9

Decatur WA, Fournier MJ (2002) rRNA modifications and ribosome function. Trends Biochem Sci 27:344–351. https://doi.org/10.1016/S0968-0004(02)02109-6

Delaunay S, Frye M (2019) RNA modifications regulating cell fate in cancer. Nat Cell Biol 21:552–559. https://doi.org/10.1038/s41556-019-0319-0

Delaunay S, Helm M, Frye M (2024) RNA modifications in physiology and disease: towards clinical applications. Nat Rev Genet 25:104–122. https://doi.org/10.1038/s41576-023-00645-2

Deng S, Zhang H, Zhu K et al (2021) M6A2Target: a comprehensive database for targets of m6A writers, erasers and readers. Brief Bioinform 22:bbaa055. https://doi.org/10.1093/bib/bbaa055

Dethoff EA, Petzold K, Chugh J et al (2012) Visualizing transient low-populated structures of RNA. Nature 491:724–728. https://doi.org/10.1038/nature11498

Dominissini D, Moshitch-Moshkovitz S, Schwartz S et al (2012) Topology of the human and mouse m6A RNA methylomes revealed by m6A-seq. Nature 485:201–206. https://doi.org/10.1038/nature11112

Dunin-Horkawicz S, Czerwoniec A, Gajda MJ et al (2006) MODOMICS: a database of RNA modification pathways. Nucleic Acids Res 34:D145–D149. https://doi.org/10.1093/nar/gkj084

Dunn DB (1961) The occurence of 1-methyladenine in ribonucleic acid. Biochim Biophys Acta 46:198–200. https://doi.org/10.1016/0006-3002(61)90668-0

Eid J, Fehr A, Gray J et al (2009) Real-time DNA sequencing from single polymerase molecules. Science 323:133–138. https://doi.org/10.1126/science.1162986

El Allali A, Elhamraoui Z, Daoud R (2021) Machine learning applications in RNA modification sites prediction. Comput Struct Biotechnol J 19:5510–5524. https://doi.org/10.1016/j.csbj.2021.09.025

Eraslan G, Simon LM, Mircea M et al (2019) Single-cell RNA-seq denoising using a deep count autoencoder. Nat Commun 10:390. https://doi.org/10.1038/s41467-018-07931-2

Flusberg BA, Webster DR, Lee JH et al (2010) Direct detection of DNA methylation during single-molecule, real-time sequencing. Nat Methods 7:461–465. https://doi.org/10.1038/nmeth.1459

Frye M, Harada BT, Behm M, He C (2018) RNA modifications modulate gene expression during development. Science 361:1346–1349. https://doi.org/10.1126/science.aau1646

Furlan M, Delgado-Tejedor A, Mulroney L et al (2021) Computational methods for RNA modification detection from nanopore direct RNA sequencing data. RNA Biol 18:31–40. https://doi.org/10.1080/15476286.2021.1978215

Furuichi Y, Shatkin AJ (2000) Viral and cellular mRNA capping: past and prospects. Adv Virus Res 55:135–184. https://doi.org/10.1016/s0065-3527(00)55003-9

Galloway A, Atrih A, Grzela R et al (2020) CAP-MAP: cap analysis protocol with minimal analyte processing, a rapid and sensitive approach to analysing mRNA cap structures. Open Biol 10:190306. https://doi.org/10.1098/rsob.190306

Gao Y, Liu X, Wu B et al (2021) Quantitative profiling of N6-methyladenosine at single-base resolution in stem-differentiating xylem of Populus trichocarpa using Nanopore direct RNA sequencing. Genome Biol 22:22. https://doi.org/10.1186/s13059-020-02241-7

Garalde DR, Snell EA, Jachimowicz D et al (2018) Highly parallel direct RNA sequencing on an array of nanopores. Nat Methods 15:201–206. https://doi.org/10.1038/nmeth.4577

Garcia DM, Baek D, Shin C et al (2011) Weak seed-pairing stability and high target-site abundance decrease the proficiency of lsy-6 and other microRNAs. Nat Struct Mol Biol 18:1139–1146. https://doi.org/10.1038/nsmb.2115

Giessing AMB, Kirpekar F (2012) Mass spectrometry in the biology of RNA and its modifications. J Proteome 75:3434–3449. https://doi.org/10.1016/j.jprot.2012.01.032

Gilbert WV, Bell TA, Schaening C (2016) Messenger RNA modifications: form, distribution, and function. Science 352:1408–1412. https://doi.org/10.1126/science.aad8711

Han D, Liu J, Chen C et al (2019a) Anti-tumour immunity controlled through mRNA m6A methylation and YTHDF1 in dendritic cells. Nature 566:270–274. https://doi.org/10.1038/s41586-019-0916-x

Han Y, Feng J, Xia L et al (2019b) CVm6A: a visualization and exploration database for m^6As in cell lines. Cells 8:168. https://doi.org/10.3390/cells8020168

Hassan D, Acevedo D, Daulatabad SV et al (2022) Penguin: a tool for predicting pseudouridine sites in direct RNA nanopore sequencing data. Methods 203:478–487. https://doi.org/10.1016/j.ymeth.2022.02.005

Hassan D, Ariyur A, Daulatabad SV et al (2024) Nm-Nano: a machine learning framework for transcriptome-wide single-molecule mapping of 2′-O-methylation (Nm) sites in nanopore direct RNA sequencing datasets. RNA Biol 21:1–15. https://doi.org/10.1080/15476286.2024.2352192

Hauenschild R, Tserovski L, Schmid K et al (2015) The reverse transcription signature of N-1-methyladenosine in RNA-Seq is sequence dependent. Nucleic Acids Res 43. https://doi.org/10.1093/nar/gkv895

Helm M, Motorin Y (2017) Detecting RNA modifications in the epitranscriptome: predict and validate. Nat Rev Genet 18. https://doi.org/10.1038/nrg.2016.169

Hendra C, Pratanwanich PN, Wan YK et al (2022) Detection of m6A from direct RNA sequencing using a multiple instance learning framework. Nat Methods 19:1590–1598. https://doi.org/10.1038/s41592-022-01666-1

Hinnebusch AG, Ivanov IP, Sonenberg N (2016) Translational control by 5'-untranslated regions of eukaryotic mRNAs. Science 352:1413–1416. https://doi.org/10.1126/science.aad9868

Holt CE, Schuman EM (2013) The central dogma decentralized: new perspectives on RNA function and local translation in neurons. Neuron 80:648–657. https://doi.org/10.1016/j.neuron.2013.10.036

Homan PJ, Favorov OV, Lavender CA et al (2014) Single-molecule correlated chemical probing of RNA. Proc Natl Acad Sci 111:13858–13863. https://doi.org/10.1073/pnas.1407306111

Hori H (2014) Methylated nucleosides in tRNA and tRNA methyltransferases. Front Genet 5:144. https://doi.org/10.3389/fgene.2014.00144

Hu B, Yang Y-CT, Huang Y et al (2017) POSTAR: a platform for exploring post-transcriptional regulation coordinated by RNA-binding proteins. Nucleic Acids Res 45:D104–D114. https://doi.org/10.1093/nar/gkw888

Hu L, Liu S, Peng Y et al (2022) m6A RNA modifications are measured at single-base resolution across the mammalian transcriptome. Nat Biotechnol 40:1210–1219. https://doi.org/10.1038/s41587-022-01243-z

Huang H, Weng H, Sun W et al (2018) Recognition of RNA N6-methyladenosine by IGF2BP proteins enhances mRNA stability and translation. Nat Cell Biol 20:285–295. https://doi.org/10.1038/s41556-018-0045-z

Huang S, Wylder AC, Pan T (2024) Simultaneous nanopore profiling of mRNA m6A and pseudouridine reveals translation coordination. Nat Biotechnol 42:1831–1835. https://doi.org/10.1038/s41587-024-02135-0

Hussain S, Aleksic J, Blanco S et al (2013) Characterizing 5-methylcytosine in the mammalian epitranscriptome. Genome Biol 14:215. https://doi.org/10.1186/gb4143

Jalkanen AL, Coleman SJ, Wilusz J (2014) Determinants and implications of mRNA Poly(A) tail size – does this protein make my tail look big? Semin Cell Dev Biol:24–32. https://doi.org/10.1016/j.semcdb.2014.05.018

Janin M, Ortiz-Barahona V, de Moura MC et al (2019) Epigenetic loss of RNA-methyltransferase NSUN5 in glioma targets ribosomes to drive a stress adaptive translational program. Acta Neuropathol 138:1053–1074. https://doi.org/10.1007/s00401-019-02062-4

Jeltsch A, Ehrenhofer-Murray A, Jurkowski TP et al (2017) Mechanism and biological role of Dnmt2 in nucleic acid methylation. RNA Biol 14:1108–1123. https://doi.org/10.1080/15476286.2016.1191737

Jenjaroenpun P, Wongsurawat T, Wadley TD et al (2021) Decoding the epitranscriptional landscape from native RNA sequences. Nucleic Acids Res 49:e7. https://doi.org/10.1093/nar/gkaa620

Johnstone TG, Bazzini AA, Giraldez AJ (2016) Upstream ORFs are prevalent translational repressors in vertebrates. EMBO J 35:706–723. https://doi.org/10.15252/embj.201592759

Jonkhout N, Tran J, Smith MA et al (2017) The RNA modification landscape in human disease. RNA 23:1754–1769. https://doi.org/10.1261/rna.063503.117

Kan RL, Chen J, Sallam T (2022) Crosstalk between epitranscriptomic and epigenetic mechanisms in gene regulation. Trends Genet 38:182–193. https://doi.org/10.1016/j.tig.2021.06.014

Ke S, Alemu EA, Mertens C et al (2015) A majority of m6A residues are in the last exons, allowing the potential for 3' UTR regulation. Genes Dev 29:2037–2053. https://doi.org/10.1101/gad.269415.115

Khoddami V, Yerra A, Mosbruger TL et al (2019) Transcriptome-wide profiling of multiple RNA modifications simultaneously at single-base resolution. Proc Natl Acad Sci USA 116:6784–6789. https://doi.org/10.1073/pnas.1817334116

Khong A, Parker R (2018) mRNP architecture in translating and stress conditions reveals an ordered pathway of mRNP compaction. J Cell Biol 217:4124–4140. https://doi.org/10.1083/jcb.201806183

Kim HS, Eun JW, Jang SH et al (2025) The diverse landscape of RNA modifications in cancer development and progression. Genes Genom 47:135–155. https://doi.org/10.1007/s13258-024-01601-y

Kowalak JA, Pomerantz SC, Crain PF, McCloskey JA (1993) A novel method for the determination of post-transcriptional modification in RNA by mass spectrometry. Nucleic Acids Res 21:4577–4585. https://doi.org/10.1093/nar/21.19.4577

Krug RM, Morgan MA, Shatkin AJ (1976) Influenza viral mRNA contains internal N6-methyladenosine and 5'-terminal 7-methylguanosine in cap structures. J Virol 20:45–53. https://doi.org/10.1128/jvi.20.1.45-53.1976

Kuznetsova SA, Petrukov KS, Pletnev FI et al (2019) RNA (C5-cytosine) Methyltransferases. Biochemistry (Mosc) 84:851–869. https://doi.org/10.1134/S0006297919080029

Łabno A, Tomecki R, Dziembowski A (2016) Cytoplasmic RNA decay pathways – enzymes and mechanisms. Biochim Biophys Acta 1863:3125–3147. https://doi.org/10.1016/j.bbamcr.2016.09.023

Lang X, Yu C, Shen M et al (2024) PRMD: an integrated database for plant RNA modifications. Nucleic Acids Res 52:D1597–D1613. https://doi.org/10.1093/nar/gkad851

Leger A, Amaral PP, Pandolfini L et al (2021) RNA modifications detection by comparative Nanopore direct RNA sequencing. Nat Commun 12:7198. https://doi.org/10.1038/s41467-021-27393-3

Legrand C, Tuorto F, Hartmann M et al (2017) Statistically robust methylation calling for whole-transcriptome bisulfite sequencing reveals distinct methylation patterns for mouse RNAs. Genome Res 27:1589–1596. https://doi.org/10.1101/gr.210666.116

Lei H-T, Wang Z-H, Li B et al (2023) tModBase: deciphering the landscape of tRNA modifications and their dynamic changes from epitranscriptome data. Nucleic Acids Res 51:D315–D327. https://doi.org/10.1093/nar/gkac1087

Li X, Xiong X, Yi C (2017) Epitranscriptome sequencing technologies: decoding RNA modifications. Nat Methods 14:23–31. https://doi.org/10.1038/nmeth.4110

Li Y, Wang Y, Vera-Rodriguez M et al (2024a) Single-cell m6A mapping in vivo using picoMeRIP-seq. Nat Biotechnol 42:591–596. https://doi.org/10.1038/s41587-023-01831-7

Li Y, Zhou H, Chen S et al (2024b) Bioorthogonal labeling and profiling of N6-isopentenyladenosine (i6A) modified RNA. Nucleic Acids Res 52:2808–2820. https://doi.org/10.1093/nar/gkae150

Limbach PA, Crain PF, McCloskey JA (1994) Summary: the modified nucleosides of RNA. Nucleic Acids Res 22:2183–2196. https://doi.org/10.1093/nar/22.12.2183

Linder B, Grozhik AV, Olarerin-George AO et al (2015) Single-nucleotide-resolution mapping of m6A and m6Am throughout the transcriptome. Nat Methods 12:767–772. https://doi.org/10.1038/nmeth.3453

Liu Q, Gregory RI (2019) RNAmod: an integrated system for the annotation of mRNA modifications. Nucleic Acids Res 47:W548–W555. https://doi.org/10.1093/nar/gkz479

Liu N, Parisien M, Dai Q et al (2013) Probing N6-methyladenosine RNA modification status at single nucleotide resolution in mRNA and long noncoding RNA. RNA 19:1848–1856. https://doi.org/10.1261/rna.041178.113

Liu N, Dai Q, Zheng G et al (2015) N(6)-methyladenosine-dependent RNA structural switches regulate RNA-protein interactions. Nature 518:560–564. https://doi.org/10.1038/nature14234

Liu L, Zhang S-W, Gao F et al (2016) DRME: count-based differential RNA methylation analysis at small sample size scenario. Anal Biochem 499. https://doi.org/10.1016/j.ab.2016.01.014

Liu H, Begik O, Lucas MC et al (2019) Accurate detection of m6A RNA modifications in native RNA sequences. Nat Commun 10:4079. https://doi.org/10.1038/s41467-019-11713-9

Liu S, Zhu A, He C, Chen M (2020) REPIC: a database for exploring the N6-methyladenosine methylome. Genome Biol 21:100. https://doi.org/10.1186/s13059-020-02012-4

Liu H, Begik O, Novoa EM (2021) EpiNano: detection of m6A RNA modifications using Oxford Nanopore direct RNA sequencing. Methods Mol Biol 2298:31–52. https://doi.org/10.1007/978-1-0716-1374-0_3

Liu H, Wang Y, Zhou X (2022) Labeling and sequencing nucleic acid modifications using bio-orthogonal tools. RSC Chem Biol 3:994–1007. https://doi.org/10.1039/d2cb00087c

Liu C, Sun H, Yi Y et al (2023) Absolute quantification of single-base m6A methylation in the mammalian transcriptome using GLORI. Nat Biotechnol 41:355–366. https://doi.org/10.1038/s41587-022-01487-9

Lorenz C, Lünse CE, Mörl M (2017) tRNA modifications: impact on structure and thermal adaptation. Biomolecules 7:35. https://doi.org/10.3390/biom7020035

Lu T, Ang CE, Zhuang X (2022) Spatially resolved epigenomic profiling of single cells in complex tissues. Cell 185:4448–4464.e17. https://doi.org/10.1016/j.cell.2022.09.035

Lu J, Sheng Y, Qian W et al (2023) scRNA-seq data analysis method to improve analysis performance. IET Nanobiotechnol 17:246–256. https://doi.org/10.1049/nbt2.12115

Lucas MC, Pryszcz LP, Medina R et al (2024) Quantitative analysis of tRNA abundance and modifications by nanopore RNA sequencing. Nat Biotechnol 42:72–86. https://doi.org/10.1038/s41587-023-01743-6

Luecken MD, Theis FJ (2019) Current best practices in single-cell RNA-seq analysis: a tutorial. Mol Syst Biol 15:e8746. https://doi.org/10.15252/msb.20188746

Luo X, Li H, Liang J et al (2021) RMVar: an updated database of functional variants involved in RNA modifications. Nucleic Acids Res 49:D1405–D1412. https://doi.org/10.1093/nar/gkaa811

Luo Z, Zhang J, Fei J, Ke S (2022) Deep learning modeling m6A deposition reveals the importance of downstream cis-element sequences. Nat Commun 13:2720. https://doi.org/10.1038/s41467-022-30209-7

Ma Q, Tatsuno T, Nakamura Y et al (2019) Immuno-detection of mRNA-binding protein complex in human cells under transmission electron microscopy. Microsc Res Tech 82:680–688. https://doi.org/10.1002/jemt.23214

Ma J, Song B, Wei Z et al (2022) m5C-Atlas: a comprehensive database for decoding and annotating the 5-methylcytosine (m5C) epitranscriptome. Nucleic Acids Res 50:D196–D203. https://doi.org/10.1093/nar/gkab1075

Marceca GP, Distefano R, Tomasello L et al (2021) MiREDiBase, a manually curated database of validated and putative editing events in microRNAs. Sci Data 8:199. https://doi.org/10.1038/s41597-021-00979-8

Matera AG, Wang Z (2014) A day in the life of the spliceosome. Nat Rev Mol Cell Biol 15:108–121. https://doi.org/10.1038/nrm3742

Mayr C (2017) Regulation by 3'-untranslated regions. Annu Rev Genet 51:171–194. https://doi.org/10.1146/annurev-genet-120116-024704

McCloskey JA, Nishimura S (1977) Modified nucleosides in transfer RNA. Acc Chem Res 10:403–410. https://doi.org/10.1021/ar50119a004

Menezes MR, Balzeau J, Hagan JP (2018) 3' RNA Uridylation in Epitranscriptomics, gene regulation, and disease. Front Mol Biosci 5:61. https://doi.org/10.3389/fmolb.2018.00061

Meyer KD (2019) DART-seq: an antibody-free method for global m6A detection. Nat Methods 16:1275–1280. https://doi.org/10.1038/s41592-019-0570-0

Meyer KD, Jaffrey SR (2017) Rethinking m6A readers, writers, and erasers. Annu Rev Cell Dev Biol 33:319–342. https://doi.org/10.1146/annurev-cellbio-100616-060758

Meyer KD, Saletore Y, Zumbo P et al (2012) Comprehensive analysis of mRNA methylation reveals enrichment in 3' UTRs and near stop codons. Cell 149:1635–1646. https://doi.org/10.1016/j.cell.2012.05.003

Meyer KD, Patil DP, Zhou J et al (2015) 5' UTR m(6)a promotes cap-independent translation. Cell 163:999–1010. https://doi.org/10.1016/j.cell.2015.10.012

Molinie B, Wang J, Lim KS et al (2016) M(6)A-LAIC-seq reveals the census and complexity of the m(6)a epitranscriptome. Nat Methods 13:692–698. https://doi.org/10.1038/nmeth.3898

Mortimer SA, Trapnell C, Aviran S et al (2012) SHAPE-Seq: high-throughput RNA structure analysis. Curr Protoc Chem Biol 4:275–297. https://doi.org/10.1002/9780470559277.ch120019

Muthukrishnan S, Moss B, Cooper JA, Maxwell ES (1978) Influence of 5'-terminal cap structure on the initiation of translation of vaccinia virus mRNA. J Biol Chem 253:1710–1715

Nachtergaele S, He C (2017) The emerging biology of RNA post-transcriptional modifications. RNA Biol 14:156–163. https://doi.org/10.1080/15476286.2016.1267096

Pan T (2013) *N*6-methyl-adenosine modification in messenger and long non-coding RNA. Trends Biochem Sci 38:204–209. https://doi.org/10.1016/j.tibs.2012.12.006

Pandey RR, Pillai RS (2019) Counting the cuts: MAZTER-Seq quantifies m6A levels using a methylation-sensitive ribonuclease. Cell 178:515–517. https://doi.org/10.1016/j.cell.2019.07.006

Parker MT, Knop K, Sherwood AV et al (2020) Nanopore direct RNA sequencing maps the complexity of Arabidopsis mRNA processing and m6A modification. elife 9:e49658. https://doi.org/10.7554/eLife.49658

Parker MT, Soanes BK, Kusakina J et al (2022) m6A modification of U6 snRNA modulates usage of two major classes of pre-mRNA 5' splice site. elife 11:e78808. https://doi.org/10.7554/eLife.78808

Passmore LA, Coller J (2022) Roles of mRNA poly(A) tails in regulation of eukaryotic gene expression. Nat Rev Mol Cell Biol 23:93–106. https://doi.org/10.1038/s41580-021-00417-y

Peer E, Rechavi G, Dominissini D (2017) Epitranscriptomics: regulation of mRNA metabolism through modifications. Curr Opin Chem Biol 41:93–98. https://doi.org/10.1016/j.cbpa.2017.10.008

Pendleton KE, Chen B, Liu K et al (2017) The U6 snRNA m6A Methyltransferase METTL16 regulates SAM Synthetase intron retention. Cell 169:824–835.e14. https://doi.org/10.1016/j.cell.2017.05.003

Picardi E, Regina TMR, Brennicke A, Quagliariello C (2007) REDIdb: the RNA editing database. Nucleic Acids Res 35:D173–D177. https://doi.org/10.1093/nar/gkl793

Picardi E, D'Erchia AM, Lo Giudice C, Pesole G (2017) REDIportal: a comprehensive database of A-to-I RNA editing events in humans. Nucleic Acids Res 45:D750–D757. https://doi.org/10.1093/nar/gkw767

Piechotta M, Naarmann-de Vries IS, Wang Q et al (2022) RNA modification mapping with JACUSA2. Genome Biol 23:115. https://doi.org/10.1186/s13059-022-02676-0

Pratanwanich PN, Yao F, Chen Y et al (2021) Identification of differential RNA modifications from nanopore direct RNA sequencing with xPore. Nat Biotechnol 39:1394–1402. https://doi.org/10.1038/s41587-021-00949-w

Qin H, Ou L, Gao J et al (2022) DENA: training an authentic neural network model using Nanopore sequencing data of Arabidopsis transcripts for detection and quantification of N6-methyladenosine on RNA. Genome Biol 23:25. https://doi.org/10.1186/s13059-021-02598-3

Ramasamy S, Sahayasheela VJ, Sharma S et al (2022) Chemical probe-based Nanopore sequencing to selectively assess the RNA modifications. ACS Chem Biol 17:2704–2709. https://doi.org/10.1021/acschembio.2c00221

Rauch S, He E, Srienc M et al (2019) Programmable RNA-guided RNA effector proteins built from human parts. Cell 178:122–134.e12. https://doi.org/10.1016/j.cell.2019.05.049

Ren X, Deng R, Zhang K et al (2021) Single-cell imaging of m6A modified RNA using m6A-specific in situ hybridization mediated proximity ligation assay (m6AISH-PLA). Angew Chem Int Ed 60:22646–22651. https://doi.org/10.1002/anie.202109118

Rieder D, Amort T, Kugler E et al (2016) meRanTK: methylated RNA analysis ToolKit. Bioinformatics 32:782–785. https://doi.org/10.1093/bioinformatics/btv647

Ries RJ, Zaccara S, Klein P et al (2019) m6A enhances the phase separation potential of mRNA. Nature 571:424–428. https://doi.org/10.1038/s41586-019-1374-1

Roundtree IA, Evans ME, Pan T, He C (2017) Dynamic RNA modifications in gene expression regulation. Cell 169:1187–1200. https://doi.org/10.1016/j.cell.2017.05.045

Rouskin S, Zubradt M, Washietl S et al (2014) Genome-wide probing of RNA structure reveals active unfolding of mRNA structures in vivo. Nature 505:701–705. https://doi.org/10.1038/nature12894

Schaefer M, Pollex T, Hanna K, Lyko F (2009) RNA cytosine methylation analysis by bisulfite sequencing. Nucleic Acids Res 37:e12. https://doi.org/10.1093/nar/gkn954

Schuberth-Wagner C, Ludwig J, Bruder AK et al (2015) A conserved histidine in the RNA sensor RIG-I controls immune tolerance to N1-2'O-methylated self RNA. Immunity 43:41–51. https://doi.org/10.1016/j.immuni.2015.06.015

Schwartz S, Motorin Y (2016) Next-generation sequencing technologies for detection of modified nucleotides in RNAs. RNA Biol 14:1124–1137. https://doi.org/10.1080/15476286.2016.1251543

Schwartz S, Bernstein DA, Mumbach MR et al (2014) Transcriptome-wide mapping reveals widespread dynamic-regulated pseudouridylation of ncRNA and mRNA. Cell 159. https://doi.org/10.1016/j.cell.2014.08.028

Shen L, Liang Z, Gu X et al (2016) N(6)-Methyladenosine RNA modification regulates shoot stem cell fate in Arabidopsis. Dev Cell 38:186–200. https://doi.org/10.1016/j.devcel.2016.06.008

Shi H, Zhang X, Weng Y-L et al (2018) m6A facilitates hippocampus-dependent learning and memory through Ythdf1. Nature 563:249–253. https://doi.org/10.1038/s41586-018-0666-1

Shi H, Wei J, He C (2019) Where, when, and how: context-dependent functions of RNA methylation writers, readers, and erasers. Mol Cell 74:640–650. https://doi.org/10.1016/j.molcel.2019.04.025

Shuman S (2015) RNA capping: progress and prospects. RNA 21:735–737. https://doi.org/10.1261/rna.049973.115

Sinnott RW, Cao Y, Dickinson BC (2024) Engineered RNA-binding proteins: studying and controlling RNA regulation. Isr J Chem 64:e202300169. https://doi.org/10.1002/ijch.202300169

Slobodin B, Han R, Calderone V et al (2017) Transcription impacts the efficiency of mRNA translation via co-transcriptional N6-adenosine methylation. Cell 169:326–337.e12. https://doi.org/10.1016/j.cell.2017.03.031

Song B, Tang Y, Chen K et al (2020a) m7GHub: deciphering the location, regulation and pathogenesis of internal mRNA N7-methylguanosine (m7G) sites in human. Bioinformatics 36:3528–3536. https://doi.org/10.1093/bioinformatics/btaa178

Song B, Tang Y, Wei Z et al (2020b) PIANO: a web server for Pseudouridine-site (Ψ) identification and functional annotation. Front Genet 11. https://doi.org/10.3389/fgene.2020.00088
Song Z, Huang D, Song B et al (2021) Attention-based multi-label neural networks for integrated prediction and interpretation of twelve widely occurring RNA modifications. Nat Commun 12: 4011. https://doi.org/10.1038/s41467-021-24313-3
Song B, Huang D, Zhang Y et al (2023) m6A-TSHub: unveiling the context-specific m6A methylation and m6A-affecting mutations in 23 human tissues. Genomics Proteomics Bioinformatics 21:678–694. https://doi.org/10.1016/j.gpb.2022.09.001
Spitale RC, Flynn RA, Zhang QC et al (2015) Structural imprints in vivo decode RNA regulatory mechanisms. Nature 519:486–490. https://doi.org/10.1038/nature14263
Squires JE, Patel HR, Nousch M et al (2012) Widespread occurrence of 5-methylcytosine in human coding and non-coding RNA. Nucleic Acids Res 40:5023–5033. https://doi.org/10.1093/nar/gks144
Starling S (2017) Fantastic beasts — cephalopod RNA recoding. Nat Rev Genet 18:329–329. https://doi.org/10.1038/nrg.2017.31
Stephenson W, Razaghi R, Busan S et al (2022) Direct detection of RNA modifications and structure using single-molecule nanopore sequencing. Cell Genom 2:100097. https://doi.org/10.1016/j.xgen.2022.100097
Stojković V, Myasnikov AG, Young ID et al (2020) Assessment of the nucleotide modifications in the high-resolution cryo-electron microscopy structure of the Escherichia coli 50S subunit. Nucleic Acids Res 48:2723–2732. https://doi.org/10.1093/nar/gkaa037
Subtelny AO, Eichhorn SW, Chen GR et al (2014) Poly(a)-tail profiling reveals an embryonic switch in translational control. Nature 508:66–71. https://doi.org/10.1038/nature13007
Sun W-J, Li J-H, Liu S et al (2016) RMBase: a resource for decoding the landscape of RNA modifications from high-throughput sequencing data. Nucleic Acids Res 44:D259–D265. https://doi.org/10.1093/nar/gkv1036
Sun L, Fazal FM, Li P et al (2019) RNA structure maps across mammalian cellular compartments. Nat Struct Mol Biol 26:322–330. https://doi.org/10.1038/s41594-019-0200-7
Tang Y, Chen K, Song B et al (2021) m6A-Atlas: a comprehensive knowledgebase for unraveling the N6-methyladenosine (m6A) epitranscriptome. Nucleic Acids Res 49:D134–D143. https://doi.org/10.1093/nar/gkaa692
Taoka M, Nobe Y, Yamaki Y et al (2018) Landscape of the complete RNA chemical modifications in the human 80S ribosome. Nucleic Acids Res 46:9289–9298. https://doi.org/10.1093/nar/gky811
Taucher M, Breuker K (2012) Characterization of modified RNA by top-down mass spectrometry. Angew Chem Int Ed 51:11289–11292. https://doi.org/10.1002/anie.201206232
Tegowski M, Flamand MN, Meyer KD (2022) scDART-seq reveals distinct m6A signatures and mRNA methylation heterogeneity in single cells. Mol Cell 82:868–878.e10. https://doi.org/10.1016/j.molcel.2021.12.038
Tegowski M, Prater AK, Holley CL, Meyer KD (2024) Single-cell m6A profiling in the mouse brain uncovers cell type-specific RNA methylomes and age-dependent differential methylation. Nat Neurosci 27:2512–2520. https://doi.org/10.1038/s41593-024-01768-3
Tomezsko PJ, Corbin VDA, Gupta P et al (2020) Determination of RNA structural diversity and its role in HIV-1 RNA splicing. Nature 582:438–442. https://doi.org/10.1038/s41586-020-2253-5
Topisirovic I, Svitkin YV, Sonenberg N, Shatkin AJ (2011) Cap and cap-binding proteins in the control of gene expression. Wiley Interdiscip Rev RNA 2:277–298. https://doi.org/10.1002/wrna.52
Tsai C-J, Sauna ZE, Kimchi-Sarfaty C et al (2008) Synonymous mutations and ribosome stalling can lead to altered folding pathways and distinct minima. J Mol Biol 383:281–291. https://doi.org/10.1016/j.jmb.2008.08.012
Tsukiyama S, Hasan MM, Deng H-W, Kurata H (2022) BERT6mA: prediction of DNA N6-methyladenine site using deep learning-based approaches. Brief Bioinform 23:bbac053. https://doi.org/10.1093/bib/bbac053

Ueda H, Dasgupta B, Yu B-Y (2023) RNA modification detection using Nanopore direct RNA sequencing and nanoDoc2. Methods Mol Biol 2632:299–319. https://doi.org/10.1007/978-1-0716-2996-3_21

van Dijk EL, Naquin D, Gorrichon K et al (2023) Genomics in the long-read sequencing era. Trends Genet 39:649–671. https://doi.org/10.1016/j.tig.2023.04.006

Vandivier LE, Anderson ZD, Gregory BD (2019) HAMR: high-throughput annotation of modified ribonucleotides. Methods Mol Biol 1870:51–67. https://doi.org/10.1007/978-1-4939-8808-2_4

Voet D, Gratzer WB, Cox RA, Doty P (1963) Absorption spectra of nucleotides, polynucleotides, and nucleic acids in the far ultraviolet. Biopolymers 1:193–208. https://doi.org/10.1002/bip.360010302

Vögele J, Duchardt-Ferner E, Kruse H et al (2023) Structural and dynamic effects of pseudouridine modifications on noncanonical interactions in RNA. RNA 29:790–807. https://doi.org/10.1261/rna.079506.122

Wahl MC, Will CL, Lührmann R (2009) The spliceosome: design principles of a dynamic RNP machine. Cell 136:701–718. https://doi.org/10.1016/j.cell.2009.02.009

Wan Y, Kertesz M, Spitale RC et al (2011) Understanding the transcriptome through RNA structure. Nat Rev Genet 12:641–655. https://doi.org/10.1038/nrg3049

Wang J, Wang L (2020) Deep analysis of RNA N6-adenosine methylation (m6A) patterns in human cells. NAR Genom Bioinform 2:lqaa007. https://doi.org/10.1093/nargab/lqaa007

Wang X, Lu Z, Gomez A et al (2014) N6-methyladenosine-dependent regulation of messenger RNA stability. Nature 505:117–120. https://doi.org/10.1038/nature12730

Wang Y, Xiao Y, Dong S et al (2020) Antibody-free enzyme-assisted chemical approach for detection of N6-methyladenosine. Nat Chem Biol 16:896–903. https://doi.org/10.1038/s41589-020-0525-x

Wang S, Sun Z, Lei Z, Zhang H-T (2022) RNA-binding proteins and cancer metastasis. Semin Cancer Biol 86:748–768. https://doi.org/10.1016/j.semcancer.2022.03.018

Wang H, Huang T, Wang D et al (2024) MSCAN: multi-scale self- and cross-attention network for RNA methylation site prediction. BMC Bioinform 25:32. https://doi.org/10.1186/s12859-024-05649-1

Wein S, Andrews B, Sachsenberg T et al (2020) A computational platform for high-throughput analysis of RNA sequences and modifications by mass spectrometry. Nat Commun 11:926. https://doi.org/10.1038/s41467-020-14665-7

Wek RC (2018) Role of eIF2α kinases in translational control and adaptation to cellular stress. Cold Spring Harb Perspect Biol 10:a032870. https://doi.org/10.1101/cshperspect.a032870

Weng H, Huang H, Wu H et al (2018) METTL14 inhibits hematopoietic stem/progenitor differentiation and promotes Leukemogenesis via mRNA m6A modification. Cell Stem Cell 22:191–205.e9. https://doi.org/10.1016/j.stem.2017.11.016

Wilkie GS, Dickson KS, Gray NK (2003) Regulation of mRNA translation by 5'- and 3'--UTR-binding factors. Trends Biochem Sci 28:182–188. https://doi.org/10.1016/S0968-0004(03)00051-3

Wilkinson E, Cui Y-H, He Y-Y (2021) Context-dependent roles of RNA modifications in stress responses and diseases. Int J Mol Sci 22:1949. https://doi.org/10.3390/ijms22041949

Workman RE, Tang AD, Tang PS et al (2019) Nanopore native RNA sequencing of a human poly(a) transcriptome. Nat Methods 16:1297–1305. https://doi.org/10.1038/s41592-019-0617-2

Wu Y, Qu R, Huang Y et al (2016) RNAex: an RNA secondary structure prediction server enhanced by high-throughput structure-probing data. Nucleic Acids Res 44:W294–W301. https://doi.org/10.1093/nar/gkw362

Wu Y, Shao W, Yan M et al (2024) Transfer learning enables identification of multiple types of RNA modifications using nanopore direct RNA sequencing. Nat Commun 15:4049. https://doi.org/10.1038/s41467-024-48437-4

Wyatt GR (1951) The purine and pyrimidine composition of deoxypentose nucleic acids. Biochem J 48:584. https://doi.org/10.1042/bj0480584

Xiao W, Adhikari S, Dahal U et al (2016) Nuclear m(6)a reader YTHDC1 regulates mRNA splicing. Mol Cell 61:507–519. https://doi.org/10.1016/j.molcel.2016.01.012

Xiao Y-L, Liu S, Ge R et al (2023) Transcriptome-wide profiling and quantification of N6-methyladenosine by enzyme-assisted adenosine deamination. Nat Biotechnol 41:993–1003. https://doi.org/10.1038/s41587-022-01587-6

Xiaodong C, Jia M, Shaowu Z et al (2016) A novel algorithm for calling mRNA m6A peaks by modeling biological variances in MeRIP-seq data. Bioinf (Oxf) 32. https://doi.org/10.1093/bioinformatics/btw281

Xiong X, Hou L, Park YP et al (2021) Genetic drivers of m6A methylation in human brain, lung, heart and muscle. Nat Genet 53:1156–1165. https://doi.org/10.1038/s41588-021-00890-3

Xu L, Seki M (2020) Recent advances in the detection of base modifications using the Nanopore sequencer. J Hum Genet 65:25–33. https://doi.org/10.1038/s10038-019-0679-0

Yan Y-B (2014) Deadenylation: enzymes, regulation, and functional implications. Wiley Interdiscip Rev RNA 5:421–443. https://doi.org/10.1002/wrna.1221

Yao H, Gao C-C, Zhang D et al (2023) scm6A-seq reveals single-cell landscapes of the dynamic m6A during oocyte maturation and early embryonic development. Nat Commun 14:315. https://doi.org/10.1038/s41467-023-35958-7

Zaccara S, Ries RJ, Jaffrey SR (2019) Reading, writing and erasing mRNA methylation. Nat Rev Mol Cell Biol 20:608–624. https://doi.org/10.1038/s41580-019-0168-5

Zaringhalam M, Papavasiliou FN (2016) Pseudouridylation meets next-generation sequencing. Methods 107:63–72. https://doi.org/10.1016/j.ymeth.2016.03.001

Zhang Y, Liu T, Meyer CA et al (2008) Model-based analysis of ChIP-Seq (MACS). Genome Biol 9. https://doi.org/10.1186/gb-2008-9-9-r137

Zhang L-S, Liu C, Ma H et al (2019a) Transcriptome-wide mapping of internal N7-Methylguanosine Methylome in mammalian mRNA. Mol Cell 74:1304–1316.e8. https://doi.org/10.1016/j.molcel.2019.03.036

Zhang Z, Chen L-Q, Zhao Y-L et al (2019b) Single-base mapping of m6A by an antibody-independent method. Sci Adv 5:eaax0250. https://doi.org/10.1126/sciadv.aax0250

Zhang L-S, Xiong Q-P, Peña Perez S et al (2021) ALKBH7-mediated demethylation regulates mitochondrial polycistronic RNA processing. Nat Cell Biol 23:684–691. https://doi.org/10.1038/s41556-021-00709-7

Zhang L-S, Ju C-W, Liu C et al (2022) m7G-quant-seq: quantitative detection of RNA internal N7-Methylguanosine. ACS Chem Biol 17:3306–3312. https://doi.org/10.1021/acschembio.2c00792

Zhang L, Chen Z, Sun G et al (2024a) Dynamic landscape of m6A modifications and related post-transcriptional events in muscle-invasive bladder cancer. J Transl Med 22:912. https://doi.org/10.1186/s12967-024-05701-x

Zhang L-S, Dai Q, He C (2024b) Base-resolution sequencing methods for whole-transcriptome quantification of mRNA modifications. Acc Chem Res 57:47–58. https://doi.org/10.1021/acs.accounts.3c00532

Zhao BS, Roundtree IA, He C (2017) Post-transcriptional gene regulation by mRNA modifications. Nat Rev Mol Cell Biol 18:31–42. https://doi.org/10.1038/nrm.2016.132

Zhou J, Wan J, Gao X et al (2015) Dynamic m(6)a mRNA methylation directs translational control of heat shock response. Nature 526:591–594. https://doi.org/10.1038/nature15377

Zhou Y, Zeng P, Li Y-H et al (2016) SRAMP: prediction of mammalian N6-methyladenosine (m6A) sites based on sequence-derived features. Nucleic Acids Res 44:e91. https://doi.org/10.1093/nar/gkw104

Zhou L, Gao G, Tang R et al (2022) m6 A-mediated regulation of crop development and stress responses. Plant Biotechnol J 20:1447–1455. https://doi.org/10.1111/pbi.13792

Zuo S, Li L, Wen X et al (2023) NSUN2-mediated m5 C RNA methylation dictates retinoblastoma progression through promoting PFAS mRNA stability and expression. Clin Transl Med 13:e1273. https://doi.org/10.1002/ctm2.1273

Chapter 11
RNA Methylation

Atsushi Shimada and Leonardo Furci

Abstract Epitranscriptomics is the field of research studying the collective epigenetic modification of RNA, affecting all aspects of the RNA fate after transcription through post-transcriptional regulation. These chemical modifications add an additional layer of information beyond the one transcribed in the genetic sequence, allowing organisms to have a higher level of fine-tuning transcriptional and biochemical responses that surpass simple gene activation or repression. Of these modifications, the deposition of a methyl group at different residues in the backbone of RNA nucleosides (RNA methylation) is the most abundant and most studied. In this chapter, we provide a state-of-the-art overview of known RNA methylation modifications in animals and plants, drawing a parallel between the two systems and highlighting the current knowledge gaps of plant models. Centering on RNA methylation in plants, this chapter will focus on the best characterized modifications to date, N6-methyladenosine and 5-methylcytidine of plant mRNAs, specifically dissecting their molecular dynamics and distribution, as well as their role in environmental adaptations, mRNA life cycle, transcription, and chromatin regulation.

Keywords RNA Methylation · Epitranscriptomics · Post-transcriptional regulation · N6-methyladenosine · 5-methylcytidine

Abbreviations

carRNA	Chromosome-associated regulatory RNAs
CDS	Coding sequence
cenRNA	Centromere-derived RNAs
lncRNA	Long noncoding RNA
M1A	1-methyladenosine

Authors Atsushi Shimada and Leonardo Furci have equally contributed to this chapter.

A. Shimada · L. Furci (✉)
Plant Epigenetics Unit, OIST, Okinawa, Japan
e-mail: Leonardo.Furci@oist.jp

L. M. Vaschetto (ed.), *Plant Transcriptomics and Epitranscriptomics*, Concepts and Strategies in Plant Sciences, https://doi.org/10.1007/978-981-95-5183-5_11

m3U	3-methyluridine
M5C	5-methylcytidine
M5U	5-methyluridine
M6A	N6-methyladenosine
m7G	7-methylguanosine
miRNA	microRNA
mRNA	messenger RNA
ncRNA	noncoding RNA
rRNA	ribosomal RNA
tRNA	transfer RNA
3′-UTRs	Three prime untranslated regions
5′-UTRs	Five prime untranslated regions

11.1 Introduction to RNA Epitranscriptomic and RNA Methylation

In biological organisms, ribonucleotides in the RNA molecule are decorated with chemical modifications. These modifications provide an additional layer of information that complements the genetic sequence and goes beyond it, determining the fate and function of different RNA molecules in the cell at the post-transcriptional level. The study of the molecular dynamics of these modifications, their biological role, and function in regulating gene and protein expression comprises the field of epitranscriptomics (Cerneckis et al. 2024). Biochemical analyses have identified more than 170 of these modifications (Boccaletto et al. 2022), with some modifications identified as early as the 1970s. With the advent of novel molecular techniques and second-generation sequencing technologies, we can now study multiple RNA modifications of the whole transcriptome of organisms/tissues/organs at single-cell resolution.

Some modifications include the incorporation of alternative nucleobases such as queuine, which forms the nucleoside queuosine in tRNAs and is involved in translation speed and efficiency (Suzuki et al. 2025), or the C5-glycoside isomer of uridine found in pseudouridine (the "fifth nucleotide"), present in different mRNA molecules and involved in gene expression and translational regulation (Cerneckis et al. 2022). On the other hand, the most abundant RNA modifications involve the addition of one or multiple chemical groups to different carbon or nitrogen positions in the backbone of RNA nucleosides (Arzumanian et al. 2022), such as methyl groups, sulfur (for instance in 2-thiouridine (Leimkühler 2024)) or acetyl group (as in N4-acetylcytidine (Jin et al. 2020). Multiple chemical modifications can occur on the same base simultaneously and subsequently in up to seven stages (identified so far) (Mathlin et al. 2020), vastly increasing the complexity of information associated with each RNA molecule.

Among these chemical modifications, methylation of different nucleosides is found to be the most abundant and, to date, the best characterized across organisms.

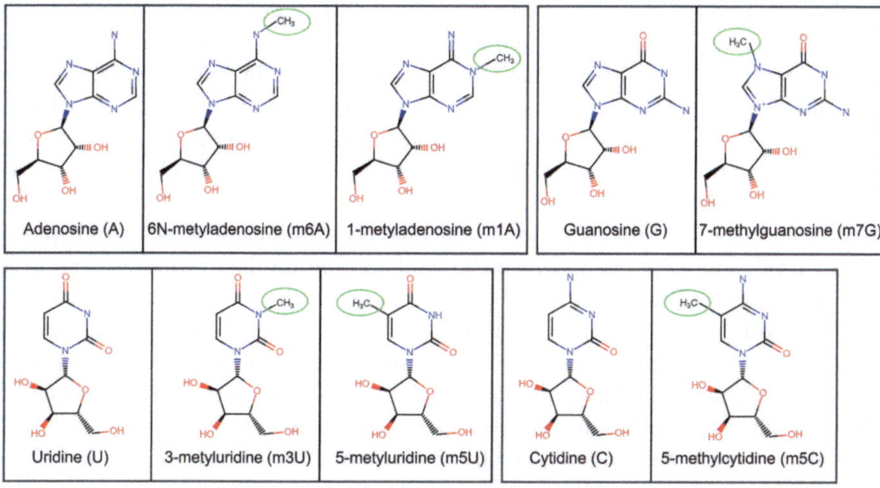

Fig. 11.1 Chemical structure of different methylation modifications of nucleosides adenosine (A), guanosine (G), uridine (U), and cytidine (C). Methyl group -CH3 is highlighted in green

Notable RNA methylation modifications include N6-methyladenosine (m6A), 1-methyladenosine (m1A), 5-methylcytidine (m5C), 7-methylguanosine (m7G), 3-methyluridine (3 mU), and 5-methyluridine (m5U) (Fig. 11.1), each found in different RNA molecules, with different distributions and different biological functions.

N6-methyladenosine (m6A) was first identified in 1974 through chemical analysis (Desrosiers et al. 1974; Salditt-Georgieff et al. 1976). However, due to the lack of robust global detection methods at the time, research focus on this modification declined. In the last two decades, the development of novel detection techniques restoked interest in m6A modification, being found to be the most abundant RNA modification of eukaryotes and bacteria, and identified in almost all RNA molecules, including mRNA, tRNA, rRNA, miRNA, snRNA, and lncRNA (Shi et al. 2019). M6A contributes to cell differentiation and cell fate determination by modulating most aspects of mRNA molecules' life cycles, from transcription to degradation, regulating splicing, poly(A) length, nucleo-cytoplasmic export rates, accumulation, translation efficiency, and half-life (Murakami and Jaffrey 2022). In plants, the modulation of mRNA lifecycle by m6A makes this modification a critical regulator of the transcriptional reprogramming occurring during physiological responses to environmental variations (phenotypic plasticity) (Miryeganeh and Saze 2020), which plants cannot avoid as sessile organisms.

1-Methyladenosine (m1A) is mostly present in tRNA and rRNA. In tRNA, this modification is conserved at position m1A58 in both prokaryotes and eukaryotes, in which it stabilizes the initiator tRNA methionine (J. Anderson et al. 1998). In human cells, m1A at position 1322 of 28S rRNA is required for the correct folding of the ribosomal unit (Waku et al. 2016). Given the structural role of m1A in fundamental biological processes, loss of this modification leads to reduced growth or cell death in animal systems (Saikia et al. 2010; Waku et al. 2016). Few studies in animal cells using both antibody-based immunoprecipitation and reverse-transcription misincorporation (m1A-MAP) methods identified m1A in mRNA molecules as well, albeit at very low frequency. This modification was found to be distributed mostly at 5′ untranslated regions (5′-UTRs) and near the stop-codon, for which it affected translation efficiency, particularly when present near the 5′ cap (Li et al. 2017b; Qi et al. 2023; Safra et al. 2017). It has to be noted, however, that one study pointed out that m1A-binding antibody can bind nonspecifically to the m7G-cap (see below), producing false positives which could account for the overrepresentation of m1A in 5′-UTRs (Grozhik et al. 2019), indicating that more studies are needed to clarify the role of this modification in mRNA life cycle regulation. In plants, only one study to date has identified m1A in mRNA from Garden petunia (*Petunia hybrida*) using antibody-based immunoprecipitation, which was mostly detected at CDS just after the translation start site (Yang et al. 2020).

Similar to m1A, 5-methylcytidine (m5C) has been documented in tRNA, rRNA, and mRNA across different life domains (Trixl and Lusser 2019), and also found in noncoding RNAs (ncRNAs) of eukaryotes (Song et al. 2022). In tRNA, m5C is found at different positions in the variable loop, anticodon loop, and T stem, and is involved in maintaining homeostasis, optimizing codon–anticodon pairing, and translation efficiency (Song et al. 2022). In eukaryotic cells, 5mC at 2 sites of 28S rRNA has been found to promote its stable folding (Bohnsack et al. 2019; Schosserer et al. 2015). In mRNA, m5C was identified 50 years ago through biochemical analysis of ribonucleotide molecules (Dubin and Taylor 1975). Like m6A, this modification is distributed across both CDS and UTRs of mRNA, but in animals is found preferentially around 5′-UTRs (Huang et al. 2019; Wei et al. 2018b), affecting stability and translation rates (Chen et al. 2019; Li et al. 2017a). In rice, m5C changes in mRNA have been linked with environmental adaptation to high temperatures (Tang et al. 2020).

3-Methyluridine (m3U) is found in 23S rRNA of archaea, 16-23S rRNA of eubacteria, and 18S-25-28S rRNA of eukaryotes. Recent studies identified a specific m3U-methylated 28S rRNA position unique to humans, frogs, and Drosophila (position U4500 in human 28S rRNA), which is located in the functional center of the large ribosomal subunit and is required for ribosomal function (Chen et al. 2024a; Taoka et al. 2018). In the model plant species Arabidopsis (*Arabidopsis thaliana*), m3U has been recently detected for the first time in 25S rRNA using a liquid chromatography-electrospray ionization-tandem mass spectrometry (LC-ESI-MS/MS) method (Tang et al. 2023), but its distribution, dynamics, and biological functions remain to date completely unknown (Sharma et al. 2023).

On the other hand, 5-methyluridine (m5U) is found mostly at position 54 in the TψC loop of bacterial and eukaryotic tRNAs, increasing the fidelity and efficiency of protein synthesis (Powell and Minczuk 2020). In recent years, a trove of machine-learning AI algorithms has been developed to identify m5U tRNA sites computationally, using both sequence-derived and physicochemical features of RNA molecules (Noor et al. 2024; Shaon et al. 2024; Xu et al. 2023). This modification has been confirmed to be present in the tRNA of both Arabidopsis and rice (*Oryza sativa*) through liquid chromatography-mass spectrometry (LC-MS/MS) (Y. Wang et al. 2018).

7-Methylguanosine (m7G) is best known for forming the 5′ cap of mature mRNA in eukaryotes. The m7G 5′ cap is deposited co-transcriptionally by the addition of guanosine nucleoside by capping enzyme (CE), which is then methylated by RNA guanosine-7 methyltransferase (RNMT), generating the m7G cap-0 form. Subsequently, the first and second nucleotides after 5′ m7G can be methylated as well, forming m7G cap-1 and cap-2 forms, respectively (Galloway and Cowling 2019). This modification is fundamental for all biological processes downstream of transcription, binding to the eukaryotic translation initiation factor elF4E for the initiation of translation and providing protection from degradation by 5′ → 3′ exonucleases (Ramanathan et al. 2016). Yet, in plant research on the mechanisms of m7G capping and its role in gene regulation in plants has been very limited. Recently, a study has found that while plants and green algae do possess m7G cap-0 structures, m7G cap-1 and cap-2 could not be identified, indicating a divergent evolution in the regulation of mRNA m7G capping (Xiao et al. 2025). Outside the 5′-cap, antibody-based immunoprecipitation techniques also identified enrichment of m7G in both 5′-UTRs, CDS, and 3′ untranslated regions (3′-UTRs) of mammalian mRNA (Malbec et al. 2019; Zhang et al. 2019). No reports of the presence of m7G outside the 5′-cap in plants' mRNAs have been published yet.

As highlighted in this introduction to RNA methylation, there is a wide gap in knowledge between animal and plant research in the field of RNA epitranscriptomics. In fact, for some of the methylated modifications listed above, only one study reports their detection in plant species. Hence, when addressing plant RNA methylation, the next sections of this chapter will focus specifically on m6A and m5C in mRNA, which to date are the best characterized in plant systems, specifically dissecting their molecular dynamics and distribution, their role in environmental adaptation, mRNA life cycle regulation, transcription, and chromatin regulation.

11.2 M6A and m5C in Plants' mRNA: Dynamics and Distribution

All organisms modulate mRNA m6A and m5C levels through a combination of RNA methyltransferase ("writers") and demethylases ("erasers"). Then, m6A- and m5C-containing mRNA molecules are recognized by specific RNA-binding proteins ("readers"), which determine the post-transcriptional fate of mRNA. In plants, this dynamic regulation of mRNA methylation allows a precise modulation of gene expression in response to developmental or environmental cues, such as different organs/tissues or (a)biotic stresses.

In the model plant *Arabidopsis thaliana*, m6A writers consist of a complex of methyltransferases, mRNA adenosine methylase A (MTA, homolog of mammalian METTL3) and methyltransferase B (MTB, homolog of mammalian METTL14), together with cofactors VIRILIZER (VIR, homolog of mammalian VIRMA), E3 ubiquitin ligase AtHAKAI (homolog of mammalian HAKAI), and FKBP12 INTERACTING PROTEIN 37 (FIP37, homolog of mammalian WTAP) (Růžička et al. 2017; Shen 2023; Wong et al. 2023). Two zinc-finger proteins, HAKAI-interacting 1 and 2 (HIZ1/2), negatively regulate HAKAI-dependent m6A deposition, suggested to be the plant equivalent of ZC3H13 in animals, m6A-METTL Associated Complex, most likely by competitive binding on mRNA (Zhang et al. 2022b). On the other hand, the removal of m6A is mediated by α-ketoglutarate-dependent dioxygenase homolog (ALKBH) eraser proteins, such as ALKBH9B and ALKBH10B (Mielecki et al. 2012; Shinde et al. 2023). Finally, members of the YTHDF family proteins called EVOLUTIONARILY CONSERVED C-TERMINAL REGION1–11 (ECT1–11) have been shown to bind m6A-modified mRNA and act as m6A readers (Nguyen and Kang 2024), with ECT2/3/4 being the best characterized (Fig. 11.1A) (Brodersen and Arribas-Hernández 2024). In Arabidopsis mRNA, m6A is deposited primarily (but not exclusively) at the DRACH/RRACH (D = A/U/C, R = A/G, H = A/U/C) consensus motif (Fig. 11.2a) (Shinde et al. 2023). Analysis of m6A distribution in Arabidopsis using both antibody-based immunoprecipitation and nanopore-based direct RNA sequencing (DRS) confirmed m6A presence in both UTRs and CDS, with a distinct density peak at 3′-UTRs near the stop codon (Luo et al. 2014; Parker et al. 2020; Wang et al. 2024; Xu et al. 2021b). Interestingly, a recent study using nanopore-DRS to achieve single-base resolution m6A calling identified the highest m6A density peak specifically at one nucleotide downstream of the CDS stop-codon (Furci et al. 2024). This is because the last two AA/GA/AG nucleotides of the stop-codon triplet can be co-opted as the first two elements of the m6A RRACH motif, leading to methylation of the next adenine when present, with preference for the TAA stop-codon trinucleotide (Furci et al. 2024).

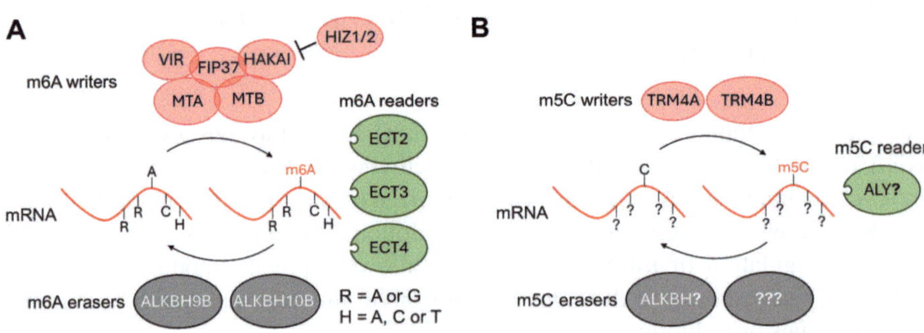

Fig. 11.2 Diagram of characterized or putative writers, erasers, and readers modulating dynamics of N6-methyladenosine (m6A, **a**) or 5-methylcytidine (m5C, **b**) in *Arabidopsis thaliana* mRNA

Regarding m5C, animal studies have shown writers include the tRNA-specific methyltransferase (TRDMT) family members, for tRNA m5C deposition, and the NOL1/NOP2/SUN domain (NSUN) family members 1–7, for mRNA 5mC deposition (H. Song et al. 2022). On the other hand, m5C erasers include Ten-Eleven Translocation 2 (TET) demethylase, which catalyzes oxidation of RNA 5mC to 5-hydroxymethyl-2′-O-methylcytidine (hm5Cm) and 5-formyl-2′-O-methylcytidine (f5Cm), with a similar mechanism to DNA 5mC demethylation (Zhang et al. 2023b). Additionally, ALKBH1 has also been shown to oxidize 5mC in human mitochondrial tRNA to hm5Cm and f5Cm (Kawarada et al. 2017), suggesting the possibility that ALKBH family proteins might target 5mC in mRNA as well. A large number of studies in animal cells characterized both ALY/REF export factor (ALYREF) as a nuclear mRNA m5C reader, involved in nucleocytoplasmic transport of modified mRNA (Y. Zhao et al. 2024), and Y-box binding protein 1 (YBX1) as a cytoplasmic mRNA m5C reader, contributing to splicing, translation, mRNA stability, and RNA sorting (Chen et al. 2019; Zhao et al. 2025). In Arabidopsis, 2 paralog genes' orthologs of mammalian NSUN2, named tRNA-specific methyltransferase 4 (TRM4) A and B, have been identified as putative m5C writers in both tRNA and mRNA, of which TRM4B showed the strongest m5C loss when removed by mutation (Cui et al. 2017; David et al. 2017). In rice, 2 orthologs of human NSUS2 have been identified through phylogenetic analysis, named OsNSUN1 and OsNSUN2, with OsNSUN2 likely being one of the major m5C methyltransferases in rice (Tang et al. 2020). For readers, the Arabidopsis mRNA export protein ALY1 (which belongs to the same protein family as ALYREF in animals) has been shown to preferentially bind to single-stranded 5mC-methylated RNA, suggesting a role of ALY proteins as potential m5C readers (Pfaff et al. 2018). On the other hand, putative m5C erasers have not been characterized at all in any plant species yet. Interestingly, plants do not possess TET enzymes, relying, for instance, on glycosylase-lyase enzymes for base excision of methylated cytidines in DNA (Zhang et al. 2022a). Removal of m5C in mRNA molecules by glycosylase-lyase enzymes is highly unlikely, given that their mechanisms of action require nucleic acid duplexes (Du et al. 2023). This raises intriguing questions on how plants achieve RNA m5C de-methylation, whether using ALKBH family proteins or a completely uncharacterized set of enzymes (Fig. 11.2b). Two studies in Arabidopsis mapped m5C at the whole-transcriptome level using different detection methods: by using bisulfite conversion of mRNA 5mC to T, David et al. showed enrichment of m5C equally distributed across CDS (David et al. 2017). By using antibody-based m5C RNA immunoprecipitation followed by deep-sequencing (m5C-RIP-seq), Cui et al. also confirmed enrichment of m5C across CDS, while also identifying an even stronger enrichment peak specifically localized around the end of CDS before the stop codon (Cui et al. 2017). On the other hand, a recent study analyzing m5C on rice mRNA using ultra-high-performance liquid chromatography-triple quadrupole mass spectrometry coupled with multiple-reaction monitoring (UHPLC-MRM-MS/MS) found overall enrichment in mRNA CDS, but with the strongest peak appearing at the beginning of CDS near the translation start site (Tang et al. 2020). It is unclear whether the differences in m5C distribution across these studies arise from the different methodologies used or denote a specificity in m5C distribution patterns between Arabidopsis and rice. Overall, all three studies confirm that m5C in plants is mostly deposited across the CDS. In animal cells, a recent study using m5C-TAC-seq, a TET-based enzymatic conversion of m5C to T enabling base-resolution m5C detection, identified the C^{m5}NGG context as a preferential m5C deposition site in mRNA (Lu et al. 2024). In plants, however, a preferential context for mRNA m5C deposition has not been identified yet (Fig. 11.2b). Identification of preferential m5C deposition contexts in plants can help resolve the apparent differences in mRNA m5C distribution identified in the literature to date, by cross-mapping m5C distribution data with the distribution of putative specific contexts across each transcript of model transcriptomes.

11.3 M6A and m5C Roles in Plants' Development and Environmental Responses

In Arabidopsis, m6A deposition is an essential epitranscriptomic modification required for correct plant development. Loss-of-function mutants of m6A deposition complex *mta*, *mtb,* and *fip37* display embryo lethality, mutants lacking cofactor *vir* show severe developmental defects and reduced size, while mutants of cofactor *hakai* have mildly reduced size (Růžička et al. 2017; Shen 2023; Wong et al. 2023). This is likely from transcriptional and post-transcriptional mis-regulation of essential genes during embryo differentiation and plant development. By complementing *mta*, *mtb,* or *fip37* mutations with embryo-specific promoters, allowing the expression of the corresponding functional protein only during embryo development, researchers could overcome the embryo lethality and obtain mutant seedlings, which still display severe developmental defects and reduced size (Bodi et al. 2012; Růžička et al. 2017; Shen 2023; Wong et al. 2023), but are sufficient for molecular and epitranscriptomic characterization. On the other hand, loss-of-function mutants for eraser *alkbh10b* mutants do not show developmental or growth defects, but possess delayed flowering due to m6A changes affecting transcript accumulation of floral transition genes (Duan et al. 2017). Similarly, reduction of mRNA m5C in the loss-of-function mutant *trm4b* did not cause major developmental defects, but the mutants displayed slightly reduced root length (Cui et al. 2017; David et al. 2017). Arabidopsis EMBRYONIC FLOWER1 (EMF1) protein associates with Polycomb Repressive Complex 1 (PRC1) for H3K27me3-dependent gene repression and has been shown to downregulate m5C writer *TRM4B* (Zhang et al. 2023a). Consistently, loss-of-function mutant *emf1–2* showed increased m5C deposition and decreased expression of photosynthesis and chloroplast genes independently of H3K27me3 deposition, resulting in defective chloroplast development and starch accumulation (Zhang et al. 2023a), though pleiotropic effects caused by *emf1* mutation might contribute to this phenotype as well.

As sessile organisms, plants cannot avoid fluctuations in environmental conditions such as temperature, water and nutrient availability, and pathogen outbreaks. To face these environmental fluctuations and overcome stresses, plants utilize global transcriptional reprogramming to achieve phenotypic plasticity and acclimation, which helps particularly in the case of recurring stresses (Charng et al. 2022; Desmedt et al. 2021). Several studies revealed a link between the transcriptional reprogramming occurring during stress and epitranscriptomic regulation: abiotic environmental cues such as salt, cold, or heat stresses, as well as pathogen attacks, can dynamically modulate m6A and m5C levels at mRNAs, whereas genetic mutants with altered m6A levels across their transcriptome show physiological changes in their responses to various (a)biotic stresses. For instance, Arabidopsis seedlings exposed to salt stress (150 mM NaCl for up to 48 h) showed transcriptome-wide loss of m6A, particularly at the 3′-UTR (Hu et al. 2021). All mutants of m6A writer components, including embryo-rescued *ABI3:MTA/mta* and partial RNA-interference (RNAi) mutant *MTB RNAi*, *vir,* and *hakai* showed salt hypersensitivity proportional to their level of m6A hypomethylation, indicating a role for m6A in salt stress tolerance (Hu et al. 2021). In another study, short-term heat stress exposure (37 °C for 3 h) also caused m6A reduction at the 3′-UTR of leaf tissues (L. Wang et al. 2022). M6A hypermethylation in floral tissue mRNAs in the M6A eraser mutant *alkbh10b-1* suppressed heat-activated genes and displayed reduced plant fertility under heat stress, indicating ALKBH10B-mediated m6A hypomethylation is involved in thermotolerance during the reproductive stage (Wang et al. 2022). In case of biotic stresses, the first study to address the role of m6A in pathogen resistance in Arabidopsis analyzed m6A-deficient mutants *ABI3: MTA/mta*, weak alleles *fip37–4* and *vir-2*, and an overexpressing line *35S: ALKBH10B* infected with bacterial pathogens *Pseudomonas syringae* (pv *syringae* and pv *maculicola*) and fungal pathogen *Botrytis cinerea*. All m6A-deficient mutants showed resistance against these pathogens, correlating with increased accumulation of defence hormones salicylic acid, jasmonic acid, and camalexin after pathogen infection (Prall et al. 2023). Correlation between global m6A distribution and transcriptional changes after treatment with *P. syringae* or its elicitor flg22 in *ABI3:MTA/mta* and *35S:ALKBH10B* mutants indicated that m6A modulation could affect abundance and internal cleavage of specific mRNA transcripts (Prall et al. 2023). A subsequent study utilized nanopore-DRS to obtain transcript-specific dynamics of m6A at single-nucleotide resolution in Arabidopsis during infection from the oomycete *Hyaloperonospora arabidopsidis* (Furci et al. 2024). In wild-type seedlings, *H. arabidopsidis* infection caused a significant reduction in global m6A ratios, correlating with the activation of m6A-modified defence gene transcripts (Furci et al. 2024). Defect of m6A deposition in the m6A writer mutant *hakai* phenocopied m6A reduction from *H. arabidopsidis* infection at around 70% m6A sites, resulting in constitutive overexpression of basal defence genes and enhanced resistance (Furci et al. 2024). Taken together, these studies demonstrate the important role of m6A dynamics in modulating Arabidopsis immune responses against pathogens. In rice, exposure to high temperature (36 °C for 48 h) resulted in both increased and decreased m5C accumulation at around 100 mRNA transcripts (Tang et al. 2020).

In addition to regulating transcript levels, mRNA m6A and m5C have been shown to enhance translation efficiency under stress conditions in Arabidopsis and rice, consistent with the findings in mammals (Wang et al. 2015). Wang et al. discovered that m6A-modified Arabidopsis mRNAs are translated efficiently under cold stress (4 °C for 24 h), even though their transcript levels are not affected (Wang et al. 2023b). Another group reported that ECT2/3/4 and FIP37 promote translation efficiency of m6A-marked mRNAs in response to elf18, a bacterial epitope that activates immune signaling (Chen et al. 2024b). Given that ECT proteins associate with PAB proteins (Due Tankmar et al. 2023), the enhanced translation may be mediated through the protection of the poly(A) tail. It is also plausible that ECT proteins help recruit translation initiation factors, as observed in mammalian systems (Wang et al. 2015). In the rice *osnsun2* mutant, mRNA m5C hypomethylation caused a negative effect on protein translation efficiency and was linked with light-dependent lesion-mimic and heat-stress hypersensitivity phenotypes (Tang et al. 2020). Further investigations are needed to elucidate the detailed mechanism by which m6A and m5C modifications modulate polysome association and translation efficiency in response to environmental stimuli.

11.4 Positive and Negative Effects of m6A on mRNA Stability

Studies across organisms have shown that m6A modification negatively affects mRNA stability. In human and budding yeast, a m6A reader protein YT521-B homology (YTH) domain-containing family protein 2 (YTHDF2) recruits Carbon Catabolite Repression 4 (CCR4), a catalytic subunit of the CCR4-NOT complex with poly(A) tail trimming activity, to promote mRNA degradation (H. Du et al. 2016; Park et al. 2019). Similar to these observations, m6A readers play a pivotal role in mRNA stability in Arabidopsis. Arabidopsis possesses YTHDF family proteins called EVOLUTIONARILY CONSERVED C-TERMINAL REGION1–11 (ECT1–11) (Ok et al. 2005), and several ECT proteins have been shown to bind to m6A and (Cai et al. 2024; Lee et al. 2023; Scutenaire et al. 2018; Song et al. 2023) destabilize their target mRNAs (Cai et al. 2024; Lee et al. 2023; Scutenaire et al. 2018; Peizhe Song et al. 2023). For example, ECT1 forms cytoplasmic condensates and promotes the decay of m6A-modified mRNA from salicylic acid-responsive genes under salicylic acid treatment, thereby preventing excessive immune activation (Lee et al. 2023). Wang et al. found ECT9 colocalizes with ECT1, and ECT9 and ECT1 together play a synergistic inhibitory role in immune responses (Wang et al. 2023a). Another study demonstrated that ECT8 interacts with DECAPPING 5 (DCP5), a component of the RNA decapping complex, in cytoplasmic processing bodies, and promotes the degradation of negative regulators involved in abiotic stress responses (Cai et al. 2024). Negative impacts of m6A on mRNA stability were also observed in m6A writer and eraser mutants. Shen et al. demonstrated that m6A modification in SHOOT MERISTEMLESS (*STM*) mRNA promotes its degradation, and the transcript levels are increased in *mta* and *fip37* mutants (Shen et al. 2016). The m6A eraser mutant *alkbh10b* exhibited reduced expression of m6A-modified transcripts under drought stress (Han et al. 2023). These findings suggest that m6A modification and its reader proteins in Arabidopsis generally promote mRNA decay under various environmental conditions.

While m6A has been shown to negatively affect mRNA stability, the opposite cases where m6A promotes mRNA stability have also been reported in Arabidopsis. Anderson et al. demonstrated that the presence of m6A in mRNA inhibits cleavage by ribonucleases, and m6A-modified transcript levels were downregulated in m6A-deficient mutant *mta* (S. J. Anderson et al. 2018). Consistent with their observations, several groups reported that m6A reader ECT proteins promote mRNA stability. Mutations in ECT2 and ECT3 lead to reduced transcript levels of their targets (Arribas-Hernández et al. 2021). Wei et al. also found that ECT2 promotes the stability of mRNAs involved in trichome development (Wei et al. 2018a). Based on these findings, one question arises: how do m6A and ECT proteins contribute to transcript stabilization? It may involve interactions with RNA-binding proteins that protect the transcript. In fact, ECT2, ECT3, and ECT4 interact with Poly(A) Binding (PAB) proteins and stabilize their target mRNAs (Peizhe Song et al. 2023). The interactions between ECT proteins and PAB proteins were confirmed in a recent study, which also reported that the overexpression of PAB proteins suppressed the leaf developmental defects of the *ect2/3/4* triple mutant, suggesting functional crosstalk between m6A-mediated regulation and poly(A) tail control (Due Tankmar et al. 2023).

On the other hand, these antagonistic biological functions of different ECT proteins also raise the question of how different m6A readers can interact with specific mRNA molecules to correctly determine their stability and fate. Preliminary studies on ECT2 showed this protein also crosslinks to U- and U/C-rich motifs such as URUAY, enriched around m6A RRACH sites in 3′-UTRs of m6-methylated mRNA (Arribas-Hernández et al. 2021; Wei et al. 2018a). This finding suggests that the sequence specificity of mRNA molecules could drive preferential binding of certain m6A readers. However, these motifs were not bound by the YTH domain of ECT2 but by its C-terminal intrinsically disordered region, prompting speculation that these motifs might not be involved in target recognition but instead act as a site of competitive binding between ECT2 and other nonspecific RNA-binding proteins (Arribas-Hernández et al. 2021). Another hypothesis is that m6A readers might be capable of binding to and distinguishing different epitranscriptomic modifications co-occurring simultaneously on the same mRNA molecules. Indeed, a recent study utilizing nanopore-DRS and a custom neural network model confirmed co-occurrence of both m6A and m5C on individual mRNA molecules as close as 5-15 nt from each other (Acera Mateos et al. 2024). While no studies in either animal or plant systems have yet confirmed the ability of m6A/m5C readers to also recognize and/or bind to different RNA methylation marks, one study reported co-localization of the m5C reader protein ALYREF with m6A-modified mRNAs in differentiated neuronal cells (Orji et al. 2025). If proven true, this hypothesis would imply the existence of an epigenetic "RNA code" complementing histone code and DNA methylation, in which combinations of different epitranscriptomic marks on mRNA molecules encode information that determines their fate post-transcriptionally.

Beyond the canonical YTHDF family, recent studies have identified additional factors that assist the m6A reader function and expand the regulatory scope of m6A. Acetylation Lowers Binding Affinity (ALBA) proteins are highly conserved DNA/RNA binding proteins (Verma et al. 2014), and were recently shown to promote mRNA stability by facilitating the binding of ECT2 to m6A-modified mRNAs (Reichel et al. 2024). Supporting this, ALBA proteins have also been implicated in mRNA stabilization under heat stress (Tong et al. 2022). These findings suggest that plants employ additional co-factors to fine-tune the effects of m6A on mRNA metabolism depending on environmental conditions. Thus, m6A exerts multifaceted effects on mRNA fate through a combinatorial network of reader and accessory proteins.

11.5 Roles of m6A in Transcription and Chromatin Regulation

It is of great importance to investigate whether m6A is implicated in nuclear processes such as transcription. Given that MTA and FIP37 are localized in the nucleus and that high levels of m6A peaks are predominantly detected in 3′-UTR regions (S. J. Anderson et al. 2018; Zhong et al. 2008), m6A modification may play a role in controlling transcription termination processes. In support of this, the protein cleavage and polyadenylation specificity factor 30-L (CPSF30-L), involved in the mRNA cleavage step during transcription termination, have been shown to recognize m6A modification through its YTHDC-type YTH domain in vitro (P. Song et al. 2021). In addition, global shifts in polyadenylation sites were observed in the *vir1* mutant, a component of the m6A writer complex (Parker et al. 2020). These findings possibly suggest that CPSF30-L and other m6A reader proteins generally regulate polyadenylation events through m6A modifications. Interestingly, Arabidopsis expresses another CPSF30 isoform called CPSF30-S, which lacks the YTH domain (P. Song et al. 2021), and it is presumed to bypass m6A signals. The presence of such isoforms would enable more flexible control of polyadenylation through m6A-dependent or -independent mechanisms. Considering that polyadenylation sites are generally altered in various environmental conditions (de Lorenzo et al. 2017; Hu et al. 2021; Kubaczka et al. 2024), it would be worth investigating how m6A and its regulatory proteins regulate environment-specific alternative polyadenylation events.

A study focusing on mouse embryonic stem cells demonstrated that m6A modification occurs in chromosome-associated regulatory RNAs (carRNA), such as enhancer RNAs, promoter-associated RNAs, and repeat RNAs (Liu et al. 2020). These m6A modifications negatively regulate the stability of carRNAs, and the half-life of carRNAs increases when m6A is depleted, thereby enhancing transcription at the carRNA-associated genes (Liu et al. 2020). Although there is limited evidence regarding whether m6A on chromatin-associated ncRNAs plays a role in chromatin regulation in Arabidopsis, Xu et al. recently found that m6A occurs on *COOLAIR*, an ncRNA that controls the epigenetic state of the flowering regulator gene *FLC* (Xu et al. 2021a). Importantly, m6A on *COOLAIR* is required for chromatin silencing of *FLC* (Xu et al. 2021a). This suggests that m6A modifications on other chromatin-associated ncRNAs in Arabidopsis may also be critical for their regulatory functions.

Beyond *COOLAIR*, Arabidopsis harbors several chromatin-associated functional ncRNAs, such as *SVALKA*, *APOLO*, *FLAIL,* and *MUSHER* (Ariel et al. 2020; Y. Jin et al. 2023; Kindgren et al. 2018; Sacharowski et al. 2025). Arabidopsis also expresses a class of chromatin-associated ncRNAs transcribed by POL V, a plant-specific RNA polymerase involved in RNA-directed DNA methylation, which regulates transcription of disease resistance genes (Yuan et al. 2024). Another such example is centromeric repeat-derived RNAs (cenRNAs) that contribute to centromere epigenome maintenance and chromosome segregation (Shimada et al. 2024). Interestingly, m6A modification of cenRNA in cancer cells ensures centromere integrity and promotes tumor growth (Kang et al. 2024), raising the possibility of a functional link between m6A and chromatin function at repetitive regions in Arabidopsis. Although it remains unclear whether these functional ncRNA harbors m6A modifications, a recent study detected m6A modifications in a broad range of Arabidopsis ncRNAs (Liang et al. 2024). Future studies focusing on how m6A influences the stability and function of ncRNAs may reveal novel mechanisms underlying chromatin regulation and transcriptional control in response to developmental and environmental cues.

11.6 Conclusions and Perspectives

Methylation of RNA nucleosides is the most abundant epitranscriptomic modification in nature, providing a structural role necessary for life, arguably more so than DNA methylation. These modifications are required for the correct folding of tRNA and rRNA molecules, which in turn are required for functional protein synthesis. For gene expression, RNA methylation has been shown to regulate both transcription and post-transcriptional processing, affecting mRNA molecule abundance and accumulation in the cytoplasm, splicing, degradation, translation rates, as well as chromatin state regulation of transcribed genes. In plants, mRNA methylation is dynamically modulated during different developmental stages or after exposure to various stress conditions, providing an additional mechanism for fine-tuning gene expression to achieve phenotypic plasticity against unavoidable environmental fluctuations.

Yet, there are still large gaps in knowledge about RNA methylation, particularly in plants, as exemplified by reports on the apparent antagonistic function of N6-adenine methylation in both stabilization and degradation of mRNA. At present, in plants, two fundamental questions are still mostly unanswered: first, what is the occurrence and distribution of all RNA methylation modifications other than N6-methyladenosine or 5-methylcytidine in all RNA molecules? Second, what are the mechanisms that decipher the information carried by each modification to produce the intended and specific biological function across different RNA molecules?

To address the first question, future immunoprecipitation-based sequencing techniques or nanopore-based direct RNA sequencing would allow the scientific community to produce rigorous atlases of the distribution of these modifications across different tissues on all different RNA molecules. First on the model plant Arabidopsis, and then on crops and commercially relevant species. Ideally, these studies would also include perturbations such as exposure to stress and/or transcription inhibitors. Coupled with transcriptomic/nascent RNA sequencing, these studies could precisely correlate the distribution of different RNA methylation modifications with physiological responses at the mRNA level.

One approach to dissect the specificity of biological function information encoded in single or co-occurring RNA modifications ("RNA code") could be through systematic identification and characterization of RNA methylation reader proteins. This can be achieved using conventional proteomic approaches or novel in-silico large-scale screenings, taking advantage of recently developed AI models, which can predict protein function and putative protein–nucleotide ligand interactions (with or without modifications), such as Alphafold 3, RoseTTAFold All-Atom, or DeepPBS.

Altogether, these strategies will generate the most accurate insights into all RNA methylation modifications distribution and functions, advancing not only fundamental research but also applied practices. Given the role of RNA methylation in modulating transcriptional responses to environmental cues in plants, key regulatory genes of (a)biotic stress responses could be precisely targeted by either RNA methylation editing or by genetic sequence editing, exchanging the methylated nucleobase for a different one (in case of protein coding genes, through synonymous mutations not to disrupt the protein sequence), to artificially modulate their expression and activity. These approaches could then be used in the near future to generate environmentally resistant crops, higher yields or longer shelf-life.

Competing Interests The authors declare no competing interests of financial or academic nature.

Ethics Approval N. A.

References

Acera Mateos P et al (2024) Prediction of m6A and m5C at single-molecule resolution reveals a transcriptome-wide co-occurrence of RNA modifications. Nat Commun 15(1):3899

Anderson J et al (1998) The essential Gcd10p-Gcd14p nuclear complex is required for 1-methyladenosine modification and maturation of initiator methionyl-tRNA. Genes Dev 12(23):3650–3662

Anderson SJ et al (2018) N(6)-Methyladenosine inhibits local Ribonucleolytic cleavage to stabilize mRNAs in Arabidopsis. Cell Rep 25(5):1146–1157

Ariel F et al (2020) R-loop mediated trans action of the APOLO long noncoding RNA. Mol Cell 77(5):1055–1065

Arribas-Hernández L et al (2021) The YTHDF proteins ECT2 and ECT3 bind largely overlapping target sets and influence target mRNA abundance, not alternative polyadenylation. elife 10: e72377

Arzumanian VA et al (2022) Epitranscriptome: review of top 25 most-studied RNA modifications. Int J Mol Sci 23(22):13851

Boccaletto P et al (2022) MODOMICS: a database of RNA modification pathways. 2021 update. Nucleic Acids Res 50(D1):D231–D235

Bodi Z et al (2012) Adenosine methylation in Arabidopsis mRNA is associated with the 3′ end and reduced levels cause developmental defects. Front Plant Sci 3:48

Bohnsack KE et al (2019) Eukaryotic 5-methylcytosine (m5C) RNA Methyltransferases: mechanisms, cellular functions, and links to disease. Genes 10(2):102

Brodersen P, Arribas-Hernández L (2024) The m6A-YTH regulatory system in plants: a status. Curr Opin Plant Biol 82:102650

Cai Z et al (2024) The m6A reader ECT8 is an abiotic stress sensor that accelerates mRNA decay in Arabidopsis. Plant Cell 36(8):2908–2926

Cerneckis J et al (2022) Decoding pseudouridine: an emerging target for therapeutic development. Trends Pharmacol Sci 43(6):522–535

Cerneckis J et al (2024) The rise of epitranscriptomics: recent developments and future directions. Trends Pharmacol Sci 45(1):24–38

Charng Y-Y et al (2022) Maintenance of abiotic stress memory in plants: lessons learned from heat acclimation. Plant Cell 35(1):187–200

Chen X et al (2019) 5-methylcytosine promotes pathogenesis of bladder cancer through stabilizing mRNAs. Nat Cell Biol 21(8):978–990

Chen J et al (2024a) The Ptch/SPOUT1 methyltransferase deposits an m3U modification on 28S rRNA for normal ribosomal function in flies and humans. Sci Adv. 10(50):eadr1743

Chen T et al (2024b) m6A modification plays an integral role in mRNA stability and translation during pattern-triggered immunity. Proc Natl Acad Sci 121(33):e2411100121

Cui X et al (2017) 5-Methylcytosine RNA methylation in Arabidopsis Thaliana. Mol Plant 10(11): 1387–1399

David R et al (2017) Transcriptome-wide mapping of RNA 5-Methylcytosine in Arabidopsis mRNAs and noncoding RNAs. Plant Cell 29(3):445–460

de Lorenzo L et al (2017) Noncanonical alternative polyadenylation contributes to gene regulation in response to hypoxia. Plant Cell 29(6):1262–1277

Desmedt W et al (2021) Chapter 5 – plant defense priming in the field: a review. In: Maienfisch P, Mangelinckx S (eds) Recent highlights in the discovery and optimization of crop protection products. Academic Press, pp 87–124

Desrosiers R et al (1974) Identification of methylated nucleosides in messenger RNA from Novikoff hepatoma cells. Proc Natl Acad Sci USA 71(10):3971–3975

Du H et al (2016) YTHDF2 destabilizes m6A-containing RNA through direct recruitment of the CCR4–NOT deadenylase complex. Nat Commun 7(1):12626

Du X et al (2023) Molecular basis of the plant ROS1-mediated active DNA demethylation. Nat Plants 9(2):271–279

Duan HC et al (2017) ALKBH10B is an RNA N(6)-Methyladenosine demethylase affecting Arabidopsis floral transition. Plant Cell 29(12):2995–3011
Dubin DT, Taylor RH (1975) The methylation state of poly A-containing messenger RNA from cultured hamster cells. Nucleic Acids Res 2(10):1653–1668
Due Tankmar M et al (2023) A YTHDF-PABP interaction is required for m6A-mediated organogenesis in plants. EMBO Rep 24(12):e57741
Furci L et al (2024) RNA N6-adenine methylation dynamics impact Hyaloperonospora arabidopsidis resistance in Arabidopsis. Plant Physiol 196(2):745–753
Galloway A, Cowling VH (2019) mRNA cap regulation in mammalian cell function and fate. Biochim Biophys Acta Gene Regul Mech 1862(3):270–279
Grozhik AV et al (2019) Antibody cross-reactivity accounts for widespread appearance of m1A in 5′UTRs. Nat Commun 10(1):5126
Han R et al (2023) ALKBH10B-mediated m6A demethylation is crucial for drought tolerance by affecting mRNA stability in Arabidopsis. Environ Exp Bot 209:105306
Hu J et al (2021) N(6) -Methyladenosine mRNA methylation is important for salt stress tolerance in Arabidopsis. Plant J 106(6):1759–1775
Huang T et al (2019) Genome-wide identification of mRNA 5-methylcytosine in mammals. Nat Struct Mol Biol 26(5):380–388
Jin G et al (2020) The processing, gene regulation, biological functions, and clinical relevance of N4-Acetylcytidine on RNA: a systematic review. Mol Ther Nucleic Acids 20:13–24
Jin Y et al (2023) LncRNA FLAIL affects alternative splicing and represses flowering in Arabidopsis. EMBO J 42(11):e110921
Kawarada L et al (2017) ALKBH1 is an RNA dioxygenase responsible for cytoplasmic and mitochondrial tRNA modifications. Nucleic Acids Res 45(12):7401–7415
Kindgren P et al (2018) Transcriptional read-through of the long non-coding RNA SVALKA governs plant cold acclimation. Nat Commun 9(1):4561
Kubaczka MG et al (2024) Light regulates widespread plant alternative polyadenylation through the chloroplast. Proc Natl Acad Sci 121(34):e2405632121
Lee KP et al (2023) The m6A reader ECT1 drives mRNA sequestration to dampen salicylic acid–dependent stress responses in Arabidopsis. Plant Cell 36(3):746–763
Leimkühler S (2024) 2-Thiouridine formation in Escherichia coli: a critical review. J Bacteriol 207: e00420–e00424
Li Q et al (2017a) NSUN2-mediated m5C methylation and METTL3/METTL14-mediated m6A methylation cooperatively enhance p21 translation. J Cell Biochem 118(9):2587–2598
Li X et al (2017b) Base-resolution mapping reveals distinct m^1A Methylome in nuclear- and mitochondrial-encoded transcripts. Mol Cell 68(5):993–1005
Liang Q et al (2024) Nanopore direct RNA sequencing reveals N6-methyladenosine and polyadenylation landscapes on long non-coding RNAs in Arabidopsis thaliana. BMC Plant Biol 24(1):1126
Liu J et al (2020) N6-methyladenosine of chromosome-associated regulatory RNA regulates chromatin state and transcription. Science 367(6477):580–586
Lu L et al (2024) Base-resolution m5C profiling across the mammalian transcriptome by bisulfite-free enzyme-assisted chemical labeling approach. Mol Cell 84(15):2984–3000
Luo G-Z et al (2014) Unique features of the m6A methylome in Arabidopsis thaliana. Nat Commun 5(1):5630
Malbec L et al (2019) Dynamic methylome of internal mRNA N(7)-methylguanosine and its regulatory role in translation. Cell Res 29(11):927–941
Mathlin J et al (2020) A census and categorization method of Epitranscriptomic Marks. Int J Mol Sci 21(13):4684
Mielecki, D., et al. (2012). Novel AlkB dioxygenases--alternative models for in silico and in vivo studies. PLoS One 7(1):e30588
Miryeganeh M, Saze H (2020) Epigenetic inheritance and plant evolution. Popul Ecol 62(1):17–27

Murakami S, Jaffrey SR (2022) Hidden codes in mRNA: control of gene expression by m6A. Mol Cell 82(12):2236–2251

Nguyen TKH, Kang H (2024) Reading m6A marks in mRNA: a potent mechanism of gene regulation in plants. J Integr Plant Biol 66(12):2586–2599

Noor S et al (2024) Deep-m5U: a deep learning-based approach for RNA 5-methyluridine modification prediction using optimized feature integration. BMC Bioinform 25(1):360

Ok SH et al (2005) Novel CIPK1-associated proteins in Arabidopsis contain an evolutionarily conserved C-terminal region that mediates nuclear localization. Plant Physiol 139(1):138–150

Orji OC et al (2025) Global co-regulatory cross talk between m6A and m5C RNA methylation systems coordinate cellular responses and brain disease pathways. Mol Neurobiol 62(4): 5006–5021

Park OH et al (2019) Endoribonucleolytic cleavage of m(6)A-containing RNAs by RNase P/MRP complex. Mol Cell 74(3):494–507

Parker MT et al (2020) Nanopore direct RNA sequencing maps the complexity of Arabidopsis mRNA processing and m6A modification. elife 9:e49658

Pfaff C et al (2018) ALY RNA-binding proteins are required for Nucleocytosolic mRNA transport and modulate plant growth and development. Plant Physiol 177(1):226–240

Powell CA, Minczuk M (2020) TRMT2B is responsible for both tRNA and rRNA m(5)U-methylation in human mitochondria. RNA Biol 17(4):451–462

Prall W et al (2023) Pathogen-induced m6A dynamics affect plant immunity. Plant Cell 35(11): 4155–4172

Qi Z et al (2023) N1-Methyladenosine modification of mRNA regulates neuronal gene expression and oxygen glucose deprivation/reoxygenation induction. Cell Death Discov 9(1):159

Ramanathan A et al (2016) mRNA capping: biological functions and applications. Nucleic Acids Res 44(16):7511–7526

Reichel M et al (2024) ALBA proteins facilitate cytoplasmic YTHDF-mediated reading of m6A in *Arabidopsis*. EMBO J 43(24):6626–6655

Růžička K et al (2017) Identification of factors required for m(6) a mRNA methylation in Arabidopsis reveals a role for the conserved E3 ubiquitin ligase HAKAI. New Phytol 215(1): 157–172

Sacharowski SP et al (2025) Chromatin retained *MUSHER* lncRNA integrates ABA and DOG1 signalling pathways to enhance Arabidopsis seeds dormancy. Nat Commun 16(1):7545

Safra M et al (2017) The m1A landscape on cytosolic and mitochondrial mRNA at single-base resolution. Nature 551(7679):251–255

Saikia M et al (2010) Genome-wide analysis of N1-methyl-adenosine modification in human tRNAs. RNA 16(7):1317–1327

Salditt-Georgieff M et al (1976) Methyl labeling of HeLa cell hnRNA: a comparison with mRNA. Cell 7(2):227–237

Schosserer M et al (2015) Methylation of ribosomal RNA by NSUN5 is a conserved mechanism modulating organismal lifespan. Nat Commun 6(1):6158

Scutenaire J et al (2018) The YTH domain protein ECT2 is an m6A reader required for Normal Trichome branching in Arabidopsis. Plant Cell 30(5):986–1005

Shaon MSH et al (2024) A robust deep learning approach for identification of RNA 5-methyluridine sites. Sci Rep 14(1):25688

Sharma B et al (2023) The diversity and functions of plant RNA modifications: what we know and where we go from here. Annu Rev Plant Biol 74(1):53–85

Shen L (2023) Functional interdependence of N6-methyladenosine methyltransferase complex subunits in Arabidopsis. Plant Cell 35(6):1901–1916

Shen L et al (2016) N(6)-Methyladenosine RNA modification regulates Shoot stem cell fate in Arabidopsis. Dev Cell 38(2):186–200

Shi H et al (2019) Where, when, and how: context-dependent functions of RNA methylation writers, readers, and erasers. Mol Cell 74(4):640–650

Shimada A, Cahn J, Ernst E et al (2024) Retrotransposon addiction promotes centromere function via epigenetically activated small RNAs. Nat Plants 10:1304–1316. https://doi.org/10.1038/s41477-024-01773-1

Shinde H et al (2023) RNA methylation in plants: an overview. Front Plant Sci 14:1132959

Song P et al (2021) Arabidopsis N(6)-methyladenosine reader CPSF30-L recognizes FUE signals to control polyadenylation site choice in liquid-like nuclear bodies. Mol Plant 14(4):571–587

Song H et al (2022) Biological roles of RNA m5C modification and its implications in cancer immunotherapy. Biomark Res 10(1):15

Song P et al (2023) m6A readers ECT2/ECT3/ECT4 enhance mRNA stability through direct recruitment of the poly(a) binding proteins in Arabidopsis. Genome Biol 24(1):103

Suzuki T et al (2025) Biogenesis and roles of tRNA queuosine modification and its glycosylated derivatives in human health and diseases. Cell Chem Biology 32(2):227–238

Tang Y et al (2020) OsNSUN2-mediated 5-Methylcytosine mRNA modification enhances Rice adaptation to high temperature. Dev Cell 53(3):272–286

Tang X-M et al (2023) Mass spectrometry profiling analysis enables the identification of new modifications in ribosomal RNA. Chin Chem Lett 34(3):107531

Taoka M et al (2018) Landscape of the complete RNA chemical modifications in the human 80S ribosome. Nucleic Acids Res 46(18):9289–9298

Tong J et al (2022) ALBA proteins confer thermotolerance through stabilizing HSF messenger RNAs in cytoplasmic granules. Nat Plants 8(7):778–791

Trixl L, Lusser A (2019) The dynamic RNA modification 5-methylcytosine and its emerging role as an epitranscriptomic mark. Wiley Interdiscip Rev RNA 10(1):e1510

Verma JK et al (2014) OsAlba1, a dehydration-responsive nuclear protein of rice (Oryza sativa L. ssp. indica), participates in stress adaptation. Phytochemistry 100:16–25

Waku T et al (2016) NML-mediated rRNA base methylation links ribosomal subunit formation to cell proliferation in a p53-dependent manner. J Cell Sci 129(12):2382–2393

Wang X et al (2015) N(6)-methyladenosine modulates messenger RNA translation efficiency. Cell 161(6):1388–1399

Wang Y et al (2018) Correction to: identification of tRNA nucleoside modification genes critical for stress response and development in rice and Arabidopsis. BMC Plant Biol 18:37

Wang L et al (2022) m6A RNA methylation impairs gene expression variability and reproductive thermotolerance in Arabidopsis. Genome Biol 23(1):244

Wang H et al (2023a) ECT9 condensates with ECT1 and regulates plant immunity. Front Plant Sci 14:1140840

Wang S et al (2023b) m6A mRNA modification promotes chilling tolerance and modulates gene translation efficiency in Arabidopsis. Plant Physiol 192(2):1466–1482

Wang G et al (2024) Quantitative profiling of m6A at single base resolution across the life cycle of rice and Arabidopsis. Nat Commun 15(1):4881

Wei LH et al (2018a) The m(6)a reader ECT2 controls Trichome morphology by affecting mRNA stability in Arabidopsis. Plant Cell 30(5):968–985

Wei Z et al (2018b) Topological characterization of human and mouse m(5)C Epitranscriptome revealed by Bisulfite sequencing. Int J Genomics 2018(1):1351964

Wong CE et al (2023) Shaping the landscape of N6-methyladenosine RNA methylation in Arabidopsis. Plant Physiol 191(3):2045–2063

Xiao C et al (2025) mRNAs of plants and green algae lack the mG cap-1 structure. New Phytol 246(2):396

Xu C et al (2021a) R-loop resolution promotes co-transcriptional chromatin silencing. Nat Commun 12(1):1790

Xu Z et al (2021b) Transcriptome-wide analysis of RNA m6A methylation and gene expression changes among two Arabidopsis ecotypes and their reciprocal hybrids. Front Plant Sci 12: 685189

Xu Z et al (2023) m5U-GEPred: prediction of RNA 5-methyluridine sites based on sequence-derived and graph embedding features. Front Microbiol 14:1277099

Yang W et al (2020) The N(1)-Methyladenosine Methylome of petunia mRNA. Plant Physiol 183(4):1710–1724

Yuan Y et al (2024) An RdDM-independent function of pol V transcripts in gene regulation and plant defence. Nat Plants 10(10):1562–1575

Zhang LS et al (2019) Transcriptome-wide mapping of internal N(7)-Methylguanosine Methylome in mammalian mRNA. Mol Cell 74(6):1304–3016

Zhang H et al (2022a) Active DNA demethylation in plants: 20 years of discovery and beyond. J Integr Plant Biol 64(12):2217–2239

Zhang M et al (2022b) Two zinc finger proteins with functions in m6A writing interact with HAKAI. Nat Commun 13(1):1127

Zhang D et al (2023a) RNA 5-Methylcytosine modification regulates vegetative development associated with H3K27 Trimethylation in Arabidopsis. Adv Sci 10(1):2204885

Zhang X et al (2023b) TET (ten-eleven translocation) family proteins: structure, biological functions and applications. Signal Transduct Target Ther 8(1):297

Zhao Y et al (2024) ALYREF (Aly/REF export factor): a potential biomarker for predicting cancer occurrence and therapeutic efficacy. Life Sci 338:122372

Zhao T et al (2025) Biological functions of 5-methylcytosine RNA-binding proteins and their potential mechanisms in human cancers. Front Oncol 15:1534948

Zhong S et al (2008) MTA is an Arabidopsis messenger RNA adenosine methylase and interacts with a homolog of a sex-specific splicing factor. Plant Cell 20(5):1278–1288

Chapter 12
Epitranscriptomic and Epigenetic Regulation of Plant–Microbe Interactions: Methylation Marks as a Case Study

Jianying Yue, Yan Lu, Régis L. Corrêa, Haijuan Wang, Fabio Pasin, and Mingmin Zhao

Abstract Recent advances in the study of nucleic acid modifications, such as N^6-methyladenosine (m^{6}A) in RNA and 5-methylcytosine (5mC) in DNA, have significantly enhanced our understanding of the complex mechanisms regulating plant–microbe interactions. The major epitranscriptomic mark m^{6}A has been implicated in immune responses against diverse pathogens, including viruses, bacteria, fungi, and oomycetes. Additionally, DNA methylation and histone modifications epigenetically control plant gene expression upon pathogen attack and during interaction with beneficial microbes. The chapter discusses the current state of knowledge on epitranscriptomic and epigenetic mechanisms that regulate plant–microbe interactions, which may ultimately lead to novel strategies for crop health management and for enhancing crop productivity.

J. Yue
College of Horticulture and Plant Protection, Inner Mongolia Agricultural University, Hohhot, China

College of Agriculture, Inner Mongolia Agricultural University, Hohhot, China

Y. Lu · H. Wang · M. Zhao (✉)
College of Horticulture and Plant Protection, Inner Mongolia Agricultural University, Hohhot, China

R. L. Corrêa
Institute of Integrative Systems Biology (I2SysBio, CSIC-UV), Paterna, Spain

Department of Genetics, Federal University of Rio de Janeiro (UFRJ), Rio de Janeiro, Brazil
e-mail: r.correa@csic.es

F. Pasin
Center for Biological Research "Margarita Salas" (CIB), Spanish National Research Council (CSIC), Madrid, Spain
e-mail: f.pasin@csic.es

L. M. Vaschetto (ed.), *Plant Transcriptomics and Epitranscriptomics*, Concepts and Strategies in Plant Sciences, https://doi.org/10.1007/978-981-95-5183-5_12

Keywords RNA and DNA methylation · Epitranscriptomics · N^6-methyladenosine · Epigenetics · 5-methylcytosine · AlkB RNA demethylase homolog (ALKBH) · RNA methyltransferase-like protein (METTL) · histone modifications · plant–microbe interactions · plant virus

Abbreviations

ac4C	N^4-acetylcytidine
AGO	ARGONAUTE
AlkB	Alkylation B
ALKBH	AlkB homolog
AMV	Alfalfa mosaic virus
BlScV	Blueberry scorch virus
BVY	Blackberry virus Y
CGMMV	Cucumber green mottle mosaic virus
CMV	Cucumber mosaic virus
CPSF	Cleavage and polyadenylation specificity factor 30 kDa
DGCR8	DiGeorge syndrome critical region 8 protein
ENMV	Endive necrotic mosaic virus
FENMV	French endive necrotic mosaic virus
FIP37	FKBP12-interacting protein of 37 kDa
Fl(2)d	Female lethal 2D
fm^6A	N^6-formyladenosine
FTO	Fat mass and obesity-associated RNA demethylase
GVA	Grapevine virus A
HCV	Hepatitis C virus
HIV-1	Human immunodeficiency virus 1
hm^6A	N^6-hydroxymethyladenosine
IGF2BPs	Insulin-like growth factor 2 mRNA-binding proteins
IME4	Inducer of meiosis 4
ISR	Induced systemic resistance
m^6A	N^6-methyladenosine
MCMV	Maize chlorotic mottle virus
MeRIP-seq	Methylated RNA immunoprecipitation sequencing
METTL	Methyltransferase-like protein
miRNA	micro RNA
MTA	mRNA adenosine methylase
MT-A1	Methyltransferase-A1
Mum2	Muddled meiosis 2
NXF1	Nuclear RNA export factor 1
PepMV	Pepino mosaic virus
Pol II	RNA polymerase II
PPV	Plum pox virus

PVY	Potato virus Y
RBM	RNA-binding motif protein
RBSDV	Rice black-streaked dwarf virus
RdDM	RNA-mediated DNA methylation
RDR	RNA-dependent RNA polymerase
RISC	RNA silencing complex
RLP43	Receptor-like protein 43
RPP7	Recognition of peronospora parasitica 7
RSV	Rice stripe virus
SAM	S-adenosylmethionine
SARS-CoV-2	Severe acute respiratory syndrome coronavirus 2
siRNA	Small interfering RNA
Slz1	Sporulation specific with a leucine zipper motif protein 1
SNC1	Suppressor of npr1, constitutive 1
snRNA	Small nuclear RNA
SRSF	Serine and arginine rich splicing factor
SV40	Simian virus 40
Sxl	Sex lethal gene
TE	Transposable elements
TMV	Tobacco mosaic virus
WTAP	Wilms' tumor 1-associating protein
XIST	Long noncoding RNA X-inactive specific transcript
XRN1	5′-3′ exoribonuclease 1
YTHDF	YTH domain-containing family protein
ZC3H13	Zinc finger CCCH domain containing protein 13
ZIKV	Zika virus
5mC	5-methylcytosine

12.1 Introduction

In eukaryotic organisms, including plants, genetic information is stored in DNA and then transmitted to protein (translation) through RNA (transcription). RNA serves as both a carrier and regulator of genetic information, thus playing a vital role in biological systems. More than 170 chemical modifications are found in RNA (Rengaraj et al. 2021), which affect almost all RNA biology aspects such as RNA processing, maturation, nuclear export, translation, and decay. RNA methylation is a main form of modification, which comprises over 60% of known RNA modification types. Among these, N^6-methyladenosine (m^6A) is the most prevalent internal modification of mRNA, whose abundance accounts for 0.4–0.7% of total adenosines in eukaryotic mRNA (Sun et al. 2023). The regulation of m^6A abundance and biological functions is a complex process that involves the coordinated action of multiple factors, including methyltransferases, demethylases, and binding proteins.

Technological advances in m^{6}A detection and site mapping in RNA are revolutionizing our understanding of the epitranscriptomic mechanisms that regulate gene expression in eukaryotic organisms, as well as their interactions with microbes.

Recent studies have shed light on the involvement of m^{6}A in plant–virus interactions (Table 12.1), revealing a complex interplay between the host plant and the invading virus (Yue et al. 2022b; Secco et al. 2024; Li et al. 2024; Ge et al. 2024). The m^{6}A regulatory components contribute to the plant immune response against infections. Furthermore, m^{6}A can be detected in genomes of plant RNA viruses, which in turn may encode homologs of Alkylation B (AlkB) RNA demethylase proteins, possibly acting as a counter-defense mechanism to protect viral genomes from methylation damage and yet-to-be-discovered antiviral effects of m^{6}A.

In addition to its role in virus infection, m^{6}A has also been implicated in plant interactions with bacteria and fungi (Table 12.1). Studies have shown that m^{6}A can regulate the mRNA stability and expression of genes involved in plant defense against fungal and bacterial pathogens, highlighting the importance of this epigenetic mark in plant immune responses. Understanding the mechanisms that govern m^{6}A homeostasis in plants is essential for the development of novel strategies to improve plant resistance to pathogens and enhance crop yields.

Nucleic acid modifications regulate plant–microbe interactions at the transcriptional level, too, and DNA methylation, mainly linked to the presence of 5-methylcytosine (5mC), is one of the best-studied classes of epigenetic marks regulating gene expression in eukaryotic organisms, including plants (Luo et al. 2018). DNA methylation, together with histone modifications, has a major influence on DNA packing around nucleosomes, resulting in epigenetic changes that control gene transcription. Epigenetic control of transcriptional activation or suppression of certain genes is rapidly being identified as critical for improved immune responses to repeated pathogen attack and for interaction of plants with beneficial microbes (Hannan Parker et al. 2022).

This chapter provides an overview of known epitranscriptomic and epigenetic regulatory mechanisms in plants, with a focus on their roles in plant–microbe interactions.

12.2 The Epitranscriptomic Mark N^6-Methyladenosine (m^{6}A)

12.2.1 m^{6}A Discovery

m^{6}A refers to the nucleobase methylation occurring at the sixth nitrogen atom of adenine, which is the most thoroughly studied type of internal mRNA modifications to date. m^{6}A is a highly conserved RNA modification, and high-throughput sequencing has revealed that it is mainly distributed near the stop codon, the 3′ untranslated region, and the long exons region of mRNA (Duan et al. 2019). m^{6}A was initially

Table 12.1 Reported epitranscriptomic regulatory roles in plant–microbe interactions

Microbe	Role[a]	References
Virus	Viral AlkB homologs from GVA, BlScV, and BVY show RNA demethylase activity in vitro	van den Born et al. (2008)
Virus	Suppression of *Arabidopsis thaliana ALKBH9B* impairs the systemic invasion of AMV	Martínez-Pérez et al. (2017)
Virus	Demethylase and methylase homolog genes are, respectively, up- and downregulated in *Nicotiana tabacum* during TMV infection	Li et al. (2018)
Virus	The global m^6A level significantly decreased in resistant watermelon after CGMMV infection	He et al. (2021)
Virus	During RSV infection of rice, expression of *OsMAT3*, *OsMAT4*, *OsALKBH9B-2*, *OsYTH1*, *OsYTH3*, *OsYTH5*, and *OsYTH7* is upregulated, whereas *OsALKBH10* and *OsYTH8* are downregulated	Zhang et al. (2021a)
Virus	During RBSDV infection of rice, expression of *OsALKBH9B-2* and *OsYTH6* is upregulated, whereas *OsALKBH10* is downregulated	Zhang et al. (2021a)
Virus	AlkB RNA demethylase homologs identified in plant RNA viruses of the *Potyvirus* genus	Pasin et al. (2022)
Virus	AlkB RNA demethylase homologs promote potyviral infection in *Nicotiana benthamiana*	Yue et al. (2022a)
Virus	MTase-like (METTL) proteins promote *N. benthamiana* resistance to potyviral infection	Yue et al. (2023)
Virus	Silencing of *ZmMTC* significantly promoted MCMV infection in maize	Xia et al. (2024)
Virus	Overexpression of the m^6A writers *MTA* and *HAKAI* inhibit the PepMV RNA accumulation	He et al. (2024)
Bacterium	m^6A regulation of *Pyrus bretschneideri* defense-responsive genes during infection of *Erwinia amylovora*	Han et al. (2021)
Bacterium	*AhALKBH15*-mediated m^6A demethylation improves resistance to *Ralstonia solanacearum* infection in peanut	Zhao et al. (2024a)
Fungus	Overexpression of *MhYTP2* enhanced the resistance of apples to powdery mildew	Guo et al. (2022)
Fungus	*Magnaporthe oryzae MTA1* plays a critical role in autophagy for fungal infection	Ren et al. (2022)
Fungus	*M. oryzae* infection upregulates the rice ac4C acetyltransferase gene *OsACYR*, enhancing jasmonic acid-dependent immunity	Lu et al. (2024)
Fungus	Inactivation of *CpALKBH* results in low *Cryphonectria parasitica* pathogenicity	Zhao et al. (2024b)
Oomycete	Inactivation of *MTA1* results in low *Phytophthora sojae* virulence	Zhang et al. (2024)

[a]See the main text for the abbreviations used

discovered in polyA-RNA of mouse (*Mus musculus*) L cells in 1974 (Perry and Kelley 1974). In the same year, Desrosiers et al. investigated the methylation of mRNA in Novikoff hepatoma cells (Desrosiers et al. 1974). Subsequently, m^6A was identified in viruses (Lavi and Shatkin 1975; Krug et al. 1976; Moss et al. 1977; Beemon and Keith 1977; Chen-Kiang et al. 1979), yeast (Clancy et al. 2002), and insects (Hongay and Orr-Weaver 2011; Guo et al. 2018).

In plants, m^6A methylation was discovered in 1979 in wheat (*Triticum aestivum*) (Kennedy and Lane 1979), followed by oat (*Avena sativa*) (Haugland and Cline 1980), and corn (*Zea mays*) (Nichols and Welder 1981). More recently, the use of m^6A-targeting antibodies in conjunction with high-throughput sequencing revealed that more than two-thirds of the transcripts are modified by m^6A in *Arabidopsis thaliana*, and m^6A peaks are enriched around the stop codon, in 3′ UTRs, and also close to the start codons (Luo et al. 2014). Transcriptome-wide m^6A high-throughput sequencing of this species revealed the differential m^6A methylation spatial patterns, and results indicate that over 80% of m^6A transcripts are shared among the leaves, flowers, and roots (Wan et al. 2015). Studies of *A. thaliana* chloroplast and mitochondrial epitranscriptomes suggest that RNA motif selection for m^6A deposition is conserved between organelles and the nucleus (Wang et al. 2017).

Apart from mRNA, m^6A also exists in tRNA, rRNA, small nuclear RNA (snRNA), and long noncoding RNA. Recently, researchers have found that approximately 23.1% of circular RNA in pancreatic ductal adenocarcinoma tissues contains m^6A modification (Ye et al. 2021). Functionally, m^6A may be involved in multiple aspects of RNA metabolism, such as mRNA splicing (Xiao et al. 2016), nuclear export (Roundtree et al. 2017), processing and transcription (Fustin et al. 2013), translation (Li et al. 2017), stability (Wang et al. 2014), and the maturation of microRNA (miRNA) (Alarcón et al. 2015b).

Regions modified by m^6A show a consensus containing the motif "RRACH" (R: G/A/U; H: U/A/C) (Duan et al. 2019), and the conserved sequence of m^6A found in yeast is "RGAC" (Schwartz et al. 2013). Additional motifs were reported, such as "URUAY" (Y: C/U) in maize (Miao et al. 2020), and "UGWAMH" (W: U/A; M: C/A), "RAGRAG," and "UGUAMM" in rice (*Oryza sativa*) (Li et al. 2014; Zhang et al. 2019). Two highly consistent motifs, UGUAY and GAACU, are significantly enriched in the m^6A peaks identified in wheat (Zhang et al. 2021b). Transcriptome-wide studies, however, indicate that not all m^6A potential motif sequences are methylated (Liu and Pan 2016).

12.2.2 m⁶A Cellular Metabolism

The m^6A deposition process in RNA uses S-adenosylmethionine (SAM) as the methyl donor, and the base adenine is methylated by an RNA methyltransferase to form m^6A. With the development of research on m^6A, it has been discovered that m^6A abundance and recognition are dynamic and reversible processes within cells, jointly regulated by methyltransferases—such as the methyltransferase-like protein

3 (METTL3), METTL14, and Wilms' tumor 1-associating protein (WTAP)—demethylases—such as fat mass and obesity-associated RNA demethylase (FTO) and AlkB homolog 5 (ALKBH5)—and some RNA-binding proteins—such as YTH domain-containing family protein 1 (YTHDF1, YTHDF2, YTHDF3)—which are referred to as "writers," "erasers," and "readers," respectively (Yue et al. 2015).

m^6A may affect a variety of cellular activities, such as mRNA transcription, maturation and splicing, nuclear export, transport, translation, degradation, and other processes (Yue et al. 2022b). However, the m^6A methylation-mediated biological functions that have been confirmed so far include YTHDF2-mediated mRNA degradation, YTHDF1-mediated mRNA translation, and YTHDF1-mediated selective splicing (Wang et al. 2014), whereas the elucidation of other potential roles awaits further studies (Liu and Pan 2016).

12.2.3 m^6A Methylases

The cellular m^6A methylase activity, also known as m^6A methyltransferase or "writing," is carried out by a protein complex composed of multiple components. In 1994, Bokar et al. isolated a m^6A methyltransferase complex from the nucleus of HeLa cells, which was further separated into methyltransferase-A1 (MT-A1; 30 kDa), MT-A2 (200 kDa), and MT-B (800 kDa) (Bokar et al. 1994). Subsequently, METTL3, a 70-kDa protein, was isolated from MT-A2, which contains the SAM binding site and the Asp-Pro-Pro-Trp (DPPW) domain with catalytic function (Bokar et al. 1997). Deletion of *METTL3* can reduce the overall m^6A level by about 30% in HeLa cells (Bokar 2005; Dominissini et al. 2012). Immunofluorescence analysis in vivo observed that METTL3 was located in nuclear speckles, which are rich in mRNA splicing factors, indicating that m^6A has a potential regulatory role in RNA metabolism (Bokar et al. 1997).

METTL3 is highly conserved in eukaryotes, and inducer of meiosis 4 (IME4), the homolog of METTL3 in *Saccharomyces cerevisiae*, regulates spore formation and meiosis in yeast (Shah and Clancy 1992). In *Drosophila melanogaster*, *IME4* is mainly expressed in the testis and ovary (Hongay and Orr-Weaver 2011), and can control sex determination by regulating the specific splicing of the sex-lethal (*Sxl*) gene (Haussmann et al. 2016). In plants, the first characterized m^6A methyltransferase is the *A. thaliana* mRNA adenosine methylase A (MTA; AT4G10760), a homolog of METTL3. MTA is mainly distributed in dividing tissues, especially in reproductive organs, stem meristems and emerging lateral roots, and its inactivation results in a reduction of m^6A in mRNA and embryo lethality (Zhong et al. 2008).

METTL14 is another subunit of the m^6A methyltransferase complex, and it exhibits homology with METTL3. METTL14 has no methyltransferase activity and mainly plays a role in binding RNA substrates and stabilizing the structure of METTL3. It can combine with METTL3 at a 1:1 ratio, significantly enhancing the catalytic efficiency of METTL3 (Liu et al. 2014). In *A. thaliana*, MTB

(AT4G09980), the homolog of METTL14, is localized within the nucleus, exhibits high expression levels in guard cells, and can interact with MTA, analogous to the interaction between METTL3 and METTL14 in other systems (Růžička et al. 2017).

WTAP, the third key protein of the mammalian m^{6}A methyltransferase complex, is located in nuclear speckles and has no catalytic domain and methyltransferase activity. It can assist METTL3 and METTL14 to form heterodimers, interact with the METTL3–METTL14 complex, and regulate their recruitment at mRNA targets, ultimately regulating m^{6}A deposition (Ping et al. 2014). The RNA-binding ability of METTL3 is greatly reduced in the absence of WTAP, whose downregulation leads to a significant reduction in total m^{6}A levels in HeLa cells. Knocking down *WTAP* in zebrafish (*Danio rerio*) can lead to defects in tissue development, and apoptosis is significantly increased after simultaneous silencing of *WTAP* and *METTL3* (Ping et al. 2014). Muddled meiosis 2 (Mum2), a homolog of WTAP in yeast, can form a complex with IME4 and sporulation-specific with a leucine zipper motif protein 1 (Slz1), which determines m^{6}A levels in mRNA during yeast meiosis (Schwartz et al. 2013). Female lethal 2D (Fl(2)d) is a homolog of WTAP in *D. melanogaster*, which can affect mRNA splicing and is highly expressed in reproductive organs (Granadino et al. 1990). In *A. thaliana*, FKBP12-interacting protein of 37 kDa (FIP37; AT3G54170) is a homolog of WTAP. It is highly expressed in bud tips, young leaves, developing flower organs and seeds, and it can interact with MTA in vivo and in vitro (Zhong et al. 2008). *FIP37* inactivation leads to an embryo-lethal phenotype, which arrests development in the spherical stage (Vespa et al. 2004). *fip37* mutants showed excessive proliferation of the stem meristem and a suppression of m^{6}A deposition in mRNA, as transcriptome-wide analysis of these mutants revealed the absence of most of the m^{6}A-rich regions identified in wild-type plants (Shen et al. 2016).

KIAA1429 (also known as VIRMA) is the fourth component of the methyltransferase complex discovered by METTL3 immunoprecipitation (Schwartz et al. 2014). KIAA1429 can promote methylation around 3′ UTRs and stop codons, thus affecting selective transcript polyadenylation (Yue et al. 2018). Similar to METTL3, METTL14, and WTAP, KIAA1429 is also located in nuclear speckles (Horiuchi et al. 2013). In human A549 cells, *KIAA1429* knockdown causes a fourfold reduction of m^{6}A peaks, which is a more pronounced decrease than that observed upon knockdown of either *METTL3* or *METTL14* (Schibler and Perry 1977). Virilizer is the homolog of KIAA1429 in *D. melanogaster*; it can interact with Fl(2)d/IME4 in the context of female-specific selective splicing (Ortega et al. 2003). The *A. thaliana* homolog of Virilizer (AT3G05680) can be immunoprecipitated by FIP37, and its functional defects lead to a dramatic reduction in m^{6}A abundance (Růžička et al. 2017).

The RNA-binding motif protein 15/15B (RBM15/15B) is a protein that interacts with WTAP, as identified by immunoprecipitation assays. Inactivation of *RBM15* and its homolog *RBM15B* can significantly reduce the m^{6}A level in mRNA. The results of immunoprecipitation showed that they interacted with METTL3 in a WTAP-dependent manner, and played an important role in X chromosome inactivation and gene silencing by mediating m^{6}A methylation of long noncoding RNA

X-inactive specific transcript (*XIST*) (Patil et al. 2016). In mice, RBM15 absence can lead to embryo lethality (Raffel et al. 2007), whereas Spenito, the *D. melanogaster* homolog of RBM15, is necessary for sex determination by controlling the alternative splicing of *Sxl* mRNA (Yan and Perrimon 2015). *Spenito* expression correlates to the level of m^6A during development, and knocking out *Spenito* leads to a significant decrease in m^6A level (Lence et al. 2016).

HAKAI was originally identified as E3 ubiquitin ligase in humans. Knocking out *HAKAI* in human cells can reduce m^6A levels of mRNA (Yue et al. 2018). The *A. thaliana* HAKAI homolog (AT5G01160) can be co-immunoprecipitated by FIP37; its inactivation results in a reduction of m^6A levels to approximately 10% of those observed in the wild type, but it does not lead to embryo lethality (Růžička et al. 2017).

The zinc finger CCCH domain-containing protein 13 (ZC3H13) plays an important role in regulating the m^6A addition to mRNA in the nucleus, and a significant reduction of the m^6A level in transcripts was detected in mouse embryonic stem cells lacking *ZC3H13*. It controls the nuclear localization of m^6A writer complex, and most of its components, WTAP, Virilizer, and HAKAI, are translocated to the cytoplasm after *ZC3H13* deletion (Wen et al. 2018). CG7358 is the homolog of ZC3H13 in *D. melanogaster* (also known as the Fl(2)d-associated complex component or Xio), which co-localizes with METTL3, METTL14, Fl(2)d, Virilizer, and Spenito in the nucleus and interacts with them, thereby stabilizing the complex and facilitating the deposition of m^6A on mRNA (Knuckles et al. 2018). Xio regulates m^6A abundance in mRNA, the selective splicing of *Sxl* in somatic and germ cells to control the level of SXL protein, and female flies develop into males when this gene is inactive (Guo et al. 2018).

METTL16 is a conserved methyltransferase that can mediate the formation of m^6A in the U6 snRNA at the "UACAGAGAA" motif and plays an important role in regulating the expression of methyl donor SAM synthesis gene *MAT2A*, thus regulating the balance of SAM in cells (Pendleton et al. 2017).

ZCCHC4 is a newly identified m^6A methyltransferase of human 28S rRNA, which contains a zinc finger structure that mediates the methylation of A4220 on 28S rRNA at the "AAC" motif. Knockout of ZCCHC4 can reduce m^6A 4220 levels in 28S rRNA, inhibit cell proliferation, and significantly reduce tumor size in xenografted mice (Ma et al. 2019).

12.2.4 *m^6A Demethylases*

Epitranscriptomic marks are reversible and can be removed by proteins with RNA demethylase activity, also known as "erasers." In humans, ALKBH5 and FTO show m^6A demethylase activity and belong to Fe(II)/α-ketoglutarate-dependent dioxygenases with homology to bacterial AlkB proteins (Fedeles et al. 2015). In mammals, the ALKBH family consists of the nine members ALKBH1 to −8, and FTO (Fu et al. 2010).

FTO, the first identified m^6A demethylase, was originally discovered as a factor associated with obesity in mice (Peters et al. 1999). FTO oxidizes m^6A to the semi-aldehyde intermediate N^6-hydroxymethyladenosine (hm^6A), which is then oxidized to N^6-formyladenosine (fm^6A) (Fu et al. 2013). *FTO* knockout in human cells leads to an increase in the level of m^6A in mRNA, while *FTO* overexpression leads to a decrease of this modification (Jia et al. 2011), and its activity can affect RNA splicing by regulating m^6A levels in mRNA precursors (Zhao et al. 2014). Notably, the transgenic expression of human FTO in potato and rice resulted in a significant increase in yield and biomass of ~50% (Yu et al. 2021), which was attributed to the stimulated proliferation of meristematic cells, as well as enhanced photosynthetic efficiency and improved drought tolerance.

After FTO, another RNA m^6A "eraser" protein, ALKBH5, was identified. In contrast to FTO, ALKBH5 can specifically bind to m^6A-modified RNA in vitro and in vivo, and directly catalyzes the demethylation of m^6A to release the unmodified nucleobase adenine without intermediate products. ALKBH5 shows a higher binding affinity to its substrate than FTO (Duan et al. 2019). In mice, *ALKBH5* inactivation significantly increased m^6A levels, which caused the apoptosis of spermatocytes and led to infertility. Knockout of *ALKBH5* in human cells can accelerate the nuclear export of poly(A)-RNA, increase the rate of synthesis and splicing of nascent mRNA, and slightly reduce the stability of total RNA (Zheng et al. 2013).

In *A. thaliana*, proteins with homology to human ALKBH5 have been identified, including AT1G48980 (encoding ALKBH9A), AT2G17970 (ALKBH9B), AT4G36090 (ALKBH9C), AT2G48080 (ALKBH10A), and AT4G02940 (ALKBH10B) (Martínez-Pérez et al. 2017). Among these, ALKBH9B is an m^6A RNA demethylase that co-localizes with small interfering RNA (siRNA) bodies and is related to P-bodies. ALKBH10B is a demethylase that has been reported to be closely involved in growth and development in *A. thaliana*, and it is capable of reversing m^6A methylation in transcripts, both in vitro and in vivo. Deletion of *ALKBH10B* increased the total m^6A level in poly(A)-RNA, affecting the mRNA stability of key flowering time regulators and delaying the floral transition. *ALKBH10B* overexpression decreased the level of m^6A in poly(A)-RNA, and the plants exhibited an early flowering phenotype (Duan et al. 2017).

12.2.5 m^6A Readers

Reader proteins are a class of proteins that can specifically bind to m^6A-modified RNA, and the downstream cellular functions of m^6A are intimately linked to them. The sequence structure of m^6A-modified RNA tends to remain single-stranded, which may facilitate the recognition and binding of these proteins (Roost et al. 2015). There are five YTH family proteins in humans that participate in RNA metabolism by recognizing m^6A, namely the cytoplasmic YTHDF1, YTHDF2,

YTHDC3, and YTHDC2, and the nuclear YTHDC1. Additionally, members of heterogeneous nuclear ribonucleoproteins, including HNRNPA2B1, HNRNPC, and HNRNPG, can also bind m^6A (Duan et al. 2019).

YTHDF2, the first identified m^6A reader, contains a proline/glutamine/asparagine-rich domain that plays an important role in recruiting the YTHDF2-RNA complex to P bodies. Targets of YTHDF2 are mostly mRNAs, but also include noncoding RNAs, with a conserved core motif "GAC" (Wang et al. 2014). YTHDF2 binds to m^6A-containing mRNA through its carboxy (C)-terminal domain, while the amino (N)-terminal domain is required for targeting of the YTHDF2-mRNA complex to cellular RNA degradation sites (Wang et al. 2014). YTHDF2 thus affects mRNA stability and translation, and can also selectively recognize m^6A-modified mRNA in vivo to regulate mRNA degradation. In YTHDF2-deficient cells, the half-life of m^6A-modified mRNA increased (Wang et al. 2014). In zebrafish, YTHDF2 accelerates the clearance of mRNA containing m^6A and promotes the maternal-to-zygote transition during embryonic development (Zhao et al. 2017). In mice, YTHDF2 regulates maternal mRNA expression, and female mice lacking YTHDF2 are infertile (Ivanova et al. 2017).

YTHDF1 selectively binds to mRNA modified with m^6A, and interacts with ribosomes and translation initiation factors to promote the ribosome loading and translation of bound mRNA (Wang et al. 2015).

YTHDF3 can also recognize mRNAs that contain m^6A, and shows many common targets with YTHDF1 and YTHDF2. It can promote mRNA translation in cooperation with YTHDF1, and enhance the degradation of mRNA mediated by YTHDF2; YTHDF3 interaction with YTHDF1 and YTHDF2 can accelerate the cytoplasmic metabolism of m^6A-modified mRNA. Downregulation of *YTHDF3* alters the specificity of YTHDF1 and YTHDF2, promoting their binding to nontarget transcripts (Shi et al. 2017).

YTHDC1 binds mRNA at sites that overlap with those known to be modified by the m^6A writer complex; it is preferentially located near the stop codon, a region where m^6A levels are also enriched. This protein can form a specific structure in the nuclear speckles and regulate mRNA splicing by recruiting serine- and arginine-rich splicing factor 3 (SRSF3) and blocking SRSF10 mRNA binding (Xiao et al. 2016). YTHDC1 can also interact with SRSF3 and nuclear RNA export factor 1 (NXF1) to promote the export of m^6A-marked transcripts from the nucleus to the cytoplasm (Roundtree et al. 2017). In addition, YTHDC1 can promote gene silencing mediated by *XIST* (Patil et al. 2016).

YTHDC2 is the largest identified member of the YTH family protein with a molecular weight of 160 kDa, while other YTH proteins typically range from 60–85 kDa (Duan et al. 2019). YTHDC2 possesses protein domains, including YTH and R3H domains and a DEAH-box helicase core spaced by ankyrin repeats. YTHDC2 can selectively bind to m^6A-modified RNA through its YTH domain, whereas its ankyrin repeats recruit the 5′-3′ exoribonuclease 1 (XRN1) in an RNA-independent manner (Kretschmer et al. 2018). In mice, knockout mutants of *YTHDC2* exhibit infertility (Hsu et al. 2017).

YT521-B in *D. melanogaster* is a homolog of human YTH proteins, which decode m^6A marks and regulate mRNA maturation. In *D. melanogaster*, defects in YT521-B mimic the deletion of *IME4* to result in a markedly reduced proportion of females (Haussmann et al. 2016).

HNRNPA2B1 is a novel m^6A-binding protein located in the nucleus. This protein has a low-complexity domain rich in glycine, which has four arginine-glycine-glycine (RGG) motifs that may help HNRNPA2B1 preferentially bind to m^6A-modified RNA in vivo and in vitro (Liu et al. 2017). Its binding site overlaps with m^6A peaks of miRNA precursors, and HNRNPA2B1 can also interact with the miRNA complex protein DiGeorge syndrome critical region 8 (DGCR8), and promote miRNA maturation (Alarcón et al. 2015a).

HNRNPC is a nuclear RNA-binding protein that can recognize secondary structure changes caused by m^6A marks in mRNA. m^6A destabilizes stem loop structures, thus enhancing mRNA accessibility and the HNRNPC-RNA interaction (Liu et al. 2015). HNRNPC binding can regulate RNA transcription, stability, and splicing, and in human cells, HNRNPC has been implicated in mediating alternative splicing, which accounts for approximately 16% of m^6A-dependent changes of pre-mRNA processing (Liu and Pan 2016).

HNRNPG can bind to m^6A-containing RNA through RGG repeats located in its C-terminal low-complexity region (Liu et al. 2017). It can further bind to the phosphorylated C-terminal domain of RNA polymerase II (Pol II), which results in transcriptome-wide modulation of alternative splicing (Zhou et al. 2019).

Insulin-like growth factor 2 mRNA-binding proteins (IGF2BPs) are conserved single-stranded RNA-binding proteins, whose RNA-binding domain can preferentially recognize "GGAC" sites modified by m^6A. IGF2BPs promote the stability, translation, and storage of their target mRNAs in normal and stress conditions in a m^6A-dependent manner, thus affecting the gene expression (Huang et al. 2018).

Homologs of mammalian m^6A readers also exist in plants, including YTH domain family proteins in *A. thaliana*, designated as ECT1 to ECT12, and cleavage and polyadenylation specificity factor 30 kDa (CPSF) (Addepalli and Hunt 2007; Scutenaire et al. 2018). Among these, ECT2 is most widely distributed in plants and shows the highest expression level. ECT2, ECT3, and ECT4 are cytoplasmic proteins localized at the shoot apex of young seedlings and in the division zone of developing leaves (Arribas-Hernández et al. 2018). ECT2 translocates to stress granules under heat shock, suggesting a regulatory role in stress responses (Arribas-Hernández et al. 2018). Through its YTH domain, ECT2 binds to the plant-specific m^6A motif "URUGY" in the 3′ UTR of target transcripts (Scutenaire et al. 2018), which may regulate processing of mRNAs in the nucleus and their stability in the cytoplasm (Wei et al. 2018). *ECT2* deletion results in an increase in trichome branching, and the *ect2/ect3* double mutant exhibits delayed leaf emergence and developmental defects, which were intensified by *ECT4* inactivation (Arribas-Hernández et al. 2018).

12.3 m^{6}A Regulation of Virus Infections

In the 1970s, m^{6}A was first detected in viral nucleic acids as part of efforts to characterize RNA modifications. After the first description in 1974 of m^{6}A in eukaryotic transcripts (Desrosiers et al. 1974), the same modification was also identified in viral RNA genomes and transcripts, including those of influenza virus (Krug et al. 1976), human immunodeficiency virus 1 (HIV-1) (Moss et al. 1977), Rous sarcoma virus (Beemon and Keith 1977; Kane and Beemon 1985), simian virus 40 (SV40) (Canaani et al. 1979; Aloni et al. 1979; Finkel and Groner 1983), and adenovirus type 2 (Chen-Kiang et al. 1979).

Due to the limitations of earlier technologies, such as chromatographic analysis of radiolabeled and enzymatically digested RNAs, research on m^{6}A progressed slowly until 2012, when the development of high-throughput approaches, including m^{6}A-specific RNA immunoprecipitation coupled with next-generation sequencing (MeRIP-seq), enabled the transcriptome-wide mapping of m^{6}A distribution (Meyer et al. 2012; Dominissini et al. 2012). m^{6}A was then shown to be conserved in genomes of diverse mammalian RNA viruses, and that virus infection can significantly alter m^{6}A levels in the host transcripts. For instance, the genomic RNA and its antisense copy of severe acute respiratory syndrome coronavirus 2 (SARS-CoV-2) are dynamically m^{6}A modified in human and monkey cells. SARS-CoV-2 infection further triggers a global increase in the host methylome, exhibiting altered localization and motifs of m^{6}A in transcripts. SARS-CoV-2 replication and transcription occur in the cytoplasm. Virus infection was shown to trigger the relocation of key m^{6}A enzymes from the nucleus to the cytoplasm, including the writers METTL3 and METTL14 and the eraser ALKBH5, which negatively and positively regulate SARS-CoV-2 infection, respectively. SARS-CoV-2 is also sensitive to the m^{6}A reader YTHDF2, which can degrade m^{6}A-labeled transcripts and negatively regulate viral replication (Liu et al. 2021).

In cells infected with HIV-1, m^{6}A detection in total RNA samples revealed a high molecular weight band, which was later identified as the viral RNA genome, indicating that HIV-1 infection induces m^{6}A methylation of viral RNA and potentially of host transcripts as well (Lichinchi et al. 2016a). Viral m^{6}A sites were mapped by photo-crosslinking-assisted m^{6}A sequencing techniques, and reported to cluster in the HIV-1 3′ UTR. The presence of m^{6}A in viral 3′ UTRs or at analogous positions in cellular transcripts strongly enhances mRNA expression in *cis* by recruiting the cellular m^{6}A reader protein YTHDF. Consistent with this notion, *YTHDF* down- or upregulation, respectively, impaired or enhanced the accumulation of viral proteins and RNA, and viral replication in human cells (Kennedy et al. 2016). Silencing of the *ALKBH5* RNA demethylase increases methylation and nuclear export of viral RNA, overall promoting viral replication. On the contrary, *METTL3* or *METTL14* silencing reduced methylation and nuclear export of viral RNA, inhibiting viral replication (Lichinchi et al. 2016a).

Several m^{6}A enrichment peaks were identified in the Zika virus (ZIKV) genomic RNA, whose abundance was controlled by the host methyltransferases METTL3 and

METTL14, as well as the demethylases ALKBH5 and FTO. Knockdown of methyltransferases increased the production of ZIKV, while demethylase silencing reduced the production of ZIKV. ZIKV replication was further enhanced by silencing *YTHDF* genes regulating the stability of m^6A-modified RNA. MeRIP-seq analysis of the host methylome showed that ZIKV infection alters the m^6A position in mRNAs, methylation motif, and methyltransferase-modified target genes (Lichinchi et al. 2016b). ZIKV infection triggered upregulation of m^6A in the CDS region of host transcripts, but downregulation of m^6A in 3′ UTRs and a reduction of m^6A peaks in most host genes, including those stimulated by interferon. m^6A-related enzymes further migrated from the nucleus to the cytoplasm after ZIKV infection (Liu et al. 2022).

Depletion of the writers METTL3 and METTL14, or the eraser FTO, respectively, increases or decreases infectious particle production of hepatitis C virus (HCV). YTHDF proteins relocalize to the sites of HCV assembly, and their downregulation was shown to increase the amount of infectious viral particles. Mapping of m^6A sites across the HCV genome showed that the reduction of m^6A levels and potential YTHDF binding sites increases viral titer without affecting viral replication (Gokhale et al. 2016).

In plants, m^6A is also involved in the regulation of antiviral responses (Table 12.1). In *Nicotiana tabacum*, tobacco mosaic virus (TMV) infection reduced the m^6A level and the expression of potential RNA methylase genes, whereas the potential demethylase XM_009801708 was upregulated at 14 and 21 days post-infection (Li et al. 2018). Martínez-Pérez et al. identified that the *A. thaliana* protein ALKBH9B is a demethylase that removes m^6A from single-stranded RNA molecules in vitro. In addition, m^6A was identified in the genomes of two members of the *Bromoviridae* family, alfalfa mosaic virus (AMV), and cucumber mosaic virus (CMV). Interestingly, the demethylation activity of ALKBH9B affected the infectivity of AMV but not of CMV, which has been linked to the AMV-specific ALKBH9B recruitment through the viral coat protein. *ALKBH9B* suppression increased the m^6A abundance in the AMV genome, impairing the viral systemic movement, whereas no effect could be detected on CMV infection (Martínez-Pérez et al. 2017). Subsequently, Martínez-Pérez et al. found that AMV spread to floral stems and roots was impaired in the ALKBH9B single mutant line, indicating that neither ALKBH9A nor ALKBH9C is required for the systemic movement, and ALKBH9B is a major determinant in the viral upload to the phloem. The absence of ALKBH9B further reduces AMV cell-to-cell movement as well as the viral replication cycle (Martínez-Pérez et al. 2021). In 2021, Alvarado-Marchena et al. identified the critical role of the ALKBH9B residues 387 and 427 for direct interaction with the AMV coat protein (Alvarado-Marchena et al. 2021).

In watermelon (*Citrullus lanatus*), the global m^6A amount significantly decreased after cucumber green mottle mosaic virus (CGMMV) infection of a resistant line. In early response to CGMMV infection, >400 differentially methylated genes were identified; most of these were hypomethylated, possibly due to the detected upregulation of the m^6A demethylase homolog *ClALKBH4B* (He et al. 2021).

In *Nicotiana benthamiana*, MeRIP-seq analysis of samples infected with plum pox virus (PPV) and potato virus Y (PVY) showed that m^6A peaks are enriched in plant transcript 3′ UTRs and in discrete internal and 3′ terminal regions of the viral RNA genomes. Downregulation of *N. benthamiana NbALKB1* and *NbALKB2* homologs of the plant-specific ALKBH9 clade resulted in a significant decrease in PPV and PVY accumulation (Yue et al. 2022a). m^6A modifications of the RNA genome of pepino mosaic virus (PepMV) were conserved in *N. benthamiana* and tomato (*Solanum lycopersicum*). Overexpression of the methyltransferases *MTA* and *HAKAI* inhibited PepMV accumulation, which was accompanied by increased viral m^6A levels, whereas deficiency of these writers decreased the viral RNA m^6A levels but enhanced virus infection (He et al. 2024).

In rice, epitranscriptomic changes were recently investigated during infection of rice stripe virus (RSV), a single-stranded RNA virus, and rice black-streaked dwarf virus (RBSDV), a double-stranded RNA virus. Rice methyltransferases *OsMAT3* and *OsMAT4* were significantly increased in RSV-infected samples, whereas no change was detected during RBSDV infection. Conversely, in both the RBSDV- and RSV-infected samples, the expression of the *OsALKBH10* demethylase was significantly suppressed, whereas *OsALKBH9B-2* expression was upregulated. In RSV-infected samples, expression of the m^6A binding protein genes *OsYTH1*, *OsYTH3*, *OsYTH5*, and *OsYTH7* was significantly upregulated, whereas *OsYTH8* was downregulated. In RBSDV-infected samples, only *OsYTH6* upregulation could be detected. m^6A modifications often occurred in the 5′ terminus of the RBSDV genome, whereas they were randomly distributed across the RSV genome (Zhang et al. 2021a).

In maize, maize chlorotic mottle virus (MCMV) infection significantly upregulated m^6A levels. MeRIP-seq detected m^6A in the viral genomic RNA, as well as its enrichment or underrepresentation in transcripts of the writer *ZmMTC*, or the reader *ZmECT18*, respectively. *ZmMTC* and *ZmECT18* played opposite functions in plant resistance against MCMV. In virus-induced gene silencing assays, *ZmECT18* was shown to be required for MCMV infection, while the *ZmMTC* downregulation significantly promoted MCMV infection in maize, possibly by m^6A-dependent regulation of the expression of resistance-related genes (Xia et al. 2024).

Several plant viruses encode AlkB protein homologs or domains, suggesting a counter-defense mechanism exerted by these viruses. Over 40 species of single-stranded RNA viruses, mainly belonging to the families *Alphaflexiviridae*, *Betaflexiviridae* and *Closteroviridae*, present intact AlkB domains (Moore and Meng 2019). More recently, AlkB domains were also identified in members of the genus *Potyvirus*, the largest group of plant RNA viruses (Pasin et al. 2022), including endive necrotic mosaic virus (ENMV) and French endive necrotic mosaic virus (FENMV) (Desbiez et al. 2017; Yue et al. 2022a). Given their homology with bacterial DNA or RNA repair components, viral AlkB domains were hypothesized to have equivalent functions protecting viral genomes from methylation damage. In 2008, experimental evidence showed that functionally active AlkB-like domains are encoded by diverse plant viruses, such as grapevine virus A (GVA), blueberry scorch virus (BlScV), and blackberry virus Y (BVY). These viral AlkB domains were found to reverse RNA damage caused by methylation, implying a possible role in removing harmful modifications from the viral genome to safeguard its integrity and stability (van den Born et al. 2008).

Our understanding of m^6A roles in plants has greatly advanced in recent years; however, the molecular mechanism(s) underlying the m^6A-dependent regulation of antiviral responses remain unknown. Yue et al. proposed a model in which m^6A coordinates plant–virus interactions (Yue et al. 2022b). On the one hand, the m^6A methylation system would act as a host defense mechanism for dampening viral infection. The viral RNA would be methylated by METTL-like proteins (Yue et al. 2023), which, together with m^6A recognition by YTH domain proteins, may affect the stability, translation, or viral particle packaging; m^6A accumulation in the viral genomes may further impair viral replication by direct inhibition of the binding and elongation activity of the virus replicase. On the other hand, viral infection may result in an m^6A imbalance that affects the methylation and demethylation status and accumulation of the host transcripts, which may indirectly influence the viral infection (Yue et al. 2022b). Additionally, m^6A modification of viral RNA genomes was proposed to impair systemic infection of plants by restricting viral upload into the phloem (Alvarado-Marchena et al. 2021).

12.4 m^6A Regulation of Plant Interactions with Fungi and Bacteria

Recently, the breadth of m^6A research in plants has expanded, and m^6A roles in the interaction between plants and pathogenic fungi have been investigated (Table 12.1). Guo et al. (2022) showed that the *Malus hupehensis* gene *MhYTP2* affected m^6A abundance, which in turn regulated the stability of *Malus domestica MdMLO19*, a susceptibility gene for powdery mildew infection. In chestnut blight fungus (*Cryphonectria parasitica*), the fungal eraser *CpALKBH* regulates *CpZap1* transcript stability, which affects the pathogen's virulence and development (Zhao et al. 2024b). The *CpALKBH* inactivation led to a significant increase in m^6A levels and a decrease in the fungal growth rate, sporulation ability, and pathogenicity. Ren et al. (2022) identified an m^6A methyltransferase in *Magnaporthe oryzae* and revealed the functional importance of m^6A in appressorium-mediated infection of this fungal pathogen; results further showed that m^6A amount negatively correlated with mRNA levels of autophagy-related genes. Beyond m^6A, N^4-acetylcytidine (ac4C) was recently shown to participate in rice resistance to fungal pathogen infection (Lu et al. 2024). Rice translatomes and epitranscriptomes upon *M. oryzae* infection uncovered the upregulation of the ac4C acetyltransferase gene *OsACYR*, which promoted transcript translation to enhance jasmonic acid-dependent immunity.

Epitranscriptomic regulation of interactions between plants and pathogenic bacteria has also been described. In peanut (*Arachis hypogaea*), the combined analysis of transcriptomes and m^6A abundance revealed that the eraser *AhALKBH15* plays a key role in resistance against bacterial wilt caused by *Ralstonia solanacearum*. Peanut m^6A levels were increased after *R. solanacearum* inoculation, and highly correlated with gene expression (Zhao et al. 2024a). Han et al. (2021) found that some defense-responsive genes were modified by m^6A methylation, which led to an increase in their expression in pear (*Pyrus bretschneideri*) during infection of the fire blight agent *Erwinia amylovora*. m^6A preferentially modifies duplicate genes rather than singleton genes, and the positive correlation with the transcript levels suggests a regulatory role for m^6A in plant immune responses.

Roles of epigenetic markers in the adaptive flexibility of oomycetes during infection are supported by recent work on *Phytophthora sojae*. Results implicated m^6A modifications in the regulation of the production of sporangia and oospores, the release of zoospores, and the virulence of this pathogen (Zhang et al. 2024), underscoring a role of m^6A-mediated RNA metabolism in regulating the development and pathogenicity of oomycetes.

12.5 Epigenetics of Plant–Microbe Interactions

It has been known for almost a century that plants can develop immunity to diseases they have already encountered (Wingard 1928). However, it was only in the last several decades that we have started to understand how plants use adaptive immunity memory. Unlike mammals, plants lack specialized mobile immune cells. In fact, when challenged or instructed by signals, all plant cells may trigger appropriate immune responses. RNA silencing mechanisms, which involve the production and amplification of mobile small RNAs (sRNAs), are commonly recognized as a form of plant adaptive immunity to viruses (Baulcombe 2022). Nonetheless, the mechanism of induced systemic resistance (ISR) in plants is the most similar to that of adaptive immunity observed in mammals (Hannan Parker et al. 2022). Plant defense genes and metabolites are pre-activated during ISR when biotic stressors are present, even in noninfected systemic tissues. When plants are challenged to the same or similar stresses repeatedly, their defenses activate faster and/or more strongly, resulting in a more effective response than when they were first exposed. This "priming," or immunological awareness state, is a complex system that requires the coordination of numerous specialized processes, such as changes to the chromatin, transcriptome, proteome, and metabolome. Understanding the components required to generate this particular priming physiological state is critical for translational research and applications. Epigenetic changes in the chromatin of certain genes are rapidly being identified as critical for improved immune responses to recurrent stress events. Epigenetic marks are divided into two categories: DNA methylation and histone modifications. Both types have a major influence on DNA packing around nucleosomes. The insertion of repressive marks promotes chromatin compaction, resulting in transcriptional suppression. In contrast, activation marks can increase gene expression. The interaction of these two types of marks during developmental programs or environmental responses determines gene expression outcomes at any given time.

12.5.1 DNA Methylation

The majority of known cases of immune genes controlled by epigenetic pathways involve the co-regulation of adjacent transposable elements (TE). Because these selfish mobile sequences have the potential to harm the genome, numerous pathways work together to restrict their expression, including at the chromatin level. The most well-known mechanism for achieving this is the RNA-mediated DNA methylation (RdDM) pathway, which targets TEs in gene-rich areas (Sigman and Slotkin 2016). The RdDM mechanism was identified in plants that carried transgenes originating from viroids, which are infectious circular long noncoding RNAs. It was discovered that viroid-derived transgenes were methylated when plants were infected with the pathogen, implying that replicating RNA sequences might guide specific DNA methylation marks on transgenes (Wassenegger et al. 1994). It is now understood that in the RdDM process, sRNAs from TEs or other repetitive sequences are generated and employed as guides that promote DNA methylation, which is followed by histone modifications, resulting in chromatin compaction. RdDM is primarily mediated by two plant-specific RNA polymerases known as Pol IV and V. These multicomponent protein complexes evolved from Pol II by gene duplication and subsequent functional divergence (Huang et al. 2015). Pol IV initiates the process by transcribing small sections of target sequences. Subsequent steps are similar to other RNA silencing mechanisms. Endogenous RNA-dependent RNA polymerases (RDR) convert Pol IV-derived single-stranded RNAs (ssRNAs) to double-stranded RNAs (dsRNAs), which are then degraded into small interfering RNAs (siRNAs) by DICER-LIKE proteins. These double-stranded siRNAs are loaded into a particular kind of RNA silencing complex (RISC), where one of the strands is selected by specialized ARGONAUTE (AGO) proteins. The same areas targeted by Pol IV are also transcribed by Pol V, resulting in RNAs with sequence complementarity to the Pol IV-generated siRNAs. This allows the loaded AGO to be recruited to target TEs, which are subsequently responsible for recruiting other protein complexes, ultimately leading to DNA methylation. Given the known ability of heterochromatin states to propagate to nearby regions, TEs present a challenge to gene expression when they are located close to genes. Activation marks added to highly expressed genes can also spread to TEs, potentially interfering with their regulation (Negi et al. 2016). Given that plant immune genes are known to be closer to TEs than other types of genes (Lai et al. 2020), it is not surprising that several associations between genes and TEs have originated over the evolutionary process.

Plants that lack RdDM-related functions have been shown to be more resistant to several diseases (López et al. 2011; Dowen et al. 2012; Luna et al. 2012; Yu et al. 2013; Le et al. 2014; López Sánchez et al. 2016; Diezma-Navas et al. 2019; Corrêa et al. 2020, 2024). This phenotype is facilitated by the direct or indirect activation of specific immunity genes in methylation-compromised lines, implying that RdDM would have a deleterious impact on immune responses. However, it has been suggested that the mechanism might have been selected in some circumstances to prevent the uncontrolled expression of defensive genes, which could otherwise compromise development. For example, when *A. thaliana* plants are challenged with the bacteria *Pseudomonas syringae*, DNA demethylase proteins are recruited to remove RdDM-promoted methylation marks from a TE located at the *RECEPTOR-LIKE PROTEIN 43* (*RLP43*) promoter region. During pathogen exposure, selective methylation erasure exposes these motifs to WRKY transcription factors, which are required to activate defensive gene expression (Halter et al. 2021). This case demonstrates the function of so-called "domesticated" transposons, which play two roles in gene expression regulation: preventing leaky defensive expression before pathogen exposure and assisting in recruiting the machinery required for appropriate expression when plants are challenged. The epigenetic regulation of TEs found inside genes can also provide cells with tools to fine-tune defense responses. For example, an element located within an intron of the *A. thaliana* immune receptor gene *RECOGNITION OF PERONOSPORA PARASITICA 7* (*RPP7*) is critical in the generation of functional alternate splicing forms (Tsuchiya and Eulgem 2013). RdDM-promoted histone repression marks deposited at a TE inside *RPP7*'s first intron restrict the usage of this splicing site, favoring the synthesis of full-length transcripts encoding the functional receptor protein. Loss of repressive histone methylation at this location causes the accumulation of a noncoding transcript, limiting the synthesis of functional RPP7 proteins. Interestingly, it was discovered that TEs placed in heterochromatic regions, i.e., away from genes, can also be essential for the appropriate expression regulation of particular defense genes, which are also close to other TEs (Cambiagno et al. 2018). In this case, the activation of heterochromatic TEs is necessary to enable the correct expression of defense genes under pathogen attack, which are often suppressed by RdDM because of adjacent TEs. This seeming paradox stems from the fact that the heterochromatic transposon transcripts have sequence similarities to those located near the key immunity genes. As a result, the heterochromatic elements begin to behave as target mimics or sponges, drawing siRNAs that typically target TEs near defense genes. This interaction causes relieving repression in defense genes, allowing for proper expression and resilencing of activated heterochromatic TEs.

12.5.2 Histone Modifications

Histone modifications also play an important role in regulating plant–microbe interactions (Kang et al. 2022). Nucleosomes, the basic structural units of chromatin are made up of octamers of histone proteins. The core nucleosome is composed of two copies of each of four histone types: H2A, H2B, H3, and H4. By wrapping DNA, nucleosome chains condense the genome to fit inside cells and protect it from nuclease degradation. Nucleosomes can also be remodeled to make the DNA available for transcription, replication, and repair as needed. This dynamic flexibility is strongly dependent on the insertion or removal of various types of epigenetic marks on the tails of histone proteins. The addition of three methyl groups to lysine 4 of the histone H3 tail, abbreviated as H3K4m3, for example, is known to cause chromatin decompaction and, as a result, gene expression. In contrast, the trimethylation of lysine 27 on the tail of the same histone (H3K27m3) promotes chromatin compaction and gene expression inhibition. Apart from methylation, additional post-translational modifications in histone tails, such as acetylation, phosphorylation, ubiquitination, and sumoylation, might influence chromatin dynamics and gene expression profiles (Haider and Farrona 2024). The binding of their associated epigenetic complexes to transcriptional factors, long noncoding RNAs, or short RNAs is typically responsible for selecting appropriate targets during developmental or environmental settings. Histone modifications have been shown to affect multiple genes involved in plant–microbe interactions (Kang et al. 2022). The expression control of the immunological receptor gene *SNC1* (*SUPPRESSOR OF npr1, CONSTITUTIVE 1*) is an interesting example of how these processes work (Yang et al. 2023). Under normal circumstances, the gene is inhibited by the combined actions of two proteins necessary to remove acetylation marks in specific histones known to be related to gene expression activation. Under infection conditions, two more modules are recruited to trigger its expression. One of them is in charge of adding different histone activation marks (H3 acetylation, H3K4m3, and H2B monoubiquitination) to enhance gene expression. The third module works independently of the others to remodel SNC1's chromatin, allowing for full expression potential.

The presence of histone variants in the chromatin of certain defense genes adds another layer of complexity to their regulation. Histone variants are specialized forms of histone proteins that differ in sequence and function from the original histones (Jiang and Berger 2017). Unlike the canonical ones, which are deposited during DNA replication, histone variants are present throughout the cell cycle and can be readily utilized to fine-tune specific responses. In plants, the most prevalent histone variants are H2A and H3. The H2A.Z variant is strongly linked to stress responses, including biotic ones (March-Díaz et al. 2008; Berriri et al. 2016; Cai et al. 2021). The conserved SWR1 chromatin-remodeling complex is required for replacing canonical H2A histones with H2A.Z (Aslam et al. 2019). The impact of H2A.Z deposition on genes is currently ambiguous. When added near a gene's transcriptional start sites, the result can be gene activation or repression. The outcome will be determined by additional alterations in nearby histones (Gómez-Zambrano et al. 2019). However, when H2A.Z is introduced to gene bodies, it is strongly related to gene repression, particularly in environmentally responsive genes (Sura et al. 2017). The removal of this variant from gene bodies is thus crucial for normal gene expression. The reintroduction of H2A.Z into the gene body of particular genes when stress is relieved may aid in priming pathogen responses, allowing for a faster and more robust response to repeated pathogen attack (Coleman-Derr and Zilberman 2012; Sura et al. 2017; Hannan Parker et al. 2022).

12.5.3 Plant Interactions with Beneficial Microbes

Aside from altering defense responses, epigenetics plays an important ecological function in the interaction of plants with beneficial microbes. Mutualistic and commensal interactions between plants and microorganisms are critical to plant health and ecosystem function. These interactions may promote plant development, nutrient absorption, and stress tolerance (Roossinck 2011; Thoms et al. 2021). These beneficial interactions are quite difficult to achieve from a molecular perspective. Immune system awareness is necessary to control microbe populations, but the system may be tightened to prevent eradicating them, which could have negative effects on the plant (Gordon and Leveau 2010). In fact, suppression of defenses by beneficial microorganisms is frequently essential to establish a beneficial interaction (Zamioudis and Pieterse 2012). A complex intercommunication of many molecular and metabolic circuits is thus required for the efficient functioning of beneficial interactions in plants. Because of epigenetics' established involvement in fine-tuning gene expression patterns, it has long been assumed that the mechanism would also apply to these types of interactions. This is, however, an area that remains largely unexplored. An intriguing case of the connection between the ectomycorrhizal fungus *Laccaria bicolor* and poplar (*Populus tremula x Populus alba*) has been documented. Poplar plants with low DNA methylation levels were unable to interact effectively with this beneficial fungus. Surprisingly, the reduction in DNA methylation in plants also resulted in lower amounts of this mark in the fungus, with implications for gene expression, suggesting an exchange of epigenetic signals between them (Vigneaud et al. 2023). In the reverse direction, some beneficial microorganisms can stimulate ISR adaptive responses in plants, making them more likely to respond properly when exposed to pathogens (Yu et al. 2022). Given that ISR is epigenetically regulated (Hannan Parker et al. 2022), it's possible that beneficial microorganisms exploit chromatin changes to prime plant defenses. Microbe–microbe interactions may also have an impact on plant health. For example, it has been demonstrated that the bacteria *P. piscium* can release a chemical that indirectly induces deacetylation and thereby suppression of virulence genes in the wheat fungus *Fusarium graminearum* (Chen et al. 2018). Wheats are thus protected from this pathogenic fungus in the presence of an epigenetic elicitor generated by beneficial microorganisms. These examples demonstrate that plant health is dependent on an intricate network of relationships. Disruptions to these sensitive ecological connections may result in impaired plant growth and/or thc introduction of new diseases. Understanding how plants and microorganisms are influenced by climate change is critical for reducing its socioeconomic consequences.

12.6 Conclusions

Recent advances in the study of nucleic acid modifications, particularly N^6-methyladenosine (m^6A) in RNA and 5-methylcytosine (5mC) in DNA, have significantly enhanced our comprehension of the complex epitranscriptomic and epigenetic mechanisms regulating plant–microbe interactions. While m^6A has been extensively investigated as a major epitranscriptomic mark, emerging evidence underscores the potential roles of other internal RNA modifications, such as N^6,2′-O-dimethyladenosine, N^1-methyladenosine, 5-methylcytidine, N^4-acetylcytidine, pseudouridine, 2′-O-methylated nucleotide, inosine, and internal N^7-methylguanosine, in controlling diverse biological processes, such as mRNA stability, translation efficiency, and immune signaling pathways in eukaryotic organisms (Wiener and Schwartz 2021; Sun et al. 2023; Lu et al. 2024). Conversely, DNA modifications beyond 5mC, such as the less abundant derivatives 5-hydroxymethylcytosine, 5-carboxylcytosine, and 5-formylcytosine, have been implicated in regulating gene expression during development and disease (Luo et al. 2018; Mattei et al. 2022). Roles of these less-abundant RNA and DNA modifications in plant immunity and plant–microbe interactions are currently unclear. However, recent technological advances for the specific mapping of RNA (and DNA) modifications at single-base resolution may soon enable to uncover the spatial and temporal patterns of m^6A and other modifications (Kong et al. 2023; Liu et al. 2023), providing valuable insights into their regulatory roles in plant immunity and stress responses. Further, evidence suggests that chromatin status may directly control m^6A abundance in mRNA by influencing Pol II elongation speed and the recruitment of the m^6A writer complex for co-transcriptional deposition of m^6A (Zaccara et al. 2019); however, investigation is required to confirm a direct connection between epigenetic and epitranscriptomic regulatory pathways.

On-demand engineering of RNA and DNA modifications holds promise for enhancing crop productivity and resistance to biotic and abiotic stress. For instance, the transgenic expression of the human RNA demethylase FTO in crops, such as potato and rice, has demonstrated remarkable improvements in growth and yield (Yu et al. 2021), highlighting the feasibility of modulating RNA methylation to achieve desirable agronomic traits. Alternatively, the use of viral vectors to deliver epitranscriptomic regulators may offer a promising strategy for transient reprogramming of plant traits (Pasin et al. 2024), potentially overcoming limitations associated with traditional transgenic approaches.

In summary, advancing our knowledge of epitranscriptomic and epigenetic regulatory mechanisms in plant immunity, through the use of cutting-edge technologies to study RNA and DNA modifications, may eventually lead to novel strategies to engineer crops with enhanced health, productivity, and adaptability, thereby contributing to global food security.

Acknowledgments R.L.C. is supported by the CIDEGENT/2021/030 grant funded by Generalitat Valenciana (Spain). F.P. is financed by the "Ramón y Cajal" program (RYC2023-045411-I) of Ministerio de Ciencia, Innovación y Universidades (Spain). M.Z. is supported by the National Natural Science Foundation of China (grants 31770165 and 31860489), a Major project of the Natural Science Foundation of Inner Mongolia of China (2021ZD06), and Talent Start funding of Inner Mongolia Agricultural University (NDGCC2016-23). Part of the text has been edited with the assistance of the artificial intelligence (AI) tool Llama-3.3-70B (Meta Platforms, Inc., 2025); every section was carefully revised and approved by the authors to ensure scientific correctness.

Competing Interests The authors declare no competing interests.

References

Addepalli B, Hunt AG (2007) A novel endonuclease activity associated with the *Arabidopsis* ortholog of the 30-kDa subunit of cleavage and polyadenylation specificity factor. Nucleic Acids Res 35:4453–4463. https://doi.org/10.1093/nar/gkm457

Alarcón CR, Goodarzi H, Lee H et al (2015a) HNRNPA2B1 is a mediator of m^{6}A-dependent nuclear RNA processing events. Cell 162:1299–1308. https://doi.org/10.1016/j.cell.2015.08.011

Alarcón CR, Lee H, Goodarzi H et al (2015b) *N*^{6}-methyladenosine marks primary microRNAs for processing. Nature 519:482–485. https://doi.org/10.1038/nature14281

Aloni Y, Dhar R, Khoury G (1979) Methylation of nuclear simian virus 40 RNAs. J Virol 32:52–60. https://doi.org/10.1128/jvi.32.1.52-60.1979

Alvarado-Marchena L, Marquez-Molins J, Martinez-Perez M et al (2021) Mapping of functional subdomains in the *at*ALKBH9B m^{6}A-demethylase required for its binding to the viral RNA and to the coat protein of alfalfa mosaic virus. Front Plant Sci 12:701683. https://doi.org/10.3389/fpls.2021.701683

Arribas-Hernández L, Bressendorff S, Hansen MH et al (2018) An m^{6}A-YTH module controls developmental timing and morphogenesis in Arabidopsis. Plant Cell 30:952–967. https://doi.org/10.1105/tpc.17.00833

Aslam M, Fakher B, Jakada BH et al (2019) SWR1 chromatin remodeling complex: a key transcriptional regulator in plants. Cells 8:1621. https://doi.org/10.3390/cells8121621

Baulcombe DC (2022) The role of viruses in identifying and analyzing RNA silencing. Annu Rev Virol 9:353–373. https://doi.org/10.1146/annurev-virology-091919-064218

Beemon K, Keith J (1977) Localization of *N*6-methyladenosine in the Rous sarcoma virus genome. J Mol Biol 113:165–179. https://doi.org/10.1016/0022-2836(77)90047-x

Berriri S, Gangappa SN, Kumar SV (2016) SWR1 chromatin-remodeling complex subunits and H2A.Z have non-overlapping functions in immunity and gene regulation in *Arabidopsis*. Mol Plant 9:1051–1065. https://doi.org/10.1016/j.molp.2016.04.003

Bokar JA (2005) The biosynthesis and functional roles of methylated nucleosides in eukaryotic mRNA. In: Grosjean H (ed) Fine-tuning of RNA functions by modification and editing. Springer, Berlin, Heidelberg, pp 141–177

Bokar JA, Rath-Shambaugh ME, Ludwiczak R et al (1994) Characterization and partial purification of mRNA *N*^{6}-adenosine methyltransferase from HeLa cell nuclei. J Biol Chem 269:17697–17704. https://doi.org/10.1016/S0021-9258(17)32497-3

Bokar JA, Shambaugh ME, Polayes D et al (1997) Purification and cDNA cloning of the AdoMet-binding subunit of the human mRNA (N^6-adenosine)-methyltransferase. RNA 3:1233–1247

Cai H, Huang Y, Chen F et al (2021) ERECTA signaling regulates plant immune responses via chromatin-mediated promotion of *WRKY33* binding to target genes. New Phytol 230:737–756. https://doi.org/10.1111/nph.17200

Cambiagno DA, Nota F, Zavallo D et al (2018) Immune receptor genes and pericentromeric transposons as targets of common epigenetic regulatory elements. Plant J 96:1178–1190. https://doi.org/10.1111/tpj.14098

Canaani D, Kahana C, Lavi S, Groner Y (1979) Identification and mapping of N^6-methyladenosine containing sequences in simian virus 40 RNA. Nucleic Acids Res 6:2879–2899. https://doi.org/10.1093/nar/6.8.2879

Chen Y, Wang J, Yang N et al (2018) Wheat microbiome bacteria can reduce virulence of a plant pathogenic fungus by altering histone acetylation. Nat Commun 9:3429. https://doi.org/10.1038/s41467-018-05683-7

Chen-Kiang S, Nevins JR, Darnell JE (1979) *N*-6-methyl-adenosine in adenovirus type 2 nuclear RNA is conserved in the formation of messenger RNA. J Mol Biol 135:733–752. https://doi.org/10.1016/0022-2836(79)90174-8

Clancy MJ, Shambaugh ME, Timpte CS, Bokar JA (2002) Induction of sporulation in *Saccharomyces cerevisiae* leads to the formation of N^6-methyladenosine in mRNA: a potential mechanism for the activity of the *IME4* gene. Nucleic Acids Res 30:4509–4518. https://doi.org/10.1093/nar/gkf573

Coleman-Derr D, Zilberman D (2012) Deposition of histone variant H2A.Z within gene bodies regulates responsive genes. PLoS Genet 8:e1002988. https://doi.org/10.1371/journal.pgen.1002988

Corrêa RL, Sanz-Carbonell A, Kogej Z et al (2020) Viral fitness determines the magnitude of transcriptomic and epigenomic reprograming of defense responses in plants. Mol Biol Evol 37:1866–1881. https://doi.org/10.1093/molbev/msaa091

Corrêa RL, Kutnjak D, Ambrós S et al (2024) Identification of epigenetically regulated genes involved in plant-virus interaction and their role in virus-triggered induced resistance. BMC Plant Biol 24:172. https://doi.org/10.1186/s12870-024-04866-3

Desbiez C, Schoeny A, Maisonneuve B et al (2017) Molecular and biological characterization of two potyviruses infecting lettuce in southeastern France. Plant Pathol 66:970–979. https://doi.org/10.1111/ppa.12651

Desrosiers R, Friderici K, Rottman F (1974) Identification of methylated nucleosides in messenger RNA from Novikoff hepatoma cells. Proc Natl Acad Sci USA 71:3971–3975. https://doi.org/10.1073/pnas.71.10.3971

Diezma-Navas L, Pérez-González A, Artaza H et al (2019) Crosstalk between epigenetic silencing and infection by tobacco rattle virus in *Arabidopsis*. Mol Plant Pathol 20:1439–1452. https://doi.org/10.1111/mpp.12850

Dominissini D, Moshitch-Moshkovitz S, Schwartz S et al (2012) Topology of the human and mouse m^6A RNA methylomes revealed by m^6A-seq. Nature 485:201–206. https://doi.org/10.1038/nature11112

Dowen RH, Pelizzola M, Schmitz RJ et al (2012) Widespread dynamic DNA methylation in response to biotic stress. Proc Natl Acad Sci USA 109:E2183–E2191. https://doi.org/10.1073/pnas.1209329109

Duan H-C, Wei L-H, Zhang C et al (2017) ALKBH10B is an RNA N^6-methyladenosine demethylase affecting Arabidopsis floral transition. Plant Cell 29:2995–3011. https://doi.org/10.1105/tpc.16.00912

Duan H, Wang Y, Jia G (2019) Dynamic and reversible RNA N^6-methyladenosine methylation. WIREs RNA 10. https://doi.org/10.1002/wrna.1507

Fedeles BI, Singh V, Delaney JC et al (2015) The AlkB family of Fe(II)/α-ketoglutarate-dependent dioxygenases: repairing nucleic acid alkylation damage and beyond. J Biol Chem 290:20734–20742. https://doi.org/10.1074/jbc.r115.656462

Finkel D, Groner Y (1983) Methylations of adenosine residues (m^6A) in pre-mRNA are important for formation of late simian virus 40 mRNAs. Virology 131:409–425. https://doi.org/10.1016/0042-6822(83)90508-1

Fu Y, Dai Q, Zhang W et al (2010) The AlkB domain of mammalian ABH8 catalyzes hydroxylation of 5-methoxycarbonylmethyluridine at the wobble position of tRNA. Angew Chem Int Ed Engl 49:8885–8888

Fu Y, Jia G, Pang X et al (2013) FTO-mediated formation of N^6-hydroxymethyladenosine and N^6-formyladenosine in mammalian RNA. Nat Commun 4:1798. https://doi.org/10.1038/ncomms2822

Fustin J-M, Doi M, Yamaguchi Y et al (2013) RNA-methylation-dependent RNA processing controls the speed of the circadian clock. Cell 155:793–806. https://doi.org/10.1016/j.cell.2013.10.026

Ge L, Pan F, Jia M et al (2024) RNA modifications in plant biotic interactions. Plant Commun 101232. https://doi.org/10.1016/j.xplc.2024.101232

Gokhale NS, McIntyre ABR, McFadden MJ et al (2016) *N6*-methyladenosine in *Flaviviridae* viral RNA genomes regulates infection. Cell Host Microbe 20:654–665. https://doi.org/10.1016/j.chom.2016.09.015

Gómez-Zambrano Á, Merini W, Calonje M (2019) The repressive role of Arabidopsis H2A.Z in transcriptional regulation depends on AtBMI1 activity. Nat Commun 10:2828. https://doi.org/10.1038/s41467-019-10773-1

Gordon TR, Leveau JHJ (2010) Plant pathology: a story about biology. Annu Rev Phytopathol 48: 293–309. https://doi.org/10.1146/annurev-phyto-080508-081919

Granadino B, Campuzano S, Sánchez L (1990) The *Drosophila melanogaster fl(2)d* gene is needed for the female-specific splicing of *Sex-lethal* RNA. EMBO J 9:2597–2602. https://doi.org/10.1002/j.1460-2075.1990.tb07441.x

Guo J, Tang H-W, Li J et al (2018) Xio is a component of the *Drosophila* sex determination pathway and RNA N^6-methyladenosine methyltransferase complex. Proc Natl Acad Sci USA 115:3674–3679. https://doi.org/10.1073/pnas.1720945115

Guo T, Liu C, Meng F et al (2022) The m^{6}A reader MhYTP2 regulates *MdMLO19* mRNA stability and antioxidant genes translation efficiency conferring powdery mildew resistance in apple. Plant Biotechnol J 20:511–525. https://doi.org/10.1111/pbi.13733

Haider S, Farrona S (2024) Decoding histone 3 lysine methylation: insights into seed germination and flowering. Curr Opin Plant Biol 81:102598. https://doi.org/10.1016/j.pbi.2024.102598

Halter T, Wang J, Amesefe D et al (2021) The Arabidopsis active demethylase ROS1 *cis*-regulates defence genes by erasing DNA methylation at promoter-regulatory regions. eLife 10:e62994. https://doi.org/10.7554/eLife.62994

Han C, Zhang F, Qiao X et al (2021) Multi-omics analysis reveals the dynamic changes of RNA N^6-methyladenosine in pear (*Pyrus bretschneideri*) defense responses to *Erwinia amylovora* pathogen infection. Front Microbiol 12:803512. https://doi.org/10.3389/fmicb.2021.803512

Hannan Parker A, Wilkinson SW, Ton J (2022) Epigenetics: a catalyst of plant immunity against pathogens. New Phytol 233:66–83. https://doi.org/10.1111/nph.17699

Haugland RA, Cline MG (1980) Post-transcriptional modifications of oat coleoptile ribonucleic acids. Eur J Biochem 104:271–277. https://doi.org/10.1111/j.1432-1033.1980.tb04425.x

Haussmann IU, Bodi Z, Sanchez-Moran E et al (2016) m^{6}A potentiates *Sxl* alternative pre-mRNA splicing for robust *Drosophila* sex determination. Nature 540:301–304. https://doi.org/10.1038/nature20577

He Y, Li L, Yao Y et al (2021) Transcriptome-wide N6-methyladenosine (m^{6}A) methylation in watermelon under CGMMV infection. BMC Plant Biol 21:516. https://doi.org/10.1186/s12870-021-03289-8

He H, Ge L, Chen Y et al (2024) m^{6}A modification of plant virus enables host recognition by NMD factors in plants. Sci China Life Sci 67:161–174. https://doi.org/10.1007/s11427-022-2377-1

Hongay CF, Orr-Weaver TL (2011) *Drosophila* inducer of MEiosis 4 (IME4) is required for notch signaling during oogenesis. Proc Natl Acad Sci USA 108:14855–14860. https://doi.org/10.1073/pnas.1111577108

Horiuchi K, Kawamura T, Iwanari H et al (2013) Identification of Wilms' tumor 1-associating protein complex and its role in alternative splicing and the cell cycle. J Biol Chem 288:33292–33302. https://doi.org/10.1074/jbc.M113.500397

Hsu PJ, Zhu Y, Ma H et al (2017) Ythdc2 is an N^6-methyladenosine binding protein that regulates mammalian spermatogenesis. Cell Res 27:1115–1127. https://doi.org/10.1038/cr.2017.99

Huang Y, Kendall T, Forsythe ES et al (2015) Ancient origin and recent innovations of RNA polymerase IV and V. Mol Biol Evol 32:1788–1799. https://doi.org/10.1093/molbev/msv060

Huang H, Weng H, Sun W et al (2018) Recognition of RNA N^6-methyladenosine by IGF2BP proteins enhances mRNA stability and translation. Nat Cell Biol 20:285–295. https://doi.org/10.1038/s41556-018-0045-z

Ivanova I, Much C, Di Giacomo M et al (2017) The RNA m^{6}A reader YTHDF2 is essential for the post-transcriptional regulation of the maternal transcriptome and oocyte competence. Mol Cell 67:1059–1067.e4. https://doi.org/10.1016/j.molcel.2017.08.003

Jia G, Fu Y, Zhao X et al (2011) *N*6-Methyladenosine in nuclear RNA is a major substrate of the obesity-associated FTO. Nat Chem Biol 7:885–887. https://doi.org/10.1038/nchembio.687

Jiang D, Berger F (2017) Histone variants in plant transcriptional regulation. Biochim Biophys Acta Gene Regul Mech 1860:123–130. https://doi.org/10.1016/j.bbagrm.2016.07.002

Kane SE, Beemon K (1985) Precise localization of m^6A in Rous sarcoma virus RNA reveals clustering of methylation sites: implications for RNA processing. Mol Cell Biol 5:2298–2306. https://doi.org/10.1128/mcb.5.9.2298-2306.1985

Kang H, Fan T, Wu J et al (2022) Histone modification and chromatin remodeling in plant response to pathogens. Front Plant Sci 13:986940. https://doi.org/10.3389/fpls.2022.986940

Kennedy TD, Lane BG (1979) Wheat embryo ribonucleates. XIII. Methyl-substituted nucleoside constituents and 5′-terminal dinucleotide sequences in bulk poly(AR)-rich RNA from imbibing wheat embryos. Can J Biochem 57:927–931. https://doi.org/10.1139/o79-112

Kennedy EM, Bogerd HP, Kornepati AVR et al (2016) Posttranscriptional m^6A editing of HIV-1 mRNAs enhances viral gene expression. Cell Host Microbe 19:675–685. https://doi.org/10.1016/j.chom.2016.04.002

Knuckles P, Lence T, Haussmann IU et al (2018) Zc3h13/Flacc is required for adenosine methylation by bridging the mRNA-binding factor Rbm15/Spenito to the m^6 a machinery component Wtap/Fl(2)d. Genes Dev 32:415–429. https://doi.org/10.1101/gad.309146.117

Kong Y, Mead EA, Fang G (2023) Navigating the pitfalls of mapping DNA and RNA modifications. Nat Rev Genet 24:363–381. https://doi.org/10.1038/s41576-022-00559-5

Kretschmer J, Rao H, Hackert P et al (2018) The m^6A reader protein YTHDC2 interacts with the small ribosomal subunit and the 5′-3′ exoribonuclease XRN1. RNA 24:1339–1350. https://doi.org/10.1261/rna.064238.117

Krug RM, Morgan MA, Shatkin AJ (1976) Influenza viral mRNA contains internal N^6-methyladenosine and 5′-terminal 7-methylguanosine in cap structures. J Virol 20:45–53. https://doi.org/10.1128/jvi.20.1.45-53.1976

Lai Y, Lu XM, Daron J et al (2020) The Arabidopsis PHD-finger protein EDM2 has multiple roles in balancing NLR immune receptor gene expression. PLoS Genet 16:e1008993. https://doi.org/10.1371/journal.pgen.1008993

Lavi S, Shatkin AJ (1975) Methylated simian virus 40-specific RNA from nuclei and cytoplasm of infected BSC-1 cells. Proc Natl Acad Sci USA 72:2012–2016. https://doi.org/10.1073/pnas.72.6.2012

Le T-N, Schumann U, Smith NA et al (2014) DNA demethylases target promoter transposable elements to positively regulate stress responsive genes in *Arabidopsis*. Genome Biol 15:458. https://doi.org/10.1186/s13059-014-0458-3

Lence T, Akhtar J, Bayer M et al (2016) m^6A modulates neuronal functions and sex determination in *drosophila*. Nature 540:242–247. https://doi.org/10.1038/nature20568

Li Y, Wang X, Li C et al (2014) Transcriptome-wide N^6-methyladenosine profiling of rice callus and leaf reveals the presence of tissue-specific competitors involved in selective mRNA modification. RNA Biol 11:1180–1188. https://doi.org/10.4161/rna.36281

Li A, Chen Y-S, Ping X-L et al (2017) Cytoplasmic m^6A reader YTHDF3 promotes mRNA translation. Cell Res 27:444–447. https://doi.org/10.1038/cr.2017.10

Li Z, Shi J, Yu L et al (2018) N^6-methyl-adenosine level in *Nicotiana tabacum* is associated with tobacco mosaic virus. Virol J 15:87. https://doi.org/10.1186/s12985-018-0997-4

Li Y, Chen J, Sun Z (2024) N6-methyladenosine (m6A) modification: emerging regulators in plant–virus interactions. Virology 110373. https://doi.org/10.1016/j.virol.2024.110373

Lichinchi G, Gao S, Saletore Y et al (2016a) Dynamics of the human and viral m^6A RNA methylomes during HIV-1 infection of T cells. Nat Microbiol 1:16011. https://doi.org/10.1038/nmicrobiol.2016.11

Lichinchi G, Zhao BS, Wu Y et al (2016b) Dynamics of human and viral RNA methylation during Zika virus infection. Cell Host Microbe 20:666–673. https://doi.org/10.1016/j.chom.2016.10.002

Liu N, Pan T (2016) N^6-methyladenosine–encoded epitranscriptomics. Nat Struct Mol Biol 23:98–102. https://doi.org/10.1038/nsmb.3162

Liu J, Yue Y, Han D et al (2014) A METTL3–METTL14 complex mediates mammalian nuclear RNA N^6-adenosine methylation. Nat Chem Biol 10:93–95. https://doi.org/10.1038/nchembio.1432

Liu N, Dai Q, Zheng G et al (2015) N^6-methyladenosine-dependent RNA structural switches regulate RNA-protein interactions. Nature 518:560–564. https://doi.org/10.1038/nature14234

Liu N, Zhou KI, Parisien M et al (2017) N^6-methyladenosine alters RNA structure to regulate binding of a low-complexity protein. Nucleic Acids Res 45:6051–6063. https://doi.org/10.1093/nar/gkx141

Liu J, Xu Y-P, Li K et al (2021) The m^6A methylome of SARS-CoV-2 in host cells. Cell Res 31: 404–414. https://doi.org/10.1038/s41422-020-00465-7

Liu Y, Li K, Xu YP et al (2022) Characterization of m^6A modifications in the contemporary Zika virus genome and host cellular transcripts. J Med Virol 94:4309–4318. https://doi.org/10.1002/jmv.27869

Liu C, Sun H, Yi Y et al (2023) Absolute quantification of single-base m^6A methylation in the mammalian transcriptome using GLORI. Nat Biotechnol 41:355–366. https://doi.org/10.1038/s41587-022-01487-9

López Sánchez A, Stassen JHM, Furci L et al (2016) The role of DNA (de)methylation in immune responsiveness of Arabidopsis. Plant J 88:361–374. https://doi.org/10.1111/tpj.13252

López A, Ramírez V, García-Andrade J et al (2011) The RNA silencing enzyme RNA polymerase V is required for plant immunity. PLoS Genet 7:e1002434. https://doi.org/10.1371/journal.pgen.1002434

Lu X, He Y, Guo J-Q et al (2024) Dynamics of epitranscriptomes uncover translational reprogramming directed by ac4C in rice during pathogen infection. Nat Plants 10:1548–1561. https://doi.org/10.1038/s41477-024-01800-1

Luna E, Bruce TJA, Roberts MR et al (2012) Next-generation systemic acquired resistance. Plant Physiol 158:844–853. https://doi.org/10.1104/pp.111.187468

Luo G-Z, MacQueen A, Zheng G et al (2014) Unique features of the m^6A methylome in *Arabidopsis thaliana*. Nat Commun 5:5630. https://doi.org/10.1038/ncomms6630

Luo C, Hajkova P, Ecker JR (2018) Dynamic DNA methylation: in the right place at the right time. Science 361:1336–1340. https://doi.org/10.1126/science.aat6806

Ma H, Wang X, Cai J et al (2019) N^6-methyladenosine methyltransferase ZCCHC4 mediates ribosomal RNA methylation. Nat Chem Biol 15:88–94. https://doi.org/10.1038/s41589-018-0184-3

March-Díaz R, García-Domínguez M, Lozano-Juste J et al (2008) Histone H2A.Z and homologues of components of the SWR1 complex are required to control immunity in Arabidopsis. Plant J 53:475–487. https://doi.org/10.1111/j.1365-313X.2007.03361.x

Martínez-Pérez M, Aparicio F, López-Gresa MP et al (2017) *Arabidopsis* m^6A demethylase activity modulates viral infection of a plant virus and the m^6A abundance in its genomic RNAs. Proc Natl Acad Sci USA 114:10755–10760. https://doi.org/10.1073/pnas.1703139114

Martínez-Pérez M, Gómez-Mena C, Alvarado-Marchena L et al (2021) The m^6A RNA demethylase ALKBH9B plays a critical role for vascular movement of alfalfa mosaic virus in Arabidopsis. Front Microbiol 12:745576. https://doi.org/10.3389/fmicb.2021.745576

Mattei AL, Bailly N, Meissner A (2022) DNA methylation: a historical perspective. Trends Genet 38:676–707. https://doi.org/10.1016/j.tig.2022.03.010

Meyer KD, Saletore Y, Zumbo P et al (2012) Comprehensive analysis of mRNA methylation reveals enrichment in 3′ UTRs and near stop codons. Cell 149:1635–1646. https://doi.org/10.1016/j.cell.2012.05.003

Miao Z, Zhang T, Qi Y et al (2020) Evolution of the RNA N^6-methyladenosine methylome mediated by genomic duplication. Plant Physiol 182:345–360. https://doi.org/10.1104/pp.19.00323

Moore C, Meng B (2019) Prediction of the molecular boundary and functionality of novel viral AlkB domains using homology modelling and principal component analysis. J Gen Virol 100: 691–703. https://doi.org/10.1099/jgv.0.001237
Moss B, Gershowitz A, Stringer HJR et al (1977) 5′-Terminal and internal methylated nucleosides in herpes simplex virus type 1 mRNA. J Virol 23:234–239. https://doi.org/10.1128/jvi.23.2.234-239.1977
Negi P, Rai AN, Suprasanna P (2016) Moving through the stressed genome: emerging regulatory roles for transposons in plant stress response. Front Plant Sci 7:1448. https://doi.org/10.3389/fpls.2016.01448
Nichols JL, Welder L (1981) Nucleotides adjacent to N^6-methyladenosine in maize poly (A)-containing RNA. Plant Sci Lett 21:75–81. https://doi.org/10.1016/0304-4211(81)90071-7
Ortega A, Niksic M, Bachi A et al (2003) Biochemical function of female-lethal (2)D/Wilms' tumor suppressor-1-associated proteins in alternative pre-mRNA splicing. J Biol Chem 278:3040–3047. https://doi.org/10.1074/jbc.M210737200
Pasin F, Daròs J-A, Tzanetakis IE (2022) Proteome expansion in the *Potyviridae* evolutionary radiation. FEMS Microbiol Rev 46:fuac011. https://doi.org/10.1093/femsre/fuac011
Pasin F, Uranga M, Charudattan R, Kwon C-T (2024) Engineering good viruses to improve crop performance. Nat Rev Bioeng 2:532–534. https://doi.org/10.1038/s44222-024-00197-y
Patil DP, Chen C-K, Pickering BF et al (2016) m^6A RNA methylation promotes *XIST*-mediated transcriptional repression. Nature 537:369–373. https://doi.org/10.1038/nature19342
Pendleton KE, Chen B, Liu K et al (2017) The U6 snRNA m^6A methyltransferase METTL16 regulates SAM synthetase intron retention. Cell 169:824–835.e14. https://doi.org/10.1016/j.cell.2017.05.003
Perry RP, Kelley DE (1974) Existence of methylated messenger RNA in mouse L cells. Cell 1:37–42. https://doi.org/10.1016/0092-8674(74)90153-6
Peters T, Ausmeier K, Rüther U (1999) Cloning of fatso (*Fto*), a novel gene deleted by the fused toes (*Ft*) mouse mutation. Mamm Genome 10:983–986. https://doi.org/10.1007/s003359901144
Ping X-L, Sun B-F, Wang L et al (2014) Mammalian WTAP is a regulatory subunit of the RNA N6-methyladenosine methyltransferase. Cell Res 24:177–189. https://doi.org/10.1038/cr.2014.3
Raffel GD, Mercher T, Shigematsu H et al (2007) Ott1 (Rbm15) has pleiotropic roles in hematopoietic development. Proc Natl Acad Sci USA 104:6001–6006. https://doi.org/10.1073/pnas.0609041104
Ren Z, Tang B, Xing J et al (2022) MTA1-mediated RNA m^6A modification regulates autophagy and is required for infection of the rice blast fungus. New Phytol 235:247–262. https://doi.org/10.1111/nph.18117
Rengaraj P, Obrdlík A, Vukić D et al (2021) Interplays of different types of epitranscriptomic mRNA modifications. RNA Biol 18:19–30. https://doi.org/10.1080/15476286.2021.1969113
Roossinck MJ (2011) The good viruses: viral mutualistic symbioses. Nat Rev Microbiol 9:99–108. https://doi.org/10.1038/nrmicro2491
Roost C, Lynch SR, Batista PJ et al (2015) Structure and thermodynamics of N^6-methyladenosine in RNA: a spring-loaded base modification. J Am Chem Soc 137:2107–2115. https://doi.org/10.1021/ja513080v
Roundtree IA, Luo G-Z, Zhang Z et al (2017) YTHDC1 mediates nuclear export of N^6-methyladenosine methylated mRNAs. eLife 6:e31311. https://doi.org/10.7554/eLife.31311
Růžička K, Zhang M, Campilho A et al (2017) Identification of factors required for m^6A mRNA methylation in *Arabidopsis* reveals a role for the conserved E3 ubiquitin ligase HAKAI. New Phytol 215:157–172. https://doi.org/10.1111/nph.14586
Schibler U, Perry RP (1977) The 5′-termini of heterogeneous nuclear RNA: a comparison among molecules of different sizes and ages. Nucleic Acids Res 4:4133–4149. https://doi.org/10.1093/nar/4.12.4133

Schwartz S, Agarwala SD, Mumbach MR et al (2013) High-resolution mapping reveals a conserved, widespread, dynamic mRNA methylation program in yeast meiosis. Cell 155:1409–1421. https://doi.org/10.1016/j.cell.2013.10.047
Schwartz S, Mumbach MR, Jovanovic M et al (2014) Perturbation of m6A writers reveals two distinct classes of mRNA methylation at internal and 5′ sites. Cell Rep 8:284–296. https://doi.org/10.1016/j.celrep.2014.05.048
Scutenaire J, Deragon J-M, Jean V et al (2018) The YTH domain protein ECT2 is an m^6A reader required for normal trichome branching in Arabidopsis. Plant Cell 30:986–1005. https://doi.org/10.1105/tpc.17.00854
Secco N, Sheikh AH, Hirt H (2024) Insights into the role of N6-methyladenosine (m6A) in plant–virus interactions. J Virol:e0159824. https://doi.org/10.1128/jvi.01598-24
Shah JC, Clancy MJ (1992) *IME4*, a gene that mediates *MAT* and nutritional control of meiosis in *Saccharomyces cerevisiae*. Mol Cell Biol 12:1078–1086. https://doi.org/10.1128/mcb.12.3.1078-1086.1992
Shen L, Liang Z, Gu X et al (2016) N^6-methyladenosine RNA modification regulates shoot stem cell fate in *Arabidopsis*. Dev Cell 38:186–200. https://doi.org/10.1016/j.devcel.2016.06.008
Shi H, Wang X, Lu Z et al (2017) YTHDF3 facilitates translation and decay of N^6-methyladenosine-modified RNA. Cell Res 27:315–328. https://doi.org/10.1038/cr.2017.15
Sigman MJ, Slotkin RK (2016) The first rule of plant transposable element silencing: location, location, location. Plant Cell 28:304–313. https://doi.org/10.1105/tpc.15.00869
Sun H, Li K, Liu C, Yi C (2023) Regulation and functions of non-m^6A mRNA modifications. Nat Rev Mol Cell Biol 24:714–731. https://doi.org/10.1038/s41580-023-00622-x
Sura W, Kabza M, Karlowski WM et al (2017) Dual role of the histone variant H2A.Z in transcriptional regulation of stress-response genes. Plant Cell 29:791–807. https://doi.org/10.1105/tpc.16.00573
Thoms D, Liang Y, Haney CH (2021) Maintaining symbiotic homeostasis: how do plants engage with beneficial microorganisms while at the same time restricting pathogens? Mol Plant-Microbe Interact 34:462–469. https://doi.org/10.1094/MPMI-11-20-0318-FI
Tsuchiya T, Eulgem T (2013) An alternative polyadenylation mechanism coopted to the *Arabidopsis RPP7* gene through intronic retrotransposon domestication. Proc Natl Acad Sci USA 110:E3535–E3543. https://doi.org/10.1073/pnas.1312545110
van den Born E, Omelchenko MV, Bekkelund A et al (2008) Viral AlkB proteins repair RNA damage by oxidative demethylation. Nucleic Acids Res 36:5451–5461. https://doi.org/10.1093/nar/gkn519
Vespa L, Vachon G, Berger F et al (2004) The immunophilin-interacting protein AtFIP37 from Arabidopsis is essential for plant development and is involved in trichome endoreduplication. Plant Physiol 134:1283–1292. https://doi.org/10.1104/pp.103.028050
Vigneaud J, Kohler A, Sow MD et al (2023) DNA hypomethylation of the host tree impairs interaction with mutualistic ectomycorrhizal fungus. New Phytol 238:2561–2577. https://doi.org/10.1111/nph.18734
Wan Y, Tang K, Zhang D et al (2015) Transcriptome-wide high-throughput deep m^6A-seq reveals unique differential m^6A methylation patterns between three organs in *Arabidopsis thaliana*. Genome Biol 16:272. https://doi.org/10.1186/s13059-015-0839-2
Wang X, Lu Z, Gomez A et al (2014) N^6-methyladenosine-dependent regulation of messenger RNA stability. Nature 505:117–120. https://doi.org/10.1038/nature12730
Wang X, Zhao BS, Roundtree IA et al (2015) N^6-methyladenosine modulates messenger RNA translation efficiency. Cell 161:1388–1399. https://doi.org/10.1016/j.cell.2015.05.014
Wang Z, Tang K, Zhang D et al (2017) High-throughput m^6A-seq reveals RNA m^6A methylation patterns in the chloroplast and mitochondria transcriptomes of *Arabidopsis thaliana*. PLoS One 12:e0185612. https://doi.org/10.1371/journal.pone.0185612
Wassenegger M, Heimes S, Riedel L, Sänger HL (1994) RNA-directed de novo methylation of genomic sequences in plants. Cell 76:567–576. https://doi.org/10.1016/0092-8674(94)90119-8

Wei J, Liu F, Lu Z et al (2018) Differential m^6A, m^6A_m, and m^1A demethylation mediated by FTO in the cell nucleus and cytoplasm. Mol Cell 71:973–985.e5. https://doi.org/10.1016/j.molcel.2018.08.011

Wen J, Lv R, Ma H et al (2018) Zc3h13 regulates nuclear RNA m^6A methylation and mouse embryonic stem cell self-renewal. Mol Cell 69:1028–1038.e6. https://doi.org/10.1016/j.molcel.2018.02.015

Wiener D, Schwartz S (2021) The epitranscriptome beyond m^6A. Nat Rev Genet 22:119–131. https://doi.org/10.1038/s41576-020-00295-8

Wingard SA (1928) Hosts and symptoms of ring spot, a virus disease of plants. J Agric Res 37:127–153

Xia Z, Zhang S, Guo H et al (2024) N^6-methyladenosine RNA modification regulates maize resistance to maize chlorotic mottle virus infection. J Agric Food Chem 72:21935–21945. https://doi.org/10.1021/acs.jafc.4c04832

Xiao W, Adhikari S, Dahal U et al (2016) Nuclear m^6A reader YTHDC1 regulates mRNA splicing. Mol Cell 61:507–519. https://doi.org/10.1016/j.molcel.2016.01.012

Yan D, Perrimon N (2015) *Spenito* is required for sex determination in *Drosophila melanogaster*. Proc Natl Acad Sci USA 112:11606–11611. https://doi.org/10.1073/pnas.1515891112

Yang L, Wang Z, Hua J (2023) Multiple chromatin-associated modules regulate expression of an intracellular immune receptor gene in Arabidopsis. New Phytol 237:2284–2297. https://doi.org/10.1111/nph.18672

Ye Y, Feng W, Zhang J et al (2021) Genome-wide identification and characterization of circular RNA m^6A modification in pancreatic cancer. Genome Med 13:183. https://doi.org/10.1186/s13073-021-01002-w

Yu A, Lepère G, Jay F et al (2013) Dynamics and biological relevance of DNA demethylation in *Arabidopsis* antibacterial defense. Proc Natl Acad Sci USA 110:2389–2394. https://doi.org/10.1073/pnas.1211757110

Yu Q, Liu S, Yu L et al (2021) RNA demethylation increases the yield and biomass of rice and potato plants in field trials. Nat Biotechnol 39:1581–1588. https://doi.org/10.1038/s41587-021-00982-9

Yu Y, Gui Y, Li Z et al (2022) Induced systemic resistance for improving plant immunity by beneficial microbes. Plants 11:386. https://doi.org/10.3390/plants11030386

Yue Y, Liu J, He C (2015) RNA N^6 -methyladenosine methylation in post-transcriptional gene expression regulation. Genes Dev 29:1343–1355. https://doi.org/10.1101/gad.262766.115

Yue Y, Liu J, Cui X et al (2018) VIRMA mediates preferential m^6A mRNA methylation in 3′UTR and near stop codon and associates with alternative polyadenylation. Cell Discov 4:10. https://doi.org/10.1038/s41421-018-0019-0

Yue J, Wei Y, Sun Z et al (2022a) AlkB RNA demethylase homologues and N^6-methyladenosine are involved in *Potyvirus* infection. Mol Plant Pathol 23:1555–1564. https://doi.org/10.1111/mpp.13239

Yue J, Wei Y, Zhao M (2022b) The reversible methylation of m6A is involved in plant virus infection. Biology 11:271. https://doi.org/10.3390/biology11020271

Yue J, Lu Y, Sun Z et al (2023) Methyltransferase-like (METTL) homologues participate in *Nicotiana benthamiana* antiviral responses. Plant Signal Behav 18:e2214760. https://doi.org/10.1080/15592324.2023.2214760

Zaccara S, Ries RJ, Jaffrey SR (2019) Reading, writing and erasing mRNA methylation. Nat Rev Mol Cell Biol 20:608–624. https://doi.org/10.1038/s41580-019-0168-5

Zamioudis C, Pieterse CMJ (2012) Modulation of host immunity by beneficial microbes. Mol Plant-Microbe Interact 25:139–150. https://doi.org/10.1094/MPMI-06-11-0179

Zhang F, Zhang Y-C, Liao J-Y et al (2019) The subunit of RNA N6-methyladenosine methyltransferase OsFIP regulates early degeneration of microspores in rice. PLoS Genet 15:e1008120. https://doi.org/10.1371/journal.pgen.1008120

Zhang K, Zhuang X, Dong Z et al (2021a) The dynamics of N^6-methyladenine RNA modification in interactions between rice and plant viruses. Genome Biol 22:189. https://doi.org/10.1186/s13059-021-02410-2

Zhang T, Wang Z, Hu H et al (2021b) Transcriptome-wide N^6-methyladenosine (m^6A) profiling of susceptible and resistant wheat varieties reveals the involvement of variety-specific m^6A modification involved in virus-host interaction pathways. Front Microbiol 12:656302. https://doi.org/10.3389/fmicb.2021.656302

Zhang F, Zhang B, Cui T et al (2024) The novel roles of RNA m^6A modification in regulating the development, infection, and oxidative DNA damage repair of *Phytophthora sojae*. PLoS Pathog 20:e1012553. https://doi.org/10.1371/journal.ppat.1012553

Zhao X, Yang Y, Sun B-F et al (2014) FTO-dependent demethylation of N6-methyladenosine regulates mRNA splicing and is required for adipogenesis. Cell Res 24:1403–1419. https://doi.org/10.1038/cr.2014.151

Zhao BS, Wang X, Beadell AV et al (2017) m^6A-dependent maternal mRNA clearance facilitates zebrafish maternal-to-zygotic transition. Nature 542:475–478. https://doi.org/10.1038/nature21355

Zhao K, Li Z, Ke Y et al (2024a) Dynamic N^6-methyladenosine RNA modification regulates peanut resistance to bacterial wilt. New Phytol 242:231–246. https://doi.org/10.1111/nph.19568

Zhao L, Wei X, Chen F et al (2024b) m^6A demethylase CpALKBH regulates *CpZap1* mRNA stability to modulate the development and virulence of chestnut blight fungus. mBio e0184424. https://doi.org/10.1128/mbio.01844-24

Zheng G, Dahl JA, Niu Y et al (2013) ALKBH5 is a mammalian RNA demethylase that impacts RNA metabolism and mouse fertility. Mol Cell 49:18–29. https://doi.org/10.1016/j.molcel.2012.10.015

Zhong S, Li H, Bodi Z et al (2008) MTA is an *Arabidopsis* messenger RNA adenosine methylase and interacts with a homolog of a sex-specific splicing factor. Plant Cell 20:1278–1288. https://doi.org/10.1105/tpc.108.058883

Zhou KI, Shi H, Lyu R et al (2019) Regulation of co-transcriptional pre-mRNA splicing by m^6A through the low-complexity protein hnRNPG. Mol Cell 76:70–81.e9. https://doi.org/10.1016/j.molcel.2019.07.005

MIX
Papier aus verantwortungsvollen Quellen
Paper from responsible sources
FSC® C105338

If you have any concerns about our products,
you can contact us on
ProductSafety@springernature.com

In case Publisher is established outside the EU,
the EU authorized representative is:
Springer Nature Customer Service Center GmbH
Europaplatz 3, 69115 Heidelberg, Germany

Printed by Libri Plureos GmbH
in Hamburg, Germany